DICTIONARY

of ENGINEERING ACRONYMS and ABBREVIATIONS

NEAL-SCHUMAN PUBLISHERS, INC.
NEW YORK   LONDON

Published by Neal-Schuman Publishers, Inc.
23 Leonard Street
New York, NY 10013

Library of Congress Cataloging-in-Publication Data

Erb, Uwe.
    Dictionary of engineering acronyms and abbreviations.

    1. Engineering—Acronyms.   2. Engineering—
Abbreviations.   I. Keller, Harald.   II. Title.
TA11.E73 1988   620'.00148   88-28840
ISBN 1-55570-028-4

To our parents
Irmgard and Konrad
and
Elfriede and Otto

Contents

Introduction

Over the past two decades the use of acronyms, abbreviations, and initialisms has increased at an almost exponential rate, and most people agree that for the specialist in a given field such linguistic shorthand is extremely convenient. However, a widespread misuse of these terms is the omission of a full explanation when they first appear in a text. As a result, a reader often encounters serious difficulties when attempting to decipher such language.

Fortunately, a number of authors have realized the need for comprehensive reference sources, and numerous excellent dictionaries on acronyms, abbreviations, and initialisms have been published, particularly in the 1970s and '80s. It appears, however, that many of these dictionaries are either too general, trying to cover all topics, from law to politics to sports; or they are too specific, dealing with only one narrow field of interest. Many other dictionaries, although fine books when they were published, are simply outdated.

With the present *Dictionary of Engineering Acronyms and Abbreviations* we present a comprehensive, up-to-date source of reference to a very special, but by no means limited, readership. As its name implies, this book is designed for all readers of literature dealing with engineering subjects. Thus, not only engineers, researchers and scientists, but also technicians, technologists, college and university students and even the layman reading technical literature will find this dictionary an extremely valuable companion when working their way through what can be a chaotic jungle of shorthand language.

Containing more than 30,000 entries, this dictionary is based on our database on engineering acronyms, abbreviations and initialisms. In order to establish the necessary skeleton for a database of this nature, we first searched through a large number of reference books, encyclopedias and textbooks pertinent to the major areas of engineering to collect the standard shorthand of the literature. Once we had built the framework for our book, we updated our database by scanning thousands of bound and current periodicals in science, engineering and technology. Since then, our database has been continually expanding—and at the same rate at which new acronyms, abbreviations and initialisms appear in the literature. Keeping the scope broad enough by including all major engineering fields and presenting a most up-to-date manuscript by the time this book was to go to press were our main objectives throughout our work.

The following alphabetical list represents the major engineering fields covered in this dictionary:

aeronautics and aerospace	government offices and programs
artificial intelligence	heating and air conditioning
associations	joining technology
auto mechanics	manufacturing
biochemistry and biophysics	marine technology
biotechnology	materials science
ceramics	mechanical engineering
chemical engineering	metallurgy
chemistry	mining engineering
civil engineering	military sciences and engineering
computer sciences	physics
drafting	plant engineering
devices and machines	processing
electrical engineering	research institutes and laboratories
electrochemistry	research programs and projects
electronics	robotics and artificial intelligence
energy	societies
environmental engineering	technical colleges and universities
experimental techniques	technical specifications
foundry technology	transportation engineering

All entries are given in letter-by-letter alphabetical order. Terms that consist of all-capital letters precede terms in lower-case letters, e.g. "AP (Access Panel)" precedes "ap (apothecaries)". Terms beginning with a capital precede those that begin with lower-case letters, e.g. "Am (Americium)" precedes "am (ante meridien) - before noon". Our policy on the usage of capital letters vs. lower-case letters was to adopt the form most commonly found in current engineering literature, with capitalization being the predominant trend. Punctuation has largely been avoided and is only used if important to the understanding of an entry. Whenever required for presenting an entry in the right context, additional information is given in brackets e.g. country of origin, translation from another language, etc. Parentheses are used to include further explanations that are not part of an acronym e.g. MIES - McMaster (University) Institute for Energy Studies. In cases where numerals form part of an entry, they are listed under the corresponding letter, e.g. 3PDT - Triple-Pole, Double-Throw - appears under "T".

With the publication of this dictionary, we hope to have served the needs of the engineering community at large, and we are looking forward to the challenge of updating this dictionary on a regular basis.

Kingston, Ontario, Canada
October 1987

A

A	Absorbance
	Acid
	Acre
	Adder
	Adenine
	Ammeter
	Ampere
	Air
	Aircraft
	Alanine
	Analogy
	Anode
	Area
	Asbestos
	Attribute
a	analog
	asymmetric
	asynchronous
Å	Angstrom [unit]
AA	Acrylic Acid
	Adjustable Angle
	All Around
	Aluminum Association
	Amino Acid
	Aminoacyl
	Angular Aperture
	Anthranilic Acid
	Anti-Aircraft
	Approximate Absolute
	Arithmetic Average
	Armature Accelerator
	Arsenic Analysis
	Auto Acquisition (Radar)
	Automobile Association
A-A	Air-to-Air
AAA	Acrylic Acid Anhydride
	Alberta Association of Architects [Canada]
	Amateur Astronomers Association
	American Airship Association
	American Automobile Association
	Anti-Aircraft Artillery
AAAA	Army Aviation Association of America
AAAI	American Association for Artificial Intelligence
AAAIP	Advanced Army Aircraft Instrumentation Program [US]
AAAR	American Association for Aerosol Research
AAAS	American Academy of Arts and Sciences
	American Association for the Advancement of Science
AAB	Advertising Advisory Board
	Aircraft Accident Board
	American Association of Bioanalysts
	Army Aviation Board [US Army]
AABE	American Association of Blacks in Energy
AABP	Acetylaminobiphenyl
	Aptitude Assessment Battery Programming
AAC	Air Approach Control
	Air Carbon-Arc Cutting
	Anti-Aircraft Cannon
	Argon Alternating Current
	Arsenic Atmosphere Czochralski
	Associate and Advisory Committee
	Automatic Aperture Control
	Automatic Approach Control
	Aviation Advisory Commission
AACB	Aeronautics and Astronautics Coordinating Board [NASA]
AACC	American Association for Contamination Control
	American Association of Cereal Chemists
	American Automatic Control Council
AACD	Antenna Adjustable Current Distribution
AACE	American Association of Cost Engineers
AACG	American Association for Crystal Growth
AACOM	Army Area Communications
AACOMS	Army Area Communications System
AACS	Airborne Astrographic Camera System
	Airways and Air Communications Service
	Army Alaska Communication System [US Air Force]
	Asynchronous Address Communications System
AACSCEDR	Associate and Advisory Committee to the Special Committee on Electronic Data Retrieval
AACT	American Association of Commodity Traders
AADA	Anti-Aircraft Defended Area
AADC	All-Applications Digital Computer
	Arnold Air Development Center [US Air Force]
AADP	Aminopyridineadeninedinucleotide Phosphate
AADS	Advanced Automated Directional Scanning
	Area Air Defense System
AAE	American Association of Engineers
	Auto-Answer Equipment
	Automatic Answering Equipment
AAEC	Australian Atomic Energy Commission
AAEE	Airplane and Armament Experimental Establishment [UK]
	American Academy of Environmental Engineers
AAEM	Acetoacetoxyethyl Methacrylate
AAES	American Association of Engineering Societies
AAESWB	Army Airborne Electronics and Special Warfare Board [US Army]
AAF	Acetylaminofluorene
	American Air Force
	Army - Air Force
	Auxiliary Air Force
AAFB	Andrews Air Force Base [US]
	Army Air Force Board [US]
AAFC	Anti-Aircraft Fire Control
AAFCE	Allied Air Forces Central Europe [NATO]
AAFNE	Allied Air Forces Northern Europe [NATO]
AAFSE	Allied Air Forces Southern Europe [NATO]
AAFSS	Advanced Aerial Fire Support System

AAG	Aeronautical Assignment Group [US]
	Acquisition Advisory Group
	Association of American Geographers
AAGS	American Association for Geodetic Surveying
AAH	Advanced Attack Helicopter
AAI	Air Aid to Intercept
AAIEE	Associate of the American Institute of Electrical Engineers
AAIM	American Association of Industrial Management
AAIMME	Associate of the American Institute of Mining and Metallurgical Engineers
AAIMS	An Analytical Information Management System
AAL	Absolute Assembly Language
	Arctic Aeromedical Laboratory [US Air Force]
	Arctic Aerospace Laboratory [US Air Force]
AALA	American Association for Laboratory Accreditation
AALAS	American Association for Laboratory Animal Science
AALC	Advanced Airborne Launch Center [US]
	Amphibious Assault and Landing Craft
	Amplified Automatic Level Control
AALMG	Anti-Aircraft Light Machine Gun
AALUE	Asymptotically Admissible Linear Unbiased Estimator
AAM	Air-to-Air Missile
	American Academy of Mechanics
	American Academy of Microbiology
AAMG	Anti-Aircraft Machine Gun
AAMGAP	Achieve as Many Goals as Possible
AAMI	Association for the Advancement of Medical Instrumentation
AAML	Arctic Aeromedical Laboratory [US Air Force]
AAMVA	American Association of Motor Vehicle Administrators
AAN	Aminoacetonitrile
AAO	Anti-Air Output
AAP	Acridinylaminopropanol
	Apollo Applications Program [NASA]
AAPG	American Association of Petroleum Geologists
AAPL	Additional Programming Language
AAPP	Auxiliary Airborne Power Plant
AAPS	American Association for the Promotion of Science
AAPSE	American Association of Professors in Sanitary Engineering
AAPT	Association of Asphalt Paving Technologists
AAPU	Auxiliary Airborne Power Unit
AAPWISM	American Association of Public Welfare Information Systems Management
AAR	Association of American Railroads
	Automotive Aftermarket Retailer
	Average Annual Rainfall
AARPS	Air Augmented Rocket Propulsion System
AARR	Argonne Advanced Research Reactor [US]

AAS	Academy of Applied Science [US]
	Advanced Antenna System
	American Astronautical Society
	American Astronomical Society
	Arithmetic Assignment Statement
	Atomic Absorption Spectrometer
	Atomic Absorption Spectroscopy
	Australian Academy of Science
	Automated Accounting System
AASB	American Association of Small Business
AASC	Association for the Advancement of Science in Canada
AASG	Association of American State Geologists
AASHO	American Association of State Highway Officials
AASHTO	American Association of State Highway and Transportation Officials
AASME	Associate of the American Society of Mechanical Engineers
AASMTC	Association of American Steel Manufacturers Technical Committee
AASO	American Association of Ship Owners
AASP	Advanced Automated Sample Processor
AASR	Airports and Airways Surveillance Radar
AASRI	Arctic and Antarctic Scientific Research Institute
AASW	American Association of Science Workers
AAT	Analytic Approximation Theory
AATB	Army Aviation Test Board [US Army]
AATBN	Aromatic Amine-Terminated Butadiene/Acrylonitrile
AATC	Anti-Aircraft Training Center [US]
	Automatic Air Traffic Control
AATCC	American Association of Textile Chemists and Colorists
AATMS	Advanced Air Traffic Management System
AA-tRNA	Aminoacyl Transfer Ribonucleic Acid
AATT	American Association for Textile Technology
AAU	Association of Atlantic Universities
AAUP	American Association of University Professors
AAV	Airborne Assault Vehicle
AAVCS	Automatic Aircraft Vectoring Control System
AAVD	Automatic Alternate Voice/Data
AAVS	Aerospace Audiovisual Service
AAW	Air-Acetylene Welding
	Anti-Air Warfare
AB	Abscisic Acid
	Adapter Booster
	Aeronautical Board
	Afterburner
	Air Blast
	Alberta [Canada]
	Alcian Blue
	Aminoazobenzene
	Anchor Bolt
	Angle Bracket
	Application Block
	Arc Brazing
	(Artium Baccalaureus) - Bachelor of Arts

ABA	Aminobutyric Acid
ABAC	A Basic Coursewriter
	Association of Balloon and Airship Constructors
ABACUS	Air Battle Analysis Center Utility System
ABAF	Alcian Blue Aldehyde Fuchsin
abamp	abampere [unit]
ABAR	Advanced Battery Acquisition Radar
ABB	Abbreviation
	American Board of Bioanalysis
ABBE	Advisory Board on the Built Environment
ABBF	Association of Brass and Bronze Founders
ABBMM	Association of British Brush Machinery Manufacturers
ABBR	Abbreviation
ABC	Abridged Building Classification
	Already Been Converted
	Approach by Concept
	Armored Bushed Cable
	Atomic, Biological and Chemical
	Automated Bleed Compensation
	Automatic Bandwidth Control
	Automatic Bias Compensation
	Automatic Brightness Control
ABCA	Association of Biological Collections Appraisers
ABCB	Air-Blast Circuit Breaker
ABCC	American Board of Clinical Chemists
ABCCTC	Advanced Base Combat Communication Training Center [US]
ABCD	Atomic, Biological, Chemical and Damage (Control)
ABCM	Association of British Chemical Manufacturers
ABCS	Advisory Board for Cooperative Systems [US]
ABCST	American-British-Canadian Standards
ABD	Adhesive Bonding
ABDC	After Bottom Dead Center
ABDL	Automatic Binary Data Link
ABEC	Annular Bearing Engineers Committee
ABEI	Aminobutylethylisoluminol
ABEND	Abnormal Ending
ABES	Aerospace Business Environment Simulator
	Automatic Burn-in and Environmental System
ABET	Accreditation Board for Engineering and Technology
ABETS	Airborne Beacon Test Set
ab ex	(ab extra) - from without
ABFA	Azobisformamide
ABG	Aural Bearing Generator
ab init	(ab inito) - from the beginning
ABK	Arrott-Belov-Kouvel (Plot)
ABL	Atlas Basic Language
	Army Biological Laboratory [US Army]
	Automated Biological Laboratory [NASA]
ABLB	Alternate Binaural Loudness Balancing
ABLE	Activity Balance Line Evaluation
ABLUE	Asymptotically Best Linear Unbiased Estimator

ABM	Anti-Ballistic Missile
	Automated Batch Mixing
	Asynchronous Balanced Mode
ABMA	American Boiler Manufacturers Association
	Army Ballistic Missile Agency [US]
ABMAC	Association of British Manufacturers of Agricultural Chemicals
ABMD	Air Ballistics Missile Division [US Air Force]
ABMDA	Advanced Ballistic Missile Defense Agency [US Army]
ABMEWS	Anti-Ballistic Missile Early Warning System
ABMIS	Anti-Ballistic Missile Intercept System
ABMS	American Bureau of Metal Standards
ABN	Airborne
ABO	Astable Blocking Oscillator
ABORT	Abnormal Termination
ABOS	Advanced Bombardment System
ABP	Actual Block Processor
	Adjustable Ball Pin
	Aminobiphenyl
ABR	Abrasive
	Abridgement
ABRC	Advisory Board for the Research Councils
ABRES	Advanced Ballistic Reentry System
ABRSV	Abrasive
ABS	Absolute
	Abstract
	Acrylonitrile-Butadiene-Styrene
	Air Break Switch
	American Biological Society
	American Bureau of Shipping
	Anti-Lock Brake System
	Association on Broadcasting Standards [US]
ABSLDR	Absolute Loader
ABSM	Alberta Bureau of Surveying and Mapping [Canada]
ABST	Abstract
ABSTI	Advisory Board on Scientific and Technological Information
ABSW	Air-Break Switch
	Association of British Science Writers
ABT	Air Blast Transformer
ABTF	Airborne Task Force
ABTICS	Abstract and Book Title Index Card Service [of ISI]
ABTS	Azinobisethylbenzthiazolinesulfonic Acid
ABTSS	Airborne Transponder Subsystem
ABU	Asian Broadcasting Union
ABV	Absolute Value
ABYC	American Boat and Yacht Council
AC	Acoustic Coupler
	Aerodynamic Center
	Air Conduction
	Air Conditioning
	Air Cooling
	Alternating Current
	Ammonium Citrate
	Analog Computer
	Anti-Coincidence Counter
	Arc Cutting
	Arctic Circle

	Armored Cable
	Armored Clad (Cable)
	Automatic Checkout
	Automatic Computer
	Automatic Control
A/C	Asbestos Cement
Ac	Actinium
	Alto-Cumulus
ACA	Adjacent Channel Attenuation
	Agricultural Computer Association
	Alberta Construction Association [Canada]
	American Communications Association
	American Commuters Association
	American Crystallographic Association
	Aminocephalosporanic Acid
	Ammonical Copper Arsenate
	Association of Commuter Airlines [US]
	Australian Council for Aeronautics
	Automatic Circuit Analyzer
	Automatic Communication Association
ACAD	Academic
	Academician
	Academy
ACAH	Acylcholine Acyl Hydrolase
ACAP	Automatic Circuit Analysis Program
ACAST	Advisory Committee on the Application of Science and Technology [of UNESCO]
ACAT	Associate Committee on Air-Cushion Technology
ACAU	Automatic Calling and Answering Unit
ACB	Access Method Control Block
	Adjusted Cost Base
	Air Circuit Breaker
	Application Control Block
ACC	Acceptance
	Accumulation
	Accumulator
	Acid Copper Chromate
	Adapter Common Card
	Advanced Carbon/Carbon
	Air Control Center
	Air Coordinating Committee
	Administrative Committee on Coordination [of UN]
	Antenna Control Console
	Application Control Code
	Area Control Center
	Automatic Contrast Control
ACCA	Accelerated Capital Cost Allowance
	Asynchronous Communications Control Attachment
ACCAP	Autocoder-to-COBOL Conversion and Program
ACCCIW	Advisory Committee to the Canadian Center for Inland Waters
ACC&CE	Association of Consulting Chemists and Chemical Engineers [also: ACCCE]
ACCD	Accelerated Construction Completion Date
ACCE	Advanced Composites Conference and Exhibition
ACCEL	Acceleration

	Automated Circuit Card Etching Layout
ACCES	Advanced Computing and Communications Educational System [Canada]
ACCESS	Accessory
	Aircraft Communication Control and Electronic Signaling System
	Automatic Computer-Controlled Electronic Scanning System
ACCG	American Conference on Crystal Growth
ACCHAN	Access Channel
	Allied Command Channel [NATO]
ACCS	Automatic Checkout and Control System
ACCT	Account
	Alliance for Coal and Competitive Transportation
ACCUM	Accumulation
ACCUMR	Accumulator
ACCW	Alternating Current Continuous Wave
ACD	Acid Citrate Dextrose
	Antenna Control Display
	Automatic Call Distributor
AC/DC	Alternating Current/Direct Current
ACDO	Air Carrier District Office [US]
ACE	Acceptance Checkout Equipment [NASA]
	Air Cushion Equipment
	Airspace Coordination Element
	Allied Command Europe [NATO]
	Altimeter Control Equipment
	Ambush Communication Equipment [Army]
	American Council on Education
	Area Control Error
	Association of Conservation Engineers
	Association for Continuing Education
	Attitude Control Electronics
	Audio Connecting Equipment
	Automated Cable Expertise
	Automated Cost Estimate
	Automatic Checkout Equipment
	Automatic Clutter Eliminator
	Automatic Computer Evaluation
	Automatic Computing Engine
	Aviation Construction Engineers
ACEAA	Advisory Committee on Electrical Appliances and Accessories
ACEC	American Consulting Engineers Council
	Association of Consulting Engineers of Canada
ACEE	Aircraft Energy Efficiency
ACEL	Aerospace Crew Equipment Laboratory [US Navy]
ACEM	Association of Consulting Engineers of Manitoba [Canada]
ACerS	American Ceramic Society
ACES	Acetamidoaminoethanesulfonic Acid
	Acoustic Containerless Experiment System
	Aerosol Containment Evacuation System
	Air Collection Engine System
	Air Collection and Enrichment System
	Association of Consulting Engineers of Saskatchewan [Canada]
	Automatic Checkout and Evaluation System [US Air Force]

ACE-S/C	Acceptance Checkout Equipment-Spacecraft
ACET	Acetylene
	Acetone
ACET A	Acetic Acid
ACF	Access Control Facility
	Advanced Communications Function
	Air Combat Fighter [US Air Force]
	Alternate Communications Facility
	$Al_2O_3 - CaO - (FeO+MgO)$
	Area Computing Facility
	Autocorrelation Function
ACFG	Automatic Continuous Function Generation
ACFT	Aircraft
ACG	Air Cargo Glider
AcGEPC	Alkyl Acetylsynglycerylphosphorylcholine
ACGIH	American Conference of Governmental Industrial Hygienists
ACGR	Associate Committee on Geotechnical Research
ACH	Acetylcholine
	Adrenal Cortical Hormone
	Association of Computers and Humanities
	Attempts per Circuit per Hour
ACHE	Acetylcholinesterase
ACI	Adjacent Channel Interference
	Air Combat Intelligence
	Alloy Casting Institute
	American Concrete Institute
	Automatic Card Identification
ACIA	Asynchronous Communications Interface Adapter
ACIB	Air Cushion Ice-Breaking Bow
ACIC	Aeronautical Chart and Information Center [US]
	Alberta Council for International Cooperation [Canada]
	Associate of the Canadian Institute of Chemistry
ACID	Association of Canadian Industrial Designers
	Automated Classification and Interpretation of Data
ACIGY	Advisory Council of the International Geophysical Year
ACIIW	American Council of the International Institute of Welding
ACIL	American Council of Independent Laboratories
ACIM	Accident Cost Indicator Model
	Axis-Crossing Interval Meter
ACIS	Association for Computing and Information Sciences
	Automated Client Information Service
ACK	Acknowledgment (Character)
ACKT	Acknowledgment
ACL	Acetal
	Aeronautical Computers Laboratory [US Navy]
	Allowable Cabin Load
	Antigen-Carrier Lipid
	Application Control Language
	Applied Color Label

	Association for Computational Linguistics
	Atlas Commercial Language
	Audit Command Language
	Automated Coagulation Laboratory
	Automatic Circuit Layout
ACLANT	Allied Command Atlantic [NATO]
ACLG	Air-Cushion Landing Gear
ACLO	Agena Glass Lunar Orbiter
ACLS	Air-Cushion Landing System
	All-Weather Carrier Landing System
	American Council of Learned Societies
	Automated Control Landing System
ACM	Active Countermeasures
	Advanced Cruise Missile
	Air Composition Monitor
	Area Contamination Monitor
	Asbestos-Covered Metal
	Associated Colleges of the Midwest [US]
	Association for Computing Machinery
	Association of Crane Makers
	Associative Communication Multiplexer
	Atmospheric Corrosion Monitor
	Automatic Clutter Mapping
ACME	Advanced Computer for Medical Research
	Association of Consulting Management Engineers
ACM/GAMM	Association for Computing Machinery/German Association for Applied Mathematics and Mechanics
ACMMM	Annual Conference on Magnetism and Magnetic Materials
ACMO	Afloat Communications Management Office [US]
ACMR	Air Combat Maneuvering Range
ACMRI	Air Combat Maneuvering Range Instrumentation
ACMRR	Advisory Committee on Marine Resource Research
ACMS	Advanced Configuration Management System
	Application Control/Management System
ACMSC	Association for Computing Machinery Standards Committee
ACN	Artificial Cloud Nucleation
	Asbestos Cloth Neck
ACNBC	Associate Committee on the National Building Code
ACNOCOMM	Assistant Chief of Naval Operations for Communications
ACO	Auto Call Originator
ACOM	Automatic Coding System
ACOPP	Abbreviated COBOL Preprocessor
ACORN	Associative Content Retrieval Network
ACOS	Automatic Control System
ACP	Acyl Carrier Protein
	Advanced Computational Processor
	Advanced Cooperative Project [NASA]
	Aerospace Computer Program [US Air Force]
	Agricultural Conservation Program
	Anodal Closing Picture

	Anticoincidence Point
	Azimuth Change Pulse
AcP	Acid Phosphatase
ACPA	Association of Computer Programmers and Analysts
ACPF	Acoustic Containerless Processing Facility
ACPM	Acoustic Chamber Position Module
	Attitude Control Propulsion Motor
ACPU	Auxiliary Computer Power Unit
ACR	Aircraft Control Room
	Airfield Control Radar
	Alaskan Communications Region [US Air Force]
	American College of Radiology
	Antenna Coupling Regulator
	Approach Control Radar
	Automatic Call Recording
	Automatic Compression Regulator
ACRE	Automatic Checkout and Readiness
ACRES	Airborne Communications Relay Station [US Air Force]
ACRI	Air Conditioning and Refrigeration Institute [US]
ACRL	Association of College and Research Libraries [US]
ACRS	Advisory Committee on Reactor Safeguards [US]
ACRT	Analysis Control Routine
ACS	Absolute Contamination Standard
	Accumulator Switch
	Acrylonitrile Chlorinated Styrene
	Adrenocortico Steroid
	Advanced Communications Service
	Advanced Computer Services
	Advanced Computer System
	Aeronautical Communications Subsystem
	Agena Control System
	Aircraft Control Surveillance
	Alaska Communication System [US]
	Alternating Current Synchronous
	Altitude Control System
	American Carbon Society
	American Cartographic Society
	American Ceramic Society
	American Chemical Society
	Anodal Closing Sound
	Application Customizer Service
	Assembly Control System
	Attitude Command System
	Attitude Control System
	Austempered Cast Steel
	Australian Computer Society
	Auto-Calibration System
	Automated Communications Set
	Automatic Checkout System
	Automatic Control System
	Auxiliary Core Storage
ACSA	Allied Communications Security Agency [NATO]
ACSB	Amplitude Compandered Sideband
ACSCE	Army Chief of Staff for Communications Electronics [US]
ACSF	Attack Carrier Striking Force
ACSM	American Congress on Surveying and Mapping
	Associate of the Cambourne School of Mines [US]
ACSN	Appalachian Community Service Network
ACSP	Advanced Control Signal Processor [NASA]
	Advisory Council on Scientific Policy [UK]
ACSR	Aluminum Cable Steel-Reinforced
	Aluminum-Clad, Steel-Reinforced
	Aluminum Conductor Steel-Reinforced
ACSR/GA	Aluminum Conductor Steel-Reinforced using Class A Zinc-Coated Steel Wire
ACSR/GB	Aluminum Conductor Steel-Reinforced using Class B Zinc-Coated Steel Wire
ACSR/GC	Aluminum Conductor Steel-Reinforced using Class C Zinc-Coated Steel Wire
ACSS	Air Combat and Surveillance System
	Analog Computer Subsystem
ACST	Access Time
	Acoustics
ACSTTO	Association of Certified Survey Technicians and Technologists of Ontario [Canada]
ACSU	Association for Computer Science Undergraduates
ACT	Action
	Activation
	Actual
	Actuation
	Adrenocortiotropin
	Advisory Council on Technology
	Air Control Team
	Air Cushion Trailer
	Algebraic Compiler and Translator
	American College Test
	Analog Circuit Technique
	Applied Circuit Technology
	Area Correlation Tracker [US Air Force]
	Association of Cytogenetic Technologists
	Asymmetric Crystal Topography
	Automatic Code Translation
ACTE	Agkistrodon Contortrix Thrombin-Like Enzyme
	Association of Commercial and Technical Employees
ACTF	Altitude Control Test Facility
ACTG	Actuating
ACTH	Adrenocortiotrophic Hormone
ACTO	Automatic Computing Transfer Oscillator
ACTOL	Air-Cushion Takeoff and Landing
ACTP	Advanced Composite Thermoplastics
	Advanced Computer Techniques Project
ACTR	Actuator
ACTRAN	Analog Computer Translator
	Autocoder-to-COBOL Translation
ACTRUS	Automatically-Controlled Turbine Run-up System
ACTS	Acoustic Control and Telemetry System
	Association of Competitive Telecommunications Suppliers
	Automatic Computer Telex Services

ACU	Address Control Unit
	Antenna Control Unit
	Arithmetic Control Unit
	Association of Computer Users
	Automatic Call Unit
ACUTE	Accountants Computer Users Technical Exchange
ACV	Air Cushion Vehicle
	Alarm Check Valve
	Armored Command Vehicle
AC&W	Aircraft Control and Warning
ACWIS	American Committee for the Weizmann Institute of Science
ACWL	Army Chemical Warfare Laboratory [US Army]
AC&WS	Air Control and Warning Station
AD	Access Door
	Advanced Design
	Air Drying
	Ampere Demand (Meter)
	Anno Domini - in the year of our lord
	Anode
	Anodal Deviation
	Antiphase Domain
	Area Drain
	As Drawn
	Automatic Detection
A/D	Altitude/Depth
	Analog-to-Digital
ADA	Acetamidoiminodiacedic Acid
	Action Data Automation
	Adenosine Deaminase
	Airborne Data Automation
	Air Defense Area [US Army]
	American Dairy Association
	Atom Development Administration
	Automatic Data Acquisition
	Automatic Distillation Analyzer
ADAC	Automated Direct Analog Computer
ADACC	Automatic Data Acquisition and Computer Complex [US Air Force]
ADAM	Advanced Data Management
	Advanced Direct-Landing Apollo Mission [NASA]
	Air Deflection and Modulation
	Automatic Distance and Angle Measurement
ADAPS	Anthropometric Design Assessment Program System
	Automatic Display and Plotting System
ADAPSO	Association of Data Processing Service Organizations [US]
ADAPT	Adaption of Automatically Programmed Tools
ADAPTICOM	Adaptive Communication
ADAPTS	Air-Delivered Antipollution Transfer System
ADAR	Advanced Design Array Radar
	Automatic Data Acquisition Routine
ADAS	Automatic Data Acquisition System
	Automatic Dialing Alarm System
ADAT	Automatic Data Accumulation and Transfer
ADATS	Air Defense Antitank System

ADB	Activated Diffusion Bonding
	Adhesive Bonding
ADBMS	Available Database Management System
ADC	Airborne Digital Computer
	Air Data Computer
	Analog-Digital Converter
	Analog-to-Digital Converter
	Anodal Duration Contraction
	Antenna Dish Control
	Automatic Data Collection
ADCA	Aminodecephalosporanic Acid
ADCAD	Airways Data Collection and Distribution
ADCC	Antibody-Dependent Cellular Cytotoxicity
	Air Defense Control Center
ADCCP	Advanced Data Communications Control Procedure
ADCE	Attitude Determination and Control Electronics
ADCI	American Die Casting Institute
ADCON	Analog-Digital Converter
ADCS	Advanced Defense Communications Satellite
	Attitude Determination and Control Subsystem
ADCSP	Advanced Defense Communications Satellite Program
ADD	Addendum
	Addition
	Aerospace Digital Development
ADDAR	Automatic Digital Data Acquisition and Recording
ADDAS	Automatic Digital Data Assembly System
ADDDS	Automatic Direct Distance Dialling System
ADDER	Automatic Digital Data Error Recorder
ADDIT	Addition
ADDL	Additional
ADDN	Automated Defense Data Network
ADDPEP	Aerodynamic Deployable Decelerator Performance Evaluation Program
ADDR	Adder
	Address
	Address Register
ADDROUT	Address Out
ADDS	Automatic Data Digitizing System
ADE	Air Density Explorer
	Armament Design Establishment [UK]
	Automated Design Engineering
	Automated Drafting Equipment
ADEE	Automated Design and Engineering for Electronics
ADEM	Adaptively Data Equalized Modem
ADEPT	Automated Direct Entry Packaging Technique
	Automatic Data Extractor and Plotting Table
ADES	Automatic Digital Encoding System
ADEU	Automatic Data Entry Unit
ADF	Aerial Direction Finding
	Airborne Direction Finder
	Automatic Direction Finder
ad fin	(ad finem) - to the end
ADFSC	Automatic Data Field Systems Command [US Army]

ADH	Adhesive
	Antidirected Hamiltonian
	Antidiuretic Hormone
ad h l	(ad hunc locum) - at the place
ad hoc	(ad hoc) - for this
ADI	Acceptable Daily Intake
	Alternating Direction Implicit
	Altitude Direction Indicator
	American Documentation Institute [now: ASIS]
	Austempered Ductile Iron
	Automatic Direction Indicator
ADIDAC	Argonne Digital Data Acquisition Computer [ANL, US]
ad inf	(ad infinitum) - to infinity
ad init	(ad initium) - at the beginning
ad int	(ad interim) - in the interim (in the meantime)
ADINTELCEN	Advanced Intelligence Center
ADIOS	Automatic Digital Input-Output System
ADIS	Advanced Data Acquisition, Imaging and Storage
	Air Defense Integrated System
	Association for the Development of Instructional Systems
	Automatic Data Interchange System
ADISP	Aeronautical Data Interchange System Panel
ADIT	Analog-Digital Integration Translator
ADJ	Adjacent
	Adjoining
	Adjunct
	Adjustment
ADL	Automatic Data Link
ad lib	(ad libitum) - freely
ADLIPS	Automated Data Link Plotting System
ad loc	(ad locum) - at the place
ADLS	Automatic Drag-Limiting System
ADM	Activity Data Method
	Adaptive Delta Modulation
	Air Decoy Missile
	Administration
	Administrator
	Analog Data Module
	Asymmetric Dimer Model
	Asynchronous Disconnected Mode
	Atomic Demolition Munition
	Automatic Drafting Machine
ADMIN	Administration
	Administrator
ADMIRAL	Automatic and Dynamic Monitor with Immediate Relocation Allocation and Loading
ADMIRE	Automatic Diagnostic Maintenance Information Retrieval
ADMS	Asynchronous Data Multiplexer Synchronizer
	Automatic Digital Message Switch
ADMSC	Automatic Digital Message Switching Center
ADMT	Air-Dried Metric Ton
ADN	Adiponitrile
ADNAC	Air Defense of the North American Continent

ADONIS	Automatic Digital On-Line Instrumentation System
ADOPE	Automatic Decisions Optimizing Predicted Estimates
ADP	Adapter
	Adenosine Diphosphate
	Airborne Data Processor
	Air Defense Position
	Airport Development Program
	Automatic Data Processing
ADPC	Automatic Data Processing Center
ADPCM	Association for Data Processing and Computer Management
ADPE	Automatic Data Processing Equipment
ADPE/S	Automatic Data Processing Equipment/System
ADPESO	Automatic Data Processing Equipment Selection Office
ADPG	Adrenosine Diphosphoglucose
ADPLAN	Advancement Planning
ADPS	Automatic Data Processing System
ADPSO	Association of Data Processing Service Organizations
ADPT	Adapter
ADPTR	Adapter
ADR	Address
	Aircraft Direction Room
	Aviation Design Research [US Navy]
ADRAC	Automatic Digital Recording and Control
ADRMP	Auto-Dialed Recorded Message Player
ADRS	Address Register
	Analog-to-Digital Data Recording System
ADRT	Analog Data Recorder Transcriber
ADS	Accessory Drive System
	Accurately-Defined System
	Activity Data Sheet
	Address
	Administrative Data System
	Advanced Data System
	Advanced Dressing Station
	Aircraft Development Services [of FAA]
	Air Defense Sector
	Alloy Descaling Salt
	Antidiuretic Substance
	Application Development System
	Asymptotic Diffuse Scattering
	Automatic Depressurization System
	Automatic Door Seal
ADSA	American Dairy Sciences Association
	Atomic Defense Support Agency
ADSAF	Automated Data Systems for the Army in the Field [US]
ADSAS	Air-Derived Separation Assurance System
ADSC	Automatic Data Service Center
	Automatic Digital Switching Center
ADSCOM	Advanced Shipboard Communication
ADSF	Automated Directional Solidification
ADSIA	Allied Data System Interoperability Agency
ADSM	Air Defense Suppression Missile
ADSS	A Diesel Supply Set
	Aircraft Damage Sensing System

	Australian Defense Scientific Service
ADST	Atlantic Daylight Saving Time
ADSTAR	Automatic Document Storage and Retrieval
ADS-TP	Administrative Data System - Teleprocessing
ADSUP	Automatic Data Systems Uniform Practices
ADT	Adenosine Triphosphate
	Adjustable Digital Thermometer
	Admission, Discharge and Transfer
	Atlantic Daylight Time
	Automatic Data Translator
ADTECH	Advanced Decoy Technology
ADTT	Average Daily Truck Traffic
ADTU	Auxiliary Data Translator Unit
ADU	Accumulation Distribution Unit
	Automatic Data Unit
	Automatic Dialing Unit
ADV	Acid Demand Value
	Advance
	Advertisement
ad val	(ad valorem) - according to value
ADVISOR	Advanced Integrated Safety and Optimizing Computer
ADVT	Advertisement
ADVTG	Advertising
ADX	Automatic (Data) Exchange
ADXRD	Angle-Dispersive X-Ray Diffraction
AE	Acoustic Emission
	Acrylic Emulsion
	Advanced ECL (– Emitter-Coupled Logic)
	Aeroelectronics
	Agricultural Engineer(ing)
	Air Escape
	Aminoethyl
	Application Engineer
	Architect Engineer
	Arithmetic Element
	Arithmetic Expression
	Atmospheric Explorer
	Atomic Energy
	Auxiliary Equation
A/E	Absorption/Emission (Ratio)
A&E	Azimuth and Elevation
AEA	Agricultural Engineers Association
	Aircraft Electronics Association
	Air-Entraining Agent
	American Electronics Association
	Atomic Energy Act
	Atomic Energy Authority
	Automotive Electric Association
AEAPS	Auger Electron Appearance Potential Spectroscopy
AEB	Atomic Energy Bureau
AEC	Airship Experimental Center [US Navy]
	American Engineering Council
	Aminoethyl Cellulose
	Aminoethyl Cysteine
	Atomic Energy Commission
AECB	Atomic Energy Control Board [Canada]
AECIP	Atlantic Energy Conservation Investment Program [Canada]

AECL	Advanced Emitter-Coupled Logic
AECT	Association for Educational Communications and Technology
AED	ALGOL Extended for Design
	Associated Equipment Distributors
	Automated Engineering Design
AEDC	Atlantic Engineering Design Competition
	Arnold Engineering Development Center [US]
AEDP	Association for Educational Data Processing
AEDS	Association for Educational Data Systems
	Atomic Energy Detection System
AEDU	Admiralty Experimental Diving Unit [UK]
AEE	Airborne Evaluation Equipment
	Association of Energy Engineers
	Atomic Energy Establishment [UK]
AEEC	Airlines Electronic Engineering Committee [US]
AEEE	Army Equipment Engineering Establishment
AEEL	Aeronautical Electronics and Electrical Laboratory [US Navy]
AEEMS	Automatic Electric Energy Management System
AeEng	Aeronautical Engineer
AEF	Aerospace Education Foundation
	Aviation Engineer Force
AEG	Active Element Group
	Association of Engineering Geologists
	Association of Exploration Geochemists
	Axial-Emission Gauge
AEI	Aerial Exposure Index
	Alternate Energy Institute
	Associated Electrical Industries [US]
	Automatic Error Interrogation
	Average Efficiency Index
AEL	Aeronautical Engine Laboratory [US]
AEM	Analytical Electron Microscopy
	Assemblability Evaluation Method
AEMB	Alliance for Engineering in Medicine and Biology
AEMS	American Engineering Model Society
AEMSA	Army Electronics Material Support Agency
AEO	Air Engineer Officer
AEOC	Aminoethylhomocysteine
AEP	American Electric Power
	Averaged Evoked Potential
AEPG	Army Electronic Proving Ground [US]
AER	Advanced Electric Reactor
	After Engine Room
AERA	American Educational Research Association
	Automotive Engine Rebuilders Association
AERCB	Alberta Energy Resources Conservation Board [Canada]
AERD	Agricultural Engineering Research and Development
AERDL	Army Electronics Research and Development Laboratory [US]
AERE	Atomic Energy Research Establishment [UK]
AERG	Advanced Environmental Research Group

AERNO	Aeronautical Equipment Reference Number
AERO	Alternate Energy Resources Organization
	Aeronautics
AERODYN	Aerodynamics
AEROSAT	Aeronautical Satellite System
AEROSPACECOM	Aerospace Communications
AES	Aerospace and Electronic Systems [of IEEE]
	Aerospace Electrical Society [US]
	Airways Engineering Society [US]
	American Electrochemical Society
	American Electronical Society
	American Electroplaters Society
	Apollo Experiment Support [NASA]
	Apollo Extension System [NASA]
	Artificial Earth Satellite [NASA]
	Atmospheric Environment Service [Canada]
	Atomic Emission Spectroscopy
	Audio Engineering Society
	Auger Electron Spectroscopy
	Auger Electron Spectrum
	Automated Environmental Station
AESA	Association of Environmental Scientists and Administrators
AESC	American Engineering Standards Committee
	Automatic Electronic Switching Center
AESE	Association of Earth Science Editors
AESF	Association for Electroplating and Surface Finishing
AESFS	American Electroplaters and Surface Finishers Society
AESG	Axial-Emission Suppression Gauge
AESMG	Axial-Emission Self-Modulating Gauge
AESOP	An Evolutionary System for On-Line Processing
	An Experimental Structure for On-Line Planning
	Automated Environmental Station for Overtemperature Probing
AET	Acoustic Emission Technology
	Aminoethylisothiouronium Bromide
AETB	Alumina-Enhanced Thermal Barrier
AETE	Aerospace Engineering Test Establishment
AEU	Amalgamated Engineering Union
AEW	Airborne Early Warning
AEW&C	Airborne Early Warning and Control
AEWES	Army Engineer Waterways Experiment Station [US]
AEWTU	Airborne Early Warning Training Unit
AF	Address Field
	Africa(n)
	Air Filter
	Aldehyde Fuchsin
	Alternating Fields
	Angle Frame
	Antiferromagnetic
	Antiferromagnetism
	Audio Frequency
	Automatic Following
A/F	Across Flat
	Air/Fuel (Ratio)
AFA	Absorption-Filtration-Adsorption (Process)
	Air Force Association [US]
	American Federation of Astrologers
	American Foundrymen's Association
AFADS	Advanced Forward Air Defense System
AFAL	Air Force Avionics Laboratory [US]
AFAPL	Air Force Aero-Propulsion Laboratory [US]
AFAR	Advanced Field Array Radar
AFA-SEF	Air Force Association - Space Education and Foundation
AFASE	Association for Applied Solar Energy [US]
AFB	Acoustic Feedback
	Air Force Base
	Antifriction Bearing
AFBC	Atmospheric Fluidized-Bed Combustion
AFBMA	Antifriction Bearing Manufacturers' Association
AFBMD	Air Force Ballistic Missile Division [US]
AFBS	Acoustic Feedback System
AFC	Air/Fuel Concentration
	Amplitude Frequency Characteristic
	Area Frequency Coordinator
	Atomic Fluid Cells
	Automatic Flatness Control
	Automatic Focus Control
	Automatic Frequency Control
AFCASI	Associate Fellow of the Canadian Aeronautics and Space Institute
AFCC	Available-Frame Capacity Count
AFCCE	Association of Federal Communications Consulting Engineers
AFCE	Automatic Flight Control Equipment
AFCEA	Armed Forces Communications and Electronics Association [US]
AFCENT	Allied Forces Central Europe [NATO]
AFCOMMSTA	Air Force Communications Station [US]
AFCRL	Air Force Cambridge Research Laboratories [US]
AFCS	Adaptive Flight Control System
	Air Force Communications System [US]
	Automatic Flight Control System
	Avionic Flight Control System
AFCSS	Air Force Communications Support System [US]
AFD	Accelerated Freeze Drying
	Automatic Forging Design
AFDASTA	Air Force Data Station [US]
AFDDA	Air Force Director of Data Automation [US]
AFEI	Americans for Energy Independence
AFELIS	Air Force Engineering and Logistics Information System [US]
AFETR	Air Force Eastern Test Range [US]
AFF	Above Finished Floor
AFFDL	Air Force Flight Dynamics Laboratory [US]
AFFF	Aqueous Film-Forming Foam
AFFP	Air-Filed Flight Plan
AFFTC	Air Force Flight Test Center [US]
AFG	Analog Function Generator
	Arbitrary Function Generator
AFGP	Antifreeze Glycoproteins
AFGWC	Air Force Global Weather Center [US]
AFHP	Advanced Flash Hydropyrolysis

AFI	Automatic Fault Isolation
AFICCS	Air Force Interim Command and Control System [US]
AFIPS	American Federation of Information Processing Societies
AFIRMS	Air Force Integrated Readiness Measurement System
AFIS	American Forces Information Service
AFIT	Air Force Institute of Technology [US]
AFL	Alberta Federation of Labour [Canada]
	American Federation of Labour
	Automatic Fault Location
AFLC	Air Force Logistics Command [US]
AFL-CIO	American Federation of Labour - Congress of Industrial Organizations
AFLCON	AFLC Operations Network [US]
AFM	Abrasive Flow Machining
	$Al_2O_3 - FeO - MgO$
	Analysis and Forecasting Mode
	Antifriction Metal
	Automatic Flight Management
AFMAG	Audio-Frequency Magnetics
AFMDC	Air Force Missile Development Center [US]
AFML	Air Force Materials Laboratory [US]
AFMR	Antiferromagnetic Resonance
AFMS	American Federation of Mineralogical Societies
AFMTC	Air Force Missile Test Center [US]
AFN	American Forces Network
AFNETSTA	Air Force Networks Station [US]
AFNORTH	Allied Forces Northern Europe [NATO]
AFOSR	Air Force Office of Scientific Research [US]
AFP	Adiabatic Fast Passage
	Association for Finishing Processes [of SME, US]
	Automatic Flow Process
AFPA	Australian Fire Protection Association
	Automatic Flow Process Analysis
AFPAM	Automatic Flight Planning and Monitoring
AFPAV	Airfield Pavement
AF/PC	Automatic Frequency/Phase Controlled
AFPL	Air Force Packaging Laboratory [US]
AFR	Africa(n)
	Alternating Frequency Rejection
	Amplitude Frequency Response
	Application Function Routine
	Automatic Field/Format Recognition
AFRCE	Air Force Regional Civil Engineer [US]
AFRPL	Air Force Rocket Propulsion Laboratory [US]
AFRRI	Armed Forces Radiobiology Research Institute [US]
AFRS	Armed Forces Radio Service
AFRSI	Advanced Flexible Reusable Surface Installation
AFRTS	Armed Forces Radio and Television Service
AFS	Aeronautical Fixed Service
	American Foundrymen's Society
	Atomic Fluorescence Spectroscopy
	Audio-Frequency Shift
	Auxiliary Fire Service

AFSAB	Air Force Science Advisory Board [US]
AFSAT	African Satellite
AFSATCOM	Air Force Satellite Communications
AFSC	Air Force Systems Command [US]
	Automatic Focus Control
AFSM	Association of Field Service Managers
AFSOUTH	Allied Forces Southern Europe [NATO]
AFSR	Argonne Fast Source Reactor [US]
AFSWC	Air Force Special Weapons Center [US]
AFT	Adaptive Ferroelectric Transformer
	Analog Facility Terminal
	Automatic Fine Tuning
AFTAC	Air Force Technical Applications Center [US]
AFTE	American Federation of Technical Engineers
AFTER	Air Force Thermionic Engineering and Research
AFTI	Advanced Fighter Technology Integration [US Air Force]
AFTN	Aeronautical Fixed Telecommunications Network
AFTRCC	Aerospace Flight Test Radio Coordinating Council
AFV	Armored Fighting Vehicle
AFVA	Armored Fighting Vehicle Association [US]
AFWAL	Air Force Wright Aeronautical Laboratories [US]
AFWETS	Air Force's Weapon Effectiveness Testing System [US]
AFWL	Air Force Weapons Laboratory [US]
AFWTR	Air Force Western Test Range [US]
AG	Advanced General-Purpose
	Air Gunner
	Armor Grating
	Arresting Gear
A-G	Air-to-Ground
Ag	(Agentum) - Silver
AGA	American Gas Association
	American Genetic Association
	Associated Geographers of America
	Astrologers' Guild of America
AGACS	Automatic Ground-to-Air Communications Systems
AGANI	Apollo Guidance and Navigation Information
AGARD	Advisory Group for Aerospace Research and Development [NATO]
AGC	Abort Guidance Computer
	Advanced Graphics Control(ler)
	Agriculture Canada
	Apollo Guidance Computer [NASA]
	Automatic Gage Control
	Automatic Gain Control
AGCA	Automatic Ground-Controlled Approach
AGCC	Air-Ground Communications Channel
AGCL	Automatic Ground-Controlled Landing
AGCR	Advanced Gas-Cooled Reactor
AGCRSP	Army Gas-Cooled Reactor Systems Program [US]
AGCY	Agency
AGD	Axial Gear Differential

AGDIC	Astro Guidance Digital Computer		Allowable Gross Weight
AGDS	American Gage Design Standard	**AH**	Analog Hybrid
AGDT	Aging Date	**A/H**	Air over Hydraulics
AGE	Aerospace Ground Equipment	**Ah**	Ampere-hour
	Automatic Ground Equipment	**AHAM**	Association of Home Appliance
AGED	Advisory Group on Electronic Devices [US]		Manufacturers [US]
AGEP	Advisory Group on Electronic Parts [US]	**AHCT**	Ascending Horizon Crossing Time
AGET	Advisory Group on Electron Tubes [US]	**AHDGA**	American Hot Dip Galvanizers Association
AGF	Army Ground Forces	**AHEM**	Association of Hydraulic Equipment
AGGR	Aggregate		Manufacturers [US]
AGI	American Geological Institute	**AHG**	Antihemophilic Globulin
AGID	Association of Geoscientists for International	**AHM**	Ampere-Hour Meter
	Development	**AHP**	Air Horsepower
AGIFORS	Airlines Group of International Federation of	**AHPL**	A Hardware Programming Language
	Operations Research Societies	**AHR**	Acceptable Hazard Rate
AGIL	Airborne General Illumination Light		Adsorptive Heat Recovery
AGL	Above Ground Level		American Heritage Radio
AGM	Air-to-Ground Missile	**AHRS**	Altitude Heading Reference System
	Annual General Meeting	**AHS**	Airborne Hardware Simulator
	Auxiliary General Missile		American Helicopter Society
AGMA	American Gear Manufacturers Association	**AHSA**	Art Historical and Scientific Association
	[US]	**AHSE**	Assembly, Handling and Shipping
AGMAP	Agricultural Market Assistance Program		Equipment
AGN	Average Group Number	**AHSR**	Air-Height Surveillance Radar
AGNIS	Apollo Guidance and Navigation Industrial		American High Speed Rail
	Support [US Army]	**AHU**	Air Handling Unit
AGNCS	Aluminum Gooseneck Clamp Strap		Antihalation Undercoat
AGO	Alkali-Graphitic-Oxide	**AHW**	Atomic Hydrogen Welding
AGOR	Auxiliary General Oceanographic Research	**AI**	Airborne Intercept(ion)
AGQ	Association of Geologists of Quebec [Canada]		Air Intensifier
AGR	Advanced Gas-Cooled Reactor		Amplifier Input
	Agriculture		Analog Input
AGRE	Army Group Royal Engineer		Anti-Icing
AGREE	Advisory Group on Reliability of Electronic		Anti-Interference
	Equipment [US]		Annoyance Index
AGRI	Agriculture		Applications Interface
AGRIC	Agriculture		Area of Intersection
AGRINTER	Inter-American Information System for the		Artificial Intelligence
	Agricultural Sciences		Atomic International
AGRIS	International Information System for		Automatic Input
	Agricultural Sciences and Technology [of		Automation Institute
	FAO]	**A&I**	Abstracting and Indexing
AGRU	Acid Gas Removal Unit	**AIA**	Aerospace Industries Association [US]
AGS	Abort Guidance System		American Institute of Architects
	Aircraft General Standards		American Inventors Association
	Alternating Gradient Synchrotron		Anthracite Industry Association
	American Geographical Society		Automatic Image Analysis
	Automatic Gain Stabilization		Automobile Industries Association
AGSAN	Astronomical Guidance System for Air	**AIAA**	Aerospace Industries Association of America
	Navigation		American Institute of Aeronautics and
AGSP	Atlas General Survey Program		Astronautics
AGT	Advanced Gas Turbine	**AIAC**	Aerospace Industries Association of Canada
	Aviation Gas Turbine	**AIAF**	American Institute of Architect Foundations
AGTERA	Advanced Gas Turbine for Engineering	**AIAG**	Automotive Industries Action Group [US]
	Research Association	**AIASA**	American Industrial Arts Student
AGU	American Geophysical Union		Association
AGV	Aniline Gentian Violet	**AIB**	Atlantic Institute of Biotechnology
	Automated Guided Vehicle	**AIBA**	American Industrial Bankers Association
AGVS	Automated Guided Vehicle System	**AIBD**	American Institute of Building Design
AGW	Actual Gross Weight	**AIBC**	Architectural Institute of British Columbia
	Advanced Graphic Workstation		[Canada]

AIBN	Azobisiobutyronitrile
AIBS	American Institute of Biological Sciences
AIC	Agricultural Institute of Canada
	Aircraft Industry Conference [US Army]
	American Institute of Chemists
	Architectural Institute of Canada
	Associate of the Institute of Chemistry
	Automatic Intercept Center
	Automatic Intersection Control
AICA	Aminoimidazole Carboxamide
AICAR	Aminoimidazole Carboxamide Ribofuranosyl
AICBM	Anti-Intercontinental Ballistic Missile
AICE	American Institute of Consulting Engineers
	Associate of the Institute of Civil Engineers
AICCP	Association of Institutes for Certification of Computer Professionals
AIChE	American Institute of Chemical Engineers
AID	Agency for International Development
	Algebraic Interpretive Dialog
	American Institute of Interior Design
	Architects Information Directory
	Attention Identification
	Attention Identifier
	Automatic Imaging Device
	Automatic Interrogation Distortion
AIDA	Automatic Instrumented Diving Assembly
	Automatic Intelligent Defect Analysis
AIDAS	Advanced Instrumentation and Data Analysis System
AIDATS	Army In-Flight Data Transmission System
AIDD	American Institute for Design and Drafting
AIDE	Airborne Insertion Display Equipment
	Aircraft Installation Diagnostic Equipment
	Army Integrated Decision Equipment [US]
	Automated Image Device Evaluation
	Automated Integrated Design Engineering
AIDJEX	Arctic Ice Dynamic Joint Experiment
AIDS	Airborne Integrated Data System
	Aircraft Integrated Data System
	Automated Integrated Debugging System
	Automated Intelligence and Data System [US Air Force]
	Automation Instrument Data Service [UK]
AIDSCOM	Army Information Data Systems Command [US]
AIEE	American Institute of Electrical Engineers [now: IEEE]
	Associate of the Institute of Electrical Engineers
AIENDF	Atomics International Evaluation Nuclear Data File
AIET	Average Instruction Executive Time
AIF	Atomic Industrial Forum
AIG	Artificial Intelligence Group [MIT, US]
AIGA	American Institute of Graphic Arts
AIGE	Association for Individually Guided Education
AIH	American Institute of Hydrology
AIHA	American Industrial Hygiene Association
AIHC	American Industrial Health Conference
AIIE	American Institute of Industrial Engineers

AIIM	Association for Information and Image Management [US]
AIL	Airborne Instruments Laboratory
	Aileron
AILAS	Automatic Instrument Landing Approach System
AILS	Advanced Integrated Landing System
	Automatic Instrument Landing System
AIM	Academy for Interscience Methodology
	Advanced Industrial Material
	Aerial Independent Model
	Age of Intelligent Machines
	Air Induction Melt
	Air Intercept Missile
	Air-Isolated Monolithic
	Alarm Indication Monitor
	American Institute of Management
	Associated Information Manager
	Automated Industrial Monitoring
	Avalanche-Induced Migration
AIMACO	Air Materiel Computer
AIMCAL	Association of Industrial Metallizers, Coaters and Laminators
AIME	American Institute of Mechanical Engineers
	American Institute of Mining, Metallurgical and Petroleum Engineers
	Applied Innovative Management Engineering
AIMechE	Associate of the Institution of Mechanical Engineers [UK]
AIMES	Automated Inventory Management Evaluation System
AIMILO	Army/Industrial Material Information Liaison Office [US]
AIMIS	Advanced Integrated Modular Instrument System
AIMM	Associate of the Institution of Mining and Metallurgy [UK]
AIMO	Audibly-Instructed Manufacturing Operation
AIMP	Anchored Interplanetary Monitoring Platform
AIMS	Advanced Intercontinental Missile System
	American Institute of Merchant Shipping
	Automated Inventory Management System
AINA	Arctic Institute of North America
AIOU	Analog Input/Output Unit
AIP	Aluminum Isopropoxide
	Aluminum Isopropylate
	American Institute of Physics
AIPE	American Institute of Plant Engineers
AIPG	American Institute of Professional Geologists
AIPS	Advanced Information Processing System
AIQS	Associate of the Institute of Quantity Surveyors [UK]
AIR	Aerospace Information Report
	Airborne Intercept Radar
	Air Injection Reaction
	Air Intercept Rocket
	American Institute of Research
	Association for Institutional Research

AI/R	Artificial Intelligence and Robotics		A (Robot-Programming) Language
AIRAC	All-Industry Research Advisory Council		Artificial Line
AIRCON	Automated Information and Reservation Computer-Oriented Network		Assembly Language
			Avionics Laboratory [US Air Force]
AIR COND	Air Conditioning	**Al**	Aluminum
AIREW	Airborne Infrared Early Warning	**ALA**	Ada Language
AIR HP	Air Horsepower		American Library Association
AIRCFT	Aircraft		Aminolevulinic Acid
AIRL	Aeronautical Icing Research Laboratory [US]	**Ala**	Alabama [US]
AIRPASS	Airborne Interception Radar and Pilot's Attack Sight System		Alanine
		ALABOL	Algorithmic and Business-Oriented Language
AIRS	Advanced Inertial Reference Sphere		
	Artificial Intelligence and Robotics Society	**ALAPCO**	Association of Local Air Pollution Control Officials
	Automatic Image Retrieval System		
	Automatic Information Retrieval System	**ALARA**	As Low As Reasonably Achievable
AIRSS	ABRES Instrument Range Safety System	**ALARM**	Air-Launched Advanced Ramjet Missile
AIS	Aeronautical Information Service		Automatic Light Aircraft Readiness Monitor
	Advanced Information System	**ALARR**	Air-Launched Air Recoverable Rocket
	Altitude Indication System	**ALART**	Army Low-Speed Air Research Tasks
	American Interplanetary Society	**ALAS**	Aminolevulinic Acid Synthetase
	Automated Information System		Asynchronous Lookahead Simulator
	Automatic Intercept System	**Alas**	Alaska [US]
	Automatic Intercity Station	**Alb**	Albumin
AISC	American Institute of Steel Construction	**ALBM**	Air-Launched Ballistic Missile
	Association of Independent Software Companies	**ALC**	Adaptive Logic Circuit
			Alcohol
AISE	Association of Iron and Steel Engineers [US]		Automatic Level(ling) Control
AISES	American-Indian Science and Engineering Society		Automatic Load Control
		ALCA	American Leather Chemists Association
AISI	American Iron and Steel Institute	**ALCAPP**	Automatic List Classification and Profile Production
AISP	Association of Information Systems Professionals		
		ALCC	Airborne Launch Control Center
AIST	Automatic Information Station	**ALCELL**	Alcohol-Cellulose
AIT	American Institute of Technology	**ALCH**	Approach Light Contact Height
	Architect-in-Training	**ALCM**	Air-Launched Cruise Missile
	Auto-Ignition Temperature	**ALCOM**	Algebraic Compiler
AITC	American Institute of Timber Construction	**ALCOR**	ARPA-Lincoln Coherent Observable Radar
AITE	Aircraft Integrated Test Equipment	**ALCS**	Aluminum Locator-Nose Clamp Strap
	Automatic Intercity Telephone Exchange	**ALD**	Analog Line Driver
AITS	Automatic Integrated Telephone System		Automated Logic Diagram
AIU	Advanced Instrumentation Unit [of NPL]		Automatic Louver Damper
AIV	Aluminum-Intensive Vehicle	**ALDEP**	Automated Layout Design Program
AIW	Auroral Intrasonic Wave	**ALDP**	Automatic Language Data Processing
	Average Industrial Wave	**ALDS**	Apollo Launch Data System
AIX	Advanced Interactive Executive	**ALE**	Atomic Layer Epitaxy
AJ	Anti-Jam(ming)	**ALERT**	Assistance for Liquid Electronic Reliability Testing
	Area Junction		
	Assembly Jig		Automted Linguistic Extraction and Retrieval Technique
AJAI	Anti-Jamming Anti-Interference		
AJD	Anti-Jam Display		Automatic Logical Equipment Readiness Tester
AJM	Abrasive Jet Machining		
	Air Jet Milling	**ALF**	Automatic Letter Facer
AK	Adenylate Kinase	**ALFA**	Air-Lubricated Free Attitude [NASA]
	Alaska [US]	**ALFTRAN**	ALGOL-to-FORTRAN Translator
	Aluminum Killed (Steel)	**ALG**	Algebra(ic)
AKM	Apogee Kick Motor	**ALGM**	Air-Launched Guided Missile
AL	Air Lock	**ALGOL**	Algorithmic Language
	Alabama [US]	**ALHT**	Apollo Lunar Hand Tool [NASA]
	Alcohol	**ALI**	Automated Logic Implementation
	Amplitude Limiter	**ALIGN**	Alignment
	Analysis Library	**ALIM**	Air-Launched Intercept Missile

ALIN	Alcoa Laboratories Information Network [US]
ALIS	Advanced Life Information System
ALIT	Association of Laser Inspection Technologies
	Automatic Line Insulation Tester
ALK	Alkaline
	Alkali
ALL	Accelerated Learning of Logic
	Application Language Liberator
ALLA	Allied Long Lines Agency [NATO]
	Automated Laboratory Liquor Analyzer
ALL-CAPS	All Capital Letters
ALLOC	Allocation
ALLOW	Allowance
ALLS	Apollo Lunar Logistic Support [NASA]
ALM	Alarm
	Applied Laboratory Method
ALMA	Alphanumeric Language for Music Analysis
	Analytical Laboratory Managers Association
ALMC	Army Logistic Management Center [US]
ALMS	Aircraft Landing Measurement System
	Analytic Language Manipulation System
ALOG	Applied Laser Optics Group
ALOR	Advanced Lunar Orbital Rendezvous
ALOT	Airborne Lightweight Optical Tracker
ALOTS	Airborne Lightweight Optical Tracking System
ALP	Assembly Language Program
	Automated Learning Process
ALPA	Alaskan Long Period Array
ALPAC	Automatic Language Processing Advisory Committee [of NAS, US]
ALPHA	Automatic Literature Processing, Handling and Analysis
ALPHANUM	Alphanumeric(s)
ALPID	Analysis of Large Plastic Incremental Deformation
ALPS	Advanced Linear Programming System
	Advanced Liquid Propulsion System
	Assembly Line Planning System
	Associated Logic Parallel System
	Automated Library Processing Service
ALRI	Airborne Long-Range Intercept
ALRR	Ames Laboratory Research Reactor [US]
ALRT	Advanced Light Rapid Transit
	Automated Light Rapid Transit
ALS	Advanced Low-Power Schottky
	Aircraft Landing System
	Alberta Land Surveyors [Canada]
	Approach Light System
	Automatic Landing System
ALSCP	Appalachian Land Stabilization and Conservation Program [US]
ALSEP	Apollo Lunar Surface Experiment Package [NASA]
ALSPEI	Association of Land Surveyors of Prince Edward Island [Canada]
ALSS	Airborne Location and Strike System
	Apollo Logistic Support System [NASA]
ALT	Activated Liquid Thromboplastin
	Airborne Laser Tracker

	Alanine Aminotransferase
	Alteration
	Alternation
	Alternator
	Altitude
ALTA	Association of Local Transport Airlines
Alta	Alberta [Canada]
ALTAC	Algebraic Transistorized Automatic Computer
	Algebraic Translator and Compiler
ALTAIR	ARPA Long-Range Tracking and Instrumentation Radar
ALTAN	Alternate Alerting Network [US Air Force]
ALTARE	Automatic Logic Testing and Recording Equipment
ALTER	Alternation
	Alternative
ALTM	Altimeter
ALTN	Alternation
ALTRAN	Algebraic Translator
ALTS	Advanced Lunar Transportation System
ALU	Advanced Levitation Unit
	Arithmetic-Logic Unit
ALUE	Admissable Linear Unbiased Estimator
ALVIN	Antenna Lobe for Variable Ionospheric Nimbus
ALY	Alloy
AM	Access Method
	Address Mark
	Airlock Module
	Air Melting
	America(n)
	Ammeter
	Amorphous
	Amplitude
	Amplitude Modulation
	Analog Monolithic
	(Anno Mundi) - in the year of the world
	(Artium Magister) - Master of Arts
	Associative Memory
	Auxiliary Memory
Am	Americium
am	(ante meridien) - before noon
AMA	Academy of Model Aeronautics
	Actual Mechanical Advantage
	Adhesives Manufacturers Association (of America)
	Air Material Area [US Air Force]
	Alberta Motor Association [Canada]
	American Management Association
	American Manufacturers Association
	American Marketing Association
	Automatic Memory Allocation
	Automatic Message Accounting
	Automobile Manufacturers Association
AMAC	Automatic Material Completion
AMACUS	Automated Microfilm Aperture Card Updating System [US Army]
AMAD	Aircraft-Mounted Accessory Drive
AMADS	Airframe-Mounted Auxiliary Drive System
AMAL	Amalgam

	Amalgamation
AMALG	Amalgamation
AMAN	Automatic Material Number
AMARS	Air Mobile Aircraft Refueling System
AMAS	Advanced Midcourse Active System
	Automatic Material Status
AMB	Amber
	Ambient
	Anti-Motor Boat
	Asbestos Millboard
AMBCS	Alaskan Meteor Burst Communication System
AMBD	Automatic Multiple Blade Damper
AMBIT	Algebraic Manipulation by Identity Translation
AMC	Acetylmethylcarbon
	Airborne Management Computer
	Aircraft Manufacturers Council
	Air Materiel Command [US Air Force]
	Alberta Microelectronics Center [Canada]
	American Mining Congress
	Army Materiel Command [US Army]
	Army Missile Command [US Army]
	Association of Management Consultants
	Automated Meter Calibration
	Automatic Message Counting
	Automatic Mixture Control
	Automatic Monitoring Circuit
	Auto Meter Correct
AMCBMC	Air Materiel Command Ballistic Missile Center [US Air Force]
AMCEC	Allied Military Communications Electronics Committee
AMCEE	Association of Media-Based Continuing Education for Engineers
AmChemSoc	American Chemical Society
AMCP	Allied Military Communications Panel
AMCS	Airborne Missile Control System
AMD	Acid Mine Drainage
	Automated Multiple Development System
AMDCB	Applied Moment Double-Cantilever Beam
AMDEL	Australian Mineral Development Laboratories
AMDSB	Amplitude Modulation Double Sideband
AME	Angle Measuring Equipment
AMEC	Aerospace Metals Engineering Committee
AME/COTAR	Angle Measuring Equipment/Correlation Tracking and Ranging
AMEG	Association for Measurement and Evaluation in Guidance
AMEIC	Associate Member of the Engineering Institute of Canada
AMER	America(n)
AMER STD	American Standard
AMES	Association of Marine Engineering Schools [UK]
AMETA	Army Management Engineering Training Agency [US]
AMETS	Artillery Meteorological System
AMF	ACE (= Allied Command Europe) Mobile Force [NATO]

	Automatic Mirror Furnace
	Automatic Mirror Facility
AMFIS	Automatic Microfilm Information Society
	Automatic Microfilm Information System
AM/FM	Amplitude Modulation/Frequency Modulation
AMHS	American Materials Handling Society
AMIA	American Metal Importers Association
AMIAS	Automated Metallurgical Image Analysis System
AMIC	Aerospace Materials Information Center [US]
AMICE	Associate Member of the Institute of Civil Engineers
AMICOM	Army Missile Command [US]
AMIEE	Associate Member of the Institute of Electrical Engineers
AMIME	Associate Member of the Institute of Mechanical Engineers
AMIMechE	Associate Member of the Institution of Mechanical Engineers
AMIMinE	Associate Member of the Institution of Mining Engineers
AMInstCE	Associate Member of the Institute of Civil Engineers
AMIS	Aircraft Movement Information Service
	Automated Management Information System
AMISC	Army Missile Command [US]
AMK	Anti-Misting Kerosene
AML	Admiralty Materials Laboratory [UK]
	Aeronautical Materials Laboratory
	Aviation Materials Laboratory [US Army]
	A Manufacturing Language
	Amplitude-Modulated Link
AMM	Ammeter
	Ammunition
	Antimissile Missile
AMMA	Acrylonitrile-Methylmethacrylate
AMMIP	Aviation Materiel Management Improvement Program [US Army]
AMMRC	Army Materials and Mechanics Research Center [now: AMTL]
AMNIP	Adaptive Man-Machine Nonarithmetical Information Processing
AM NIT	Ammonium Nitrate
AMOR	Amorphous
AMORP	Amorphous
AMOS	Acoustic Meteorological Oceanographic Survey
	Associative Memory Organizing System
	Automatic Meteorological Observing System
AMP	Adaptation Mathematical Processor
	Advanced Manned Penetrator
	Adenosine Monophosphate
	American Melting Point
	Aminomethylpropanol
	Amperage
	Ampere [also: amp]
	Associative Memory Processor
AMPCO	Association of Major Power Consumers in Ontario [Canada]

AMPD	Aminomethylpropanediol
AMPH	Amphibian
	Amphibious
amp-hr	ampere-hour
AMPIC	Atomic and Molecular Processes Information Center
AMPL	Amplification
	Amplifier
AMPLG	Amplidyne Generator
AMPLMG	Amplidyne Motor Generator
AMPNS	Asymmetric Multiple Position Neutron Source
AMPR	Aeronautical Manufacturers Planning Report
AMPS	Advanced Mobile Phone Service
	Automatic Message Processing System
AMPSO	Dimethylmethylaminohydroxypropanesulfonic Acid
AMPSS	Advanced Manned Precision Strike System
AMR	Advance Material Request
	Atlantic Missile Range [US Air Force]
	Automated Management Report
	Automatic Message Registering
	Automatic Message Routing
AMRA	Army Materials Research Agency [US]
AMRAC	Anti-Missile Research Advisory Council [US]
AMRC	Automotive Market Research Council [US]
AMRF	Automated Manufacturing Research Facility [NBS, US]
AMRNL	Army Medical Research and Nutrition Laboratory [US]
AMS	Access Method Services
	Advanced Memory System
	Administrative Management Society
	Aeronautical Material Specification
	Aeronautical Mobile Services
	Aerospace Material Specification [of SAE]
	Air Mail Service
	American Mathematical Society
	American Meteorological Society
	American Microchemical Society
	American Microscopical Society
	Army Map Service [US]
	Asymmetric Multiprocessing System
	Attitude Measurement Sensor
AMSA	Advanced Manned Strategic Aircraft
	American Metal Stamping Association
AMSAT	Amateur Satellite
AMSC	Army Mathematics Steering Committee [US]
AMSE	Association of Muslim Scientists and Engineers
AMSL	Above Mean Sea Level
AMSOC	American Miscellaneous Society
AMSSB	Amplitude Modulation, Single-Sideband
AMST	Advanced Medium STOL (= Short Takeoff and Landing) Transport
AMT	Alternative Minimum Tax
	Amount
	Audio Magnetotellurics

AMTA	Antenna Measurement Techniques Association
AMTB	Anti-Motor Torpedo Boat
AMTCL	Association for Machine Translation and Computational Linguistics [US]
AMTDA	American Machine Tool Distributors Association
AMTEC	Association for Media and Technology in Education in Canada
AMTI	Area Moving Target Indicator
	Automatic Moving Target Indicator
AMTIDE	Aircraft Multipurpose Test Inspection and Diagnostic Equipment
AMTL	Army Materials Technology Laboratory [formerly: AMMRC]
AMTRAC	Amphibian Tractor
AMTRAN	Automatic Mathematical Translator
AMTU	Advanced Materials Technology Unit [Canada]
AMU	Associated Midwestern Universities [US]
	Association of Minicomputer Users
	Astronaut Maneuvering Unit [NASA]
	Atomic Mass Unit
AMV	Advanced Marine Vehicle
AMVER	Automated Merchant Vessel Reporting
AN	Acorn Nut
	Acrylonitrile
	Advanced Navigator
	Air Force - Navy
	Aminonitrogen
	Ammonium Nitrate
	Army-Navy
ANA	Antinuclear Antibody
	Army Navy Aeronautical
	Automatic Network Analyzer
ANACHEM	Association of Analytical Chemists
ANACOM	Analog Computer
ANAL	Analysis
ANALYT	Analytical
ANARE	Australian National Antarctic Research Expedition
ANATRAN	Analog Translator
ANBLS	Association of New Brunswick Land Surveyors [Canada]
ANC	All-Number Calling
	Air Navigation Commission
	Air Navigation Conference
	Automatic Nutation Control
ANCA	Allied Naval Communications Agency [NATO]
ANCAR	Australian National Committee for Antarctic Research
ANCS	American Numerical Control Society
ANCU	Airborne Navigation Computer Unit
AND	Air Force-Navy Design
	Army-Navy Design
ANDB	Air Navigation Development Board [US]
A-nDNA-A	Antinative DNA Antibody
ANDREE	Association for Nuclear Development and Research in Electrical Engineering
ANDZ	Anodize

ANEC	American Nuclear Energy Council
ANF	Anti-Nuclear Factor
	Atlantic Nuclear Force
ANFO	Ammonium Nitrate and Fuel Oil
ANG	Air National Guard [US]
ANG-CE	Air National Guard - Civil Engineering [US]
ANH	Anhydrous
ANHYD	Anhydrous
ANI	Annular Isotropic
	Automatic Number Identification
ANIM	Association of Nuclear Instrument Manufacturers
ANIP	Army-Navy Instrumentation Program [US]
	Army-Navy Integrated Presentation
ANIS	Annular Isotropic Source
ANL	Anneal
	Argonne National Laboratory [US]
	Automatic Noise Limiter
ANLG	Annealing
ANLOR	Angle Order
ANMC	American National Metric Council
ANN	Anneal
	Annual
	Annuniciator
ANO	Alphanumerical Output
ANOD	Anodize
ANOVA	Analysis of Variants
	Analysis of Variation
ANP	Aircraft Nuclear Propulsion
ANPO	Aircraft Nuclear Propulsion Office [of AEC]
ANPOD	Antenna Positioning Device
ANPP	Army Nuclear Power Program [US]
ANPRM	Advanced Notice of Proposed Rule Making
ANPS	American Nail Producers Society
ANPT	Aeronautical National (Taper) Pipe Thread
ANR	Alphanumeric Replacement
ANRAC	Aids Navigation Radio Control
ANS	Academy of Natural Sciences
	Air Navigation School
	American National Standards
	American Nuclear Society
	Anilinenaphthalene Sulfonate
	Astronautical Netherlands Satellite
ANSA	Anilinonaphthalenesulfonic Acid
ANSAM	Antinuclear Surface-to-Air Missile
ANSCAD	Associate of Nova Scotia College of Art and Design [Canada]
ANSI	American National Standards Institute
ANSL	American National Standard Label
ANSLS	Association of Nova Scotia Land Surveyors [Canada]
ANSVIP	American National Standard Vocabulary for Information Processing
ANSW	Antinuclear Submarine Warfare
ANT	Antenna
ANTAC	Air Navigation Tactical Control System
ANTC	Antichaff Circuit
ANTEC	Annual Technical Conference
ANTI	Anticoincidence
antilog	antilogarithm
AN/TN	Aminonitrogen/Total Nitrogen (Ratio)

ANTS	Airborne Night Television System
ANTU	Alphanaphthylthiourea
ANU	Airplane Nose Up
ANVIS	Aviator's Night Vision Imaging System
ANYST	Analyst
ANZAAS	Australian and New Zealand Association for the Advancement of Science
AO	Access Opening
	Amplifier Output
	Atomic Orbital
	Automated Operator
AOAA	Aminooxyacetic Acid
AOAC	Association of Official Agricultural Chemists (of North America)
	Association of Official Analytical Chemists [US]
AOB	Address Operation Block
AOC	Airport Operators Council
	Automatic Output Control
	Automatic Overload Circuit
	Automatic Overload Control
	Oxygen-Arc Cutting
AOCI	Airport Operators Council International
AOCR	Advanced Optical Character Reader
AOCS	Alpha Omega Computer System
	American Oil Chemists Society
AOD	Argon-Oxygen Decarburization
AOEW	Airplane Operating Empty Weight
AOF	American Optometric Foundation
AOI	And - Or Invert
	Automated Operator Interface
AOL	Application-Oriented Language
	Atlantic Oceanographic Laboratory
AOLO	Advanced Orbital Launch Operation
AOLS	Association of Ontario Land Surveyors [Canada]
AOML	Atlantic Oceanographic and Meteorological Laboratories [NOAA]
AOPA	Aircraft Owners and Pilots Association
AOQ	Average Outgoing Quality
AOQL	Average Outgoing Quality Level
	Average Outgoing Quality Limit
AOR	Angle of Reflection
	Atlantic Ocean Region
AORB	Aviation Operational Research Branch
AORE	Army Operational Research Establishment
AOS	Acquisition of Signal
	Angle of Sight
	Automated Office System
	Azimuth Orientation System
AOSERP	Alberta Oil Sands Environmental Research Program [Canada]
AOSO	Advanced Orbiting Solar Observatory
AOSP	Automatic Operating and Scheduling Program
AOSTRA	Alberta Oil Sands Technology and Research Authority [Canada]
AOT	Alignment Optical Telescope
AOTV	Aero-Assisted Orbital Transfer Vehicle
AOU	Apparent Oxygen Utilization
	Automated Offset Unit

AP	Access Panel
	Adenosine Monophosphate
	After Peak
	After Perpendicular
	Air Pollution
	Airport
	Alkaline Permanganate
	Alkaline Phosphatase
	All-Pass (Filter)
	American Patent
	Annealing Point
	Application Program
	Applications Processor
	Applied Physics
	April
	Argument Programming
	Arithmetic Progression
	Armor Piercing
	Array Processor
	Assembly of Parties
	Associated Press
	Associative Processor
	Atom Probe
	Attached Processor
	Automatic Polisher
	Autopilot
A/P	Attached Processor
ap	apothecaries
APA	Amalgamated Printers Association
	American Plywood Association
	American Polygraph Association
	Aminopenicillanic Acid
	Automobile Protection Association
	Axial Pressure Angle
ApA	Adenylyladenosine
APAC	Alkaline Permanganate Ammonium Citrate
APAD	Acetylpyridine Adenine Dinucleotide
APACS	Airborne Position and Altitude Camera System
APADS	Automatic Programmer and Data System
APAG	Atlantic Policy Advisory Group [NATO]
APAM	Array Processor Access Method
APAP	Acetyl-P-Aminophenol
ApApC	Adenylyladenylylcytidine
APAR	Authorized Program Analysis Report
	Automatic Programming and Recording
APAS	Automatic Performance Analysis System
APATS	Automatic Programming and Testing System
APB	Air Portable Bridge
	Antiphase Boundary
	As-Purchased Basis
APC	Active Path Corrosion
	Advanced Pocket Computer
	Advanced Processor Card
	Advanced Programming Course
	Aeronautical Planning Chart
	Air Pollution Control
	Air Purification Control
	Amplitude Phase Conversion
	Approach Control

	Area Position Control
	Armored Personnel Carrier
	Armor Piercing Capped
	Aromatic Polymer Composite
	Automatic Particle Counter
	Automatic Position Control
	Automatic Phase Control
	Automatic Pressure Control
	Autoplot Controller
ApC	Adenylylcytidine
APCA	Air Pollution Control Association [US]
APCC	Advanced Physical Coal Cleaning
APCHE	Automatic Programmed Checkout Equipment
APCI	Armor Piercing Capped Incendiary
	Atmospheric Pressure Chemical Ionization
APCI/MS	Atmospheric Pressure Chemical Ionization/Mass Spectrometry
APCI-T	Armor Piercing Capped Incendiary with Tracer
ApCpC	Adenylylcytidylylcytidine
APCS	Associative Processor Computer System
	Automatic Program Control System
APC-T	Armor Piercing Capped with Tracer
APCVD	Atmospheric Pressure Chemical Vapor Deposition
APD	Aerospace Power Division [US Air Force]
	Alloy Phase Diagram
	Amplitude-Phase Diagram
	Amplitude Probability Distribution
	Angular Position Digitizer
	Antiphase Domain
	Avalanche Photodiode
	Avalanche Photodiode Detector
APDC	Ammonium Pyrrolidinedithiocarbamate
APDIC	Alloy Phase Diagram International Commission
APDS	Armour Piercing Discarding Sabot
APDSMS	Advanced Point Defense Surface Missile System [US Navy]
APDTC	Ammonium Pyrrolidinedithiocarbamate
APE	Abbreviated Plain English (Language)
	Aminopropyl Epoxy
	Antenna Positioning Electronics
	Association of Professional Engineers
APEA	Association of Professional Engineers of Alberta [Canada]
APEBC	Association of Professional Engineers of British Columbia [Canada]
APEC	All-Purpose Electronic Computer
	Atlantic Provinces Economic Council [Canada]
	Automated Procedures for Engineering Consultants
APEG	Alkaline Polyethyleneglycol
APEGGA	Association of Professional Engineers, Geologists and Geophysicists of Alberta [Canada]
APEGGNWT	Association of Professional Engineers, Geologists and Geophysicists of the Northwest Territories [Canada]

APEL	Aeronautical Photographic Experimental Laboratory
APEM	Association of Professional Engineers of Manitoba [Canada]
APEN	Association of Professional Engineers of Newfoundland [Canada]
APEO	Association of Professional Engineers of Ontario [Canada]
APEPEI	Association of the Professional Engineers of Prince Edward Island [Canada]
APERS	Antipersonnel
APES	Association of Professional Engineers of Saskatchewan [Canada]
APEX	Assembler and Process Executive
APEYT	Association of Professional Engineers of the Yukon Territory [Canada]
APF	Advanced Printer Function
	Asphalt Plank Floor
	Atomic Packing Factor
	Authorized Program Facility
	Automatic Program Finder
	Autopilot Flight Director
APFA	American Pipe Fittings Association
APFCS	Automatic Power Factor Control System
APFIM	Atom Probe Field Ion Microscopy
APG	Aminopropyl Glass
	Automatic Priority Group
	Automatic Program Generator
	Azimuth Pulse Generator
ApG	Adenylylguanosine
APGA	American Personnel Guidance Association
APGC	Air Proving Ground Center
APGGQ	Association of Professional Geologists and Geophysicists of Quebec [Canada]
ApGpU	Adenylylguanylyluridine
APHA	American Public Health Association
APHI	Association of Public Health Inspectors
API	Addition(-Reaction) Polyimide
	Air Position Indicator
	American Petroleum Institute
	American Paper Institute
	Armor Piercing Incendiary
	Application Programming Interface
	Automatic Priority Interrupt
APIC	Apollo Parts Information Center [NASA]
APICS	American Production and Inventory Control Society
	Atlantic Provinces Inter-University Council on the Sciences [Canada]
APIP	Alberta Petroleum Incentive Program [Canada]
APIS	Army Photographic Interpretation Section [US]
API-T	Armor Piercing Incendiary with Tracer
APK	Amplitude Phase Keyed
APKS	Amplitude Phase Keyed System
APL	Aero-Propulsion Laboratory [US Air Force]
	Airplane
	Applied Physics Laboratory [US]
	A Progamming Language
	Association of Programmed Learning

	Associative Programming Language
	Automatic Phase Lock
	Average Picture Level
APLE	Association of Public Lighting Engineers
APM	Amplitude and Phase Modulation
	Analog Panel Meter
	Antenna Positioning Mechanism
	Associative Principle for Multiplication
	Asynchronous Packet Manager
	Atom Probe Microanalysis
APMA	Automotive Parts Manufacturers Association
APMC	Alberta Petroleum Marketing Commission [Canada]
APMI	Area Precipitation Measurement Indicator
	American Powder Metallurgy Institute
APNIC	Automatic Programming National Information Center [UK]
APOA	Arctic Petroleum Operators Association
APOS	Advanced Polar Orbiting Satellite
APP	Apparatus
	Appendix
	Application
	Arctic Pilot Project
	Auxiliary Power Plant
APPA	American Public Power Association
APPAR	Apparatus
APPC	Advanced Program-to-Program Communication
APPD	Approved
APPECS	Adaptive Pattern Perceiving Electronic Computer System
APPG	Adjacent Phase Pulse Generator
APPHS	Auger Peak-to-Peak Heights
APPI	Advanced Planning Procurement Information
APPL	Appliance
	Application
APPLE	Associative Processor Programming Language Evaluation
APPN	Advanced Peer-to-Peer Networking
APPR	Approximately
APPROX	Approximate(ly)
APPS	Adenosine Phosphate Phosphosulfate
APPX	Appendix
APQ	Available Page Queue
APR	Airborne Profile Recorder
	Alternate Path Retry
	April
APRA	Automotive Parts Rebuilders Association
APRF	Army Pulse Radiation Facility [US]
APRFR	Army Pulse Radiation Facility Reactor [US]
APRIL	Aqua Planning Risk Indicator for Landing
	Automatically Programmed Remote Indication Logged
APRO	Aerial Phenomena Research Organization
APRS	Automatic Position Reference System
APRT	Adenine Phosphoribosyltransferase
APS	Adenosine Phosphosulfate
	Alphanumeric Photocomposer System
	American Physical Society
	American Physiological Society

	American Polar Society		Arithmetic Register
	Aminopolystyrene		Arkansas [US]
	Appearance Potential Spectroscopy		As Received
	Assembly Programming System		As Required
	Atmospheric Plasma Spraying		Assembly and Repair
	Automatic Patching System		Associative Register
	Auxiliary Power System		Attention Routine
	Auxiliary Program Storage		Autoregression
APSA	Automatic Particle Size Analyzer		Aviation Radionavigation
APSE	Ada Program Support Environment		Avionic Requirements
APSM	Association for Physical and System Mathematics	**Ar**	Argon
APSP	Array Processor Subroutine Package	**ARA**	Aerial Rocket Artillery
APT	Advanced Passenger Train		Aircraft Replaceable Assemblies
	Advanced Passenger Transport		Aircraft Research Association [US]
	Ammonium Paratungstate		Aluminum Recycling Association
	Apartment		Amateur Rocket Association [US]
	Association for Preservation Technology		Angular Rate Assembly
	Augmented Programming Training		Associates for Radio Astronomy
	Automatically Programmed Tool		Automotive Retailers Association
	Automatic Picture Taking	**Ara-A**	Arabinofuranosyl Adenine
	Automatic Picture Transmission	**ARABSAT**	Arab Countries Regional Communications Satellite
	Automation Planning and Technology	**ARAC**	Aerospace Research Applications Center [US]
	Automatic Position Telemetering		
AP-T	Armor Piercing with Tracer	**ARAD**	Airborne Radar and Doppler
APTA	Atlantic Provinces Trucking Association [Canada]	**ARAF**	Air Reserve Augmentation Flights
		ARAL	Automatic Record Analysis Language
APTC	Atlantic Provinces Transport Commission [Canada]	**ARALL**	Aramid-Aluminum Laminate
		ARB	Air Registration Board
APTS	Automatic Picture Transmission System		American Research Bureau
APTT	Activated Partial Thromboplastin Time		Arbitration
APU	Acid Purification Unit	**ARC**	Advanced Reentry Concept
	Audio Playback Unit		Aeronautical Research Council [UK]
	Automatic Power Up		Agricultural Research Council
	Auxiliary Power Unit		Aiken Relay Calculator
ApU	Adenylyluridine		Alberta Research Council [Canada]
ApUpG	Adenylyluridylylguanosine		Altitude Rate Command
ApUpU	Adenylyluridylyluridine		Ames Research Center [US]
APVD	Approved		Amplitude and Rise-Time Compensation
APW	Augmented Plane Wave		Antireflection Coating
	Average Piece Weight		Argonne Reactor Computation
APX	Appendix		Attached Resource Computer
AQ	Achievement Quotient		Augmentation Research Center [US]
	Aminoquinoline		Automatic Relay Calculator
	Any Quality		Automatic Remote Control
	(aqua) - water		Auxiliary Roll Control
	Aqueous		Average Response Computer
AQAP	Allied Quality Assurance Publication	**ARCAS**	Automatic Radar Chain Acquisition System
AQL	Acceptable Quality Level	**arccos**	arccosine
AQO	Aminoquinoline Oxide	**arccot**	arccotangent
AQ REG	Aqua Regia	**arccsc**	arccosecant
AQT	Acceptable Quality Test	**ARCE**	Amphibious River Crossing Equipment
AR	Acid-Resisting	**ARCH**	Architect
	Acoustic Reflex		Architectural
	Acquisition Radar		Architecture
	Address Register		Articulated Computing Hierarchy [UK]
	Aerial Refueling	**ARCHE**	Architectural Engineer(ing)
	Air Resistance	**ARCOMSAT**	Arab League Communications Satellite
	Anti-Reflection	**ARCRL**	Agricultural Research Council Radiobiological Laboratory
	Annual Review		
	Arabia(n)	**ARCS**	Air Resupply and Communication Service

	Associate of the College of Science
	Autonomous Remote Controlled Submersible
arcsec	arcsecant
arcsin	arcsine
arctan	arctangent
ARC/W	Arc Weld
ARD	Airborne Respirable Dust
	Association of Research Directors
ARDA	Agriculture and Rural Development Act
	American Railway Development Association
	Analog Recording Dynamic Analyzer
ARDC	Air Research and Development Command [US Air Force]
	Armament Research and Development Center
ARDE	Armament Research and Development Establishment [now: RARDE]
ARDIS	Army Research and Development Information System [US]
ARDS	Advanced Remote Display Station
ARE	Activated Reactive Evaporation
	Asymptotic Relative Efficiency
	Automated Responsive Environment
AREA	American Railway Engineering Association
ARELEM	Arithmetic Element (Program)
AREP	Automated Reliability Estimation Program
ARFA	Allied Radio Frequency Agency [NATO]
ARG	Argument
Arg	Arginine
ARGUS	Automatic Routine Generating and Updating System
ARI	Airborne Radio Instrument
	Air-Conditioning and Refrigeration Institute [US]
ARIA	Advanced Range Instrumented Aircraft
ARIBA	Associate of the Royal Institute of British Architects
ARIC	Associate of the Royal Institute of Chemistry
ARICS	Associate of the Royal Institute of Chartered Surveyors
ARIES	Advanced Radar Information Evaluation System
ARIP	Automatic Rocket Impact Predictor
ARIS	Advanced Range Instrumented Ship
ARITH	Arithmetic(s)
Ariz	Arizona [US]
Ark	Arkansas [US]
ARL	Acceptable Reliability Level
	Admiralty Research Laboratory [UK]
	Aeronautical Research Laboratory
	Aerospace Research Laboratory [UK]
	Applied Research Laboratories
	Arctic Research Laboratory [US]
	Army Radiation Laboratory [US Army]
	Association of Research Libraries [US]
	Average Run Length
ARLIS	Arctic Research Laboratory Ice Station [US Navy]
ARM	Accumulator Read-in Module
	Anhysteretic Remanent Magnetization

	Anti-Radar Missile
	Anti-Radiation Missile
	Armature
	Armor
	Articulated Remote Manual
	Asynchronous Response Mode
	Atomic Resolution Microscope
ARMA	American Records Management Association
ARMAN	Artificial Methods Analyst
ARMC	Automotive Research and Management Consultants
ARMD	Armored
ARMMS	Automated Reliability and Maintainability Measurement System
ARM-PL	Armor Plate
ARMS	Advanced Receiver Model System
	Aerial Radiological Measuring Survey
	Amateur Radio Mobile Society [US]
ARMT	Armament
ARO	Air Radio Officer
	Army Research Office [US]
AROD	Airborne Range and Orbit Determination
ARODS	Airborne Radar Orbital Determination System
AROM	Alterable Read-Only Memory
AROU	Aviation Repair and Overhaul Unit
ARP	Advanced Reentry Program
	Aeronautical Recommended Practice
	Airborne Radar Platform
	Aramid-Reinforced Plastics
	Azimuth Reset Pulse
ARPA	Advanced Research Projects Agency [USDOD]
ARPES	Angle-Resolved Photoemission Spectroscopy
ARPS	Aerospace Research Pilot School [US Air Force]
ARQ	Automatic Repeat Request
	Automatic Request for Repetition
	Automatic Response Query
ARR	Anti-Repeat Relay
	Arrangement
	Arrest
	Arrestor
	Arrival
ARRL	Aeronautical Radio and Radar Laboratory
	American Radio Relay League [US]
ARRS	Aerospace Rescue and Recovery Service [US Air Force]
ARS	Active Repeater Satellite
	Advanced Reconnaissance Satellite
	Advanced Record System
	American Radium Society [US]
	American Rocket Society
	Asbestos Roof Shingles
	Automatic Recovery System
ARSM	Associate of the Royal School of Mines
ARSME	All-Round Shape Memory Effect
ARSP	Aerospace Research Satellite Program [US Air Force]
	Aerospace Research Support Program [US Air Force]

ARSR	Air Route Surveillance Radar
	Arrestor
ART	Admissable Rank Test
	Advanced Reactor Technology
	Advanced Research and Technology
	Airborne Radiation Thermometer
	Article
	Artificial
	Artificial Resynthesis Technology
	Automated Rotor Test (System)
	Automatic Range Tracker
	Automatic Reporting Telephone
	Average Retrieval Time
ARTC	Aircraft Research and Testing Committee
	Air Route Traffic Control
ARTCC	Air Route Traffic Control Center [US]
ARTE	Admiralty Reactor Test Establishment [UK]
ARTFL	Artificial
ARTG	Azimuth Range and Timing Group
ARTI	Arab Regional Telecommunications Institute
ARTIC	A Real-Time Interface Coprocessor
ARTOC	Army Tactical Operations Center [US]
ARTRAC	Advanced Range Testing, Reporting and Control
	Advanced Real-Time Range Control
ARTS	Automated Radar Terminal System
	Advanced Radar Traffic-Control System
ARTU	Automatic Range Tracking Unit
ARTY	Artillery
ARU	Audio Response Unit
ARUBIS	Angle-Resolved Ultraviolet Bremsstrahlung Isocromate Spectroscopy
ARUPS	Angle-Resolved Ultraviolet Photoelectron Spectroscopy
ARV	Aeroballistic Reentry Vehicle
	Armored Recovery Vehicle
AS	Acetanisidine
	Advanced Schottky
	Aeronautical Standards
	Aerospace Standards
	Air Screw
	Alongside
	Ammeter Switch
	Anthranilate Synthetase
	Antistatic
	Antisubmarine
	Applied Science
	Area Surveillance
	Australian Standard
	Automatic Sprinkler
	Automatic Switching
	Auxiliary Storage
A/S	Ascent Stage
As	Arsenic
	Altostratus
ASA	Acoustical Society of America
	Acrylonitrile Styrene Acrylonitrile
	American Standards Association [now: ANSI]
	American Statistical Association

	Antistatic Additive
	Army Security Agency
	Atomic Sphere Approximation
ASAE	Advanced School of Automobile Engineering
	American Society for Agricultural Engineers
	American Society of Association Executives
ASAP	Aerospace Supplier Accreditation Program
	American Society of Aerospace Pilots
	Analog System Assembly Pack
	Anti-Submarine Attack Plotter
	Applied Systems and Personnel
	Army Scientific Advisory Panel
	As Soon As Possible
	Automated Shipboard Aerological Program
	Automated Statistical Analysis Program
ASAT	Antisatellite
ASB	Aluminum sec-Butoxide
	Asbestos
	Association of Shell Boilermakers
ASBC	American Society of Biological Chemists
	American Society of Brewing Chemists
	American Standard Building Code
ASBD	Advanced Sea-Based Deterrent
ASBE	American Society of Bakery Engineers
	American Society of Body Engineers
ASBI	Advisory Service for the Building Industry
ASBO	Association of School Business Officials
ASC	Adhesive and Sealant Council [US]
	Advanced Scientific Computer
	Aeronautical Systems Center
	Agricultural Stabilization and Cultivation
	Air-Powered Swing Clamp
	American Society of Cartographers
	American Society for Cybernetics
	American Standard Code
	Associated School of Construction [US]
	Associate in Science [also: ASc]
	Association of Systematics Collections
	Associative Structure Computer
	Atlantic Systems Conference
	Automatic Sensitivity Control
	Automatic Switching Center
	Automatic System Controller
	Auxiliary Switch, (Normally) Closed
ASc	Associate in Science
ASCA	Automated Spin Chemistry Analyzer
	Automatic Subject Citation Alerting
ASCAC	Anti-Submarine Classification Analysis Center [US Navy]
ASCATS	Apollo Simulation Checkout and Training System [NASA]
ASCB	American Society for Cell Biology
ASCC	Air Standardization Coordinating Committee
	American Society of Concrete Constructors
	Army Strategic Communications Command [US Army]
	Automatic Sequence Controlled Calculator
ASCE	American Society of Civil Engineers
ASCEND	Advanced System for Communications and Education in National Development

ASCENT	Assembly System for Central Processor [UNIVAC]
ASCET	American Society of Certified Engineering Technicians
ASCG	Automatic Solution Crystal Growth
ASCII	American Standard Code for Information Interchange
ASCMA	American Sprocket Chain Manufacturers Association
ASCO	Automatic Sustainer Cutoff
ASCRT	Association for the Study of Canadian Radio and Television
ASCS	Aluminum Straight Clamp Strap
	Area Surveillance Control System
	Automatic Stabilization and Control System
ASD	Access Storage Device
	Aeronautical Systems Division [US Air Force]
	Aerospace Systems Division [US Air Force]
	Atmospheric Sciences Department
ASDE	Airport Surface Detection Equipment
	American Society of Danish Engineers
ASDG	Aircraft Storage and Disposition Group [US Air Force]
ASDI	Automatic Selective Dissemination of Information
ASDIC	Antisubmarine Detection Investigation Committee [also: Asdic]
ASDIRS	Army Study Documentation and Information Retrieval System [US]
ASDL	Automated Ship Data Library
ASDR	Airport Surface Detection Radar
ASDSRS	Automatic Spectrum Display and Signal Recognition System
ASDSVN	Army Switched Data and Secure Voice Network
ASE	Airborne Search Equipment
	Allowable Steering Error
	Alternative Sources of Energy
	Amplified Spontaneous Emission
	Association of Scientists and Engineers
	Amalgamated Society of Engineers
	Automatic Spectroscopic Ellipsometry
	Automatic Support Equipment
ASEA	American Solar Energy Association
ASEAN	Association of Southeast Asian Nations
ASEB	Aeronautics and Space Engineering Board [US]
	American Society for Experimental Biology
ASEC	American Standard Elevator Code
ASECH	Acetylselenocholine [also: ASECh]
ASEE	American Society for Engineering Education
ASEM	American Society for Engineering Management
ASES	American Solar Energy Society
ASESA	Armed Services Electronic Standards Agency [US]
ASESS	Aerospace Environment Simulation System
ASET	Aeronautical Services Earth Terminal
ASF	Ampere per Square Feet
	Army Service Forces

ASFDO	Antisubmarine Fixed Defense Office
ASFE	Association of Soil and Foundation Engineers
ASFG	Atmospheric Sound-Focusing Gain
ASFIR	Active-Swept Frequency Interferometer Radar
ASFTS	Airborne Systems Functional Test Stand
ASG	Aeronautical Standards Group
	Advanced Study Group
ASGE	American Society of Gas Engineers
ASGLS	Advanced Space-Ground Link Subsystem
ASGMT	Assignment
ASH	Assault Support Helicopter
ASHAE	American Society of Heating and Air Conditioning Engineers
ASHE	American Society for Hospital Engineering
ASHG	American Society of Human Genetics
ASHRAE	American Society of Heating, Refrigeration and Air Conditioning Engineers
ASHVE	American Society of Heating and Ventilating Engineers
ASI	Advanced Scientific Instrument
	Air Speed Indicator
	Altimeter Setting Indicator
	American Society of Inventors
	American Standards Institute [US]
A-Si	Amorphous Silicon
ASIC	American Society of Irrigation Consultants
	Application-Specific Integrated Circuit
ASID	Address Space Identifier
	American Society for Interior Designers
ASIDIC	Association of Information and Dissemination Centers
ASII	American Science Information Institute [US]
ASIP	Aircraft Structural Integrity Program
ASIRC	Aquatic Sciences Information Retrieval Center [US]
ASIS	Abort Sensing and Instrumentation System
	American Society for Information Sciences
ASIST	Advanced Scientific Instruments Symbolic Translator
ASIWPCA	Association of State and Interstate Water Pollution Control Administrators [US]
ASJ	All-Service Jacket
ASK	Amplitude Shift Keying
ASKA	Automatic System for Kinematic Analysis
ASKS	Automatic Station Keeping System
ASL	Aeronautical Structures Laboratory [US Navy]
	Association of Symbolic Logic
	Atmospheric Sciences Laboratory [US]
	Available Space List
	Average Signal Level
ASLBM	Air-to-Ship Launched Ballistic Missile
ASLE	American Society of Lubrication Engineers
ASLE&F	Amalgamated Society of Locomotive Engineers and Firemen
ASLIB	Association of Special Libraries [US]
ASLO	American Society of Limnology and Oceanography

ASLT	Advanced Solid Logic Technology
ASM	Advanced Surface-to-Air Missile
	Air-to-Surface Missile
	American Society for Metals
	American Society for Microbiology
	Apollo Service Module [NASA]
	Apollo Systems Manual [NASA]
	Assembler
	Association for Systems Management
	Asynchronous State Machine
ASMB	Acoustical Standards Management Board
ASMC	Aviation Surface Material Command [US Army]
	Automatic Systems Management and Control
ASME	American Society of Mechanical Engineers
ASMEA	American Society of Mechanical Engineers, Auxiliary
ASMER	Association for the Study of Man-Environment Relations
ASMFER	American Society for Metals - Foundation for Education and Research
ASMI	Airfield Surface Movement Indicator
ASM-MSD	American Society for Metals - Materials Science Division
ASMS	Advanced Surface Missile System
	American Society of Mass Spectroscopy
ASMT	Assortment
ASN	Atlantic Satellite Network
	Average Sample Number
Asn	Asparagine
ASNE	American Society of Naval Engineers
ASNT	American Society for Nondestructive Testing
ASO	Auxiliary Switch, (Normally) Open
ASODDS	ASWEPS Submarine Oceanographic Digital Data System
ASOP	Automatic Scheduling and Operating Program
	Automatic Structural Optimization Program
ASOS	Airport Surface Observing System
ASP	Alcoa Smelting Process
	Alloy Steel Plant
	Air-Speeded Post
	American Selling Price
	American Society for Photobiology
	American Society of Photogrammetry
	Anti-Segregation Process
	Asea-Stora Process
	Asymmetric Multiprocessing System
	Attached Support Processor
	Automatic Schedule Procedure
	Automatic Servo Plotter
	Automatic Synthesis Program
	Average Selling Price
Asp	Asparagine Acid
ASPARTAME	Aspartylphenylalanine Methylester
ASPE	American Society of Plumbing Engineers
	American Society of Professional Ecologists
ASPEN	Advanced System for Process Engineering
	Arctic Ship Probability Evaluation Network
ASPEP	Association of Scientists and Professional Engineering Personnel

ASPH	Asphalt
ASPI	American Society for Performance Improvement
ASPJ	Advanced Self-Protective Jammer
ASPN	American Society of Precision Nailmakers
ASPO	Apollo Spacecraft Project Office [NASA]
ASPP	Alloy-Steel Protective Plating
ASPR	Average Specific Polymerization Rate
ASQC	American Society for Quality Control
ASR	Accumulator Shift Right
	Airborne Surveillance Radar
	Airport Surveillance Radar
	Air Search Radar
	Air-Sea Rescue
	Automatic Send and Receive
	Automatic Speech Recognition
	Automatic Sprinkler Riser
	Available Supply Rate
ASRA	Automatic Stereophonic Recording Amplifier
ASRE	Admiralty Signal and Radar Establishment [UK]
	American Society of Refrigeration Engineers [now: ASHRAE]
ASRL	Aeroelastic and Structures Research Laboratory [MIT, US]
ASROC	Antisubmarine Rocket
ASRPA	Army Signal Radio Propagation Agency [US]
ASRS	Amalgamated Society of Railway Servants
AS/RS	Automated Storage and Retrieval System [also: ASRS]
ASRWPM	Association of Semi-Rotary Wing Pump Manufacturers
ASS	Aerospace Support System
	Aerospace Surveillance System
	Assistant
ASSE	American Society of Safety Engineers
	American Society of Sanitary Engineering
	American Society of Swedish Engineers
ASSEM	Assembly
ASSESS	Analytical Studies of Surface Effects of Submerged Submarines
ASSET	Aerothermodynamic, Structural Systems Environment Test
	American Society of Scientific and Engineering Translators
ASSIM	Assimilation
ASSN	Association
ASSOC	Associate
AssocSc	Associate in Sciences
Assoc Prof	Associate Professor
ASSP	Acoustics, Speech, and Signaling Processing Group [of IEEE]
ASSR	Airborne Sea and Swell Recorder
	Autonomous Soviet Socialist Republic
ASST	Advanced Supersonic Transport
	Assistant
Asst Prof	Assistant Professor
ASSU	Air Support Signal Unit
ASSY	Assembly

AST	Advanced Supersonic Technology
	Aerospace Technology [NASA]
	Army Satellite Tracking Center [US]
	Atlantic Standard Time
	Automatic Shop Tester
	Auxiliary Segment Table
ASTA	Aerial Surveillance and Target Acquisition
	Association of Short-Circuit Testing Authorities
ASTAS	Antiradar Surveillance and Target Acquisition System
ASTC	Airport Surface Traffic Control
ASTD	American Society for Training and Development
ASTE	American Society of Test Engineers
	American Society of Tool Engineers
ASTEC	Antissubmarine Technical Evaluation Center
ASTEM	Analytical Scanning Transmission Electron Microscopy
ASTF	Aero-Propulsion System Test Facility
ASTI	Applied Science and Technology Index
	Association for Science, Technology and Innovation
ASTIA	Armed Services Technical Information Agency
ASTIS	Arctic Science and Technology Information System
ASTM	American Society for Testing and Materials
ASTME	American Society of Tool and Manufacturing Engineers
ASTMS	Association of Scientific, Technical and Managerial Staffs
ASTOR	Anti-Ship Torpedo
ASTP	Apollo-Soyuz Test Project [NASA]
ASTR	Astronomer
	Astronomical
	Astronomy
ASTRA	Application of Space Techniques Relating to Aviation
	Automatic Scheduling with Time-Integrated Resource Allocation
ASTRAC	Arizona Statistical Repetitive Analog Computer
ASTRAL	Analog Schematic Translator to Algebraic Language
	Assurance and Stabilization Trends for Reliability by Analysis of Lots
ASTREC	Atomic Strike Recording System [US Air Force]
ASTRO	Advanced Spacecraft Transport Reusable Orbiter
	Aerodynamic Spacecraft Two-Stage Reusable Orbiter
	Airspace Travel Research Organization [US]
ASTRON	Astronomer
	Astronomical
	Astronomy
ASU	Arizona State University [US]
	Automatic Switching Unit
ASUN	Associated Students of the University of Nevada [US]

ASUS	Arts and Science Undergraduate Society
ASV	Angle Stop Valve
	Automatic Self-Verification
ASVIP	American Standard Vocabulary for Information Processing
ASW	Antisubmarine Warfare
	Antisubmarine Weapon
	Applications Software
	Automatic Socket Weld
	Auxiliary Switch
ASWE	Admiralty Surface Weapon Establishment [UK]
ASWEPS	Anti-Submarine Warfare Environmental Prediction System [US Navy]
AS&WG	American Steel and Wire Gage [also: ASWG]
ASWORG	Antisubmarine Warfare Operations Research Group [US Navy]
ASW/SCCS	ASW (= Antisubmarine Warfare)/Ship Command and Control System
ASWSPO	Antisubmarine Warfare Systems Project Office
ASXRED	American Society for X-Ray and Electron Diffraction
ASYM	Asymmetry
ASYNCH	Asynchronous
ASZD	American Society for Zero Defects
AT	Acceptance Test
	Acoustical Tile
	Address Translation
	Address Translator
	Air Temperature
	Airtight
	Air Transport
	Alignment Telescope
	Aminotriazole
	Ampere Turn
	Antitank
	Anti-Thrombin
	Antitorpedo
	Atomic
	Atomic Time
	Audit Trail
	Automatic Ticketing
At	Astatine
AT %	Atomic Percent
ATA	Aeronautical Telecommunications Agency
	Air Transport Association (of America)
	Alberta Trucking Association [Canada]
	American Teleport Association
	American Transit Association
	American Trucking Association
	Atlantic Treaty Association [NATO]
ATAA	Air Transport Association of America
ATAC	Air Transport Advisory Council [US]
	Air Transport Association of Canada
	Army Tank Automotive Center [US Army]
AT/AC	Alignment Telescope/Autocollimator
ATAR	Antitank Aircraft Rocket
ATARC	Advanced Technology Alcoa Reduction Cell
ATB	Access Type Base
	Acetylene-Terminated Bisphel

	Advanced Technology Bomber
	Air Transport Board
	All Trunks Busy
	Aluminum tert-Butoxide
	Asphalt Tile Base
ATBM	Advanced Tactical Ballistic Missile
ATC	Acoustic Tile Ceiling
	Aircraft Technical Committee
	Air Traffic Control(ler)
	Air Transport Command
	Air Transport Committee
	Alloy-Tin Couple
	American Technical Ceramics
	Applied Technology Council
	Armored Troop Carrier
	Automatic Temperature Compensation
	Automatic Through Center
	Automatic Toggle Clamp
	Automatic Train Control
	Automation Training Center
ATCA	Air Traffic Control Association [US]
	Allied Tactical Communications Agency [NATO]
	Automatic Tuned Circuit Adjustment
ATCAA	Automatic Tuned Circuit Adjustment Amplitude
ATCAC	Air Traffic Control Advisory Committee
ATCC	American-Type Culture Collection
ATCAP	Air Traffic Control Automation Panel
ATCBI	Air Transport Control Beacon Interrogator
ATCC	Air Traffic Control Center
ATCCC	Advanced Technical Command Control Capability
ATCE	Automatic Test and Checkout Equipmemt
ATCOS	Atmospheric Composition Satellite [NASA]
ATCRBS	Air Traffic Control Radar Beacon System
ATCS	Advanced Train Control System
	Air Traffic Control Service
ATCSS	Air Traffic Control Signaling System
ATCT	Air Traffic Control Tower
ATD	Admission, Transfer and Discharge (System)
	Aerospace Technology Division
	Along Track Distance
	Asynchronous Time Dilation
ATDA	Augmented Target Docking Adapter [NASA]
ATDC	Advanced Technology Development Center
	After Top Dead Center
ATDM	Asynchronous Time-Division Multiplexing
ATDS	Airborne Tactical Data System
	Automatic Telemetry Decommutation System
ATE	Advanced Technology Engine
	Air-Turbo Exchanger
	Automatic Test Equipment
ATEA	Army Transportation Engineering Agency [US]
ATEC	Automated Technical Control
	Automatic Test Equipment Complex
ATEE	Acetyl Tyrosine Ethylester
ATEGG	Advanced Turbine Engine Gas Generator
ATEM	Analytical Transmission Electron Microscopy

ATEWS	Advanced Tactical Early Warning System
ATF	Acid-Treated Florisil
	Advanced Tactical Fighter (Plane)
	Advanced Technology Fighter (Plane)
	Automatic Track Finding
	Automatic Transmission Fluid
	Aviation Turbine Fuel
ATFR	Automatic Terrain-Following Radar
ATFT	Additive Thin Film Technology
ATG	Advanced Technology Group
	Air-to-Ground
	Air Transport Group [Canadian Forces]
	Air Turbine Generator
ATGAR	Antitank Guided Air Rocket
ATGM	Antitank Guided Missile
ATH	Alumina Trihydrate
	Aluminum Trihydrate
ATI	Air Technical Intelligence [US Air Force]
	Associate of the Textile Institute [UK]
	Association of Technical Institutions
ATIS	Automatic Terminal Information Service
	Automatic Traffic Information System
ATJ	Automatic Through Junction
ATJS	Advanced Tactical Jamming System
ATL	Aeronautical Turbine Laboratory [US Navy]
	Analog-Threshold Logic
	Appliance Testing Laboratory
	Anti-Thrust Law
	Artificial Transmission Line
	Atlantic
ATLAS	Abbreviated Test Language for Avionics Systems
	Adaptive Test and Logic Analysis System
	Antitank Laser Assisted System
	Automated Tape Lay-Up System
	Automatic Tabulating, Listing and Sorting System
ATLB	Air Transport Licensing Board
ATLIS	Army Technical Libraries and Information Systems [US]
ATLSS	Advanced Technology for Large Structural Systems
ATM	Advanced Test Module
	Air Turbine Motor
	Apollo Telescope Mount [NASA]
	Automated Teller Machine
	Auxiliary Tape Memory
atm	(standard) atmosphere [unit]
ATMAM	Analytical and Test Methodologies for Design with Advanced Materials
ATME	Automatic Transmission Measuring Equipment
ATMI	American Textile Manufacturers Institute
ATMOS	Atmosphere
ATMS	Advanced Text Management System
	Automatic Transmission Measuring System
ATN	Augmented Transition Network
AT NO	Atomic Number
ATO	Assisted Takeoff
	Automatic Trunk Office
ATOL	Assisted Takeoff and Landing

ATOLL	Acceptance Test or Launch Language		Attention
ATOM	Apollo Telescope Orientation Mount [NASA]		Automatic Testing Technology
ATOMDEF	Atomic Defense	**ATTAC**	Advanced Technologies Testing Aircraft
ATOMIC	Automatic Test of Monolithic Integrated Circuits	**ATTC**	American Towing Tank Conference
		ATTEN	Attenuation
ATOMS	Automated Technical Order Maintenance Sequences		Attenuator
		ATTI	Arizona Transportation and Traffic Institute [US]
ATP	Acceptance Test Procedure		
	Accepted Test Procedure	**ATTITB**	Air Transport and Travel Industry Training Board
	Adenosine Triphosphate		
	Advanced Turboprop Airliner	**ATTM**	Attachment
	Arsenical Tough Pitch (Copper)	**ATTN**	Attention
	Association of Technical Professionals	**ATTP**	Advanced Transport Technology Program
	Asynchronous Transaction Processing	**ATTS**	Automatic Telemetry Tracking System
	Automated Test Plan		Auto(matic) Tape Time Select
ATPG	Automatic Test Pattern Generator	**ATU**	Aerial Tuning Unit
	Automatic Test Program Generator		Automatic Tracking Unit
ATPISO	Acetyl-Terminated Polyimidesulfone	**ATV**	All-Terrain Vehicle
ATR	Advanced Terminal Reactor		Automatic Threshold Variation
	Air Transport Radio	**ATW**	Automatic Tube (Butt) Weld
	Air Turbo Rocket	**AT/W**	Atomic Hydrogen Weld
	Anti-Transmit-Receive	**AT WT**	Atomic Weight
	Attenuated Total Reference	**AU**	Acousto-Ultrasonics
	Attenuated Total Reflection		Arithmetic Unit
	Attenuated Total Reflectance		Amplifier Unit
	Attenuated Reflection		Astronomical Unit
	Automatic Target Recognition		Angstrom Unit
ATRA	Advanced Transit Association	**Au**	(Aurum) - Gold
ATRAN	Automatic Terrain Recognition and Navigation	**AUA**	Association of University Architects
		AUC	Alberta Universities Commission [Canada]
ATRAX	Air Transportable Communications Complex	**AUCC**	Association of Universities and Colleges of Canada
ATRC	Antitracking Control		
ATREM	Average Time Remaining	**AUCS**	Atlantic University Computer Study
ATRID	Automatic Target Recognition Identification and Detection	**AUD**	Audibility
			Audio
ATRS	Automated Temporary Roof Support	**AUDAR**	Autodyne Detecting and Ranging
ATRT	Anti-Transmit-Receive Tube	**AUDREY**	Audio Reply
ATS	Acetylene-Terminated Sulfone	**AUG**	Add-On's and Upgrades
	Acquisition and Tracking System		August
	Administrative Terminal System	**AUFIS**	Automated Ultrasonic Flaw Imaging System
	Advanced Technology Satellite	**AUL**	Above Upper Limit
	Advanced Test System	**AUM**	Air-to-Underwater Missile
	Air Traffic Service [of FAA]	**AUNT**	Automatic Universal Translator
	Air Traffic System	**AUR**	Association of University Radiologists [US]
	American Technical Society	**AURA**	Association of Universities Research in Astronomy
	Analog Test System		
	Applications Technology Satellite [NASA]	**AUS**	Australia(n)
	Applied Test System	**AUSS**	Advanced Unmanned Search System
	Astronomical Time Switch		Automatic Ultrasonic Scanning System
	Automated Telemetry System	**AUSSAT**	Australian Satellite Network
	Automatic Transport System	**AUST**	Austria(n)
	Auxiliary Tug Service	**AUSTRAL**	Australia(n)
ATSD	Airborne Traffic Situation Display	**AUT**	Advanced User Terminal
ATSIT	Automatic Techniques for the Selection and Identification of Targets	**AUTEC**	Atlantic Undersea Test and Evaluation Center [US]
ATSR	Along-Track Scanning Radiometer	**AUTECS**	Automated Eddy Current Inspection System
ATSS	Acquisition Tracking Subsystem	**AUTH**	Authorization
ATSU	Air Traffic Service Unit	**AUTO**	Automatic
	Association of Time-Sharing Users	**AUTOCALL**	Automatic Calling
ATT	Advanced Transport Technology	**AUTOCOM**	Automotive Composites Conference and Exhibition
	Attach(ment)		

AUTO CV	Automatic Check Valve
AUTODIN	Automatic Digital Information Network
AUTODOC	Automated Documentation
AUTOMAP	Automatic Machining Program
AUTOMAST	Automatic Mathematical Analysis and Symbolic Translation
AUTONET	Automatic Network Display
AUTOPIC	Automatic Personal Identification Code
AUTOPROMPT	Automatic Programming of Machine Tools
AUTOPSY	Automatic Operating System
AUTO-QC	Automatic Quality Control
AUTO RECL	Automatic Reclosing
AUTOSATE	Automated Data Systems Analysis Technique
AUTOSCRIPT	Automated Systems for Composing, Revising, Illustrating and Phototypesetting
AUTOSEVCOM	Automatic Secure Voice Communications
AUTOSPOT	Automatic System for Positioning Tools
AUTOSTRAD	Automated System for Transportation Data
AUTOSTRT	Automatic Starter
AUTOSTRTG	Automatic Starting
AUTOTESTCON	Automobile Testing Conference [US]
AUTO TR	Auto-Transformer
AUTO-TRIP	Automatic Transportation Research Investigation Program
AUTOVON	Automatic Voice Network
AUTRAN	Automatic Target Recognition Analysis
	Automatic Utility Translator
AUUA	American UNIVAC Users Association
AUV	Armored Utility Vehicle
	Autonomous Underwater Vehicle
AUVS	Association for Unmanned Vehicle Systems
AUW	All-Up-Weight
	Anti-Underwater Warfare
AUWE	Admiralty Underwater Weapons Establishment [UK]
AUX	Auxiliary
AUXIL	Auxiliary
AV	Added Value
	Angular Velocity
	Arctic Vessel
	Audio-Visual
	Authorized Version
	Avenue
	Average
	Average Variability
av	avoirdupois [unit]
AVA	Adventitious Viral Agent
	Automated Vision Association
	Azimuth versus Amplitude
AVAS	Automatic VFR (= Visual Flight Rules) Advisory Service
AVASI	Abbreviated Visual Approach Slope Indicator
AVBL	Armored Vehicle Bridge Launcher
AVC	Automatic Voltage Compression
	Automatic Volume Control
AVCG	Automatic Vapor Crystal Growth

AVCOM	Aviation Materiel Command [US]
AVCS	Advanced Vidicon Camera System
AVD	Alternate Voice/Data
	Aluminum Vacuum Degassing
	Automatic Voice/Data
avdp	avoirdupois [unit]
AVE	Aerospace Vehicle Electronics
	Automatic Voltammetric Electrode
	Automatic Volume Expansion
	Avenue
AVEC	Amplitude Vibration Exciter Control
AVERT	Association of Volunteer Emergency Radio Teams
AVF	Azimuthally Varying Field
AVG	Aminoethoxyvinylglycine
	Average
AVGAS	Aviation Gasoline
AVI	Airborne Vehicle Identification
	Automated Visual Inspection
	Aviation
AVID	Advanced Visual Information Display
AVIS	Automatic Video Inspection System
AVL	Adel'son-Vel'skii and Landis Trees
	Automatic Vehicle Location
AVLB	Armored Vehicle Launch Bridge
AVLC	Automatic Vehicle Location and Control System
AVLSI	Analog Very Large-Scale Integration
AVM	Airborne Vibration Monitor(ing)
	Automatic Vehicle Monitor(ing)
	Automatic Vibration Monitor(ing)
AVMRI	Arctic Vessel and Marine Research Institute
AVNL	Automatic Video Noise Limiting
AVO	Advanced Video Option
AVOID	Airfield Vehicle Obstacle Indication Device
avoir	avoirdupois [unit]
avoir oz	avoirdupois ounce [unit]
AVOLO	Automatic Voice Link Observation
AVOSS	Added Value Operating Support System
AVR	Automatic Volume Recognition
AVRS	Automated Vehicle Roading System
AVS	Advanced Vacuum System
	Aerospace Vehicle Simulation
	American Vacuum Society
AvSat	Aviation Satellite [also: AVSAT]
AVSF	Advanced Vertical Strike Fighter
AVTA	Automatic Vocal Transaction Analysis
AW	Above Water
	Acid-Washed
	Air Warning
	Arc Welding
	Arming Wire
	Automatic Weapon
A/W	Actual Weight
AWA	Alliance of Women in Architecture
	American Wire Association
AWACS	Advanced Warning and Control System
	Airborne Warning and Control System
AWADS	All-Weather Aerial Delivery System
AWAR	Area Weighted Average Resolution
AWARS	Airborne Weather and Reconnaissance System

AWAT	Area Weighted Average T-Number
AWC	Association for Women in Computing
AWCLS	All-Weather Carrier Landing System
AWCS	Air Weapons Control System
AWEA	American Wind Energy Association
AWG	American Wire Gage
	Arbitrary Wave Generator
	Association for Women Geoscientists
AWIC	Association for Women in Computing
AWIM	Association for Women in Mathematics
AWIS	Association for Women in Science
	Aviation Weather Information Service
AWLS	All-Weather Landing System
AWM	Association for Women in Mathematics
AWN	Automatic Weather Network [US Air Force]
AWOP	All-Weather Operations Panel [of ICAO]
AWOS	Airport Weather Observing System
	Automated Weather Observing System
AWPA	American Wood Preservers Association
AWPB	American Wood Preservers Bureau
AWRA	American Water Resources Association
AWRE	Atomic Weapon Research Establishment [UK]
AWS	Air Warning System
	Air Weapon System
	Air Weather Service
	American War Standards
	American Welding Society
	Association of Women Scientists
AWT	Actual Work Time
	Automatic Weld Tube
AWTP	Advanced Waste Treatment Process
AWU	Atomic Weight Unit
AWWA	American Water Works Association
AX	Attack Experimental
	Automatic Transfer
AXAA	Australian X-Ray Analytical Association
AXD	Auxiliary Drum
AXFMR	Automatic Transformer
AXP	Axial Pitch
AZ	Arizona [US]
	Azimuth
	Azure
AZAS	Adjustable Zero Adjustable Span
AZBN	Azobisisiobutyronitrile
AZEL	Azimuth Elevation
AZON	Azimuth Only
AZRAN	Azimuth and Range
AZS	Automatic Zero Set
AZT	Azinothymidine

B

B	Bachelor
	Bale
	Bandwidth
	Base
	Batch
	Bay
	Beam
	Bit
	Block
	Bonded
	Braid
	Boron
	Brazing
	Breadth
	British
	Broadcasting
	Bus
	Byte
b	barn [unit]
	bel
	binary
BA	Bachelor of Arts
	Basal Area
	Bayard-Alpert (Gauge)
	Benzanthracene
	Benzla Acetone
	Bend Allowance
	Binary Add(er)
	Box Annealing
	Breathing Apparatus
	Bromo-Acetone
	Buffer Amplifier
B/A	Boron/Aluminum
Ba	Barium
BAA	Bachelor of Applied Arts
	British Acetylene Association
	British Airports Authority
	Broadband Antenna Amplifier
BAAR	Board for Aviation Accident Research [US Army]
BAArch	Bachelor of Arts in Architecture
BAAS	British Association for the Advancement of Science
BAB	Babbitt
BABS	Blind Approach Beacon System
BAC	Background Analysis Center
	Barometric Altitude Control
	Binary Asymmetric Channel
	Blood Alcohol Concentration
BACAIC	Boeing Airplane Company Algebraic Interpretive Computing System
BACAN	British Association for the Control of Aircraft Noise
BACE	Basic Automatic Checkout Equipment
	British Association of Consulting Engineers
BACG	British Association for Crystal Growth
BADAS	Binary Automatic Data Annotation System
BADC	Binary Asymmetric Dependent Channel
BADGE	Base Air Defense Ground Experiment
BADIC	Biological Analysis Detection Instrumentation and Control
BAdmin	Bachelor of Administration
BAE	Beacon Antenna Equipment
BAED	Bachelor of Arts in Environmental Design
BAEE	Benzoyl Arginine Ethyl Ester
BAeE	Bachelor of Aeronautical Engineering

BAF	Baffle
BAgE	Bachelor of Agricultural Engineering
BAID	Bachelor of Interior Design
BAINS	Basic Advanced Integrated Navigation System
BAIR	Berkeley Artificial Intelligence Research
BAIT	Bacterial Automated Identification Technique
BAK	Backup
BAL	Balance
	Basic Assembly Language
	British Anti-Lewisite
BALGOL	Burroughs Algorithmic Language
BALL	Ballast
BALLAST	Balanced Loading via Automatic Stability and Trim
BALLOTS	Bibliographic Automation of Large Library Operations using Time Sharing
BALLUTE	Balloon Parachute
BALMI	Ballistic Missile
BALPA	British Airline Pilots Association
BALS	Balancing Set
	Blind Approach Landing System
BALUN	Balance-to-Unbalance
BAM	Ballistic Missile
	Basic Access Method
	Bituminous Aggregate Mixture
	Broadcasting Amplitude Modulation
BAMBI	Ballistic Missile Boost Intercept
BAME	Benzoylarginine Methylester
BAMIRAC	Ballistic Missile Radiation Analysis Center [US]
BAMO	Bureau of Aeronautic Materials Officer
BAMS	Bachelor of Arts, Master of Science
BAMTM	British Association of Machine Tool Merchants
BAN	Best Asymptotically Normal
BANA	Benzoyl Arginine Naphthylamide
BAND	Bandolier
BANS	Bright Alphanumeric Subsystem
BAP	Band Amplitude Product
	Basic Assembly Program
	Biologically Active Peptides
BAPNA	Benzoyl Argenine Nitroanilide
BAPTA	Bearing and Power Transfer Assembly
BAR	Barometer
	Barometric
	Base Address Register
	Browning Automatic Rifle
	Buffer Address Register
	Bureau of Aeronautics Representative
bar	barrel [unit]
BArch	Bachelor of Architecture
BArchEng	Bachelor of Architectural Engineering
BARE	Biased-Activated Reactive Evaporation
BARR	Bureau of Aeronautics Resident Representative
BARS	Backup Attitude Reference System
	Ballistic Analysis Research System
BARSTUR	Barking Sands Tactical Underwater (Test) Range [Hawaii]

BART	Bay Area Rapid Transit [US]
BARTD	Bay Area Rapid Transit District [US]
BARV	Beach Armored Recovery Vehicle
BARZREX	Bartok Archives Z-Symbol Rhythm Extraction
BAS	Bachelor in Agricultural Science
	Basic Airspeed
	Blind Approach System
	British Acoustical Society
BASc	Bachelor of Applied Science
BASCOM	BASIC Compiler
BASEC	British Approval Service for Electric Cables
BASEEFA	British Approval Service for Electrical Equipment in Flammable Atmospheres
BASIC	Basic Algebraic Symbolic Interpretative Compiler
	Basic Automatic Stored Instruction Computer
	Beginner's All-Purpose Symbolic Instruction Code
BASICPAC	BASIC Processor and Computer
BASICS	Battle Area Surveillance and Integrated Communication System
BASIS	Bank Automated Service Information System
	Burroughs Applied Statistical Inquiry System
BASJE	Bolivian Air Shower Joint Experiment
BAS NET	Basic Network
BASRA	British Amateur Scientific Research Association
BASW	Bell Alarm Switch
BASYS	Basic System
BAT	Batch
	Basic Assurance Test
	Battalion Anti-Tank
	Battery
	Bond - Assembly - Test
BATS	Basic Additional Teleprocessing Support
BATT	Batten
	Battery
BATTY	Battery
BAUA	Business Aircraft Users Association
BAUD	Baudot Code
BAY CAND DC	Bayonette Candelabra, Double Contact
BB	Ball Bearing
	Baseband
	Best Best
	Block Brazing
	Broadband
	Building Block
B-B	Butane-Butane
B&B	Bell and Bell
BBA	Bachelor of Business Administration
BBB	Bisbenzimidazobenzophenanthroline
BB/B	Body Bound Bolts
BBC	Bachelor of Building Construction
	Broadband Conducted
BBD	Benzylaminonitrobenzoxadiazole
	Bubble Bath Detector
	Bucket-Brigade Device

BBDC	Before Bottom Dead Center
BBEA	Brewery and Bottling Engineers Association
BBF	Baseband Frequency
BBL	Barrel
	Basic Business Language
	Benzimidazobenzophenanthroline
bbl	barrel [unit]
BBM	Beam Brightness Modulation
	Break Before Make
BBMRA	British Brush Manufacturers Research Association
BBN	Bolt, Beranek and Newman
	Borabicyclononane
BBO	Bisbiphenylyloxazole
BBOT	Bisbutylbenzoxazolylthiophene
BBP	Butylbenzylphthalate
BBQ	Benzimidazobenzisoquinoline
BBR	Bend-Bend-Roll
	Broadband Radiated
BBRG	Ball Bearing
BBT	Bombardment
BBU	Baseband Unit
BC	Back Connected
	Basic Circle
	Basic Control
	Bathyconductograph
	Between Centers
	Binary Code
	Binary Counter
	Bolt Circle
	Bonded (Single) Cotton (Wire)
	Bottom Chord
	Bottom Contour
	British Columbia [Canada]
	Broadcast Control
	Bus Controller
	Buried Channel
BCA	Battery Control Area
BCAA	Branched-Chain Amino Acid
BCABP	Bureau of Competitive Assessment and Business Policy [US]
BCAC	British Columbia Aviation Council [Canada]
	British Conference on Automation and Computation
BCAS	British Compressed Air Society
BCB	Broadcast Band
	Brown Cardboard
BCC	Biological Council of Canada
	Block Check Character
	Body-Centered Cubic
	British Color Council
BCCA	British Columbia Construction Association [Canada]
BCCF	British Cast Concrete Federation
BCD	Binary-Coded Decimal
	Burst Cartridge Detection
BCDC	Binary-Coded Decimal Counter
BCDIC	Binary-Coded Decimal Interchange Code
BCE	Bachelor of Civil Engineering
BCET	Biochemical Engineering Technology
BCF	Billion Cubic Feet
	Bromochlorofluoromethane
	Bulked Continuous Fiber
BCFCA	British Columbia Floor Covering Association
BCFL	British Columbia Federation of Labour [Canada]
BCFMCA	British Columbia Frequency Modulation Communications Association [Canada]
BCGLO	British Commonwealth Geographical Liaison Office
BCH	Binary-Coded Hollerith
	Block Control Header
	Bose-Chaudhuri-Hocquenguem [SPADE]
BCH	Bunch
BChem	Bachelor of Chemistry
BCI	Battery Council International
	Binary-Coded Information
	Broadcast Interference
BCIP	Belgium Center for Information Processing
BCIRA	British Cast Iron Research Association
BCISC	British Chemical Industrial Safety Council
BCIT	British Columbia Institute of Technology [Canada]
BCKA	Branched Chain A-Keto Acid
BCKCD	Back-Order Code
BCL	Basic Contour Line
	Binary Cutter Location
	Biological Computer Laboratory [US]
	Burroughs Common Language
BCLMA	British Columbia Lumber Manufacturers Association [Canada]
BCM	Back Course Marker
	Beyond Capability of Maintenance
	Bromochloromethane
BCME	Bischloromethylether
BC-MOSFET	Buried-Channel Metal-Oxide Semiconductor Field-Effect Transistor
BCMTA	British Columbia Motor Transport Association [Canada]
BCN	Beacon
BCNI	Business Council on National Issues
BCO	Binary-Coded Octal
	Bridge Cutoff
BCOA	Bituminous Coal Operators' Association
BCOM	Burroughs Computer Output to Microfilm
BCompS	Bachelor of Computer Science
BCP	Best Current Practices
	Byte Control Protocol
BCPL	Basic Combined Programming Language
BCPMA	British Chemical Plant Manufacturers Association
BCPS	Beam Candle Power Seconds
BCR	Bituminous Coal Research [US]
BCRA	British Ceramic Research Association
	British Coke Research Association
BCRC	British Columbia Research Council [Canada]
BCRNL	Bituminous Coal Research National Laboratories
BCRU	British Committee on Radiological Units
BCRUM	British Committee on Radiation Units and Measurements
BCS	Bachelor of Computer Science

	Bardeen-Cooper-Schrieffer (Theory)
	Best Cast Steel
	Biomedical Computing Society [US]
	Block Check Sequence
	Boeing Computer Services [US]
	British Computer Society
BCSA	British Constructional Steelwork Association
BCSP	Board of Certified Safety Professionals
BCT	Body-Centered Tetragonal
	Bushing Current Transformer
BCTD	Building and Construction Trades Department
BCU	Binary Counting Unit
	Block Control Unit
BCUC	British Coal Utilities Commission
BCURA	British Coal Utilization Research Association
BCW	Buffer Control Word
	Burst Codewords
BCWWA	British Columbia Water and Waste Association [Canada]
BD	Band
	Base Detonating
	Binary Decoder
	Blocker Deflector
	Blocking Device
	Board
	Bond
	Bottom Down
	Butadiene
B/D	Binary-to-Decimal
BDA	Booster-Distributor Amplifier
	Bright-Dipped and Anodized
BDAM	Basic Direct Access Method
BDC	Binary Decimal Counter
	Bonded Double Cotton (Wire)
	Bottom Dead Center
BDCS	Butyldimethylchlorosilane
BDD	Binary-to-Decimal Decoder
BDE	Baseband Distribution Equipment
	Batch Data Exchange
BD ELIM	Band Elimination
BDES	Batch Data Exchange Service
BDes	Bachelor of Design
BDF	Baseband Distribution Frame
	Base Detonating Fuse
bd ft	board foot [unit]
BDGH	Binding Head
BDH	Bearing Distance and Heading
BDHI	Bearing Distance and Heading Indicator
BDI	Bearing Deviation Indicator
BDIA	Base Diameter
BDIAC	Battelle Defender Information Analysis Center [BMI]
BDL	Bundle
BDM	Bomber Defense Missile
BDMA	Benzyldimethylamine
BDN	Blocked Deoxynucleoside
BDNA	Benzyldinonylame
BDP	Bioenergy Development Program [Canada]
	Bonded, Double Paper (Wire)

	Business Data Processing
BDPA	Bisdiphenylenephenylallyl
	Black Data Processing Associates
BDPEC	Bureau of Disease Prevention and Environmental Control [US]
BDRY	Boundary
BDS	Benzenediazothioether
	Bonded Double Silk (Wire)
BDSA	Business and Defense Services Administration [US]
BDT	Binary Deck to Binary Tape
BDU	Baseband Distribution Unit
	Basic Device Unit
	Basic Display Unit
BDWTT	Battelle Drop Weight Tear Test
BDY	Boundary
BE	Bachelor of Engineering
	Back End
	Backscattered Electron
	Base Ejection
	Baume [unit]
	Bell End
	Benchmark Experiment
	Breaker End
Be	Beryllium
BEA	British Electrical Authority
	Business Education Association
BEAB	British Electrical Approvals Board
BEAC	Boeing Engineering Analog Computer
	British Export Advisory Council
BEACON	British European Airways Computer Network
BEAIRA	British Electrical and Allied Industries Research Association
BEAM	Building Equipment, Accessories and Materials (Program) [Canada]
	Burroughs Electronic Accounting Machine
BEAMA	British Electrical and Allied Manufacturers Association
BEBC	Big European Bubble Chamber
BEC	Bioenergy Council
	British Engineers Club
BECO	Booster-Engine Cutoff
BECS	Building Energy Conservation Sector
BECTO	British Electric Cable Testing Organization
BED	Bachelor of Environmental Design
	Bridge Element Delay
BEDA	Bachelor of Environmental Design in Architecture
BEDC	Building Economic Development Committee [UK]
BEDes	Bachelor of Environmental Design
BEE	Bachelor of Electrical Engineering
BEEC	Binary Error-Erasure Channel
BEEE	Bachelor of Electrical and Electronics Engineering
BEEF	Business and Engineering Enriched FORTRAN
BEF	Band Elimination Factor
	Band Elimination Filter
	Blunt End Forward

	Buffered Emitter Follower
BEFAP	Bell Laboratories FORTRAN Assembly Program
BEI	Backscattered Electron Image
BEL	Bell Character
BELG	Belgian
	Belgium
BEM	Bachelor of Engineering, Mining
	Boundary Element Method
BEMA	Business Equipment Manufacturers Association [US]
BEMAC	British Exports Marketing Advisory Committee
BEMS	Bioelectromagnetics
BEng	Bachelor of Engineering
BEngSc	Bachelor of Engineering and Science
BEP	Bachelor of Engineering Physics
	Bureau of Engraving and Printing [US]
BEPC	British Electrical Power Convention
BEPO	British Experimental Pile Operation
BER	Bit Error Rate
BERA	Biomass Energy Research Association
BER CONT	Bit Error Rate Continuous
BERT	Basic Energy Reduction Technology
BERTS	Bit Error Rate Test Set
BERTH	Berthing
BES	Bachelor of Engineering Science
	Biological Engineering Society
	Biomedical Engineering Society
	Bisaminoethanesulfonic Acid
	Black Enamel Slate
	Boundary Element Solver
BESA	British Engineering Standards Association
BESS	Bessemer
	Binary Electromagnetic Signal Signature
BEST	Ballastable Earthmoving Sectionized Tractor
	Ballistic Electron Schottky-Gate Transistor
	Battery Energy Storage Test
	Boehler Electroslag Topping
	Business EDP (= Electronic Data Processing) Systems Technique
	Business Electronic Systems Technique
	Business Equipment Software Technique
BET	Best Estimate of Trajectory
	Biological Engineering Technology
	Brunauer-Emmett-Teller
BETA	Business Equipment Trade Association [US]
BETT	Buildings Energy Technology Transfer
BEU	Basic Encoding Unit
BEV	Bevel(led)
BeV	Billion Electronvolts
BEV-BD	Bevel Board
BEX	Broadband Exchange
BF	Backface
	Back Feed
	Band Filter
	Base Fuse
	Beat Frequency
	Blast Furnace
	Blocking Factor
	Boiler Feed

	Boldface
	Both Faces
	Bottom Face
	Bright Field
	British Forces
B&F	Bell and Flange
BFBS	British Forces Broadcasting System
BFCO	Band Filter Cutoff
BFD	Back Focal Distance
BFE	Beam Forming Electrode
BFER	Base Field Effect Register
BFG	Binary Frequency Generator
BFL	Back Focal Length
BFMA	Business Forms Management Association
BFMIRA	British Food Manufacturing Industries Research Association
BFN	Beam Forming Network
BFO	Beat Frequency Oscillator
BFP	Back Focal Plane
	Boiler Feed Pump
BFPDDA	Binary Floating Point Digital Differential Analyzer
bFSH	Bovine Follicle-Stimulating Hormone
BFSK	Binary Frequency Shift Keying
BFW	Boiler Feed Water
BG	Back Gear
	Background
	Birmingham Gauge
	Board of Governors
	Bottom Grille
	British (Standard) Gauge
BGA	Blue-Green Algae
	Brilliant Green Agar
BG/BARC	Board of Governor's Budget and Accounts Review Committee [INTELSAT]
BGIRA	British Glass Industry Research Association
BGMA	British Gear Manufacturers Association
BGP	Becker, Green and Pearson (Equation)
BG/PC	Board of Governor's Planning Committee [INTELSAT]
BGRR	Brookhaven's Graphite Research Reactor [BNL, US]
BGRV	Boost Glide Reentry Vehicle
BGS	Beta Gamma Signal
	Bosson, Gutmann and Simmons (Equation)
BGWF	British Granite and Whinstone Federation
BH	Bake Hardening
	Block Handler
	Boiler House
	Brinell Hardness
B-H	Binary-to-Hexadecimal
BHA	Base Helix Angle
	Bottomhole Assembly
	Butylated Hydroxyanisole
BHC	Benzene Hexachloride
BHD	Bulkhead
BHET	Bishydroxyethyl Terephthalate
BHFP	Bottomhole Flowing Pressure
BHN	Brinell Hardness Number
BHP	Brake Horsepower
	Boiler Horsepower

	Bottomhole Pressure
	Bulk Handling Plant
BHPhr	Brake Horsepower-Hour
BHR	Biotechnology and Human Research
	Block Handling Routine
BHRA	British Hydromechanics Research Association
BHSL	Basic Hytran Simulation Language
BHT	Blowdon Heat Transfer
	Butylated Hydroxytoluene
	Butylated Hydroxytoluol
BHW	Boiling Heavy Water
BI	Base Ignition
	Basicity Index
	Black Iron
	Blanking Input
	Bus Interconnect
B&I	Base and Increment
Bi	Bismuth
BIA	Boost, Insertion and Abort
	Brick Institute of America
BIAA	British Industrial Advertising Association
BIAC	Bio-Instrumentation Advisory Council [of AIBS]
	Business and Industry Advisory Committee [OECD]
BIAS	Battlefield Illumination Airborne System
BIATA	British Independent Air Transport Association
BIB	Burn-In-Board
BIBD	Balanced Incomplete Block Design
BIBL	Bibliography
BIBLIO	Bibliography
BIBO	Bounded Input, Bounded Output
BIBRA	British Industrial Biological Research Association
BIC	Business Information Center
BICEMA	British Internal Combustion Engine Manufacturers Association
BICERI	British Internal Combustion Engine Research Institute
BICINE	Bishydroxyethylglycine
BICROS	Binaural Contralateral Routing of Signal
BICS	Building Industry Consulting Service
BICTA	British Investment Casters Technical Association
BID	Bachelor of Industrial Design
	Blocker Initial Design
	Blocker Initial-Guess Device
BIDAP	Bibliographic Data Processing
BIDCO	Built-in Digital Circuit Observer
BIE	Bachelor of Industrial Engineering
	Boundary Integral Equation
	British Institute of Engineers
BIEE	British Institute of Electrical Engineers
BIIA	British Institute of Industrial Art
BIIL	Basic Impulse Isolation Level
BIL	Block Input Length
	Built-In Logic
BILA	Battelle Institute Learning Automation [BMI]

BILBO	Built-in Logic Block Observability
	Built-in Logic Block Observer
BILD	Business Industrial Leadership Development
BIM	Beginning of Information Marker
	Best in Match
	Branch 'If' Multiplexer
BIMAC	Bistable Magnetic Core
BIMCAM	British Industrial Measuring and Control Apparatus Manufacturers
BIMOS	Bipolar Metal-Oxide Semiconductor
BIMRAB	BuWeps-Industry Materiel Reliability Advisory Board [US Navy]
BIN	Binary
BINAC	Binary Automatic Computer
BINAP	Binaphthyl
BInd	Bachelor of Industry
BINOMEXP	Binomial Expansion
BINR	Basic Intrinsic Noise Ratio
BIntArch	Bachelor of Interior Architecture
BIntDes	Bachelor of Interior Design
BIO	Bedford Institute of Oceanography [Canada]
BIOAL	Bioastronautics Laboratory
BIOALRT	Bioastronautics Laboratory Research Tool
BIOCHEM	Biochemical
	Biochemist
	Biochemistry
BIOL	Biological
	Biologist
	Biology
BIONESS	Zooplankton Net Sampling System
BIONICS	Biological Electronics
BIOPHYS	Biophysical
	Biophysicist
	Biophysics
BIOR	Business Input-Output Rerun
BIORED	Biological Resources Development
BIOS	Basic Input/Output System
	Biological Investigations of Space
	British Intelligence Objectives Subcommittee
BIOSIS	Biosciences Information Service [US]
BIP	Balanced in Plane
	Binary Image Processor
	Bismuth Iodoform Paraffin
BIPD	Bi Parting Doors
BIR	British Institute of Radiology
BIRDIE	Battery Integration and Radar Display Equipment
BIRE	British Institute of Radio Engineers
BIRES	Broadband Isotropic Real-Time Electric-Field Sensor
BIRS	Basic Indexing and Retrieval System
BIS	Basic Interchange System
	British Information Service
	British Interplanetary Society
	Business Information System
BISAM	Basic Indexed Sequential Access Method
Bis-AMP	Bisaminomethylpropanol
BISF	British Iron and Steel Federation
BISFA	British Industrial and Scientific Film Association

BisGMA	Bisphenolglycidyldimethacrylate
BISITS	British Industrial and Scientific International Translation Service
BISL	British Information Service Library
BISMRA	Bureau of Inter-Industrial Statistics and Multiple Regression Analysis [US]
Bis-MSB	Bismethylstyrylbenzene
Bis-NAD	Bisnicotinamide Adenine Dinucleotide
BISPA	British Independent Steel Producers Association
BISPAD	Bissilylated Phenylaminodiol
BISRA	British Iron and Steel Research Association
BIS-TRIS	Bisiminotrismethane
BISYNCH	Binary Synchronous
BIT	Bipolar Integrated Technology
	Built-in Test
bit	binary digit
BITA	British Industrial Truck Association
BITE	Built-in Test Equipment
bit/s	bits per second
BIU	Basic Information Unit
BIV	Best Image Voltage
BIX	Binary Information Exchange
BJ	Bar Joist
BJT	Bipolar Junction Transistor
BK	Bar Knob
	Black
	Block
	Book
	Brake
Bk	Berkelium
BKGRD	Background
BKR	Breaker
BKSP	Backspace
BKT	Basket
	Bracket
BL	Bale
	Base Line
	Bend Line
	Blotting
	Blue
	Bottom Layer
	Breadth-Length
	Breech Loading
	Building Line
B/L	Bill of Lading
bl	barrel [unit]
BLAC	British Light Aviation Center
BLADE	Basic Level Automation of Data through Electronics
BLADS	Bell Laboratories Automatic Design System
BLAM	Ballistically-Launched Aerodynamic Missile
BLC	Board Level Computer
	Boundary Layer Control
	British Lighting Council
BLD	Beam-Lead Device
	Binary Load Dump
BLDG	Building
BLE	Bombardment-Induced Light Emission
BLESSED	Bell Little Electronic Symbolic System for the Electrodata

BLEU	Blind Landing Experimental Unit [of RAE]
BLEVE	Boiling Liquid Expanding Vapor Explosion
BLF	Band Limiting Filter
BLG	Benzyl-L-Glutamate
	Breech Loading Gun
BLI	Basic Learning Institute
BLIP	Background Limited Infrared Photoconductor
BLIS	Bell Laboratories Interpretative System
BLK	Black
	Blank
	Block
	Bulk
BLK CAR	Bulk Carrier
BLKG	Blocking
BLL	Below Lower Limit
BLLDZR	Bulldozer
BLM	Basic Language Machine
	Bureau of Land Management [US]
BLMRA	British Leather Manufacturers Research Association
BLO	Blower
BLODI	Block Diagram
BLP	Ball Lock Pin
	Bypass Label Processor
BLR	Base Line Restorer
	Boiler
	Breech Loading Rifle
BLS	Bureau of Labour Statistics [US]
BLSTG	Blasting
BLSTG PWD	Blasting Powder
BLT	Bachelor of Laboratory Technology
	Basic Language Translator
	Borrowed Light
BL&T	Blind Loaded and Traced
BLTC	Bottom Loading Transfer Cast
BLU	Basic Link Unit
	Basic Logic Unit
BLUE	Best Linear Unbiased Estimator
BLVD	Boulevard
BLWA	British Laboratory Ware Association
BM	Ballistic Missile
	Beam
	Bench Mark
	Bellows Metering (Valve)
	Bill of Material
	Binary Multiply
	Board Measure
	Boundary Marker
	Bureau of Mines [US]
	Business Machine
B/M	Bill of Material
BMAR	Ballistic Missile Acquisition Radar
BMath	Bachelor of Mathematics
BMAW	Bare Metal-Arc Welding
BMB	British Metrication Board
BMBT	Bismethoxybenzylidenebitoluidine
BMC	Bulk Molding Compound
BMCS	Bureau of Motor Carrier Safety
BME	Bachelor of Mechanical Engineering
	Bachelor of Mining Engineering

	Biomedical Engineering		Blanking Oscillator
BMEC	Ball Bearing Manufacturers Engineers Committee		Blocking Oscillator
			Blow-off
BMEP	Brake Mean Effective Pressure		Branch Office
BMES	Biomedical Engineering Society	B-O	Binary-to-Octal
BMet	Bachelor of Metallurgy	BOA	British Optical Association
BMetE	Bachelor of Metallurgical Engineering	BOB	Bobbin
BMETO	Ballistic Missiles European Task Organization	BOBMA	British Oil Burners Manufacturers Association
BMEWS	Ballistic Missile Early Warning System	BOC	Blowout Coil
BMF	Bulk Mail Facility		Butoxycarbonyl
BMI	Battelle Memorial Institute	BOCA	Building Officials Conference of America
	Bismaleimide	BOCOL	Basic Operating Consumer-Oriented Language
BMIC	Bachelor of Microbiology	BOC-ON	Butoxycarbonyloxyiminophenylacetonitrile
BMILS	Bottom-Mounted Impact Location System	BOCS	Bendix Optimum Configuration Satellite
BMKR	Boilermaker	BOD	Beneficial Occupancy Date
BMO	Ballistic Missile Office		Biochemical Oxygen Demand
BMOC	Ballistic Missile Orientation Course		Biological Oxygen Demand
BMOM	Base Maintenance and Operations Model		Bottom of Duct
BMP	Batch Message Processing	BOE	Beginning of Extent
BMR	Bipolar Magnetic Region	BOF	Basic Oxygen Furnace
	Brookhaven's Medical Reactor [BNL, US]	B of B	Back of Board
BMS	Ballistic Missile Ship [US Navy]	BOFS	Black Oil Finish Slate
	Basic Mapping Support	BOFT	Board of Foreign Trade
	Battle Management System	BOL	Beginning of Life
	Boeing Materials Specification	BOLD	Bibliographic On-line Library Display
	Boranemethylsulfide	BOLS	Bolster
BMTA	British Mining Tools Association	BOLT	Beam of Light Transistor
BMTD	Ballistic Missile Terminal Defense	BOM	Basic Operating Monitor
BMTFA	British Malleable Tube Fittings Association		Bill of Materials
BMTS	Ballistic Missile Test System		Binary Order of Magnitude
BMTT	Buffered Magnetic Tape Transport	BoM	Bureau of Mines [US]
BMW	Beam Width	BOMEX	Barbados Oceanographic and Meteorological Experiment
BN	Battalion	BONAC	Broadcasting Organization of the Nonaligned Countries
	Binary Number		
BNB	Bullet Nose Bushing	BOOL	Boolean (Algebra)
BNC	Baby 'N' Connector	BOOST	Booster
BNCHBD	Benchboard	BOP	Basic Operator Panel
BNCM	British National Committee on Materials		Basic Oxygen Process
BNCOR	British National Committee for Oceanographic Research		Binary Output Program
			Blowout Preventer
BNCS	Board on Nuclear Codes and Standards	BOPD	Barrels of Oil per Day
	British Numerical Control Society	BOPET	Biaxially-Oriented Polyethylene Terephthalate
BNCSR	British National Committee on Space Research		
BND	Benzoylated Naphthoylated DEAE	BOM	Bureau of Mines [US]
	Bullet Nose Dowel	BORAM	Block-Oriented Random Access Memory
BNF	Backus Naur Form	BOS	Background Operating System
	Bomb Nose Fuse		Back of Slip
BNFMRA	British Nonferrous Metals Research Association		Basic Operating System
			Basic Oxygen Steel(making)
BNH	Burnish		Bundle of Sticks
BNL	Brookhaven National Laboratory [US]	BOSOR	Buckling of Shell of Revolution
BNO	Backus Normal Form	BOSS	Behaviour of Off-Shore Structures (Conference)
BNS	Bachelor of Natural Sciences		
	Bachelor of Naval Sciences		Boeing Operational Supervisory System
	Binary Number System		Book of SEMI Standards
BNSC	British National Space Center		Business Opportunities Sourcing System
BNT	(Salt) Bath Nitriding	BOST	Basic Offshore Survival Training
BNW	Battelle Northwest (Laboratories)	BOT	Basic Offshore Training
BO	Beat Oscillator		

	Beginning of Tape	
	Board of Trade	
	Bottom	
BP	Back Pressure	
	Ball Plunger	
	Bandpass	
	Base Plate	
	Batch Processing	
	Between Perpendiculars	
	Blueprint	
	Boiling Point	
	Bonded, (Single) Paper (Wire)	
	Bypass	
B/P	Blueprint	
BPA	Bauhinia Purpurea	
	Biological Photographic Association [US]	
	Bonneville Power Administration [US]	
BPAM	Basic Partitioned Access Method	
BPB	Bromophenol Blue	
BPBIRA	British Paper and Board Industry Research Association	
BPC	Back Pressure Control	
	Boost Protective Cover [NASA]	
	British Productivity Council	
BPCD	Barrels per Calendar Day	
BPCF	British Precast Concrete Federation	
BPD	Barrels per Day	
	Beam Positioning Drive	
	Bushing Potential Device	
BPE	Boiling Point Elevation	
BPEA	Bisphenylethynylanthracene	
BPEC	Building Products Executives Conference	
BPEN	Bisphenylethynylnaphthacene	
BPetE	Bachelor of Petroleum Engineering	
BPF	Bandpass Filter	
	Bottom Pressure Fluctuation	
	British Plastics Foundation	
BPGMA	British Pressure Gauge Manufacturers Association	
BPH	Barrels per Hour	
BPhys	Bachelor of Physics	
BPI	Boost Phase Intercept	
bpi	bits per inch	
BPKT	Basic Programming Knowledge Test	
BPL	Beta-Propiolactone	
BPM	Batch Processing Monitor	
BPMA	British Photographic Manufacturers Association	
	British Printing Manufacturers Association	
BPMC	Butylphenylmethylcarbamate	
BPMD	Battelle's Project Management Division [US]	
BPO	Biphenylylphenyloxazole	
	British Post Office	
BPPF	Basic Program Preparation Facility	
BPPMA	British Power Press Manufacturers Association	
BPR	Bureau of Public Roads	
BPRF	Bullet-Proof	
BPRTHM	Bureau of Public Roads Transport Highway Mobilization [Canada]	
BPS	Basic Programming System	

	Batch Processing System	
	Biophysical Society	
	Bulk Processing Stream	
	Bulk Processing System	
bps	bits per second	
BPSD	Barrels per Stream Day	
BPSK	Binary Phase-Shift Keying	
	Biphase Shift Keyed	
BPST	Benzothiazolylstyrylphtalhydrazidyltetra-zolium	
BPT	Blade Passage Tones	
	Borderline Pumping Temperature	
BPWR	Burnable Poison Water Reactor	
BQ	Brine Quenching	
BQL	Basic Query Language	
BR	Base Register	
	Bellows Regulating (Valve)	
	Bend Radius	
	Biological Reactivation	
	Boil-Resistant	
	Bottom Register	
	Branch	
	Brand	
	Breeder Reactor	
	Britain	
	British	
	British Railways	
	Bronze	
	Brown	
	Brush	
	Butadiene Rubber	
Br	Bromine	
BRA	British Refrigeration Association	
	British Robot Association	
BRANE	Bombing Radar Navigation Equipment	
BRASS	Bottom Reflection Active Sonar System	
BRAZ	Brazil	
	Brazilian	
BRB	British Railways Board	
BRBMA	Ball and Roller Bearing Manufacturers Association	
BRC	Branch Conditional	
BRD	Binary Rate Divider	
	Board	
	Braid	
BRDG	Bridge	
BRE	Building Research Establishment [UK]	
	Bureau of Research and Engineering	
BRECOM	Broadcast Radio Emergency Communication [US Air Force]	
BREL	Boeing Radiation Effects Laboratory [US]	
BREMA	British Radio Equipment Manufacturers Association	
BRG	Beacon Reply Group	
	Bearing	
BRGW	Brake Release Gross Weight	
BRH	Bureau of Radiological Health [US]	
BRI	Building Research Institute	
BRICS	Bureau of Research Information Control Systems [of USOE]	
BRIT	Britain	

	British
BRK	Break
	Brick
BRKR	Breaker
BRKT	Bracket
BRL	Ballistics Research Laboratory [US]
	Bureau of Research and Laboratories [Philippines]
BRLESC	Ballistics Research Laboratories Electronic Scientific Computer
BRLP	Burlap
BRLS	Barrier Ready Light System
BRN	Brown
BRO	Broach
BROOM	Ballistic Recovery of Orbiting Man
BRP	Beacon Ranging Pulse
BRR	Battelle Research Reactor
BRRL	British Road Research Laboratory
BRS	Bureau of Railroad Safety [US]
BRS	Brass
BR STD	British Standard
BRT	Binary Run Tape
	Brightness
BRU	Basic Resolution Unit
	Branch Unconditional
	Brown University [US]
BRUNET	Brown University Network [US]
BRVMA	British Radio Valve Manufacturers Association
BRWM	Board on Radioactive Waste Management
BRZ	Bronze
BRZG	Brazing
BS	Bachelor of Science
	Backscattered
	Backspace (Character)
	Bandeiraea Simplicifolia
	Basic Sediment
	Below Slab
	Bill of Sale
	Binary Subtract
	Biochemical Society
	Biological Society [US]
	Biometric Society
	Bismuth Sulphite
	Bonded (Single) Silk (Wire)
	Both Sides
	Bottom Sediment
	Breaking Strength
	British Standard
	Broadcasting Satellite
	Bureau of Standards
	Bursting Strength
B/S	Bill of Sale
B&S	Beams and Stringers
	Bell and Spigot
	Brown and Sharpe (Gauge)
b/s	bits per second
BSA	Bachelor of Science and Agriculture
	Bearing Specialists Association [US]
	Bimetal Steel-Aluminum
	Biophysical Society of America

	Bistrimethylsilylacetamide
	Body Surface Area
BSAD	Bachelor of Science in Architectural Design
BSAE	Bachelor of Science in Aeronautical Engineering
	Bachelor of Science in Architectural Engineering
BSAgrEng	Bachelor of Science in Agricultural Engineering
BSAM	Basic Sequential Access Method
BSArch	Bachelor of Science in Architecture
BSArchTech	Bachelor of Science in Architectural Technology
BSB	British Standard Beam
BSBC	Bachelor of Science in Building Construction
BSBldgSci	Bachelor of Science in Building Science
BSBP	British Standard Bulb Plate
BSC	Binary Synchronous Communications
	British Standard Channel
BSc	Bachelor of Science
BSCA	Binary Synchronous Communications Adapter
BScAgrEng	Bachelor of Science in Agricultural Engineering
BScArch	Bachelor of Science in Architecture
BScBldgDes	Bachelor of Science in Building Design
BScCE	Bachelor of Science in Civil Engineering
	Bachelor of Science in Construction Engineering
BScCS	Bachelor of Science in Computer Science
BScDA	Bachelor of Science in Data Analysis
BScE	Bachelor of Science in Engineering
BScEng	Bachelor of Science in Engineering
BScFE	Bachelor of Science in Forestry Engineering
BSCh	Bachelor of Science in Chemistry
BSCN	Bit Scan
BSConE	Bachelor of Science in Construction Engineering
BSConEng	Bachelor of Science in Construction Engineering
BSConTech	Bachelor of Science in Construction Technology
BSCRA	British Steel Castings Research Association
BScTech	Bachelor of Science and Technology
BSD	Ballistic Systems Division [US Army]
BSDC	Binary Symmetric Dependent Channel
	British Standard Data Code
BSDL	Boresight Datum Line
BSE	Bachelor of Science in Engineering
	Bachelor of Science in Sanitary Engineering
	Backscattered Electron
BSEA	British Standard Equal Angle
BSED	Bachelor of Science in Environmental Design
BSEE	Bachelor of Science in Electrical Engineering
BSEM	Bachelor of Science in Engineering of Mines
BSEng	Bachelor of Science in Engineering
BSERC	British Science and Engineering Research Council
BSF	Bachelor of Science in Forestry
	British Standard Fine (Thread)

BSFA	British Steel Founders Association
BSG	British Standard Gauge
B&SG	Brown and Sharpe Gauge
BSGeog	Bachelor of Science in Geography
BSH	Benzenesulfonylhydrazide
bsh	bushel [unit]
BSI	British Standards Institution
BSIE	Bachelor of Science in Industrial Engineering
BSIntArch	Bachelor of Science in Interior Architecture
BSIRA	British Scientific Instrument Research Association
BSKT	Basket
BSME	Bachelor of Science in Mechanical Engineering
	Bachelor of Science in Mining Engineering
BSMet	Bachelor of Science in Metallurgy
BSMT	Basement
BSNDT	British Society for Nondestructive Testing
BSO	Blue Stellar Object
BSP	British Standard Pipe (Thread)
	Bromosulphalein
	Burroughs Scientific Processor
BSPT	Benzothiazolylstyrylphthalhydrazidyltetra-zolium
BSR	Blip-Scan Ratio
BSRA	British Ship Research Association
BSRP	Biological Sciences Research Paper
BSS	Balanced Salt Solution
	British Standard Specification
	Broadcasting Satellite Service
BSSG	Biological Sciences Study Group
BSSM	British Society for Strain Measurement
BSSRS	British Society for Social Responsibility in Science
	Bureau of Safety and Supply Radio Services [US]
BST	Beam-Switching Tube
	Block the SPADE Terminal (Command)
	Bovine Somatostatin
	British Standard Tee
	British Standard Thread
	British Summer Time
BSTAN	Beta Solution Treated and Annealed
BSTD	Bastard
BSTFA	Bistrifluoroacetamide
	Bistrimethylsilyltrifluoroacetamide
BSTOA	Beta Solution Treated and Overaged
BSTR	Booster
BSU	Baseband Switching Unit
BSUA	British Standard Unequal Angle
BSW	British Standard Whitworth (Thread)
BS&W	Bottom Sediment and Water
BSWM	Bureau of Solid Waste Management
BSXF	Burst Sync Failure [SPADE]
BT	Backtracker
	Bathythermograph
	Bent
	Benzotriazole
	Branch Tee
	Bus Tie

BTA	Benzoyltrifluoroacetone
	Boring-Trepanning Association
BTAM	Basic Telecommunications Access Method
BTB	Bus Tie Breaker
BT-BASIC	Backtracker-BASIC
BTCC	Benzothiazolocarbon Cyanine
	Board of Transportation Commissioners of Canada
BTD	Binary-to-Decimal
	Bomb Testing Device
BTDA	Benzophenonetetracarboxylic Dianhydride
BTDC	Before Top Dead Center
BTDL	Back Transient Diode Logic
	Basic Transient Diode Logic
BTDU	Benzylthiodihydrouracil
BTE	Bachelor of Textile Engineering
	Baldwin, Tate and Emery (Theory)
	Battery Terminal Equipment
	Bench Test Equipment
	Business Terminal Equipment
BTech	Bachelor of Technology
BTEE	Benzoyltyrosineethylester
BTF	Bomb Tail Fuse
BTFLY VLV	Butterfly Valve
BTH	Basic Transmission Header
BTI	Bridged Tap Isolator
	British Telecommunications International
B-TIP	Battelle Technical Inputs to Planning
BTL	Beginning Tape Level
	Bell Telephone Laboratories
BTM	Basic Time-Sharing Monitor
BTMA	British Typewriter Manufacturers Association
	Busy-Tone Multiple Access
BTMSA	Bistrimethylsilylacetylene
BTO	Bombing through Overcast
B to B	Back to Back
BTR	Better
	Bit Time Recovery
	Broadcast and Television Receivers [IEEE Group]
BTS	Barium Titanate/Stannate
	Batch Terminal Simulation
	Bias Temperature Stress
	British Tunneling Society
	Bisynch Terminal Support
BTSS	Basic Time Sharing System
BT ST	Billet Steel
BTT	Business Transfer Tax
BTTP	British Towing Tank Panel
BTU	Basic Transmission Unit
	British Thermal Unit
BU	Baseband Unit
	Basic User
	Binding Unit
	Bottom Up
	Bureau
bu	bushel [unit]
BUAER	Bureau of Aeronautics [US Navy]
BUCLASP	Buckling of Laminated Stiffened Plates
BUDC	Before Upper Dead Center

BUE	Built-up Edge
BUEC	Back-up Emergency Communications
BUF	Buffer
BUGS	Brown University Graphic System
BUI	Buildup Index
BUIC	Back-up Interceptor Control
BUILD	Base for Uniform Language Definition
BUIS	Barrier Up Indicator System
BULL	Bullet
	Bulletin
BULL PRF	Bullet-Proof
BUM	Back-up Mode
	Break-up Missile
BUN	Blood Urea Nitrogen
BUORD	Bureau of Ordinance [US Navy] [also: BuOrd]
BUR	Bureau
BURB	Back-Up Roll Bending
BUR ST	Bureau of Standards
BUS	Broken-up Structure
	Business
bus	bushel [unit]
BUSH	Bushing
BUSHIPS	Bureau of Ships [US] [also: BuShips]
BU STD	Bureau of Standards
BUT	Broadband Unbalanced Transformer
	Button
BUV	Backscatter Ultraviolet
BUWEPS	Bureau of Naval Weapons [US]
BUZ	Buzzer
BV	Back View
	Bochumer Verein [Vacuum Degassing Process]
	Breakdown Voltage
BVC	Black Varnish Cambric (Insulation)
BVE	Bivariate Exponential (Distribution)
BVMA	British Valve Manufacturers Association
BVO	Brominated Vegetable Oil
BVSV	Bimetal Switching Valve
BVU	Brightness Value Unit
BW	Bacteriological Warfare
	Bandwidth
	Biological Warfare
	Bloch Wall
	Both Ways
	Braided Wire
BWA	Backward Wave Amplifier
	British Waterworks Association
BWC	Backward Wave Converter
BWD	Barrels of Water per Day
BWG	Birmingham Wire Gauge
	British (Imperial) Wire Gauge
BWH	Barrels of Water per Hour
BWM	Backward-Wave Magnetron
	Barrels of Water per Minute
BWMA	British Woodwork Manufacturers Association
BWO	Backward-Wave Oscillator
BWOC	By Weight of Cement
BWP	Brown Wrapping Paper
BWPA	Backward-Wave Power Amplifier

	British Wood Preserving Association
BWPD	Barrels of Water per Day
BWR	Bandwidth Ratio
	Boiling Water Reactor
BWRA	British Welding Research Association
BWTSDS	Base Wire and Telephone System Development Schedule
BWV	Back Water Valve
BX	Box
BYP	Bypass
BZ	Brillouin Zone
Bz	Benzene
	Benzoyl

C

C	Candle
	Capacitance
	Capacitor
	Carbide
	Carbon
	Cast(ing)
	Celsius
	Centigrade
	Clock
	Coefficient
	Collector
	Commercial
	Computer
	Concentration
	Conductor
	Control
	Controller
	Cotton
	Counter
	Coulomb
	Coulombmeter
	Course
	Cysteine
	Cytosine
c	calorie
	carry
	candle
	carat
	clear
	crystalline
	cubic
CA	Cable
	California [US]
	Candle
	Cathode
	Cellulose Acetate
	Central America
	Centrifugal Accelerator
	Chaney Adapter
	Chromic Acid
	Cinnamic Acid
	Citric Acid

Clamp Assembly
Clear Aperture
Close Annealing
Compression Axis
Computer Application
Computers and Automation
Continue-Any (Mode)
Continuous Annealing
Control Area
Controlled Atmosphere
Conventional Alloy

C/A Close Annealing
Ca Calcium
ca (circa) - approximately; about
CAA Canadian Automobile Association
Civil Aeronautics Administration [US]
Civil Aviation Authority [UK]
Clean Air Act [US]
Computer-Assisted Accounting
Computer-Assisted Analysis
CAARC Commonwealth Advisory Aeronautical
Research Council
CAAS Ceylon Association for the Advancement of
Science
Computer-Aided Approach Sequencing
CAAT College of Applied Arts and Technology
CAB Cabinet
Cabriolet
Canadian Accreditation Board
Canadian Association of Broadcasters
Captive Air Bubble
Cellulose Acetobutyrate
Civil Aeronautics Board [US]
Commonwealth Agricultural Bureau
CABATM Civil Aeronautics Board Air Transport
Mobilization
CABB Captured Air Bubble Boat
CaBP Calcium Binding Protein
CAC Carbon-Arc Cutting
CACA Canadian Agricultural Chemicals Association
CACAC Civil Aircraft Control Advisory Committee
[UK]
CACB Compressed-Air Circuit Breaker
CACDP California Association of County Data
Processors [US]
CACHE Computer-Controlled Automated Cargo
Handling Envelope
CACM Communications of the Association for
Computing Machinery
CACP Cartridge-Actuated Compaction Press
CACRS Canadian Advisory Council on Remote
Sensing
CACTS Canadian Air Cushion Technology Society
CACUL Canadian Association of College and
University Libraries
CAD Cabling Diagram
Cartridge-Activated Device
Compensated Avalanche Diode
Computer Access Device
Computer-Aided Design
Computer-Aided Detection

CADA Cellulose Acetate Diethylaminoacetate
Computer-Aided Design Analysis
CADAM Computer Graphics Augmented Design and
Manufacturing
CADAPSO Canadian Association of Data Processing
Service Organizations
CADAR Computer-Aided Design Analysis and
Reliability
CADC Cambridge Automatic Digital Computer
Central Air Data Computer
CAD/CAM Computer-Aided Design/Computer-Aided
Manufacturing
CADD Coding and Analysis of Drillhole Data
Computer-Aided Design and Drafting
Computer-Aided Design Drawing
CADDIF Computer-Aided Design Data Interchange
Format
CADE Canadian Association of Drilling Engineers
Computer-Aided Design Engineering
Computer-Aided Design Evaluation
CADEP Computer-Aided Design of Electronic
Products
CADET Computer-Aided Design Experimental
Translator
CADETS Classroom-Aided Dynamic Educational
Time-Sharing System
CADF CRT (= Cathode-Ray Tube) Automatic
Direction Finding
Commutated Antenna Direction Finder
CADFISS Computation and Data Flow Integrated
Subsystem
CADI Computer-Aided Diagnostic Information
CADIC Chemical Analysis Detection
Instrumentation Control
Computer-Aided Design of Integrated
Circuits
CADIF Computer-Aided Design Instructional
Facility [US]
CADRE Current Awareness and Document Retrieval
for Engineers
CADSS Combined Analog-Digital Systems Simulator
CAE Chemical Engineering Abstract
Compare Alpha Equal
Computer-Aided Education
Computer-Aided Engineering
Computer-Assisted Engineering
Computer-Assisted Enrollment
Constant Analyzer Energy
CAED Canadian Association of Equipment
Distributors
CAEM Controlled Atmosphere Electron Microscopy
CAER Community Awareness and Emergency
Response
CAF Cost, Assurance, and Freight
CaF Calcium-Free
CAFCG Constant Amplitude Fatigue Crack Growth
CAFE Computer-Aided Film Editor
Corporate Average Fuel Economy
CAG Canadian Air Group [of RCAF]
Canadian Association of Geographers
Civil Aviation Group

	Computer Applications Group [US Air Force]
CAGC	Clutter Automatic Gain Control
	Coded Automatic Gain Control
	Continuous Access Guided Communication
CAGE	Compiler and Assembler by General Electric
	Computerized Aerospace Ground Equipment
CAGI	Compressed Air and Gas Institute [US]
CAGS	Canadian Arctic Gas Study
CAI	Canadian Aeronautical Institute
	Computer-Administered Instruction
	Computer-Aided Instruction
	Computer Analog Input
	Computer-Assisted Instruction
CAIBE	Chemically-Assisted Ion Beam Etching
CAINS	Carrier Aircraft Inertial Navigation System
CAINT	Computer-Assisted Interrogation
CAI/O	Computer Analog Input/Output
CAI/OP	Computer Analog Input/Output
CAIRS	Canadian Institute for Radiation Safety
CAIS	Canadian Association for Information Science
	Central Abstracting and Indexing Service
	Computer-Aided Instruction System
CAISIM	Computer-Assisted Industrial Simulation [US Army]
CAIT	Coalition for the Advancement of Industrial Technology
CAJ	Caulked Joint
CAK	Command Access Key
	Command Acknowledgment
CAL	Caliber
	Calibration
	Computer-Aided Learning
	Computer Animation Language
	Computer-Assisted Learning
	Confined Area Landing
	Continuous Annealing Line
	Conservation Analytical Laboratory [US]
	Conversational Algebraic Language
	Cornell Aeronautical Laboratory [US]
Cal	California
cal	calorie [unit]
CALC	Calculation
	Cargo Acceptance and Load Control
CaLC	Calcia-Containing Lanthanum Chromite
CALM	Collected Algorithms for Learning Machines
	Computer-Assisted Library Mechanization
CALOGSIM	Computer-Assisted Logistics Simulation [US Army]
CALS	Computer-Aided Logistics Support
CAM	Cellulose Acetate Metacrylate
	Cement Aggregate Mixture
	Central Address Memory
	Checkout and Maintenance
	Commercial Air Movement
	Computer-Address Matrix
	Computer-Aided Manufacturing
	Content-Addressable Memory
	Custom Application Module
	Cybernetic Anthropomorphous Machine

CAMA	Canadian Appliance Manufacturers Association
	Centralized Automatic Message Accounting
CAMAC	Center for Applied Mathematics and Advanced Computation
	Computer-Automated Measurement and Control
CAMA-ONI	Centralized Automatic Message Accounting - Operator Number Identification
CAMAR	Common-Aperture Multifunction Array Radar
CAMCOS	Computer-Assisted Maintenance Planning and Control System
CAMEL	Collapsible Airborne Military Equipment Lifter
CAMESA	Canadian Military Electronic Standards Agency
CAMI	Computer Applications in the Mining Industry
CAM-I	Computer-Aided Manufacturing - International
CAMM	Computer-Aided Manufacturing Management
	Computer-Aided Modeling Machine
CAMMD	Canadian Association of Manufacturers of Medical Devices
CAMMS	Computer-Assisted Material Management System
CAMP	Cabin Air Manifold Pressure
	Compiler for Automatic Machine Programming
	Computer-Assisted Mathematics Program
	Computer-Assisted Movie Production
	Continuous Air Monitoring Program
	Controls and Monitoring Processor
cAMP	Cyclic Adenosine Monophosphate
CAMRAS	Computer-Assisted Mapping and Records Activities System
CAMS	Cybernetic Anthropomorphous Machine System
CAMSIM	Computer-Assisted Maintenance Simulation [US Army]
CAN	Canada
	Canadian
	Cancel (Character)
	Ceric Ammonium Nitrate
	Correlation Air Navigation
CANAPES	Canadian Acoustic Parabolic Equation System
CANAS	Canadian Naval Air Station
CANBIOCON	Canadian Biotechnology Conference (and Exhibition)
CANC	Cancellation
CAN-CAM	Canadian Congress of Applied Mechanics
CANCEE	Canadian National Committee for Earthquake Engineering
CAND	Candelabra
CANDU	Canada Deuterium Uranium
CANDUBLW	Canada Deuterium Uranium Boiling Light Water
CANDUPHW	Canada Deuterium Uranium Pressurized Heavy Water

CANEWS	Canadian Electronic Warfare System
CAN-LAW	Canadian Computer-Assisted Legal Research System
CANMARC	Canadian Machine-Readable Cataloging
CANMET	Canada Center for Mineral and Energy Technology
CAN/OLE	Canadian On-Line Enquiry [CISTI]
CANQUA	Canadian Quaternary Association
CANS	Computer-Assisted Network Scheduling System
CANSARP	Canadian Search and Rescue Planning
CAN/SDI	Canadian Service for the Selective Dissemination of Information
CAN/SIS	Canadian Soil Information System
CAN/SND	Canadian Service for Scientific Numeric Databases
CAN/TAP	Canadian Technical Awareness Program
CANTAT	Canadian Transatlantic Telephone
CANTRAN	Cancel in Transmission
CAN-UK	Canada - United Kingdom
CANUNET	Canadian University Computer Network
CANUSE	Canada-United States Eastern Power Complex
CANWEC	Canadian National Committee of the World Energy Conference
CAO	Completed as Ordered
CAODC	Canadian Association of Oilwell Drilling Contractors
CAOS	Completely Automatic Operational System
CAP	Card Assembly Program
	Canadian Association of Physics
	Capacitance
	Capacitor
	Capital (Letter)
	Carburizing Atmosphere Process
	Cellulose Acetate Propionate
	Chloramphenicol
	Cleaner Air Package
	Computer Access Panel
	Computer-Aided Planning
	Computer-Aided Publishing
	Computer-Aided Purchasing
	Computer Analysts and Programmers
	Computer-Assisted Processing
	Computer-Assisted Production
	Consolidation by Atmospheric Pressure (Process)
	Controlled Atmosphere Packaging
	Council on Advanced Programming
	Cryotron Associative Processor
	Customer Access Panel
CAPA	Commission on Asian and Pacific Affairs [of ICC]
CAPAL	Computer and Photographically Assisted Learning
CAPCC	Computer Application Process Control Committee
CAPCOM	Capsule Communications
CAPD	Computer-Aided Process Design
CAPDAC	Computer-Aided Piping Design and Construction

CAPDM	Canadian Association of Physical Distribution Management
CAPE	Communication Automatic Processing Equipment
	Computer Assisted Program Evaluation
	Computer-Assisted Project Execution
CAPERT	Computer-Assisted Program Evaluation Review Technique
CAPERTSIM	Computer-Assisted Program Evaluation Review Technique Simulation
CAPL	Continuous Annealing and Processing Line
CAPM	Computer-Aided Patient Management
CAPR	Catalogue of Programs
CAPRI	Coded Address Private Radio Intercom
	Computerized Advance Personnel Requirements and Inventory
CAPS	Capital Letters
	Center for Advanced Purchasing Studies
	Computer-Assisted Problem Solving
	Control and Auxiliary Power Supply System
	Courtauld's All-Purpose Simulator
	Cyclohexylaminopropanesulfonic Acid
CAP SCR	Cap Screw
CAPSO	Cyclohexylaminohydroxypropanesulfonic Acid
CAPST	Capacitor Start
CAP'Y	Capacity
CAQ	Computer-Aided Quality Control
CAQC	Computer-Aided Quality Control
CAR	Canadian Association of Radiologists
	Cargo
	Channel Address Register
	Channel Assignment Record
	Community Antenna Relay
	Computer-Aided Repair
	Computer-Assisted Research
	Conditioned Avoidance Response
	Corrective Action Request
CARA	Cargo and Rescue Aircraft
	Combat Aircrew Rescue Aircraft
CARAM	Content-Addressable Random-Access Memory
CARB	California Air Resources Board [US]
	Carburetor
	Carburization
CARBINE	Computer-Automated Real-Time Betting Information Network
CARC	Canadian Agricultural Research Council
	Chemical Agent Resistant Coating
CARCA	Computer-Assisted Rocking Curve Analysis
CARD	Channel Allocation and Routing Data
	Compact Automatic Retrieval Device
	Compact Automatic Retrieval Display
CARDE	Canadian Armament Research and Development Establishment
CARE	Ceramic Application in Reciprocating Engines
	Computer-Aided Reliability Estimation
CARF	Central Altitude Reservation Facility
CARI	Canadian Association of Recycling Industries

CARN	Conditional Analysis for Random Networks
CARP	Computed Air Released Point
CARP	Carpenter
CARPTR	Carpenter
CARR	Carrier
CARRS	Close-In Automatic Route Restoral System [NORAD]
CARS	Center for Atomic Radiation Studies
	Coherent antiStokes Raman Scattering
	Common Antenna Relay System
	Computer-Aided Routing System
	Computerized Automotive Reporting Service
CART	Cartographer
	Cartography
	Cartridge
	Central Automated Replenishment Technique
	Complete Automatic Reliability Testing
	Computerized Automatic Rating Technique
CARTIS	Computer-Aided Real-Time Inspection System
CAS	Calculated Air Speed
	Calibrated Air Speed
	Chemical Abstracts Service [of ACS]
	Center of Atmospheric Studies [US]
	Central Amplifier Station
	Circuits and Systems [IEEE Society]
	Close Air Support
	Collision Avoidance System [FAA]
	Commission for Atmospheric Sciences [NAS]
	Computer-Assisted Steelmaking
	Control Augmentation System
	Controlled Airspace
	Customer Administrative Service
CASA	Computer and Automated Systems Association [of SME]
CASALS	Congress on Advances in Spectroscopy and Laboratory Sciences
CASB	Canadian Aviation Safety Board
CASD	Computer-Aided System Design
CASDAC	Computer-Aided Ship Design and Construction
CASE	Citizens Association for Sound Energy
	Committee on Academic Science and Engineering [of FCST]
	Computer-Aided Software Engineering
	Computer Automated Support Equipment
	Coordinating Agency for Suppliers Evaluation [US]
	Council for Advancement and Support of Education
CASI	Canadian Aeronautics and Space Institute
CASING	Crosslinking by Activated Species of Inert Gases
CAS NUT	Castle Nut
CASP	Canadian Atlantic Storms Program
CASPAR	Cambridge Analog Simulator for Predicting Atomic Reactions
CASPAR	Cushion Aerodynamics System Parametric Assessment Research
CASS	Consolidated Automatic Support System

	Commanded Active Sonobuoy System
	Copper-Accelerated Salt Spray (Test)
CASSA	Coarse Analog Sun Sensor Assembly
CASSANDRA	Chromatogram Automatic Soaking Scanning and Digital Recording Apparatus
CASSI	Chemical Abstracts Service Source Index [of ACS]
CAST	Canadian Air-Sea Transportable (Brigade Group)
	Clearinghouse Announcements in Science and Technology
	Computer-Aided Solidification Technique
	Computerized Automatic System Tester
CASW	Council for the Advancement of Science Writing
CASWS	Close Air Support Weapon System
CAT	Carburetor Air Temperature
	Catalogue
	Catalysis
	Catalyst
	Catapult
	Clear Air Turbulence
	Civil Air Transport
	College of Advanced Technology
	Compile and Test
	Computer-Aided Teaching
	Computer-Aided Testing
	Computer-Aided Translation
	Computer-Aided Typesetting
	Computer-Assisted Tomography
	Computer Average Transient
	Computerized Axial Tomography
	Constant Analyzer Transmission
	Controlled Attenuator Timer
CATA	Canadian Advanced Technology Association
CATAL	Catalogue
CATC	Canadian Air Transport Commission
	Circular-Arc-Toothed Cylindrical
	Commonwealth Air Transport Council
CATCC	Canadian Association of Textile Colorists and Chemists
CATCH	Computer-Aided Testing and Checking
CATE	Center for Advanced Technology Education
	Ceramic Applications in Turbine Engines
CATE	Center for Advanced Technology Education
CATH	Cathode
CATI	Computer-Assisted Television Instruction
CATLG	Catalogue
CAT NO	Catalogue Number
CATO	Compiler for Automatic Teaching Operation
CATS	Center for Advanced Television Studies
	Central Automated Transit System
	Computer-Aided Teaching System
	Computer-Aided Testing System
	Computer Automated Test System
CATV	Cable Television
	Common Antenna Television
CAU	Compare Alpha Unequal
CAUML	Computers and Automation Universal Mailing List

CAURA	Canadian Association of University Research Administrators
CAUSE	College and University System Exchange
CAV	Cavity
	Component Analog Video
	Composite Analog Video
	Constant Angular Velocity
	Constant Arc Voltage
CAVORT	Coherent Acceleration and Velocity Observations in Real Time
CAVU	Ceiling and Visibility Unlimited
CAW	Canadian Autoworkers (Union)
	Carbon-Arc Welding
	Channel Address Word
CAW-G	Gas Carbon-Arc Welding
CAWS	Common Aviation Weather System
CAW-S	Shielded Carbon-Arc Welding
CAW-T	Twin Carbon-Arc Welding
CAX	Community Automatic Exchange
CB	Carrier-Based
	Catch Basin
	Cellulose Butyrate
	Center of Buoyancy
	Circuit Board
	Circuit Breaker
	Citizens' Band (Radio)
	Coated Back
	Common-Base
	Common Battery
	Component Board
	Conduction Band
	Construction Ball
	Continuous Blowdown
	Contract Bolt
	Convergent Beam
	Coupled Biquad
CBA	Chemical Bond Approach
	Computer-Based Automation
	Cost Benefit Analysis
CBAL	Counterbalance
CBARC	Conference Board of Associated Research Councils
CBAS	Chemical Bond Approach Study
CBAT	Central Bureau for Aeronomical Telegrams
CB/ATDS	Carrier-Based/Airborne Tactical Data System
CBC	Chemically-Bonded Ceramics
	Continuity Bar Connector
CBCT	Customer-Bank Communication Terminal
CBD	Constant Bit Density
	Convergent Beam Diffraction
CBDP	Convergent Beam Diffraction Pattern
CBDS	Circuit Board Design System
CBE	Circuit Breaker for Equipment
	Computer-Based Education
	Council of Biology Editors
CBED	Convergent-Beam Electron Diffraction
CBEDP	Convergent-Beam Electron-Diffraction Pattern
CBEMA	Canadian Business Equipment Manufacturers Association

	Computer and Business Equipment Manufacturers Association [US]
CBFC	Copper and Brass Fabricators Council
CBI	Charles Babbage Institute
	Computer-Based Instruction
	Computer-Based Instrumentation
CBIE	Canadian Bureau of International Education
CBIS	Computer-Based Instructional System
CBKP	Convergent Beam Kikuchi Pattern
CBL	Cement Bond Log
	Computer-Based Learning
CBM	Conduction Band Minimum
	Containerized Batch Mixer
CBMA	Canadian Battery Manufacturers Association
	Canadian Business Manufacturers Association
CBMIS	Computer-Based Management Information System
CBMM	Council of Building Materials Manufacturers
CBMPE	Council of British Manufacturers of Petroleum Equipment
CBMS	Chill Block Melt Spinning
	Conference Board of the Mathematical Sciences
CBMU	Current Bit Motor Unit
CBMUA	Canadian Boiler and Machinery Underwriters Association
CBN	Cubic Boron Nitride
CBNM	Central Bureau for Nuclear Measurements
CBOM	Current Break-Off and Memory
CBORE	Counterbore
CBOSS	Count Back Order and Sample Select
CBOTA	Cape Breton Offshore Trade Association [Canada]
CBP	Construction Ball Pad
	Convergent Beam Pattern
CBR	California Bearing Ratio
	Chemical, Biological, Radiological (Warfare)
	Community Bureau of Reference
CBRA	Chemical, Biological, and Radiological Agency [US]
CBRS	Chemical, Biological, and Radiological Section [US Military]
CBS	Call Box Station
	Canadian Biochemical Society
	Columbia Broadcasting System
CBT	Core Block Table
CBTA	Canadian Business Telecommunications Alliance
CBW	Chemical and Biological Warfare
	Chemical and Biological Weapons
CBX	Cam Box
	C-Band Transponder
	Computerized (Private) Branch Exchange
CBZ	Carbobenzyloxy
CC	Calorimetry Conference
	Canada Council
	Capsule Communicator
	Central Computer
	Centrifugal Casting

	Channel Command
	Circuit Closing
	Close-Coupled
	Closing Coil
	Cluster Controller
	Color Code
	Color Compensation
	Color Correction
	Combined Carbon
	Command Chain
	Common Collector
	Communication Center
	Communication Controller
	Comparison Circuit
	Computer Community
	Condition Code
	Constant Current
	Continuous Casting
	Continuous Current
	Control Center
	Control Computer
	Control Counter
	Cotton Count
	Cross Couple
C/C	Carbon/Carbon (Composite)
C-C	Center to Center
C&C	Command and Control
cc	carbon copy
	complex conjugate
	cubic centimeter
CCA	Cam Clamp Assembly
	Canadian Center of Architecture
	Canadian Chemical Association
	Canadian Construction Association
	Capital Cost Allowance
	Carrier-Controlled Approach
	Cement and Concrete Association
	Central Computer Accounting
	Chromated Copper Arsenate
	Coagulation Control Abnormal
	Common Communication Adapter
	Consumer and Corporate Affairs [Canada]
CCAB	Canadian Circulation Audit Board
CCAC	Consumer and Corporate Affairs Canada
CCACD	Canadian Center for Arms Control and Disarmament
CCAM	Canadian Congress on Applied Mechanics
CCAP	Communications Control Application Program
CCASM	Canadian Council of the American Society for Metals
CCB	Circuit Concentration Bay
	Command Control Block
	Configuration Control Board
	Convertible Circuit Breaker
CCBDA	Canadian Copper and Brass Development Association
CCC	Canadian Chamber of Commerce
	Canadian Computer Conference
	Central Computer Complex
	Chemical Coal Cleaning

	Chlorocholine Chloride
	Communications Control Console
	Computer Communication Console
	Computer Control Complex
	Convert Character Code
	Copy Control Character
	Copyright Clearance Center
	Customs Cooperation Council
C³	Command, Control and Communication
	Computer, Communications and Components
CC&C	Command Control and Communications
CCCESD	Council of Chairmen of Canadian Earth Science Departments
C³I	Command, Control, Communication and Intelligence
C³L	Complementary Constant-Current Logic
CCCS	Command, Control and Communications System
CCD	Capacitor-Charged Device
	Charge-Coupled Device
	Cold Cathode Discharge
	Computer-Controlled Display
	Controlled Current Distribution
	Core Current Driver
	Counter-Current Decantation
CCDC	Construction Contract Development Association
CCDP	Canadian Continental Drilling Program
CCDSO	Command and Control Defense Systems Office
CCE	Certified Cost Engineer
	Chief Construction Engineer
CCECA	Consultative Committee on Electronics for Civil Aviation [US]
CCEE	Committee of Concerned Electrical Engineers
CCEPC	Canadian Civil Engineering Planning Committee
CCES	Canadian Congress of Engineering Students
CCETT	Canadian Council of Engineering Technicians and Technologists
CCF	Central Computing Facility [NASA]
	Central Communications Facility [US Air Force]
	Cross Correlation Function
	Cumulative Cost per Foot
CCFM	Cryogenic Continuous-Film Memory
CCFR	Constant Current Flux Reset (Test Method)
	Coordinating Committee on Fast Reactors
CCFT	Controlled Current Feedback Transformer
CCG	Canadian Coast Guard
	Computer-Controlled Goniometer
	Constant Current Generator
CCGE	Cold Cathode Gauge Experiment
CCGS	Canadian Coast Guard Service
CCGT	Closed-Cycle Gas Turbine
CCH	Channel-Check Handler
	Connections per Circuit per Hour
CCHS	Cylinder-Cylinder-Head Sector
CCI	Canadian Conservation Institute

	Chamber of Commerce, International
	Consultative Committee International
	Continuous Computation of Impact Points
CCIA	Console Computer Interface Adapter
CCIC	Chapter Chairmen's Invitational Conference [of ASM]
CCIF	Consultative Committee on International Frequencies
CCIIW	Canadian Council of the International Institute of Welding
CCIR	Consultative Committee International on Radio [of ITU]
CCIS	Command, Control and Information System
	Common Channel Interoffice Signaling
CCITT	Consultive Committee International on Telegraph and Telephone
CCIW	Canada Center for Inland Waters
CCKF	Continuity Checktone Failure [SPADE]
CCL	Coating and Chemical Laboratory
	Command Control Language
	Common-Carrier Line
	Core Current Layer
CCLS	Canadian Council of Land Surveyors
CCLW	Counterclockwise
CCM	Center for Composite Materials
	Chain Crossing Model
	Communications Controller Multichannel
	Counter Countermeasures
ccm	cubic centimeter
CCMD	Continuous Current Monitoring Device
CCMTC	Cape Canaveral Missile Test Center [now: KSC]
CCMS	Committee on Challenges of Modern Society
CCNDT	Canadian Council for Nondestructive Testing
CCNG	Computer Communications Network Group [Canada]
CCNR	Current-Controlled Negative Resistance
CCO	Crystal-Controlled Oscillator
	Current-Controlled Oscillator
CCOHS	Canadian Center for Occupational Health and Safety
CCOP	Constant-Control Oil Pressure
C-CORE	Center for Cold Ocean Resources Engineering [Canada]
CCOT	Cycling Clutch Orifice Tube
CCP	Communication Control Package
	Communication Control Panel
	Communication Control Program
	Configuration Control Panel
	Console Control Package
	Critical Compression Pressure
	Cubic Close-Packed
C&CP	Corrosion and Cathodic Protection [IEEE Committee]
CCPA	Canadian Chemical Producers Association
	Cemented Carbide Producers Association
CCPE	Canadian Council of Professional Engineers
CCPES	Canadian Council of Professional Engineers and Scientists
CCPIT	China Council for the Promotion of International Trade [PR China]

CCPO	Conference on Charged Particle Optics
CCR	Computer Cassette Recorder
	Control Contactor
	Council for Chemical Research
	Critical Compression Ratio
CCRL	Combustion and Carbonization Research Laboratory [of CANMET]
CCRMP	Canadian Certified Reference Materials Project
CCROS	Card Capacitor Read-Only Storage
CCRS	Canada Center for Remote Sensing
CCS	Canadian Ceramic Society
	Cartesian Coordinate System
	Center Core Signal
	Central Computer Station
	Change Control System
	Commercial Communications Satellite [Japan]
	Common Command Set
	Communication Control System
	Continuous Commercial Service
	Controlled Combustion System
	Conversational Compiling System
	Custom Computer System
	Hundred Call Seconds
CC&S	Central Computer and Sequencer [NASA]
CCSA	Common-Control Switching Arrangement
CCSG	Computer Components and Systems Group
CCSL	Comparative Current Sinking Logic
	Compatible Current Sinking Logic
CCST	Center for Computer Sciences and Technology [of NBS]
CCT	Center-Cracked Tension (Specimen)
	Communications Control Team
	Computer Compatible Tape
	Constant Current Transformer
	Continuity Check Transceiver
	Continuous Cooling Transformation
	Correlated Color Temperature
	Creosote Coal Tar
	Crystal-Controlled Transmitter
CCTP	Clean Coal Technology Program
CCTS	Coordinating Committee on Satellite Communications
CCTV	Closed Circuit Television
CCU	Camera Control Unit
	Canadian Commission for UNESCO
	Communications Control Unit
	Computer Control Unit
	Contaminant Collection Unit
	Coupling Control Unit
CCV	Control Configured Vehicle
CCVS	COBOL Compiler Validation System
CCW	Channel Command Word
	Counterclockwise
CCWBAD	Counterclockwise Bottom Angular Down
CCWBAU	Counterclockwise Bottom Angular Up
CCWBH	Counterclockwise Bottom Horizontal
CCWDB	Counterclockwise Down Blast
CCWTAD	Counterclockwise Top Angular Down
CCWTAU	Counterclockwise Top Angular Up

CCWTH	Counterclockwise Top Horizontal		**CDI**	Carbodiimide
CCWUB	Counterclockwise Up Blast			Collector Diffusion Isolation
CD	Cable Duct			Community Development Index
	Cage Dipole			Compact Disk - Interactive
	Capacitor Diode			Comprehensive Dissertation Index
	Carrier Detect			Control Data Institute [US]
	Cathodoluminescence			Control Deviation Indicator
	Ceiling Diffuser			Control Direction Indicator
	Center Distance			Course Deviation Indicator
	Check Digit		**CD-I**	Compact Disk - Interactive
	Circuit Description		**CDIF**	Component Development and Integration Facility
	Coherent Detector			Crystal Data Identification File
	Cold Drawing		**CDL**	Common Display Logic
	Collision Detection			Computer Description Language
	Cord			Computer Design Language
	Common Digitizer		**CDMA**	Code-Division Multiple Access
	Compact Disk		**CDMS**	COMRADE Data Management System
	Contracting Definition		**CDN**	Canadian
	Critical Dimension		**cDNA**	Complimentary Deoxyribonucleic Acid
	Crystal Driver		**CDO**	Community Dial Office
	Current Density		**CDOM**	Center for Development of Materials
C&D	Control and Display		**CDP**	Central Data Processor
Cd	Cadmium			Centralized Data Processing
cd	candela [unit]			Certificate in Data Processing
	cord [unit]			Checkout Data Processor
CDA	Canada			Communications Data Processor
	Command and Data Acquisition			Compact Disk Player
	Commercial Development Association			Compressor Discharge Pressure
	Comprehensive Development Area			Correlated Data Processor
	Copper Development Association			Cytidine Diphosphate
CDAP	Cyanodimethylaminopyridinium		**CDPC**	Central Data Processing Computer
CDAPSO	Canadian Data Processing Service Organization			Commercial Data Processing Center
CDAS	Central Data Acquisition System		**CDPIR**	Crash Data Position Indicator Recorder
	Command and Data Acquisition Station		**CD PL**	Cadmium Plate
CDB	Current Data Bit		**CDPS**	Computing and Data Processing Society
CDBN	Column-Digit Binary Network		**CD-PROM**	Compact Disk - Programmable Read-Only Memory
CDC	Call Directing Code		**CDR**	Card Reader
	Canadian Dairy Commission			Command Destruct Receiver
	Code Directing Character			Composite Damage Risk
	Computer Display Channel			Current Directional Relay
	Configuration Data Control		**CDRA**	Canadian Drilling Research Association
CDCE	Central Data Conversion Equipment			Committee of Directors of Research Associations
CDCR	Center for Documentation and Communication Research [US]		**CDRE**	Chemical Defense Research Establishment [UK]
CDCU	Communications Digital Control Unit		**CDRAP**	Collaborative Diesel Research Advisory Panel
CDD	Common Data Dictionary		**CDRILL**	Counterdrill
	Conference on Dual Distribution		**CDRL**	Contractual Data Requirements List
CDE	Canadian Depletion Expense		**CD-ROM**	Compact Disk - Read-Only Memory
	Certificate in Data Education		**CDS**	Communications and Data Subsystem
	Control and Display Equipment			Cold-Drawn Steel
CDEE	Chemical Defense Experimental Establishment [of MOD]			Compatible Duplex System
CDF	Carrier Distribution Frame			Comprehensive Display System
	Centered Dark Field			Computer Duplex System
	Combined Die Forging			Configuration Data Set
	Combined Distribution Frame			Container Delivery System
	Confined Detonating Fuse			(Single) Cotton Double Silk (Wire)
	Cumulative Distribution Function		**CDSE**	Computer-Driven Simulation Environment
cd ft	cord foot [unit]			
CDG	Capacitor Diode Gate			

CDSF	COMRADE Data Storage Facility		Combination Emission Control
C&DSS	Communication and Data Subsystem		Commission of European Communities
CDST	Central Daylight Saving Time		Commonwealth Engineering Conference
CDT	Central Daylight Time		Consulting Engineers Council
	Control Data Terminal		Coordinating European Council
CDTA	Cyclohexylenedinitrilotetraacetic Acid		Cryogenic Engineering Conference
CDU	Central Display Unit	**CECA**	Consumers Energy Council of America
	Control Display Unit	**CED**	Capacitance Electronic Disk
	Coolant Distribution Unit		Carbon-Equivalent-Difference
	Coupling Display Unit		Cohesive Energy Density
CDV	Compact Disk Video		Committee for Economic Development
	Component Digital Video		Constant Energy Difference
CDW	Capacitor Discharge Welding	**CEDA**	Canadian Electrical Distributors Association
	Charge Density Wave	**CEDO**	Consulting and Engineering Design Organization
	Computer Data Word		
CDX	Control Differential Transmitter	**CEDPA**	California Educational Data Processing Association [US]
CE	Carbon Equivalence		
	Carbon Equivalent	**CEE**	Canadian Exploration Expense
	Channel End	**CEDC**	Canadian Engineering Design Competition
	Chemical Engineer(ing)	**CEF**	Carrier Elimination Filter
	Chief Engineer		Critical Experiments Facility
	Civil Engineer(ing)	**CEFAC**	Civil Engineering Field Activities Center [UK]
	Common-Emitter		
	Communications - Electronics	**CEFNO**	Carbethoxyformonitrile
	Commutator End	**CEGB**	Central Electricity Generating Board [UK]
	Computer Engineer(ing)	**CEI**	Committee for Environmental Information
	Conducted Emission		Communication Electronics Instructions
	Coulomb Excitation		Computer Extended Instruction
	Current Efficiency		Council of Engineering Institutions [UK]
	Customer Engineer		Cycle Engineers Institute [US]
Ce	Cerium	**CEIC**	Canada Employment and Immigration Commission
CEA	Cambridge Electron Accelerator		
	Canadian Electrical Association	**CEIF**	Council of European Industrial Federations
	Canadian Export Association	**CEIL**	Ceiling
	Carcinoembryonic Antigen	**CEL**	Carbon-Equivalent, Liquidus
	Circular Error Average		Central European Line (– Pipeline)
	Combustion Engineering Association		Cryogenic Engineering Laboratory [NBS]
	Communications-Electronics Agency [US]	**CELL**	Cellulose
	Competitive Equipment Analysis	**Cels**	Celsius
CEAC	Central European Analysis Commission	**CEM**	Cement
	Committee for European Airspace Coordination [NATO]		Channel Electron Multiplier
			Communications-Electronics-Meteorological
	Consulting Engineers Association of California [US]		Conventional Electron Microscope
			Counter Electromotive Cell
CEADI	Colored Electronic Altitude Director Indicator	**CEMA**	Canadian Electrical Manufacturers Association
CEAL	Canadian Explosive Atmospheres Laboratory [of CANMET]		Conveyor Equipment Manufacturers Association [US]
CEANAR	Commission on Education in Agriculture and Natural Resources [NRC, US]	**CEM A**	Cement Asbestos
		CEM AB	Cement Asbestos Board
CEAP	Canada Energy Audit Program	**CEMAP**	Cotton Export Market Acreage Program
CEARC	Computer Education and Applied Research Center	**CEMC**	Canadian Engineering Manpower Council
		CEMF	Counter Electromotive Force
CEB	Central Electricity Board	**CEM FL**	Cement Floor
CEBC	Consulting Engineers of British Columbia [Canada]	**CEMI**	Canadian Engineering Manpower Inventory
		CEMIRT	Civil Engineering Maintenance, Inspection, Repair and Training Team [US Air Force]
CEBM	Corona, Eddy Current, Beta Ray, Microwave		
CEC	Canada Employment Center	**CEM MORT**	Cement Mortar
	Canadian Electrical Code	**CEMON**	Customer Engineering Monitor
	Cation Exchange Capacity	**CEM PLAS**	Cement Plaster
	Ceramic Educational Council	**CEM PC**	Cement Plaster Ceiling

CEMREL	Central Midwest Regional Educational Laboratory [US]
CEMS	Central Electronic Management System
	Conversion Electron Moessbauer Scattering
CEN	Central
CENB	Consulting Engineers of New Brunswick [Canada]
CENT	Centigrade
	Central
	Centrifugal
cent	(centum) - hundred
CENTCON	Centralized Control Facility
CENTL	Central
CENTO	Central Treaty Organization
CEO	Chief Executive Officer
	Consulting Engineers of Ontario [Canada]
CEOA	Central Europe Operating Agency [NATO]
CEP	Central European Pipeline
	Circle of Equal Probability
	Circular Error Probability
	Citizens' Energy Project
	Civil Engineering Package
	Command Executive Procedures
	Computer Entry Punch
	Cotton Equalization Program
	Cylinder Escape Probability
CEPA	Civil Engineering Programming Applications
	(Society for) Computer Applications in Engineering, Planning and Architecture
CEPACS	Customs Entry Processing and Control System
CEPC	Canada Civil Engineering Planning Committee
CEPD	Council for Economic Planning and Development
CEPE	Central Experimental and Proving Establishment [Canada]
CEPEX	Controlled Ecosystem Pollution Experiment
CEPO	Central European Pipeline Office [NATO]
CEPPC	Central European Pipeline Policy Committee [NATO]
CEPR	Center for Energy Policy and Research
CEPS	Central Europe Pipeline System
	Command Module Electrical Power System
	Cornish Engines Preservation Society
CEQ	Committee on Environmental Quality [of FCST]
	Corporation of Engineers of Quebec [Canada]
CER	Ceramics
	Civil Engineering Report
	Constant Extension Rate
	Coordinated Experimental Research
	Cost Estimating Relationship
CERA	Canadian Electronic Representatives Association
	Civil Engineering Research Association [now: CIRIA]
CERC	Coastal Engineering Research Center [US Army]
	Coastal Engineering Research Council [US]

CERDIP	Ceramic Dual In-Line Package [also: Cerdip]
CerE	Ceramics Engineer(ing)
CERG	Chemical Ecology Research Group
CERI	Canadian Energy Research Institute
CERL	Cambridge Electronics Research Laboratory
	Canadian Explosives Research Laboratory [of CANMET]
	Central Electricity Research Laboratories
CERMET	Ceramic Metal [also: Cermet]
	Ceramic-to-Metal [also: Cermet]
CERS	Cornell Electron Storage Ring
CERT	Certificate
	Combined Environmental Reliability Test
	Constant Extension Rate, Tensile
	Constant Extension Rate Test
CES	Constant Elasticity of Substitution
	Conjugated Estrogens
	Coordinate Evaluation System
CESA	Canadian Engineering Standards Association
CESAC	Communications-Electronics Scheme Accounting and Control [US Air Force]
CESD	Composite External Symbol Directory
CESEMI	Computer Evaluation of Scanning Electron Microscope Images
CESI	Council for Elementary Science International
CESO	Canadian Executive Service Overseas
CESR	Conduction Electron Spin Resonance
CESSE	Council of Engineering and Scientific Society Executives
CESSS	Council of Engineering and Scientific Society Secretaries
CET	Cement Evaluation Tool
	Center for Educational Technology [US]
	Certified Engineering Technologist
	Chemical Engineering Technology
	Civil Engineering Technology
	Corrected Effective Temperature
	Cumulative Elapsed Time
CETDC	China External Trade Development Council [Taiwan]
CETR	Center for Explosives Technology Research [US]
CETS	Conference on European Telecommunications Satellites
CEU	Communications Expansion Unit
	Continuous Education Unit
	Construction Engineering Unit
	Coupler Electronics Unit
CEV	Combat Engineer Vehicle
	Corona Extinction Voltage
CEVM	Consumable Electrode Vacuum Melting
CEW	Coextrusion Welding
CEX	Central Excitation
CEY	Consulting Engineers of the Yukon Territory [Canada]
CF	Canadian Forces
	Carbofuchsin
	Cathode Follower
	Cationized Ferritin

	Center Frequency
	Center of Flotation
	Centrifugal Force
	Centripetal Force
	Cerium-Free
	Circuit Finder
	Coated Front
	Cold Finish
	Combined Function
	Concept Formulation
	Constant Fraction
	Contemporary Force
	Context Free
	Conversion Factor
	Corn Flour
	Correlation Function
	Corrosion Fatigue
	Cresolformaldehyde
	Crude Fiber
	Counterfire
C&F	Cost and Freight
Cf	Californium
cf	(confer) - compare
CFA	Canadian Foundry Association
	Council of Iron Founders Associations
	Council on Fertilizer Application [US]
	Cross-Field Amplifier
CFADC	Canadian Forces Air Defense Command
CFAE	Contractor-Furnished Aircraft Equipment
CFAP	Copenhagen Frequency Allocation Plan
CFAR	Constant False Alarm Rate
CFAV	Canadian Forces Arctic Vessel
CFB	Coated Front and Back
CFBS	Canadian Federation of Biological Sciences
	Canadian Federation of Biological Societies
CFC	Capillary Filtration Coefficient
	Central Fire Control
	Chlorofluorocarbon
	Colony Forming Cells
	Consolidated Freight Classification
	Crossed-Film Cryotron
CFD	Computational Fluid Dynamics
	Constant Fraction Discriminator
CFDA	Current File Disk Address
CFDM	Companded Frequency-Division-Multiplex
CFDM/FM	Companded Frequency-Division-Multiplex/Frequency Modulation
CFE	Chlorotrifluoroethylene
	Continuous Flow Electrophoresis
	Continuous Fuel Economizer
	Contractor-Furnished Equipment
CFER	Collector Field Effect Register
CFES	Continuous Flow Electrophoresis System
CFF	Critical Flicker Frequency
	Critical Fusion Frequency
CFFTP	Canadian Fusion Fuels Technology Project
CFH	Cubic Feet per Hour
CFHQ	Canadian Forces Headquarters
CFI	Canadian Film Institute
	Cost, Freight, and Insurance
CFIA	Component Failure Impact Analysis

CFIB	Canadian Federation of Independent Business
CFIEI	Canadian Farm and Industrial Equipment Institute
CFIF	Continuous Flow Isoelectric Focusing
CFL	Calibrated Focal Length
	Canadian Federation of Labour
	Clear Flight Level
	Context-Free Language
CFM	Cathode Follower Mixer
	Cerium-Free Mischmetall
	Center Frequency Modulation
	Companded Frequency Modulation
	Continuous Film Memory
	Continuous Flow Manufacturing
	Cubic Feet per Minute
CFMR	Center for Fundamental Materials Research
CFMS	Chained File Management System
CFN	Canadian Forces Network
CFO	Critical Flashover
CFPA	Canadian Fluid Power Association
	Chlorophenyltrifluoromethyl Phenoxyacetate
CFPHT	Constant Fraction of Pulse Height Trigger
CFPO	Canadian Forces Post Office
CFPP	Canadian Federation of Pulp and Paper Workers
CFR	Carbon Fiber Reinforced
	Carbon-Film Resistor
	Center for Field Research
	Code of Federal Regulations
	Commercial Fast Reactor
	Contact Flight Rules
	Coordinating Fuel Research
	Cumulative Failure Rate
CFRC	Continuous Fiber-Reinforced Composites
CFRP	Carbon Fiber-Reinforced Polymers
	Carbon Fiber-Reinforced Plastics
CFS	Canadian Forestry Service
	Carrier Frequency Shift
	Constant Final State
	Container Freight Station
	Continuous Forms Stacker
	Cubic Feet per Second
CFSG	Cometary Feasibility Study Group [of ESRO]
CFSSB	Central Flight Status Selection Board
CFSTI	Clearinghouse for Federal Scientific and Technical Information [USDOC]
CFSTR	Continuous Flow Stirred Tank Reactor
CFT	Complement-Fixation Test
	Constant Fraction Trigger
	Controlled Finishing Temperature
	Cubic Feet per Ton
CFTA	Committee of Foundry Technical Associations
CFTD	Constant Fraction Timing Discriminator
CFU	Colony Forming Units
	Current File User
CG	Center of Gravity
	Chain Grate
	Coarse Grain

	Coincidence Gate
	Compacted Graphite (Cast Iron)
	Computer Graphics
	Crystal Growth
cg	centigram [unit]
CGA	Canadian Gas Association
	Color Graphics Adapter
	Compressed Gas Association [US]
	Contrast Gate Amplifier
CGB	Ceramics and Graphite Branch [US Air Force]
	Convert Gray to Binary
CGC	Canadian Geoscience Council
	Clock Generator Controller
	Compact Glass Column
CGE	Carriage
CGF	Carrier Gas Fusion
CGG	Continuous Galvanizing Grade (of Zinc)
CGHAZ	Coarse-Grained Heat-Affected Zone
CGI	Ground Control Interceptor
	Computer Graphics Interface
CGIAR	Consultative Group on International Agricultural Research [US]
CGIC	Ceramics and Graphite Information Center [US Air Force]
CGIS	Canadian Geographic Information Service
CGM	Character Generation Module
	Commission for the Geological Map of the World
	(International Conference on Joining of) Ceramics, Glass and Metal
CGMID	Character Generation Module Identifier
CG/MOI	Center of Gravity/Moment of Inertia
CGP	Color Graphics Printer
CGPA	Canadian Gas Processors Association
CGPSA	Canadian Gas Processors and Suppliers Association
CGRAM	Clock Generator Random-Access Memory
CGS	Canadian Geographical Society
	Canadian Geotechnical Society
	Centimeter-Gram-Second System
	Control Guidance Subsystem
CGSB	Canadian General Standards Board
	Canadian Government Specification Board
CGSE	GS Electrostatic System
CGSM	GS Electromagnetic System
CGSU	CGS Unit
CGT	Current Gate Tube
CGU	Canadian Geophysical Union
CH	Calcium Hydroxide
	Case Hardening
	Ceiling Height
	Central Heating
	Chain
	Chain Home (Radar)
	Channel
	Chapter
	Choke
	Coal Hook
	Cold Work Hardening
	Conductor Head

CHA	Concentric Hemispherical Analyzer
CHABA	Committee on Hearing and Bioacoustics
CHAD	Code for Handling Angular Data
CHAG	Chain Arrester Gear
CHAM	Chamfer
CHAMPS	Computerized History and Maintenance Planning System
CHAN	Channel
CHANCOM	Channel Committee [NATO]
CHAP	Chapter
CHAPS	Cholamidopropyldimethylammoniopropanesulfonate
CHAR	Character
CHARA	Center for High Angular Resolution Astronomy [US]
CHART	Computerized Hierarchy and Relationship Table
CHBA	Canadian Home Builders Society
CHC	Chemical Hydrogen Cracking
	Choke Coil
	Cyclohexylamine Carbonate
CHD	Chord
CHDM	Cyclohexanedimethanol
CHE	Channel End
ChE	Chemical Engineer
	Cholinesterase
CHEC	Channel Evaluation and Call
CheE	Chemical Engineer
CHEM	Chemical
	Chemist
	Chemistry
ChemE	Chemical Engineer
CHEM-MET	Chemical Metallurgy
CHES	Cyclohexylaminoethanesulfonic Acid
CHESNAVFAC	Chesapeake Naval Division Facilities Engineering Command [US Navy]
CHESS	Cornell High Energy Synchrotron Source
CHEVMA	Canadian Heat Exchanger and Vessel Manufacturers Association
CHF	Critical Heat Flux
CHG	Change
	Charge
CHI	Computer - Human Interaction
CHIEF	Controlled Handling of Internal Executive Functions
CHIL	Current Hogging Injection Logic
CHILD	Cognitive Hybrid Intelligent Learning Device
	Computer Having Intelligent Learning and Development
ChIME	Chemical Industry for Minorities in Engineering
CHK	Check
CHKPT	Checkpoint
CHKR	Checker
CHL	Chloroform
Chl	Chlorophyll
CHM	Chairman
	Chemical Machining
	Chemical Milling
CHMN	Chairman

CHMT Components, Hybrids and Manufacturing
Technology Society
CHN Carbon - Hydrogen - Nitrogen
CHNL Channel
CHO Carbohydrate
Carbon - Hydrogen - Oxygen
CHORD Chief of Office of Research and
Development
CHPAE Critical Human Performance and Evaluation
CHR Character
c-hr candle-hour [unit]
CHS Calder Hall Reactor [UK]
Canadian Hydrographic Service
CHT Charactron Tube
Chemical Heat Treatment
Cycloheptatriene
CHU Centigrade Heat Unit
Channel Unit [SPADE]
CHY Chymotrypsinogen
CI Call Indicator
Card Input
Cast Iron
Ceiling Average
Chemical Inspectorate [MOD]
Circuit Interrupter
Colloidal Iron
Combustion Institute
Control Interval
Critical Item
Cubic Inch [also: ci]
C/I Carrier-to-Interference (Ratio)
C&I Control and Indication
Ci Curie [unit]
CIA Central Intelligence Agency [US]
Chemical Industries Association [UK]
Chemiluminescence Immunoassay
Communications Industry Association
Communications Interrupt Analysis
Computer Industry Association
Computer Interface Adapter
CIAJ Communications Industry Association of
Japan
CIAM Computerized Integrated and Automated
Manufacturing
CIAP Climatic Impact Assessment Program
CIAR Canadian Institute for Advanced Research
CIAS CCRS Image Analysis System
CIB Centralized Intercept Bureau
CIC Canada Immigration Center
Chemical Industry Council [US]
Chemical Institute of Canada
Cogeneration Coalition
Combat Information Center
Common Interface Circuit
Communication Intelligence Channel
Computer Intelligence Corps [US Army]
Construction Industry Commission
Cover Integrated Cell
CICA Chicago Industrial Communications
Association [US]
CICAT Center for International Cooperation and
Appropriate Technology

CICC Custom Integrated Circuit Conference [of
IEEE]
CICD Computer-Integrated Circuit Design
CICS Canadian Industrial Computer Society
Committee for Index Cards for Standards
Customer Information Control System
CICT Commission on International Commodity
Trade
CID Center for Information and Documentation
Charge Injection Device
Collision-Induced Decomposition
Commercial Item Description
Communication Identifier
Component Identification
Cubic Inch Displacement
Current-Image Diffraction (Test)
CIDA Canadian International Development Agency
Current Input Differential Amplifier
CIDAS Conversational Interactive Digital/Analog
Simulator
CIDB Canadian International Development Board
CIDF Control Interval Definition Field
CIDIN Common ICAO Data Interchange Network
CIDNP Chemically-Induced Dynamic Nuclear
Polarization
CIDS Chemical Information and Data System [US
Army]
Concrete Island Drilling System
CIDST Committee for Information and
Documentation on Science and
Technology [UK]
CIE Chinese Institute of Engineers
CIEDS Computer-Integrated Electrical Design
Series
CIEF Continuous Isoelectric Focusing
CIEP Canadian Institute for Economic Policy
CIE-USA Chinese Institute of Engineers - USA
CIF Central Integration Facility
Cold-Insoluble Fibrinogen
Computer Integrated Factory
Cost, Insurance, and Freight
CIFC Cost, Insurance, Freight and Commission
CIFST Canadian Institute of Food Science and
Technology
CIG Cryogenic-in-Ground
CIGGT Canadian Institute for Guided Ground
Transport
CIGTF Central Inertial Guidance Test Facility
CIIA Canadian Information Industry Association
CIIT Chemical Industry Institute of Toxicology
CIL Carbon-in-Leach
CIM Canadian Institute of Mining and
Metallurgy [also: CIMM]
Charge Imaging Matrix
Communications Improvement
Memorandum
Computer Input from Microfilm
Computer Input Multiplexer
Computer-Integrated Manufacturing
Continuous Image Microfilm
Custom Injection Molding

CIMA	Construction Machinery Manufacturers Association
CIMAC	Computer Integrated Manufacturing Control
CIMCO	Card Image Correction
CIME	Chartered Institute of Marine Engineers [US]
CIMM	Canadian Institute of Mining and Metallurgy [also: CIM]
CIMS	Chemical Ionization Mass Spectroscopy
	Coordination and Interference Management System [INTELSAT]
CIMT	Center for Instructional Media and Technology
CIMTC	Construction and Industrial Machinery Technical Committee [of SAE]
CIN	Carrier Input
	Communication Identification Navigation
CINC	Commander-in-Chief
CIND	Computer Index Neutron Data
CINS	CENTO Institute of Nuclear Science
CIO	Central Input/Output
	Chief Information Officer
	Congress of Industrial Organizations
CIOCS	Communications Input/Output Control System
CIOM	Communications Input/Output Multiplexer
CIOU	Custom Input/Output Unit
CIP	Carbon-in-Pulp
	Cast-in-Place
	Cast Iron Pipe
	Cleaned-in-Place
	Cold Isostatic Pressing
	Compatible Independent Peripherals
	Composite Interlayer Bonding
	Current Injection Probe
CIPASH	Committee on International Programs in Atmospheric Sciences and Hydrology [of NAS/NRC]
CIPEC	Canadian Industry Program for Energy Conservation
CIPH	Canadian Institute of Plumbing and Heating
CIPM	Council for International Progress in Management
CIPOM	Computers, Information Processing and Office Machines [of CSA]
CIPPOCS	Cast-In-Place, Push-Out Cylinders
CIPS	Canadian Information Processing Society
CIR	Canada India Reactor
	Characteristic Instants of Restitution
	Circle
	Circular
	Circulate
	Circumference
	Cost Information Report
	Current Instruction Register
CIRA	Committee on Industrial Research Assistance
	COSPAR International Reference Atmosphere
CIRB	Canadian Industrial Renewal Board
CIRC	Centralized Information Reference and Control

Circle	Circulate
	Circumference
CIRCAL	Circuit Analysis
CIRCLE	Cylindrical Internal Reflection Cell
CIRCUM	Circumference
CIRES	Communication Instruction for Reporting Enemy Sighting
CIRIA	Construction Industry Research and Information Association [UK]
CIRIS	Complete Integrated Reference Instrumentation System
cir mils	circular mils [unit]
CIRP	Canadian Industrial Renewal Program
CIRRA	Canadian Industrial Relations Research Association
CIS	Canadian Institute of Surveying
	Characteristic Isochromate Spectroscopy
	Commercial Instruction Set
	Communication Information System
	Communication and Instrumentation System
	Constant Initial State Spectroscopy
	Course Information System
	Cue Indexing System
	Current Information Selection
CISA	Canadian Industrial Safety Association
	Casting Industry Suppliers Association
CISC	Canadian Institute for Steel Construction
	Complex Instruction Set Computer
CISCO	Compass Integrated System Compiler
CISIR	Ceylon Institute of Scientific and Industrial Research
CISP	Canadian Institute of Surveying and Photogrammetry
CISR	Center for International Systems Research [US]
CISTI	Canada Institute for Scientific and Technical Information
CIT	California Institute of Technology [US]
	Canadian Import Tribunal
	Carnegie Institute of Technology [US]
	Citation
	Compressor Inlet Temperature
	Computer-Integrated Testing
CITAB	Computer Instruction and Training Assistance for the Blind
CITB	Construction Industry Training Board
CITC	Canadian Institute of Timber Construction
	Canadian International Trade Classification
	Construction Industry Training Center
CITE	Capsule-Integrated Test Equipment
	Cargo-Integrated Test Equipment
	Compression, Ignition and Turbine Engine
	Council of Institute of Telecommunications Engineers [US]
CITEC	Contractor-Independent Technical Effort
CITL	Canadian Industrial Transportation League
CITS	Computer-Integrated Testing System
CITT	Canadian Institute of Traffic and Transportation
CIU	Computer Interface Unit

	Control Indicator Unit
CIV	Civil
CIW	Carnegie Institute of Washington [US]
CJ	Chuck Jaw
CK	Cask
	Creatine Kinase
CKD	Completely Knocked Down
	Count-Key-Data Device
CKAFS	Cape Kennedy Air Force Station [US]
CKMTA	Cape Kennedy Missile Test Annex [now: Cape Canaveral, NASA]
CKSNI	Cape Kennedy Space Network [NASA]
CKS	Cell Kinetics Society
cks	centistokes [unit]
CKT	Circuit
CKW	Counterclockwise
CL	Cable Line
	Cable Link
	Carload
	Cathodoluminescence
	Center Line
	Chemiluminescence
	Class
	Clearance
	Closing
	Clutch
	Command Language
	Computational Linguistics
	Condenser Lens
	Confidence Limits
	Connecting Line
	Control Language
	Control Leader
	Conversion Loss
	Current Layer
	Current Logic
	Cutter Location
	Cutting Lubricant
C/L	Carload
Cl	Chlorine
cL	centiliter [also: cl]
CLA	Canadian Labour Association
	Canadian Lumbermen's Association
	Centerline Average
	Clear and Add
	Communication Line Adapter
	Communication Link Analyzer
CLAC	Closed Loop Approach Control
CLAIRA	Chalk Lime and Allied Industries Research Association [now: WHRA]
CLAM	Chemical Low-Altitude Missile
CLARA	Cornell Learning and Recognizing Automaton
CLAS	Computer Library Applications Service
CLASP	Closed Line Assembly for Single Particles
	Composite Launch and Spacecraft Program
	Computer Language for Aeronautics and Space Programming [NASA]
CLASS	Classification
	Closed Loop Accounting for Store Sales
	Composite Laminate Analysis Systems

	Computer-Based Laboratory for Automated School Systems
CLASSMATE	Computer Language to Aid and Stimulate Scientific, Mathematical, and Technical Education
CLAT	Communications Line Adapter for Teletype
CLAW	Clustered Atomic Warhead
CLB	Cone Locator Bushing
CLC	Canadian Labour Congress
	Central Logic Control
	Column Liquid Chromatography
	Constant Light Compensating
	Course Line Computer
	Cost of Living Council [US]
CLCOLL	Clinch Collar
CLCR	Controlled Letter Contract Reduction
CLCS	Chinese Language Computer Society
	Current Logic, Current Switching
CLD	Central Library and Documentation Branch [of ILO]
	Constant Level Discriminator
CLD	Cooled
CLDAS	Clinical Laboratory Data Acquisition System
CLDS	Canada Land Data System
CLEA	Conference on Laser Engineering and Applications [IEEE]
CLEAR	Compiler, Executive Program, Assembler Routines
	Components Life Evaluation and Reliability
CLEFT	Cleavage of Lateral Epitaxial Film for Transfer
CLEM	Cargo Lunar Excursion Module
	Closed Loop Ex-Vessel Machine
	Composite for the Lunar Excursion Module
CLEO	Clear Language for Expressing Orders
	Conference on Lasers and Electrooptics
CLF	Capacitive Loss Factor
CLG	Catalogue
	Ceiling
	Controlled Lead Grade (of Zinc)
CLI	Cathodoluminescence Imaging
	Coin Level Indicator
	Command Language Interpreter
CLIP	Compiler Language for Information Processing
	Contributions to Laboratory Investigations Program
CLIRA	Closed Loop In-Reactor Assembly
CLIST	Command List
CLK	Clock
CLKG	Caulking
CLL	Central Light Loss
CLM	Coal-Liquid Mixture
C/L min	Carload, Minimum Weight
CLML	Chicago Linear Music Language
CLO	Closet
CLOS	Closure
CLP	Clamp
	Cone Locator Pin
	Contractor Laboratory Program
	Current Line Pointer

ClPh	Chlorophenyl
CLR	Clear
	Combined Line and Recording
	Computer Language Recorder
	Computer Language Research
	Constant Load Rupture
	Coordinating Lubricants Research
	Council of Library Resources [US]
	Current Limiting Resistor
CLRB	Canada Labour Relations Board
CLRI	Central Leather Research Institute [UK]
CLRU	Cambridge Language Research Unit
CLS	Canada Land Surveyor
	Canadian Lumber Standard
	Cathodoluminescence Spectroscopy
	Clear and Subtract
	Common Language System
	Comparative Systems Laboratory [US]
	Concept Learning System
	Constant Level Speech
	Control Launch Subsystem
CLSR	Computer Law Service Reporter
CLSS	Communication Link Subsystem
CLT	Central Limit Theorem
	Certificate of Laboratory Technology
	Communications Line Terminal
	Computer Language Translator
	Constant Load Tensile
CLU	Central Logic Unit
	Creusot-Loire-Uddeholm (Process)
	Circuit Line-Up
CLUT	Computer Logic Unit Tester
CLV	Constant Linear Velocity
CLV	Clevis
CLW	Clockwise
CM	Calibrated Magnification
	Carboxymethyl
	Cell Membrane
	Center Matched
	Center of Mass
	Central Memory
	Chemical Metallurgy
	Chemical Milling
	Circular Measure
	Circular Mil
	Color Monitor
	Command Module
	Common Mode
	Communications Multiplexer
	Configuration Management
	Construction and Machinery
	Consumable Electrode Remelted
	Controlled Mine Field
	Core Memory
	Countermeasure
	Cross Modulation
	Cylinder Mount
C/M	Communications Multiplexer
Cm	Curium
cm	centimeter
c m	circular mil [unit]

CMA	Canadian Manufacturers Association
	Canadian Metric Association
	Cast Metals Association [US]
	Chemical Manufacturers Association [US]
	Computer-Aided Microanalyzer
	Contact-Making Ammeter
	Copper-Magnesium-Aluminum (Alloy)
	Cylindrical Mirror Analyzer
CMACL	Composite Mode Adjective Checklist
CMANY	Communications Managers Association of New York [US]
CMB	Carbolic Methylene Blue
	Concrete Median Barrier
CMBES	Canadian Medical and Biological Engineering Society
CMBG	Canadian Mechanized Brigade Group
CMC	Canadian Meteorological Center
	Carboxymethyl Cellulose
	Ceramic Matrix Composite
	Code for Magnetic Characters
	Command Module Computer
	Communications Mode Control
	Contact-Making Clock
	Coordinating Manual Control
	Critical Micelle Concentration [also: cmc]
	Cyclohexylmorpholinoethylcarbodiimide
CMCA	Canadian Masonry Contractors Association
CMCP	Cruise Missile Conversion Project
CMCR	Continuous Melting, Casting and Rolling
CMCSA	Canadian Manufacturers of Chemical Specialties Association
CMCTL	Current Mode Complementary Transistor Logic
CMD	Carboxymuconolactone Decarboxylase
	Condensed Matter Division
	Core Memory Driver
	Command
CMDAC	Current Mode Digital-to-Analog Converter
CMDR	Command Reject
CMDS	Centralized Message Data System
CME	Chemically-Modified Electrode
	Chloromethyl Ether
	Crucible Melt Extraction
CMEA	Council of Mutual Economic Aid
CMERI	Central Mechanical Engineering Research Institute [India]
CMF	Cartesian Mapping Function
	Center for Metals Fabrication [of BMI]
	Coherent Memory Filter
cmf	circular mil-foot [unit]
CMfgE	Certified Manufacturing Engineer
CMfgT	Certified Manufacturing Technologist
CMG	Computer Management Group
	Control Moment Gyro
CMHEC	Carboxymethylhydroxyethyl Cellulose
cm Hg	centimeters of mercury [unit]
CMI	Casting Metals Institute
	Computer-Managed Instruction
	Cultured Marble Institute
CMIG	Canadian MAP (= Manufacturing Automation Protocol) Interest Group

cmil	circular mil
CMITP	Canada Manpower Industrial Training Program
CMJ	Canadian Mining Journal
CML	Chemical
	Commercial
	Current-Mode Logic
CMLCENCOM	Chemical Corps Engineering Command [US]
CMLE	Carboxymuconate Lactonizing Enzyme
CMM	Commission for Maritime Meteorology [of WMO]
	Computerized Modular Monitoring
	Coordinate Measuring Machine
CMMA	Concrete Mixer Manufacturers Association
CMME	Chloromethyl Methyl Ether
CMMI	Council of Mining and Metallurgy Institutions
CMMP	Carnegie Multi-Mini Processor
	Commodity Management Master Plan
CMMS	Cation Micro Membrane Suppressor
CMN	Cerous Magnesium Nitrate
	Commission
CMND	Command
CMOD	Crack Mouth Opening Displacement
CMOP	Canadian Market Opportunities Program
CMOS	Canadian Meteorological and Oceanographic Society
	Complementary Metal-Oxide Semiconductor
CMOS-RAM	Complementary Metal-Oxide Semiconductor - Random Access Memory
CMOS/SOS	Complimentary Metal-Oxide Semiconductor/Silicon on Sapphire
CMP	Canadian Mineral Processors
	Center for Metal Production
	Ceramic Mold Process
	Compounded Mobilized Planes (Theory)
	Cooperative Marketing Partner
	Cooperative Marketing Program
	Cooperative Market Participant
	Cytidine Monophosphate
CMPM	Computer-Managed Parts Manufacture
cmps	centimeter per second
CMPSS	Conference on Molecular Processes on Solid Surfaces
CMPT	Component
CMPTR	Computer
CMR	Carbon Magnetic Resonance
	Committee on Manpower Resources
	Common-Mode Rejection
	Communications Monitoring Report
CMRA	Chemical Marketing Research Association
CMRR	Common-Mode Rejection Ratio
CMS	Calcium-Magnesium Silicate
	Cambridge Monitor System
	Canadian Micrographic Society
	Center for Materials Science [of NBS]
	Centrifuge Melt Spinning
	Chapter Management Seminar [of ASM]
	Chloromethylated Polystyrene
	Clay Minerals Society [US]

	Code Management System
	Coincidence Moessbauer Spectroscopy
	Cold Melt Steel
	Compiler Monitor System
	Conservation Materials and Services
	Continuous Mining System
	Conversational Monitor System
	Current-Mode Switching
C&MS	Consumer and Marketing Service
CMT	Code Matching Technique
	Computer-Managed Training
	Conversational Mode Terminal
	Corrugating Medium-Flat Test
CMTA	Constant Momentum Transfer Averaging
CMTC	Coupled Monostable Trigger Circuit
CMTDA	Canadian Machine Tool Distributors Association
CMTI	Celestial Moving Target Indicator
CMTP	Canadian Manpower Training Program
	Continuum Mechanics of Textured Polycrystals
CMTM	Communications and Telemetry
CMTS	Canadian Machine Tool Show (and Exhibition)
CMU	Carnegie-Mellon University [US]
	Computer Memory Unit
CMV	Common-Mode Voltage
	Contact-Making Voltmeter
CMVM	Contact-Making Voltmeter
CMVPCB	California Motor Vehicle Pollution Control Board [US]
CMVSA	Canadian Motor Vehicle Safety Act
CMVSS	Canadian Motor Vehicle Safety Standard
CN	Canadian National Railways
	Carbon Number
	Cascade Nozzle
	Caudate Nucleus
	Cellulose Nitrate
	Center Notch
	Cetane Number
	Chain
	Chance Notice
	Close Nipple
	Coordination Number
	Cyanogen
C/N	Carrier-to-Noise Ratio
cn	cosine of the amplitude
CNA	Canadian Nuclear Association
	Center for Naval Analyses [US Navy]
	Copper Nickel Alloy
	Cosmic Noise Absorption
CNAA	Council for National Academic Awards [UK]
CNAD	Conference on National Armament Directors [NATO]
CNAS	Civil Navigation Aids System
CNBMDA	Canadian National Building Materials Distribution Association
CNBr	Cyanogen Bromide
CNC	Canadian National Committee
	Computer Numerical Control
CN-CA	Cellulose Nitrate - Cellulose Acetate

CNC/IAPS	Canadian National Committee for the International Association on the Properties of Steam
CNC/IUTAM	Canadian National Committee for the International Union of Theoretical and Applied Mechanics
CNCRM	Canadian National Committee on Rock Mechanics
CNCTRC	Concentric
CND	Conduit
CNDCTN	Conduction
CNDCTR	Conductor
CNDP	Communication Network Design Program
	Continuing Numerical Data Project
CNDS	Condensate
CNE	Canadian National Exhibition
	Charge Neutrality Equation
	Compare Numeric Equal
CNET	Communications Network
CNEt	Cyanoethyl
CNG	Compressed Natural Gas
CNGA	California Natural Gas Association [US]
CNI	Communication, Navigation and Identification
CNL	Circuit Net Loss
CNLA	Council on National Library Associations
C-NMR	Carbon Nuclear Magnetic Resonance
CNOP	Conditional Non-Operational
CNR	Canadian National Railway
	Carrier-to-Noise Ratio
	Composite Noise Rating
CNS	Canadian Navigation Society
	Canadian Nuclear Society
CNSS	Center for National Space Study
CNT	Canadian National Telecommunications
	Celestial Navigation Trainer
	Count
CNTL	Control
CNTP	Committee for a National Trade Policy
CNTR	Container
	Counter
CNTU	Confederation of National Trade Unions
CNU	Compare Numeric Unequal
CNV	Contingent Negative Variation
CNVR	Conveyor
CO	Carbonyl
	Central Office
	Change Order
	Circuit Order
	Cleanout
	Close-Open Operation
	College of Aeronautics
	Colorado [US]
	Combined Operations
	Commanding Officer
	Communications Officer
	Company
	Crystal Oscillator
	Cut Out
Co	Cobalt
c/o	care of

CO$_2$-EOR	Carbon Dioxide Enhanced Oil Recovery
COA	College of Aeronautics [UK]
	Conversion of Acetyl
COACH	Canadian Organization for Advancement of Computers in Health
COAL PRO	Coal Research Project
COAM	Customer Owned and Maintained
COAM	Coaming
COAT	Coherent Optical Array Technique
	Corrected Outside Air Temperature
COAX	Coaxial (Cable) [also: coax]
COBESTCO	Computer-Based Estimating Technique for Contractors
COBI	Coded Biphase
COBIS	Computer-Based Instruction System
COBLIB	COBOL Library
COBLOC	CODAP Language Block-Oriented Compiler
COBOL	Common Business-Oriented Language
COC	Carbon Oxide Chemisorption
	Coded Optical Character
COCODE	Compressed Coherency Detection
COCOM	Coordinating Committee
COD	Carrier Onboard Delivery
	Carrier-Operated Device
	Cash on Delivery
	Chemical Oxygen Demand
	Clean-Out Door
	Crack Opening Displacement
CODAC	Coordination of Operating Data by Automatic Computer
CODAG	Combined Diesel and Gas
CODAN	Carrier-Operated Device Antinoise
	Coded Weather Analysis [US Navy]
CODAP	Control Data Assembly Program
CODAR	Coherent Display Analyzing and Recording
	Correlation Display Analyzing and Recording
CODAS	Control and Data Acquisition System
	Customer-Oriented Data Acquisition System
CODASYL	Conference on Data System Languages
CODATA	Committee on Data for Science and Technology [of ICSU]
CODC	Canadian Oceanographic Data Center
CODEC	Coder-Decoder
CODED	Computer Design of Electronic Devices
CODES	Computer Design and Evaluation System
	Computer Design and Evaluation System
CODF	Crystal(lite) Orientation Distribution Function
CODIC	Color Difference Computer
	Computer-Directed Communications
CODIL	Control Diagram Language
CODIPHASE	Coherent Digital Phased Array System
CODIS	Controlled Digital Simulator
CODIT	Computer Direct to Telegraph
CODOG	Combined Diesel or Gas
CODORAC	Coded Doppler Radar Command
CODSIA	Council of Defense and Space Industry Associations [US]
COE	Cab over Engine
COED	Char Oil Energy Development

	Computer-Operated Electronic Display
COEF	Coefficient
COF	Coefficient of Friction
	Correct Operation Factor
COFIL	Core File
COG	Commentary Graphics
	Composite Group
	Computer Operations Group
COGAG	Combined Gas and Gas
COGD	Circulator Outlet Gas Duct
COGENE	Committee on Genetic Engineering [of ICSU]
COGENT	Compiler and Generalized Translator
COGEO	Coordinate Geometry
COGLA	Canada Oil and Gas Lands Administration
GOGPE	Canadian Oil and Gas Property Expense
COGS	Continuous Orbital Guidance Sensor
	Continuous Orbital Guidance System
COGSME	Composite Group of SME (= Society of Manufacturing Engineers)
COHO	Coherent Oscillator
COI	Communication Operation Instructions
COIA	Canadian Ocean Industries Association
COIC	Canadian Oceanographic Identification Center
COID	Council of Industrial Design
COIME	Committee on Industry, Minerals and Energy
COIN	Coal Oxygen Injection Process
	Committee on Information Needs [US]
	Complete Operating Information
	Computer and Information
	Computerized Ontario Investment Network [Canada]
COINC	Coincidence
COINS	Computer and Information Sciences
	Counter-Insurgency
	Computerized Information System
	Control in Information Systems
COL	Column
	Computer-Oriented Language
COLA	Cost-of-Living Allowance
COLASL	Compiler/Los Alamos Scientific Laboratory [US]
COLIDAR	Coherent Light Detecting and Ranging
COLINET	College Libraries Information Network
COLINGO	Compile On-Line and Go
COLL	Collator
	Collector
	College
Colo	Colorado [US]
colog	cologarithm
COLLOQ	Colloquium
COLT	Communication Line Terminator
	Computerized On-Line Testing
	Computer Oriented Language Translator
	Control Language Translator
COM	Command
	Commerce
	Commissioner
	Commission

	Communication
	Computer Output Microfiche
	Computer Output Microfilm
	Computer Output Microfilmer
	Conference of Metallurgists [of CIM]
COMA	Canadian Oilfield Manufacturers Association
COMAC	Continuous Multiple Access Collator
COMANSEC	Computation and Analysis Section [of DRB]
COMAPS	Commons Automated Publishing System [Canadian Parliament]
COMAR	Committee on Man and Radiation [of IEEE]
COMAT	Computer-Assisted Training
COMB	Combination
	Combustion
	Console-Oriented Model Building
COMCM	Communication Countermeasures
COMDA	Canadian Office Machinery Dealers Association
COMEINDORS	Composite Mechanized Information and Documentation Retrieval System
COMEPP	Cornell Manufacturing Engineering and Productivity Program
COMER	College of Mineral and Energy Resources
COMESA	Committee on Meteorological Effects of Stratospheric Aircraft [UK]
COMET	Computer-Operated Management Evaluation Technique
COMETS	Comprehensive On-Line Manufacturing and Engineering Tracking System
COMEX	Commodity Exchange
	Computer and Office Machinery Exhibition
COMIT	Compiler, Massachusetts Institute of Technology [US]
	Computing System, Massachusetts Institute of Technology [US]
COMITEXTIL	Coordinating Committee for the Textile Industry [European Economic Community]
COML	Commercial
COMLAB	Commerce Laboratory
COMLO	Compass Locator
COMLOGNET	Combat Logistics Network
	Communications Logistics Network
COMM	Commercial Mission
	Communication
	Communications Society [of IEEE]
	Commutator
COMMEN	Compiler Oriented for Multiprogramming and Multiprocessing Environments
COMMEND	Computer-Aided Mechanical Engineering Design System
COMM'L	Commercial
COMP	Comparison
	Compatibility
	Compensation
	Composition
	Compound
	Compression
	Computer
	Computer Society [of IEEE]
COMPAC	Commonwealth Pacific Cable

	Computer Program for Automatic Control
COMPACT	Compatible Algebraic Compiler and Translator
	Computer Planning and Control Technique
	Computer-Programmed Automatic Checkout and Test System
COMPANDER	Compressor Expander
COMPARE	Computerized Performance and Analysis Response Evaluator
	Console for Optical Measurement and Precise Analysis of Radiation from Electronics
COMPAS	Committee on Physics and Society [AIP, US]
COMPASS	Compiler-Assembler
	Computer-Assisted Classification and Assignment System
COMPAY	Computerized Payroll System
COMPCON	Computer Convention [of IEEE]
COMPEL	Compute Parallel
COMPL	Completion
COMPO Image	Composition Image [microscopy]
COMPOOL	Communications Pool
COMPR	Compressor
COMPRESS	Computer Research, Systems and Software
COMPROC	Command Processor
	Computer Program
COMPS	Commercial Materials Processing Support Facility
COMPSO	Computer Software and Peripherals Show
COMPT	Compartment
COMRADE	Computer-Aided Design Environment
COMSAT	Communications Satellite
COMSEC	Communications Security
COMSL	Communication System Simulation Language
COMSOAL	Computer Method of Sequencing Operations for Assembly Lines
COMSOC	Communications Spacecraft Operation Center
COMT	Catechol-O-Methyltransferase
COMTRAN	Commercial Translator
COMZ	Communication Zone
CON	Conclusion
CONBAT	Converted Battelion Antitank
CONC	Concrete
	Concentration
	Concentric(ity)
CONC B	Concrete Block
CONC C	Concrete Ceiling
CONC F	Concrete Floor
COND	Condenser
	Condition
	Conduct
	Conductivity
	Conductor
CONELRAD	Control of Electromagnetic Radiation
CONF	Conference
CONFIDAL	Conjugate Filter Data Link
CONFLEX	Conditional Reflex
CONHAN	Contextural Harmonic Analysis

CONLIS	Committee on National Library Information Systems [US]
CONN	Connection
	Connector
Conn	Connecticut [US]
CONN DIA	Connection Diagram
CONRAD	Committee on Nuclear Radiology
CONS	Carrier-Operated Noise Suppression
	Console
CONSORT	Conversation System with On-Line Remote Terminals
CONST	Construction
	Constant
CONSUL	Control Subroutine Language
CONT	Contact
	Content
	Continent
	Continuation
	Contract
	Control(ler)
CONTAM	Contamination
CONT'D	Continued
CONT HP	Continental Horsepower
CONTR	Container
	Contract
	Contraction
	Contractor
	Controller
CONTRAN	Control Translator
CONTRANS	Conceptual Thought, Random Net Simulation
CONTREQS	Contingency Transportation Requirements System
CONTR O	Contracting Officer
CONUS	Continental United States
CONV	Conversion
	Converter
COOL	Checkout-Oriented Language
	Control-Oriented Language
CO-OP	Cooperative [also: Co-op]
COORD	Coordinate
COOSRA	Canadian Offshore Oilspill Research Association
COP	Coefficient of Performance
	Computer Optimization Package
	Copper
COPA	Canadian Office Products Association
COPAC	Continuous Operation Production Allocation and Control
COPAG	Collision Prevention Advisory Group [of FAA]
COPAN	Command Post Alerting Network
COPE	Communications-Oriented Processing Equipment
	Console Operator Proficiency Examination
COPEP	Committee on Public Engineering Policy [of NAE]
COPES	Conceptually-Oriented Program in Elementary Science
COPI	Computer-Oriented Programmed Instruction
	Cooperative Projects with Industry

COPIS	Communication-Oriented Production Information System
COPOL	Copolarization [also: CO-POL]
COP PL	Copper Plate
COPS	Command Operations
COPSS	Committee of Presidents of Statistical Societies
COPUOS	Committee on the Peaceful Uses of Outer Space
COR	Canadian Ownership Regulation
	Carrier-Operated Relay
	Conditioned Orientation Reflex
	Corner
	Correction
	Correspondence
CORA	Coherent Radar Array
	Conditioned Reflex Analog
	Conditioned Response Analog (Machine)
CORAD	Correlation Radar
CORAL	Class-Oriented Ring Associated Language
	Command Radio Link
	Computer On-Line Real-Time Applications Language
	Correlated Radio Link
CORC	Cornell Computing Language
CORDIC	Coordinate Rotation Digital Computer
CORDIS	Cold Reflex Discharge Ion Source
CORDPO	Correlated Data Printout
CORDPO-SORD	Correlated Data Printout - Separation of Radar Data
CORDS	Coherent-On Receive Doppler System
CORE	Canadian Offshore Resources Exposition
	Coherent-On-Receive
	Cold Ocean Resources Engineering
	Computer-Oriented Reporting Efficiency
CORF	Committee on Radio Frequencies
CORNET	Corporation Network
CORP	Corporation
CORR	Correction
	Correspondence
	Corrugation
CORREGATE	Correctable Gate
CORS	Canadian Operational Research Society
CORSA	Cosmic Radiation Satellite [Japan]
CORTEX	Communications-Oriented Real-Time Executive
CORTS	Convert Range Telemetry Systems
COS	Compatible Operating System
	Concurrent Operating System [UNIVAC]
	Contactor Starting
	Corporation for Open Systems [US Consortium]
	Cosmic Rays and Trapped Radiation Committee [of ESRO]
cos	cosine
COSAG	Combination of Steam and Gas
COSAM	Conservation of Strategic Aerospace Materials
COSA NOSTRA	Computer-Oriented System and Newly Organized Storage-to-Retrieval Apparatus
COSATI	Committee on Scientific and Technical Information [of FCST]
COSBA	Computer Services and Bureau Association [UK]
COSEC	Culham On-Line Single Experimental Console
cosec	cosecant
cosh	hyperbolic cosine
COSI	Committee on Scientific Information [US]
COSIE	Commission on Software Issues in the Eighties
COSINE	Committee on Computer Science in Electrical Engineering Education [USACM]
COSIP	College Science Improvement Program [NSF, US]
COSMAR	Committee on Surface Mining and Reclamation
COSMAT	Committee on the Survey of Materials Science and Engineering
COSMIC	Computer Software Management and Information Center [US]
COSMON	Component Open/Short Monitor
COSMOS	Coastal Multidisciplinary Oceanic System
	Computer-Oriented System for Management Order Synthesis
	Console-Oriented Statistical Matrix Operator System
	Consortium of Selected Manufacturers Open Systems
	Cornell Simulator of Manufacturing Operations
COS/MOS	Complementary-Symmetry Metal-Oxide Semiconductor
COSOS	Conference on Self-Operating Systems
COSP	Canadian Oil Substitution Program
COSPAR	Committee on Space Research [of ICSU]
COSPEAR	Committee on Space Programs for Earth Observation [NASA]
COSPUP	Committee on Science and Public Policy [of NAS]
COSRIMS	Committee on Research in the Mathematical Sciences [of NAS]
COSS	Computer-Optimized Storehouse System
COSY	Compiler System
	Correction System
COT	Cotter
	Cotton
	Cyclooctatetraene
cot	cotangent
COTAR	Correlation Tracking and Ranging [also: Cotar]
COTAR-AME	COTAR Angle Measuring Equipment
COTAR-DAS	COTAR Data Acquisition System
COTAR-DME	COTAR Distance Measuring Equipment
COTAT	Correlation Tracking and Triangulation
coth	hyperbolic cotangent
COTRAN	COBOL-to-COBOL Translator
COT WEB	Cotton Webbing
COU	Council of Ontario Universities [Canada]
COV	Covariance

	Coefficient of Variation
	Cover
COVER	Cut-Off Velocity and Range
covers	coversed sine
COVOA	Canadian Offshore Vessel Operators Association
COW	Crude Oil Washing
COWAR	Committee on Water Resources [of ICSU]
COWL	Cowling
CP	Call Processor
	Canadian Pacific
	Candlepower
	Card Punch
	Cellulose Propionate
	Cement Points
	Center of Pressure
	Center Punch
	Central Processor
	Cesspool
	Chamber Pressure
	Channel Program
	Character Printer
	Chemical Preparation
	Chemical Properties
	Chromatographically-Purified
	Circuit Package
	Circular Pitch
	Circular Polarization
	Clamping Pin
	Clock Pulse
	Coefficient of Performance
	Coherent Potential
	Cold Punch(ed)
	Collision Probability
	Color Print(ing)
	Command Privilege
	Command Processor
	Command Pulse
	Commercial Purity
	Communications Processor
	Conference Paper
	Connection Point
	Constant Pressure
	Continuous Path
	Continuous Phase
	Control Panel
	Control Point
	Control Processor
	Control Program
	Cracking Pressure
	Crack Propagation
	Cross-Ply
	Current Paper
	Customized Processor
cp	chemically pure
CPA	Canadian Petroleum Association
	Canadian Pharmaceutical Association
	Closest Point of Approach
	Coherent Potential Approximation
	Commutative Principle for Addition
	Concurrent Photon Amplification

	Cracking Pressure Adjusted (Valve)
	Critical Path Analysis
CpA	Cytidylyladenosine
CPAR	Cooperative Pollution Abatement Research
CPAWS	Computer-Planning and Aircraft-Weighing Scales
CPB	Channel Program Block
CPC	Calling Party Control
	Card-Programmed Calculator
	Cerium Pyridinium Chloride
	Channel Program Command
	Chemical Protective Clothing
	Clock Pulse Control
	Committee for Program and Coordination
	Computerized Plasma Control
	Computer Program Component
	Computer Programming Concepts
	Controlled-Potential Coulometry
	Cycle Program Control
	Cycle Program Counter
CpC	Cytidylylcytidine
CPCEI	Computer Program Contract End Item
CPCFA	Council of Pollution Control Financing Agencies
CPCI	Canadian Prestressed Concrete Institute
CPCS	Cheque Processing Control System
CPD	Carboxypeptidase
	Citrate Phosphate Dextrose
	Compound
	Consolidated Programming Document
	Contact Potential Difference
	Critical Point Drier
	Critical Point Drying
	Cumulative Probability Distribution
CPDS	Computer Program Design Specification
CPDU	Continuous Process Development Unit [of CANMET]
CPE	Central Processing Element
	Central Programmer and Evaluator
	Chlorinated Polyethylene
	Circular Probable Error
	Computer Premises Equipment
	Continuous Particle Electrophoresis
	Cross-Roll Piercing Elongation
	Customer Premises Equipment
	Custom Power Engineering
CPEA	Concentrated Phospate Export Association [US]
CPED	Continuous Particle Electrophoresis Device
CPEM	Conference on Precision Electromagnetic Measurements
CPEQ	Corporation of Professional Engineers of Quebec [Canada]
C-PET	Crystallized Polyethylene Terephthalate
CPEUG	Computer Performance Evaluation Users Group
CPF	Calcium Phosphate Free
CPFMS	COMRADE Permanent File Management System
CPG	Clock Pulse Generator
	Controlled Pore Glass

CpG	Cytidylylguanosine
CPH	Close-Packed Hexagonal
CPI	Canadian Plastics Institute
	Carbon-Number Predominance
	Characters per Inch
	Chemical Processing Industries
	Computer Prescribed Instruction
	Consumer Price Index
	Council for Professional Interest
	Crash Position Indicator
	Condensation-Reaction Polyimide
CPIA	Chemical Propulsion Information Agency [US]
CPIB	Chlorophenoxyisobutyrate
CPID	Computer Program Integrated Document
CPILS	Correlation Protected Instrument Landing System
CPIP	Computer Pneumatic Input Panel
	Computer Program Implementation Process
CPK	Creatine Phosphokinase
CPL	Canadian Plastics
	CAST Programming Language
	Combined Programming Language
	Computer Program Library
	Critical Path Length
CPLEE	Changed Particle Lunar Environment Experiment
CPLG	Coupling
CPM	Carbide Powdered Metal
	(Punched) Cards per Minute
	Central Processor Module
	Commutative Principle of Multiplication
	Connection Point Manager
	Control Path Method
	Counts per Minute [also: cpm]
	Critical Path Method
	Crucible Particle Metallurgy
	Cycles per Minute [also: cpm]
CP/M	Control Program/Microcomputer
CPO	Canada Post Office
	Catalytic Partial Oxidation
	Code Practice Oscillator
	Command Pulse Output
	Concurrent Peripheral Operations
CPP	Card Punching Printer
	Coal Preparation Plant
	Controllable Pitch Propeller
	Cyclopentenophrenanthrene
CPPA	Canadian Pulp and Paper Association
	Coal Preparation Plant Association
CPPO	Certified Public Purchasing Officer
CPPS	Critical Path Planning and Scheduling
CPR	Cam Plate Output
	Canadian Pacific Railway [now: CP]
	Coal Pile Runoff
	Committee on Polar Research [of NAS/NRC]
	Component Pilot Rework
	Corrosion Penetration Rate
CPRG	Computer Personnel Research Group
CPS	Cathode Potential Stabilized

	Central Processing System
	C-Frame Profile Scanner
	Characters per Second
	Chinese Petroleum Society
	Circuit Package Schematic
	Coils per Slot
	Commission on the Patent System [US]
	Controlled Path System
	Conversational Programming System
	Counts per Second [also: cps]
	Critical Path Scheduling
	Customer Premises System
	Cycles per Second [also: cps]
CPSC	Consumer Product Safety Commission [US]
CPSE	Counterpoise
CPSG	Common Power Supply Group
CPSK	Coherent Phase-Shift Keying
CPSM	Critical Path Scheduling Method
CPSR	Controlled Process Serum Replacement
CPSS	Committee of Presidents of Statistical Societies
	Common Programming Support System
CPT	Canadian Pacific Telecommunications
	Control Power Transformer
	Critical Path Technique
CPTA	Computer Programming and Testing Activity
CPTR	Computer
CPTS	Collidine-P-Toluenesulfonate
CPTU	Continuous Process Development Unit
CPU	Canadian Paperworkers Union
	Central Processing Unit
	Collective Protection Unit
	Computer Peripheral Unit
	Computer Processor Unit
CpU	Cytidylyluridine
CPUID	Central Processing Unit Identification (Number)
CPUNCH	Counterpunch
CPVC	Chlorinated Polyvinyl Chloride
CPVO	Chlorinated Polyvinyl Chloride
CPW	Commercial Projected Window
	Coplanar Waveguide
CQ	Chloroquine
	Commercial Quality
	Controlled Quality
CQAK	Commercial Quality Aluminum-Killed (Steel)
CQD	Customary Quick Dispatch
CQRS	Commercial Quality Rimmed Steel
CQS	Common Queue Space
CR	Call Request
	Card Reader
	Carriage Return (Character)
	Carrier Recovery
	Cathode Ray
	Ceiling Register
	Change Request
	Channel Request
	Chemical Report
	Chloroprene Rubber

	Clamp Rest
	Cold Rolled
	Command Register
	Communications Register
	Complete Round
	Conference Report
	Constant Rate
	Contract Report
	Control Relay
	Control Routine
	Controlled Rectifier
	Cooling Rate
	Conference Report
	Control Relay
	Corrosion Resistance
	Crew
	Crystal Rectifier
	Current Rate
	Current Relay
	Chloroprene Rubber
C/R	Chamfer or Radius
Cr	Chromium
CRA	California Redwood Association [US]
	Catalogue Recovery Area
	Clear Rear Access
	Composite Research Aircraft
	Corona Australis
CRAC	Careers Research and Advisory Center
CRAD	Committee for Research into Apparatus for the Disabled
CRAFT	Changing Radio Automatic Frequency Transmission
	Computerized Relative Allocation of Facilities Technique
CRAGS	Chemistry Records and Grading System
CRAM	Card Random Access Memory
	Computerized Reliability Analysis Method
	Conditional Relaxation Analysis Method
CRAN	Cross Scan
CRB	Corona Borealis
CRBL	Charles River Breeding Laboratories [US]
CRC	Carriage Return Contact
	Communications Research Center [Canada]
	Control and Reporting Center
	Coordinating Research Council [US]
	Copy Research Council
	Corporate Research Center
	Cumulative Results Criterion
	Cyclic Redundancy Check
CRCA	Cold-Rolled, Close-Annealed
CRCC	Cyclic Redundancy Check Character
CRCMF	Circumference
CRCP	Continuously-Reinforced Concrete Pavement
CRD	Capacitor-Resistor Diode
	Card Reader
CRDF	Canadian Radio Direction Finder
	Cathode-Ray Direction Finding
CRDL	Chemical Research and Development Laboratories [US Army]
CRDME	Committee for Research into Dental Materials and Equipment
CRDSD	Current Research and Development in Scientific Documentation
CRE	Corrosion Resistance
	Current Ring End
CREATE	Chalk River Experiment to Assess Tritium Emission [Canada]
CREQ	Center for Research and Environmental Quality
CRES	Corrosion-Resistant Steel
CRESP	Coding Region Expression Selection Plasmid
CRESS	Center for Research in Experimental Space Science
	Combined Reentry Effort in Small Systems
	Computerized Reader Enquiry Service System
CRESTS	Courtauld's Rapid Extracting, Sorting and Tabulating System
CRETC	Combined Radiation Effects Test Chamber
CRF	Capital Recovery Factor
	Control Relay Forward
	Corticotropin Releasing Factor
	Cross-Reference Facility
	Cryptographic Repair Facility
CRG	Carriage
	Classification Research Group
CRH	Channel Reconfiguration Hardware
CR-hi	Channel Request, High Priority
CRI	Carbohydrate Research Institute [Canada]
	Color Rendering Index
	Committee for Reciprocity Information
CRIFO	Civilian Research, Interplanetary Flying Objects
CRIM	Composite Reaction Injection Molding
CRIS	Command Retrieval Information System
	Current Research Information System [USDA]
CRITIC	Chalk River In-Reactor Tritium Instrumented Capsule [Canada]
CRITICOMM	Critical Intelligence Communications System [US Air Force]
CRJE	Conversational Remote Job Entry
CRK	Crank
CRKC	Crankcase
CRL	Coal Research Laboratories
	Coherent Radiation Laboratory [US]
	Communications Research Laboratory
CR/LF	Carriage Return/Line Feed
CR-lo	Channel Request, Low Priority
CRLR	Chemical and Radiological Laboratories [US Army]
CRM	Comprehensive Resource Management
	Confusion Reflector Material
	Control and Reproducibility Monitor
	Counter-Radar Measures
	Counter-Radar Missile
	Cross-Reacting Material
CRMA	Canadian Research Managers Association
	Canadian Rock Mechanics Association
CR-med	Channel Request, Medium Priority
CRMG	Calgary Rock-Mechanics Group [Canada]
CR MOLY	Chrome Molybdenum

CRMR	Continuous-Reading Meter Relay
CRN	Continuous Random Network
CRNL	Chalk River Nuclear Laboratories [Canada]
CRO	Cathode-Ray Oscillograph
	Cathode-Ray Oscilloscope
CR-OFC	Corrosion-Resistant Oxygen-Free Copper
CROM	Capacitive Read-Only Memory
	Control and Read-Only Memory
CROS	Contralateral Routing-of-Signal
CROSS	Computerized Rearrangement of Subject Specialties
CROW	Counter-Rotating Optical Wave
CRP	Constant Rate of Penetration
	Control and Reporting Post
	Corrosion-Resistant Pump
	C-Reactive Protein
CRPL	Central Radio Propagation Laboratory [now: ITSA]
CR PL	Chromium Plate
CRPM	Communication Registered Publication Memorandum
CRR	Center for Renewable Resources
	Churchill Research Range [US Air Force]
	Constant Relative Resolution
	Customer Response Representative
CRREL	Cold Regions Research and Engineering Laboratory [US Army]
CRS	Center for Resource Studies [Canada]
	Clamp Rest Screw
	Cold-Rolled Steel
	Commercially Rapidly Solidified
CRSR	Center for Radiophysics and Space Research [US]
CRSS	Critical Resolved Shear Stress
CRT	Cathode-Ray Tube
	Charactron Tube
	Circuit Requirement Table
	Constant Rate of Traverse
CR TAN LTHR	Chromium Tanned Leather
CRTC	Canadian Radio-Television and Telecommunications Commission
CRTPB	Canadian Radio Technical Planning Board
CRTS	Constant Return to Scale
	Controllable Radar Target Simulator
CRTU	Combined Receiving and Transmitting Unit
CRTV	Composite Reentry Test Vehicle
CRUIS	Cruising
CR VAN	Chrome Vanadium
CRV DWG	Curve Drawing
CRW	Community Radio Watch
CRWPC	Canadian Radio Wave Propagation Committee
CRY	Crystalline
CRYO	Cryogenics
CRYST	Crystalline
CRYPTO	Cryptography
CS	Casein
	Carbon Steel
	Cast Steel
	Center Section
	Channel Status

	Check Sorter
	Chemical Society
	Citrate Synthase
	Clamp Strap
	Coal Store
	Coblentz Society (for Spectroscopy)
	College of Science
	Color Specification
	Commercial Standard
	Communications Satellite
	Communications System
	Community Service
	Computer Science
	Concrete Slab
	Conducted Susceptibility
	Continue-Specific (Mode)
	Control Section
	Control Switch
	Control System
	Core Shift
	Crystalline Solid
	Current Strength
	Cutting Speed
	Cycle Shift
	Cyclo-Stationary
	(Single) Cotton (Single) Silk (Wire)
C/S	Call Signal
C&S	Computers and Systems
Cs	Cesium
c/s	cycles per second [also: cps or CPS]
CSA	Canadian Science Association
	Canadian Standards Association
	Cast Section Angle
	Common Service Area
	Common System Area
	Computer Sciences Association
	Computer Systems Association
	Cross-Sectional Area
	Cryogenic Society of America
CSAE	Canadian Society of Agricultural Engineers
CSAM	Conventional Scanning Acoustic Microscope
CSAO	Construction Safety Association of Ontario [Canada]
CSB	Coherent Surface Bremsstrahlung
	Concrete Splash Block
CSC	Cadmium-Sulfide Cell
	Canadian Society for Chemistry
	Cartridge Storage Case
	Centrifugal Shot Casting
	Chief Sector Control
	Common Signaling Channel
	Commonwealth Scientific Committee
	Communications Systems Center
	Computer Society of Canada
	Construction Specification Canada
	Course and Speed Computer
	Customer Support Center
csc	cosecant
CSCC	Canadian Steel Construction Council
	Cumulative Sum Control Chart
CSCD	Common Signaling Channel Demodulator

CSCE	Canadian Society for Chemical Engineers
	Canadian Society for Civil Engineering
	Communications Subsystem Checkout Equipment
	Conference on Security and Cooperation in Europe [NATO]
CSChE	Canadian Society for Chemical Engineering
csch	hyperbolic cosecant
CSCM	Common Signaling Channel Modem
CSCS	Common Signaling Channel Synchronizer
CSCT	Canadian Society for Chemical Technology
CSD	Charge State Distribution
	Computer Science Division
	Constant Speed Drive
	Controlled-Slip Differentials
CSE	Commission on Science Education
	Computerized Shrinkage Evaluation
	Computer Science and Engineering
	Containment Systems Experiment
	Control Systems Engineering
	Core Storage Element
CSEA	California State Electronics Association [US]
CSECT	Control Section
CSEE	Canadian Society for Electrical Engineers
CSEF	Current Switch Emitter Follower
CSEG	Canadian Society of Exploration Geophysicists
CSEM	Conventional Scanning Electron Microscope
	Conventional Scanning Electron Microscopy
CSF	Catalytic Seed Fund
	Central Switching Facility
	Colony Stimulating Factor
	Cumulative Size Selection Function
CSFGR	Canadian Society for Fifth Generation Research
CSFS	Computer-Simulated Fracture Surface
CSG	Casing
	Constructive Solid Geometry
CSH	Calcium - Silicate - Hydrate
CSI	Chlorosulfonyl Isocyanate
	Colloquium Spectroscopicum Internationale [International Colloquium on Spectroscopy]
	Construction Specification Institute [US]
CSIA	Canadian Solar Industries Association
CSIC	Computer System Interface Circuits
CSICC	Canadian Steel Industries Construction Council
CSIM	Compact Short-Channel IGFET Model
CSIR	Council for Scientific and Industrial Research
CSIRA	Canadian Steel Industry Research Association
CSIRAC	Commonwealth Scientific and Industrial Research Automatic Computer
CSIRO	Commonwealth Scientific and Industrial Research Organization [Australia]
CSK	Countersink
CSK-O	Countersink other Side
CSL	Code Selection Language
	Coincidence Site Lattice

	Comparative Systems Laboratory [US]
	Computer Sensitive Language
	Computer Simulation Language
	Constant Scattering Length
	Continuous Sheet Lamination
	Control and Simulation Language
	Coordinated Science Laboratory [US]
CSLM	Confocal Scanning Laser Microscope
CSLMA	Canadian Softwood Lumber Manufacturers Association
CSLT	Canadian Society of Laboratory Technologists
CSM	Canadian Society of Microbiologists
	Chinese Society of Metals
	Colorado School of Mines [US]
	Command Service Module
	Commission for Synoptic Meteorology
	Communications System Monitoring
	Constant System Monitor
	Continuous Sheet Memory
CSMA	Carrier-Sense Multiple-Access
	Chemical Specialties Manufacturers Association
	Communications Systems Management Association
CSMA/CD	Carrier-Sense Multiple-Access with Collision Detection
CSMCS	Continuous Space Monte Carlo Simulation
CSME	Canadian Society of Mechanical Engineering
	Communications System Monitoring Equipment
CSMITH	Coppersmith
CSML	Continuous Self Mode Locking
CSMMFRA	Cotton, Silk and Man-Made Fibers Research Association [UK]
CSMP	Continuous System Modelling Program
CSN	Cable Satellite Network
	Circuit Switching Network
	Conductive Solids Nebulizer
CSNDT	Canadian Society for Nondestructive Testing
CSO	Chained Sequential Operation
CSP	Certified Safety Professionals
	Coder Sequential Pulse
	Commercial Subroutine Package
	Continuous Sampling Plan
	Control Signal Processor
	Control Switching Point
	Council for Scientific Policy
	Crystallographic Shear Plane
CSPG	Canadian Society of Petroleum Geologists
CSPI	Corrugated Steel Pipe Institute
CSPO	Communications Satellite Project Office
CSR	Control Status Register
CSRF	Canadian Synchrotron Radiation Facility [AEC]
CSRG	Chemical Sciences Research Group
CSRO	Compositional Short-Range Ordering
CSRP	Chemical Sciences Research Paper
CSRS	Cooperative State Research Service [of USDA]

CSRT	Canadian Society of Radiological Technicians		Communications Terminal
CSS	Canada Standard Size		Compact Tension
	Cognitive Science Society		Compact Type (Specimen)
	Communications Subsystem		Computerized Tomography
	Computer Special System		Computer Technology
	Computer Subsystem		Computer Tomography
	Computer System Simulator		Connecticut [US]
CSSA	Computer Society of South Africa		Continuous Transformation
CSSB	Compatible Single-Sideband		Continuous Time
CSSBI	Canadian Sheet Steel Building Institute		Control Transformer
CSSCI	Canadian Steel Service Center Institute		Cooling Tower
CSSE	Canadian Society of Safety Engineers		Counter
	Conference of State Sanitary Engineers [US]		Counter Tube
CSSG	Chemical Sciences Study Group		Current Transformer
CSSL	Continuous System Simulation Language	**C/T**	Carrier-to-Noise Temperature (Ratio)
CSSP	Canadian Space Station Program	**ct**	(centum) - one hundred
	Council on Scientific Society Presidents	**CTA**	California Trucking Association [US]
CSSS	Canadian Society of Soil Science		Canadian Trucking Association
CSST	Computer System Science Training		Compatibility Test Area
CST	Central Standard Time		Constant-Time Anemometer
CSTG	Casting		Cystine Trypticase Agar
CSTAR	Classified Scientific and Technical Aerospace Reports [NASA]	**CTAB**	Canadian Technology Accreditation Board
			Cetyltrimethylammonium Bromide
CSTEX	Cost Extension		Commerce Technical Advisory Board [US]
CSTF	Continuous Stirred Tank Fermentator	**CTB**	Coherent Twin Boundary
CSTI	Committee on Scientific and Technological Information [US]		Concentrator Terminal Buffer
		CTBN	Curboxyl-Terminated Butadiene Acrylonitrile
CSTL	Computer System Terminal Log	**CTC**	Canadian Transport Commission
CSTPC	Cost Price		Carbon Tetrachloride
CSTR	Canister		Channel-to-Channel
	Committee on Solar Terrestrial Research [US]		Chlorotetracycline
			Compact Transpiration Cooling
	Continuous Stirred Reactor	**CTCA**	Ceramic Tile Contractors Association
	Continuous Stirred Tank Reactor		Channel-to-Channel Adapter
CSTS	Combined System Test Stand	**CTCS**	Consolidated Telemetry Checkout System
	Computer Systems Technical Support	**CTD**	Combination Thermal Drive
CSTUN	Cost per Unit		Conductivity, Temperature and Density
CSU	Central Switching Unit		Cross Track Distance
	Channel Service Unit	**CTDH**	Command and Telemetry Data Handling
	Circuit Switching Unit	**CTDS**	Code Translation Data System
	Colorado State University [US]	**C/TDS**	Count/Time Data System
	Combined Shaft Unit	**CTE**	Coefficient of Thermal Expansion
	Crude Steel Unit	**CTEA**	Channel Transmission and Engineering Activation
	Customer Setup		
CSUA	Computer Science Undergraduate Association [US]		Cyanotriethylammonium Tetrafluoroborate
		CTEM	Conventional Transmission Electron Microscope
CSV	Corona Start Voltage		
CSW	Channel Status Word	**CTF**	Chlorine Trifluoride
	Crude Steel Weight		Colloquium on Thin Films
	Control Switch		Controlled Temperature Furnace
CSWA	Canadian Science Writers Association		Controlled Thermonuclear Fission
CSZ	Calcia-Stabilized Zirconia		Cottonseed Flour
CT	Cell Culture Tested	**CTFE**	Chlorotrifluoroethylene
	Center Tap	**CTFM**	Continuous-Transmission, Frequency-Modulated
	Charge Transfer (Complex)		
	Ciguatoxin	**CTG**	Cartridge
	Code Telegram	**CTI**	Comparative Tracking Index
	Colloidal Thorium		Configuration and Tuning Interface
	Commercial Translator	**CTIS**	Computerized Transport Information System
	Communications Technician		

CTL	CAGE Test Language		**CTST**	Critical Trade Skills Training
	Checkout Test Language		**CTT**	Central Trunk Terminal
	Cincinnati Testing Laboratories [US]			Character Translation Table
	Complementary Transistor Logic		**CTTC**	Canadian Trade and Tariffs Committee
	Constant Time Loci		**CTTF**	Continuous Thermal Treatment Facility
	Constructive Total Loss		**CTTL**	Complimentary Transistor-Transistor Logic
	Core Transistor Logic		**CTTMA**	Canadian Truck and Trailer Manufacturers
CTMA	Canadian Tooling Manufacturers			Association
	Association		**CTU**	Centigrade Thermal Unit
CTMC	Communication Terminal Module Controller			Central Terminal Unit
	[UNIVAC]			Channel Testing Unit
CTMT	Combined Thermomechanical Treatment			Communications Terminal Unit
CTN	Cable Termination Network			Compatibility Test Unit
	Carton		**CTUC**	Committee on Tunneling and Underground
CTO	Charge Transforming Operator			Construction
CTOA	Crack Tip Opening Angle		**CTV**	Canadian Television
C to C	Center to Center			Commercial Television
CTOD	Crack Tip Opening Displacement			Control Test Vehicle
C to F	Center to Face		**CTWT**	Counterweight
CTOL	Conventional Takeoff and Landing		**CU**	Columbia University [US]
CTP	Central Transfer Point			Community User
	Character Table Pointer			Concordia University [US]
	Character Translation Pointer			Construction Unit
	Charge Transforming Parameter			Control Unit
	Command Translator and Programmer			Cornell University [US]
	Coordinated Test Plan			(Piezoelectric) Crystal Unit
	Curboxyl-Terminated Polybutadiene			Cubic
	Cytidine Triphosphate		**Cu**	(Cuprum) - Copper
CTPB	Curboxyl-Terminated Polybutadiene Binder		**CUA**	Circuit Unit Assembly
CTR	Center(ing)			Compugraphics Users Association
	Certified Test Requirement		**CUAS**	Computer Utilization Accounting System
	Collective Television Reception		**CUBE**	Concertation Unit for Biotechnology in
	Contour			Europe [also: Cube]
	Controlled Thermonuclear Reactor [US]			Cooperating Users of Burroughs Equipment
	Controlled Thermonuclear Research		**CUBOL**	Computer Usage Business-Oriented
	Counter			Language
CTRA	Coal Tar Research Association [UK]		**cu cm**	cubic centimeter
CTRF	Canadian Transport Research Forum		**CUD**	Craft Union Department [of AFL-CIO]
CTRL	Control		**CUDOS**	Continuously-Updated Dynamic Optimizing
CTRS	Contrast			System
CTS	Cable Transmission System		**CUE**	Computer Updating Equipment
	Carrier Test Switch			Control Unit End
	Chemically-Treated Steel			Cooperating Users Exchange
	Clear-to-Send		**CUEA**	Coastal Upwelling Ecosystem Analysis
	Common Test Subroutines		**CUEBS**	Commission on Undergraduate Education in
	Communications and Tracking Subsystem			the Biological Sciences [US]
	Communications Technology Satellite [US-		**CUES**	Computer Utility Educational System
	Canada]		**CUFT**	Center for the Utilization of Federal
	Computer Telewriter Systems [US]			Technology
	Communications Terminal Synchronous		**cu ft**	cubic foot
	Contralateral Threshold Shift		**cu in**	cubic inch
	Contrast Transfer Function		**CUJT**	Complementary Unijunction Transistor
	Controlled Thermal Severity		**CULP**	Computer Usage List Processor
	Conversational Terminal System		**CuLPCN**	Copper Leucophalocyanine
	Conversational Time Sharing		**cu m**	cubic meter
	Counter-Timer System		**CUMA**	Canadian Urethane Manufacturers
	Courier Transfer Station			Association
	Curved Track Simulator		**CUMM**	Council of Underground Machinery
CTS/RTS	Clear-to-Send/Request-to-Send			Manufacturers
CTSS	Compatible Time Sharing System		**CUMREC**	College and University Machine Record
	Conversational Time Sharing System			Conference

cu mm	cubic millimeter		Communication Valve Development [UK]
cu mu	cubic micron		Compact Video Disk
CUNY	City University of New York [US]		Coupled Vibration-Dissociation
CUP	Canadian University Press		Current-Voltage Diagram
	Communications Users Program	**CVDV**	Coupled Vibration-Dissociation-Vibration
CUPID	Cornell University Program for Integrated Devices	**CVF**	Controlled Visual Flight
		CVI	Chemical Vapor Infiltration
CUPM	Committee on the Undergraduate Program in Mathematics [MAA, US]	**CVIC**	Conditional Variable Incremental Computer
		CVIS	Computerized Vocational Information System
CU PL	Copper Plate	**CVM**	Cluster Variation Method
CUPTE	Canadian Union of Professional and Technical Employees		Consumable-Electrode Vacuum Remelted
		CVN	Charpy V-Notch (Test)
CUR	Complex Utility Routine	**CVR**	Cockpit Voice Recorder
	Current		Continuous Video Recorder
CURE	Color Uniformity Recognition Equipment	**CVS**	Chinese Vacuum Society
	Council for Unified Research and Education		Constant Volume Sampling
CURES	Computer Utilization Reporting System		Cyclic Voltametric Stripping
CURTS	Common User Radio Transmission System	**CVSG**	Channel Verification Signal Generator
CURV	Cable-Controlled Underwater Recovery Vehicle	**CVT**	Chemical Vapor Transport
			Continuously Variable Transmission
	Cable-Controlled Underwater Research Vehicle		Crystal Violet Tetrazolium
		CVTR	Carolians Virginia Tube Reactor
CUS	Coordinatively Unsaturated Sites	**CVU**	Constant Voltage Unit
CUSCL	Customer Class	**CW**	Call Waiting
CUSEC	Canada-United States Environmental Council		Carrier Wave
			Cell Wall
CUSIP	Committee on Uniform Security Identification Procedures		Chemical Warfare
			Clockwise
CUSO	Canadian Universities Service Overseas		Cold Water
CUSP	Central Unit for Scientific Photography [of RAE]		Cold Welding
			Cold Working
CUSRPG	Canada-United States Regional Planning Group		Composite Wave
			Continuous Wave
CUST	Customer		Control Word
CUT	Circuit Under Test		C-Washer
	Control Unit Tester	**CWA**	Communication Workers of America [union]
CUTS	Cassette User Tape System	**CWAR**	Continuous Wave Acquisition Radar
	Computer-Utilized Turning System	**CWARC**	Canadian Workplace Automation Research Center
CUW	Committee on Undersea Warfare [USDOD]		
cu yd	cubic yard	**CWAS**	Contractor Weighted Average Share
CV	Capacitance versus Voltage	**CWB**	Canadian Welding Bureau
	Check Valve	**CWBAD**	Clockwise Bottom Angular Down
	Coefficient of Variation	**CWBAU**	Clockwise Bottom Angular Up
	Computerized Vision	**CWC**	Canadian Wood Council
	Computer Vision		Compound Water Cyclone
	Condensing Vacuole	**CWDB**	Clockwise Down Blast
	Constant Arc Voltage	**CWDC**	Canadian Wood Development Council
	Constant Value	**CWDI**	Canadian Welding Development Institute
	Constant Voltage	**CWED**	Cold Weld Evaluation Device
	Continuously Variable	**CWEI**	Canadian Wood Energy Institute
	Counter Voltage	**CWF**	Coal-Water-Fuel (Technology)
	Curriculum Vitae	**CWFM**	Continuous Wave/Frequency Modulation [also: CW/FM]
	(Single) Cotton Varnish (Wire)		
C/V	Capacitance versus Voltage	**CWG**	Committee for Women in Geophysics
CVA	California Vehicle Act [US]	**CWIF**	Continuous Wave Intermediate Frequency
	Computerized Vibration Analysis	**CWL**	Continuous Working Level
CVC	Continuously Variable Crown (Technology)	**CWLS**	Canadian Well Logging Association
	Current Voltage Characteristics	**CWME**	Canadian Waste Materials Exchange
CV/CC	Constant Voltage/Constant Current	**CWMTU**	Cold-Weather Material Test Unit
CVCM	Collected Volatile Condensable Materials	**CWO**	Continuous Wave Oscillator
CVD	Chemical Vapor Deposition	**CWP**	Circulating Water Pump

	Cold-Work Peak		Drop
CWRA	Canadian Water Resources Association	d	date
CWRC	Chemical Warfare Review Council		day
CWRU	Case Westen Reserve University [US]		decompose
CWS	Canadian Welding Society		degree
	Caucus for Women in Statistics		delete
	Chemical Warfare Service		depth
	Cooperative Wholesale Society		derivative
CWSF	Coal-Water Slurry Fuels		(inside) diameter
CWT	Cooperative Wind Tunnel		dyne
	Critical Water Temperature	DA	Data Administrator
	Hundredweight [also: cwt]		Data Area
CWTAD	Clockwise Top Angular Down		Data Available
	Hundredweight, Air Dry		Decimal Add(er)
CWTAU	Clockwise Top Angular Up		Decimal-to-Analog
CWTH	Clockwise Top Horizontal		Define(d) Area
CWTN	Hundredweights Net		Delay Amplifier
CWUB	Clockwise Up Blast		Demand Assignment
CWV	Continuous Wave Video		Department of the Army [US]
CWWA	Canadian Water Well Association		Design Automation
CX	Character Transfer		Destination Address
	Compatible Expansion		Dielectric Absorption
	Composite		Differential Analyzer
	Control Transmitter		Digital-to-Analog
CX-HLS	Cargo Experimental, Heavy Logistic		Direct Access
	Support [US Air Force]		Directory Assistance
CXR	Carrier		Disaccommodation
CY	Cycle		Discharge Afloat
	Cyclophoshamide		Discrete Address
CYBORG	Cybernetic Organism		Dissolved Acetylene
CYL	Cylinder		Dopamine
CYL L	Cylinder Lock		Double Amplitude
Cyn	Cyanide		Duplex Annealed
Cys	Cysteine	D/A	Digital-to-Analog
CZ	Canal Zone [Panama]	dA	Deoxyadenosine
CZC	Chromated Zinc Chlorate	DAA	Data Access Arrangement
CZE	Compare Zone Equal		Direct Access Arrangement
CZR	Central Zone Remelting	DAAM	Diacetone Acrylamide
CZU	Compare Zone Unequal	DAB	Diaminobenzidine
		DABCO	Diazabicyclooctane
		DABL	Daisy Behavioural Language
		DABS	Discrete Address Beacon System
		DABSYL	Dimethylaminoazobenzenesulfonyl

D

		DAC	Data Acquisition
D	Asparagine Acid		Data Acquisition Center
	Data		Data Analysis Computer
	December		Delayed Atomization Cuvette
	Decimal		Design Augmented by Computer
	Density		Development Assistance Committee [of
	Destination		OECD]
	Deuterium		Digital Amplitude Curve
	Diameter (Outside)		Digital-to-Analog Converter
	Diffusion		Display Analysis Console
	Digit(al)		Distance Amplitude Compensation
	Diode		Distance Amplitude Correction
	Diopter	DACC	Direct Access Communications Channel
	Displacement	DACM	Dimethylamino Methylcoumarinyl
	Dividend		Maleimide
	Division	DACOM	Data Communications
	Doctor	DACON	Data Controller
	Domain		Digital-to-Analog Converter
	Drain		

DACOR	Data Correction
	Data Correlator
DACS	Data Acquisition Control System
	Directorate of Aerospace Combat Systems
DACU	Digitizing and Control Unit
DAD	Diode Array Detector
	Direct Access Device
	Drum and Display
DADB	Data Analysis Database
DADEE	Dynamic Analog Differential Equation Equalizer
DADPS	Diaminodiphenylsulfone
DADS	Data Acquisition Display System
	Digital Air Data System
	Digitally-Assisted Dispatch System
DADSM	Direct Access Device Space Management
DADT	Durability and Damage Tolerance
DADTA	Durability and Damage Tolerance Assessment
DAEMON	Data Adaptive Evaluator and Monitor
DAF	Data Acquisition Facility
	Destination Address Field
	Discard at Failure
	Dissolved-Air Flotation
	Dry and Ash-Free
DAFC	Digital Automatic Frequency Control
DAFM	Discard-at-Failure Maintenance
DAFS	Deepwater Actively Frozen Seabed
	Department of Agriculture and Forest Service
DAFT	Digital/Analog Function Table
DAG	Distance Amplitude Gate
DAGC	Delayed Automatic Gain Control
DAI	Dissertation Abstracts International
DAIM	Digital Acceleration Integration Module
DAIR	Direct Altitude and Identity Readout [FAA]
	Driver Air, Information and Routing
	Dynamic Allocation Interface Routine
DAIRS	Dial Access Information Retrieval System
DAIS	Defense Automatic Integrated Switch(ing System)
	Digital Avionics Information System
DAISY	Data Acquisition and Interpretation System
	Double-Precision Automatic Interpretative System
DAK	Decision Acknowledgement
DAL	Data Address List
	Diallyl Phthalate
daL	decaliter
DAM	Damage
	Data Addressed Memory
	Data Association Message
	Descriptor Attribute Matrix
	Digital-to-Analog Multiplier
	Direct Access Memory
	Direct Access Method
	Dual Absorption Model
dam	dekameter
DAMA	Demand-Assignment Multiple Access
DAM CON	Damage Control
DAMN	Diaminomaleonitrile

DAMP	Downrange Anti-Missile Program [US Army]
DAMPS	Data Acquisition Multiprogramming System
DAMS	Defense against Missile Systems
DANSYL	Dimethylaminonaphthalenesulfonyl
DAP	Deformation of Aligned Phases
	Diallyl Phthalate
	Diammonium Phosphate
	Diammonium Phosphate Plant
	Digital Assembly Program
	Dihydroxyacetone Phosphate
	Distributed Array Processor
	Division of Air Pollution [DOE]
DAPA	Diaminodipropylamine
DAPCA	Development and Production Costs for Aircrafts
DAPI	Diamidinophenylindole
DAPP	Data Acquisition and Processing Program
DAPN	Diaminopropanol
DAPR	Digital Automatic Pattern Recognition
DAPS	Direct Access Programming System
DApSc	Doctor of Applied Science
DAR	Damage Assessment Routine
	Defense Acquisition Radar
	Directorate of Atomic Research
DARC	Direct Access Radar Channel
DArch	Doctor of Architecture
DARE	Document Abstract Retrieval Equipment
	Documentation Automated Retrieval Equipment
	Doppler and Range Evaluation
	Doppler Automatic Reduction Equipment
DARES	Data Analysis and Reduction System
DARLI	Digital Angular Readout by Laser Interferometry
DARME	Directorate of Armament Engineering
DARMS	Digital Alternate Representation of Music Symbols
DARPA	Defence Advanced Research Projects Agency [US]
DARS	Digital Adaptive Recording System
	Digital Attitude and Rate System
DART	Daily Automatic Rescheduling Technique
	Data Acquisition in Real Time
	Data Analysis Recording Tape
	Digital Automatic Readout Tracker
	Direct Access Radio Transceiver
	Director and Response Tester
	Dual Axis Rate Transducer
	Dynamic Acoustic Response Trigger
DARTS	Digital Azimuth Range Tracking System
DAS	Diacetylsulfide
	Data Acquisition System
	Data Analysis System
	Data Automation System
	Dendrite-Arm Spacing
	Digital Analog Simulator
	Digital Attenuator System
DASA	Defense Atomic Support Agency [USDOD]
DASC	Defense Automotive Supply Center
	Direct Air Support Center

DASc	Doctor of Applied Science
DASCOTAR	Data Acquisition System Correlating Tracking and Ranging
DASD	Direct-Access Storage Device
DASH	Drone Anti-Submarine Helicopter
DASHER	Dynamic Analysis of Shells of Revolution
DASM	Direct-Access Storage Media
DASS	Demand-Assignment Signaling and Switching (Device)
DAST	Diethylaminosulfur Trifluoride
DAT	Datum
	Designation - Acquisition - Track
	Digital Audio Tape
	Director of Advanced Technology
	Dynamic Address Translation
	Dynamic Address Translator
DATA	Development and Technical Assistance
DATAC	Data Analog Computer
	Digital Automatic Tester and Classifier
	Digital Autonomous Terminal Access Communication
DATACOL	Data Collection System
DATACOM	Data Communication
DATAGEN	Data File Generator
DATAN	Data Analysis
DATANET	Data Network
DATAR	Digital Automatic Tracking and Ranging [also: Datar]
	Digital Auto-Transducer and Recorder
DATCOM	Data Support Command [US Army]
DATDC	Data Analysis and Technique Development Center
DATICO	Digital Automatic Tape Intelligence Checkout
DATR	Design Approval Test Report
DATRIX	Direct Access to Reference Information
DATS	Despun Antenna Test Satellite
	Dynamic Accuracy Test Set
DATSA	Depot Automatic Test System for Avionics
DAU	Data Adapter Unit
	Data Acquisition Unit
	Disposable Adsorbent Unit
DAV	Data above Voice
	Data Valid
DAVC	Delayed Automatic Volume Control
DAVI	Dynamic Antiresonant Vibration Isolator
DAVIE	Digital Alphanumeric Video Insertion Equipment
DAW	Data Address Word
DAWNS	Design of Aircraft Wing Structures
DAX	Data Acquisition and Control
DAZD	Double Anode Zener Diode
DB	Database
	Data Bus
	Dead Band
	Dichlorophenoxybutyric Acid
	Diffusion Bonding
	Dip Brazing
	Distribution Box
	Doppler-Broadening
	Double-Biased

	Double Bottom
	Double Braid (Wire)
	Double-Break
	Dry Bulb
D-B	Decimal-to-Binary
dB	Decibel
DBA	Database Administration
	Database Administrator
	Design Basis Accident
	Dibenzanthracene
dBA	Adjusted Decibel
DBAO	Digital Block AND-OR (Gate)
DBB	Detector Back Bias
	Detector Balanced Bias
DBC	Data Bus Controller
	Diameter Bolt Circle
	Digital-to-Binary Converter
dBC	C-Scale Sound Level in Decibels
DBCP	Dibromochloropropane
DBD	Database Definition
	Database Description
	Double-Base Diode
DBDA	Dibenzyldodecylamine
DB/DC	Database/Data Communications
DBF	Demodulator Band Filter
DBFS	Dull Black Finish Slate
DBH	Diameter at Breast Height
	Dopamine Beta Hydroxylase
DBHI	Dopamine Beta Hydroxylase Inhibitor
DBI	Differential Bearing Indicator
dBJ	Decibel, Jerrold
dB/K	Decibels per degree Kelvin
DBL	Double
DBLR	Doubler
DBM	Database Management
	Data Buffer Module
	Data Bus Monitor
	Dead-Burnt Magnesia
	Direct Bombardment Mode
dBm	Decibels relative to 1 mW
dB/m^2	Decibels above 1 mW per m^2
dBm/m^2/Mhz	Decibels above 1 mW per m^2 per megahertz
DBMC	Data Bus Monitor/Controller
dBm0p	dBm at 0 (= zero) transmission level, psophometrically weighted
dB/mi	Decibels per mile
DBMS	Database Management Software
	Database Management System
DBN	DECnet Business Network [Canada]
	Diazabicyclononene
DBOS	Disk-Based Operating System
DBP	Dibutylphthalate
DBPCB	Database Program Communication Block [also: DB/PCB]
DBR	Directorate of Biosciences Research
	Division of Building Research [NRC]
dBRAP	Decibels above Reference Acoustical Power
dBRN	Decibels above Reference Noise
dBRNC	Decibels above Reference Noise, C-Message Weighted
DBS	Direct Broadcast Satellite

	Direct Broadcasting System
	Division of Biology Standards
	Dodecylbenzene Sulfonate
DBSP	Double-Based Solid Propellant
DBTG	Database Task Group
DBTT	Ductile-Brittle Transition Temperature
	Ductile-Brittle Transition Transformation
DBU	Diazabicycloundecene
	Digital Buffer Unit
DBUT	Database Update Time
DBV	Doppler Broadening Velocity
dBV	Decibels relative to 1 Volt
DBW	Design Bandwidth
dBW	Decibels relative to 1 Watt
dBx	Decibels above Reference Coupling
DC	Damage Control
	Data Channel
	Data Check
	Data Classifier
	Data Code
	Data Collection
	Data Communications
	Dead Center
	Decimal Classification
	Depolarization Current
	Deposited Carbon
	Device Control (Character)
	Digital Comparator
	Digital Computer
	Disk Controller
	Dimensional Coordination
	Diphenylarsine Cyanide
	Direct Connection
	Direct Current [also: dc]
	Directional Coupler
	Dispatcher Console
	Display Console
	District of Columbia [US]
	Disk Cartridge
	Double Channel
	Double Column
	Double Contact
	Double Cotton (Wire)
	Driver Control
	Drought Code
	Duty Cycle
D&C	Display and Control
D/C	Down-Converter
dC	Deoxycytidine
DCA	Decade Counting Assembly
	Defense Communications Agency [US]
	Device Cluster Adapter
	Dichloroacetic Acid
	Dichloroaniline
	Digital Command Assembly
	Digital Computer Association
	Diploma in Computer Application
	Direct Colorimetric Analysis
	Document Content Architecture
	Driver Control Area
DCAA	Defense Contract Audit Agency [US]

DCA/DIA	Document Content Architecture/Document Interchange Architecture
DCASA	Dyers and Colorists Association of South Africa
DCAOC	Defense Communications Agency Operations Center [US]
DCAS	Deputy Commander Aerospace System
DCB	Data Control Block
	Defense Communications Board [US]
	Define Control Block
	Double-Cantilever Beam
	Drawout Circuit Breaker
DCBD	Define Control Block Dummy
DCBRE	Defense Chemical, Biological and Radiation Establishment [of DRB]
DCBRL	Defense Chemical, Biological and Radiation Laboratories [now: DCBRE]
DCC	Data Communication Channel
	Defense Construction Canada
	Device Control Character
	Device Cluster Controller
	Dicyclohexylcarbodiimide
	Digital Cross Correct
	Direct Computer Control
	Direct Contact Condensation
	Display Channel Complex
	District Communications Center
	Double Cotton Covered (Wire)
	Droplet Counter-Current Chromatograph
DCCA	Design Change Cost Analysis
DCC-MSF	Direct Contact Condensation - Multistage Flash
DCCS	Defense Communications Control System [US Air Force]
	Digital Command Communications System
DCCU	Data Communications Control Unit
DCD	Digital Coherent Detector
	Double Channel Duplex
	Double Crystal Diffractometer
	Dynamic Computer Display
DC/DC	Direct Current to Direct Current
DCDMA	Diamond Core Drill Manufacturers Association [US]
DCDS	Digital Control Design System
	Double Cotton Double Silk (Wire)
DCE	Data Circuit-Terminating Equipment
	Data Communications Equipment
DCEN	Direct Current Electrode Negative
DCEO	Defense Communications Engineering Office
DCEP	Direct Current Electrode Positive
DCF	Data Count Field
	DeCarb-Free (Product)
	Discounted Cash Flow
	Disk Control Field
	Droplet Combustion Facility
DCFROR	Discounted Cash Flow Rate of Return
DCG	Diode Capacitor Gate
DCGEM	Directorate of Clothing, General Engineering and Maintenance
DCH	Data Channel
DCHP	Dicyclohexylphthalate

DCI	Data Communication Interrogate	**DCT**	Destination Control Table
	Defense Computer Institute [US]		Direct Current Test
	Dichloroisoproterenol	**DCTL**	Direct-Coupled Transistor Logic
DCIB	Data Communication Input Buffer	**DCU**	Data Collection Unit
DCKP	Direct Current Key Pulsing		Data Command Unit
DCL	Declaration		Data Control Unit
	Digital Command Language		Decade Counting Unit
	Digital Computer Laboratory		Decimal Counting Unit
	Dual Current Layer		Device Control Unit
DCM	Digital Capacitance Meter		Dichlorourethane
	Digital Command Language		Digital Control Unit
	Direction Cosine Matrix		Digital Counting Unit
DCME	Dichloromethyl Ether		Display and Control Unit
DCN	Drawing Change Notice	**DCUTL**	Direct-Coupled Unipolar Transistor Logic
DCOS	Data Communication Output Selector	**DCV**	Direct Current, Volts
DCP	Data Collection Platform		Double Cotton Varnish (Wire)
	Data Communication Processor	**DCW**	Data Communication Write
	Design Criteria Plan	**DCWV**	Direct Current Working Voltage
	Dicetyl Phosphate	**DD**	Data Definition (Statement)
	Dicumyl Peroxide		Data Dictionary
	Digital Computer Processor		Data Display
	Digital Computer Programming		Data Division
	Digital Control(ler) Programmer		Decimal Divide
	Direct-Current Plasma		Deep Drawing
	Directly-Coupled Plasma		Degree Day
DCPA	Dichloropropionaniline		Density Dependent
DC-PBH	Double-Channel Planar-Buried		Dewey Decimal
	Heterostructure		Dichloropropane-Dichloropropane
DCPG	Defense Communications Planning Group		Digital Data
	[US]		Digital Display
DCPIP	Dichlorophenolindophenol		Disconnecting Device
DCPS	Digitally-Controlled Power Source		Drawing Deviation
DCPSK	Differentially-Coherent Phase Shift Keying		Duplex Drive
DCR	Data Communication Read	**D/D**	Digital-to-Digital
	Data Conversion Receiver	**DDA**	Data Differential Analyzer
	Data Coordinator and Retriever		Digital Differential Analyzer
	Decrease		Direct Differential Analyzer
	Digital Conversion Receiver		Dodecenyl Acetate
	Direct Conversion Reactor		Dynamics Differential Analyzer
	Direct-Current Restorer	**DD&A**	Depreciation, Depletion and Amortization
	Double Cold Reducing	**DDAS**	Digital Data Acquisition System
	Drawing Change Request	**DDB**	Dimethoxybisdimethylaminobutane
DCRP	Direct Current Reverse Polarity		Double Declining Balance
DCRT	Data Collection Receive Terminal	**DDC**	Deck Decompression Chamber
DCS	Data Collection System		Defence Documentation Center [USDOD]
	Data Communication Subsystem		Dewey Decimal Classification
	Data Communication System		Die Casting Development Council
	Data Conditioning System		Digital Data Converter
	Data Control Service		Digital Display Converter
	Defense Communication System		Direct Digital Control(ler)
	Design Control Specifications		Dual Diversity Comparator
	Digital Command System	**DD/C**	Dual Down-Converter
	Digital Communication System	**DDCE**	Digital Data Conversion Equipment
	Direct-Coupled System	**DDCS**	Digital Data Calibration System
	Distributed Computer System	**DDD**	Dichlorophenyl Dichlorophenyl
	Distributed Control System		Dichloroethane
DCSC	Defense Construction Supply Center		Direct Distance Dialing
	[USDOD]	**DDDA**	Dodecadienyl Acetate
DCSP	Defense Communications Satellite Program	**DDDOL**	Dodecadienol
	Defense Communications Satellite Project	**DDE**	Dichlorophenyl Dichloroethylene
	Direct Current Straight Polarity		Direct Design Engineering

DDF	Design Disclosure Format		Digital Debugging Tape
DDG	Digital Data Generator		Dithiotreitol
	Digital Data Group		Doppler Data Translator
DDGE	Digital Display Generator Element		Dynamic Debugging Technique
DDI	Depth Deviation Indicator	**DDTE**	Digital Data Terminal Equipment
	Direct Dial-In	**DDTESM**	Digital Data Terminal Equipment Service Module
	Direct Digital Interface		
	Dislocation-Dislocation Interaction	**DDTS**	Digital Data Transmission System
DDIS	Development Drilling Incentive System	**DDTV**	Dry Diver Transport Vehicle
	Document Data Indexing Set	**DDU**	Digital Distribution Unit
DDL	Data Definition Language	**DDV**	Deck Drain Valve
	Data Description Language	**DDXF**	DISOSS Document Exchange Facility
	Digital Data Link	**DDZ**	Dimethyldimethoxybenzyloxycarbonyl
	Dispersive Delay Line	**DE**	Data Entry
DDM	Data Demand Module		Delaware [US]
	Department of Data Management [US]		Deposition Efficiency
	Derived Delta Modulation		Design Engineering
	Design, Drafting and Manufacturing		Design Evaluation
	Difference in Depth Modulation		Device End
	Digital Display Makeup		Diesel-Electric
DDMC	Design and Drafting Management Council		Differential Equation
DDMS	Department of Defense Manned Spaceflight [US]		Digital Element
			Digital Encoder
DDN	Defense Data Network		Display Electronics
DDNAME	Data Definition Name		District Engineer
DDNP	Diazodinitrophenol		Doctor of Engineering
DDOCE	Digital Data Output Conversion Element	**DEA**	Display Electronics Assemblies
DDOL	Dodecenol	**DEAB**	Diethylamine Borane
DDP	Dichlorodiammineplatinum	**DEACON**	Direct English Access and Control
	Digital Data Processor	**DEAD**	Diethyl Azodicarboxylate
	Distributed Data Processing	**DEAE**	Diethylaminoethyl
DDPE	Digital Data Processing Equipment	**DEAL**	Decision Evaluation and Logic
DDPL	Demand Deposit Program Library	**DEB**	Data Extension Block
DDPU	Digital Data Processing Unit	**DEC**	December
DDQ	Deep Drawing Quality		Decimal
	Dichlorodicyanobenzoquinone		Declination
DDR	Digital Data Receiver		Decoder
	Dynamic Device Reconfiguration		Decomposition
DDR&E	Directorate of Defense Research and Engineering [USDOD]		Decrease
			Decrement
DDRS	Digital Data Recording System		Dynamic Environmental Conditioning
DDS	Data Dependent System	**DECAL**	Decalcomania
	Data Display System		Desk Calculator
	Dataphone Digital Service		Digital Equipment Corporation's Adaptation of Algorithmic Language
	Deep-Diving System		
	Deployable Defense System	**DECB**	Data Event Control Block
	Diaminodiphenylsulfone	**DECEO**	Defense Communication Engineering Office
	Differential and Derivative Spectrophotometry	**DECLAB**	Digital Equipment Corporation Laboratory [US]
	Digital Data Service	**DECM**	Defense Electronic Countermeasures System [US Army]
	Digital Data System		
	Digital Display Scope	**DECOM**	Decommutator
	Digital Dynamics Simulator	**DECONTN**	Decontamination
	Doppler Detection System	**DECOR**	Digital Electronic Continuous Ranging
DDSA	Digital Data Service Adapter	**DECR**	Decrease
	Dodecylsuccinic Anhydride		Decrement
DDT	Data Description Table	**DECUS**	Digital Equipment Computer Users Society
	Defect Detection Trial	**DED**	Dedendum
	Deflagration to Detonation Transition		Design Engineering Directorate
	Dichlorodiphenyltrichloroethane		Doctor of Environmental Design
	Digital Data Transmission	**DEDB**	Data Entry Database

DEDS	Data Entry and Display System
DEDUCOM	Deductive Communicator
DEE	Digital Evaluation Equipment
	Digital Events Evaluator
DEER	Directional Explosive Echo Ranging
DEF	Defect
	Definition
DEFL	Deflection
DEFT	Definite Time
	Dynamic Error-Free Transmission
DEFTR	Digital's Ethernet Frequency Translator
DEG	Degree
	Diethylene Glycol
DEGA	Diethylene Glycol Adipate
DEGB	Diethylene Glycol Benzoate
DEGS	Diethylene Glycol Succinate
DEHA	Diethylhydroxylamine
DEI	Development Engineering Inspection
	Double Electrically Isolated
DEJF	Double End Jig Feet
DEL	Delete (Character)
	Delineation
	Delivery
Del	Delaware [US]
DELDT	Delivery Date
DELFIA	Dissociation-Enhanced Lanthanide Fluoroimmunoassay
DELIQ	Deliquescence
DELNI	Digital's Local Network Interconnect
DELOS	Division for Experimentation and Laboratory-Oriented Studies [of ASEE]
DELTIC	Delay-Line Time Compression
DEM	Demodulator
DEMIST	Design Methodology Incorporating Self-Test
DEML	Demolition
DEMOD	Demodulator
DEMON	Decision Mapping via Optimum Network
DEMPR	Digital's Ethernet Multiport Repeater
DEMPS	Directorate of Engineering and Maintenance Planning Standardization
DEMS	Development Engineering Management System
DEMUX	Demultiplexer [also: Demux]
DEN	Density
	Diethylnitrosamine
	Double Edge Notched
DENALT	Density Altitude
DENISE	Dense Negative Ion-Beam Surface Experiment
DENS	Density
DEP	Department
	Diethyl Phthalate
	Diethyl Pyrocarbonate
DEPA	Defense Electric Power Administration [US]
DEPC	Diethyl Pyrocarbonate
DEPE	Double Escape Peak Efficiency
DEPI	Differential Equations Pseudocode Interpreter
DEPREC	Depreciation
DEPSK	Differential-Encoded Phase Shift Keying
DEPT	Department

DEQ	Dequeue
DE/Q	Design Evaluation/Qualification
DEQUE	Double-Ended Queue [also: deque]
DER	Declining Error Rate
	Derivation
	Derivative
	Diesel Engine Reduction Drive
DEREP	Digital's Ethernet Repeater
DERIV	Derivation
	Derivative
DERV	Diesel Engine Road Vehicle
DES	Data Encryption Standard
	Data Entry System
	Department of Education and Science
	Department of Employment Security
	Design
	Desoxycholate
	Diethylstilbestrol
	Differential Equation Solver
	Digital Expansion System
	Dynamic Environment Simulator
DESC	Defense Electronics Supply Center [USDOD]
	Digital Equation Solving Computer
	Directorate of Engineering Standardization
DESCR	Description
DESG	Design
DESGN	Design
DESGR	Designer
DESIG	Designation
DESO	Double End Shut-Off
DEST	Destination
DESTA	Digital's Ethernet Station Adapter
DET	Detail
	Detector
	Determinant
	Detonator
	Diethyltryptamine
	Divorced Eutectic Transformation
	Double Exposure Technique
	Divorced Eutectic Transformation
DETA	Dielectric Thermal Analyzer
	Diethylenetriamine
DETAB	Decision Table
	Design Table
DETAB-X	Design Table, Experimental
DETC	Diethylthiacarbocyanine
DETP	Data Entry and Teleprocessing
DETWAD	Divorced Eutectic Transformation With Associated Deformation
DEU	Data Entry Unit
	Data Exchange Unit
DEUA	Diesel Engines and Users Association
DEUCE	Digital Electronic Universal Calculating Engine
	Digital Electronic Universal Computing Machine
DEV	Development
	Deviation
DEVEL	Development
DEVN	Deviation

DEVR	Distortion-Eliminating Voltage Regulator		**DFRL**	Differential Relay
DEVSIS	Development Sciences Information System [IDRC, Canada]		**DFS**	Dynamic Flight Simulator
			DFSR	Directorate of Flight Safety Research [US]
DEW	Digital Electronic Watch		**DFSU**	Disk File Storage Unit
	Distant Early Warning		**DFT**	Deaerating Feed Tank
DEWIZ	Distant Early Warning Identification Zone			Design for Testability
DEX	Deferred Execution			Diagnostic Function Test
DEXAN	Digital Experimental Airborne Navigator			Digital Facility Terminal
DF	Dark Field			Discrete Fourier Transform
	Data Field			Draft
	Decimal Fraction		**DFTG**	Drafting
	Decontamination Factor		**DFTPP**	Decafluorotriphenylphosphine
	Deflection Factor		**DFTSMN**	Draftsman
	Degrees of Freedom		**DFU**	Data File Utility
	Describing Function			Disposable Filter Unit
	Destination Field		**DFW**	Diffusion Welding
	Device Flag			Disk File Write
	Digital Fluoroscopy		**DFZ**	Dislocation-Free Zone
	Dilution Factor [also: Df]		**DG**	Dangerous Goods
	Direction Finder			Datagram
	Direction Finding			Detonation Gun
	Disappearing Filament			Differential Generator
	Disk File			Diglyme
	Dissipation Factor			Diode Gate
	Distribution Feeder			Directional Gyro(scope)
	Double Feeder			Director General
	Drinking Fountain			Double Girder (Crane)
	Drive Fit			Double Glass
	Drop Forging			Double Groove (Insulator)
DFAG	Double-Frequency Amplitude Grating			Diacylglycerol
DFAW	Direct Fire Antitank Weapon		**Dg**	Grain Density [crude oil]
DFB	Diffusion Brazing		**dG**	Deoxyguanosine
	Distributed Feedback		**dg**	decigram
	Distribution Fuse Board		**DGB**	Disk Gap Bond
DFBT	Dynamic Functional Board Tester		**DGBC**	Digital Geoballistic Computer
DFC	Data Flow Control		**DGCI&S**	Directorate General of Commercial Intelligence and Statistics [India]
	Diagnostic Flowchart		**DGDME**	Diethyleneglycol Dimethylether
	Direct Field Costs		**DGDP**	Double Groove, Double Petticoat (Insulators)
	Disk File Check		**DGEBA**	Diglycidyl Ether of Bisphenol A
DFCU	Disk File Control Unit		**DGES**	Division of Graduate Education in Science [NSF]
DFD	Digital Flight Display			
DFDR	Digital Flight Data Recorder		**DGIR**	Department of Scientific and Industrial Research [UK]
DFDSG	Direct-Fired Downhole Steam Generator			
DFDT	Difluorodiphenyltrichloroethane		**DGS**	Data Generation System
DFHP	Dislocation-Free High-Purity			Data Ground Station
DFI	Developmental Flight Instrumentation			Degaussing System
	Disk File Interrogate			Data Gathering System
	Display Formatting Language		**DGT**	Directorate General of Telecommunication [France]
DFLD	Data Field			
DFMSR	Directorate of Flight and Missile Safety Research [US Air Force]		**DH**	Dead Head
				Denavit-Hartenberg (Process)
DFO	Department of Fisheries and Oceans [Canada]			Document Handling
				Dortmund-Hoerder Huettenunion (Vacuum Treatment Process)
	Director of Flight Operations			Double Heterojunction
DFOLS	Depth of Flash Optical Landing System			
DFP	Diisopropyl Fluorophosphate		**D-H**	Decimal-to-Hexadecimal
DFPG	Double-Frequency Phase Grating		**DHA**	Dehydrated Humulinic Acid
DFR	Decreasing Failure Rate			Dihydroanthracene
	Disk File Read		**DHAP**	Dihydroxyacetone Phosphate
	Dornreay Fast Reactor		**DHBA**	Dihydroxybenzylamine
DFRA	Drop Forging Research Association [UK]			

DHC	Data Handling Center
DHCF	Distributed Host Command Facility
DHD	Double Heat-Sink Diode
DHE	Data Handling Equipment
	Down-Hole Emulsification
	Dump Heat Exchanger
DHEA	Dehydroepiandrosterone
DHG	Dihydroxyethylglycinate
DHP	Deoxidized, High-Residual Phosphorus (Copper)
	Dihydropyridine
DHS	Data Handling System
DHW	Domestic Hot Water
	Double Hung Windows
DHX	Dump Heat Exchanger
DI	Data Input
	Demand Indicator
	Device Independence
	Digital Input
D/I	Distinctness of Image
DIA	Defense Intelligence Agency [USDOD]
	Diameter
	Document Interchange Architecture
DIAC	Defense Industry Advisory Control
	Diode AC (= Alternating Current) Switch
DIAD	Diisopropyl Azodicarboxylate
	Drum Information Assembler and Dispatcher
DIAG	Diagonal
	Diagram
DIAL	Differential-Absorption Lidar
	Display Interactive Assembly Language
	Draper Industrial Assembly Language
	Drum Interrogation, Alteration and Loading System
DIALGOL	Dialect of ALGOL
DIAM	Data Independent Architecture Model
DIAM	Diameter
DIAN	Digital - Analog
DIANE	Digital Integrated Attack Navigation Equipment
	Direct Information Access Network for Europe
DIANS	Digital Integrated Attack Navigation System
DIAPH	Diaphragm
DIAS	Dynamic Inventory Analysis System
DIB	Diiodobutane
DIBAL	Diisobutylaluminum
DIBAL-H	Diisobutylaluminum Hydride
DIBOL	Digital Interactive Business-Oriented Language
DIC	Data Insertion Converter
	Detailed Interrogation Center
	Differential Interference Contrast
	Digital Interchange Code
DICBM	Defense Intercontinental Ballistic Missile
DICO	Dissemination of Information through Cooperative Organization
DICON	Digital Communication through Orbiting Needles
DICORAP	Directional-Controlled Rocket-Assisted Projectile

DICORS	Diver Communication Research System
DICT	Dictionary
DID	Device-Independent Display
	Digital Information Display
	Direct Inward Dialing
	Division of Isotopes Development [of AEC]
	Drum Information Display
DIDACS	Digital Data Communication System
DIDAD	Digital Data Display
DIDAP	Digital Data Processor
DI/DO	Data Input/Data Output
DIDOC	Desired Image Distribution fusing Orthogonal Constraints
DIDOCS	Device-Independent Display Operator Console Support and Status
DIDP	Diisodecylphthalate
DIDS	Digital Information Display System
	Diisothiocyanatostilbene Disulfonic Acid
DIE	Document of Industrial Engineering
DIEA	Diisoprophylethylamine
DIEGME	Diethylene Glycol Monomethyl Ether
DIF	Data Interchange Format
	Differential Interference Microscopy
DIFET	Double-Injection Field-Effect Transistor
DIFF	Difference
	Differential
DIFFTR	Differential Time Relay
DIFKIN	Diffusion Kinetics
DIFP	Diisopropyl Fluorophosphate
DIFPEC	Differentially Pumped Environmental Chamber
DIG	Digit
	Digital Input Gate
DIGACC	Digital Guidance and Control Computer
DIGCOM	Digital Computer
DIGICOM	Digital Communications
DIGITAC	Digital Airborne Computer
	Digital Tactical Automatic Control
DIGM	Diffusion-Induced Grain Boundary Migration
DIGS	Data and Information Gathering System
	Disorder-Induced Gap States
DIL	Dilution
	Dual-in-Line
DILS	Doppler Inertial LORAN System
DIM	Device Interface Module
	Dimension
DIMATE	Depot-Installed Maintenance Automatic Test Equipment
DIMES	Defense-Integrated Management Engineering System
DIMPLE	Deuterium-Moderated Pile Low Energy
DIMUS	Digital Multibeam Steering
DIN	Deutsches Institut fuer Normung [German Institute for Standards]
DINA	Direct Noise Amplification
DINAP	Digital Network Analysis Program
DINP	Diisononylphthalate
DIO	Diode
	Direct Input/Output
DIOB	Digital Input/Output Buffer

DIOP	Diisopropylidene Dihydroxy Bisdiphenylphosphinobutane
DIOS	Distribution, Information and Optimizing System
DIP	Defense Industry Productivity (Program) [US and Canada]
	Display Information Processor
	Dual-in-line Package
	Dual-in-line Pin
DIPD	Double Inverse Pinch Device
DIPEC	Defense Industrial Plant Equipment Center [USDOD]
DipEng	Diploma in Engineering
DipEngTech	Diploma in Engineering Technology
DIPP	Defense Industry Productivity Program
DIPS	Development Information Processing System
DIPSO	Dihydroxyethylamino Hydroxypropanesulfonic Acid
DIPT	Diisopropyltartrate
DIR	Development Inhibitor Releasing
	Diffusion-Induced Recrystallization
	Direction
	Director
	Directory
DIRCOL	Direction Cosine Linkage
DIR-CONN	Direct-Connected
DIRP	Defense Industrial Research Program
DIS	Dissertation Inquiry Service
	Ductile Iron Society
DISAC	Digital Simulator and Computer
DISC	Differential Isochronous Self-Collimating Counter
	Disconnection
	Discovery
DISCH	Discharge
DISCOM	Digital Selective Communications
DISCOS	Disturbance Compensation System
DISCR	Discriminator
DISD	Data and Information Systems Division
dis/min	disintegration per minute [unit]
DISOSS	Distributed Office Support System
DISP	Dispatcher
	Displacement
	Display
	Disposal
DISS	Digital Interface Switching System
DIST	Distance
	Distribution
	District
DISTR	Distribution
DISTRAM	Digital Space Trajectory Measurement System
DISTRO	Distribution Rotation
DITC	Department of Industry, Trade and Commerce
	Diisothiocyanate
DITEC	Digital Television Communication System
DITRAN	Diagnostic FORTRAN
DIU	Data Interchange Utility
DIV	Data in Voice

	Digital Input Voltage
	Diverter
	Division
	Divergence
	Dividend
DIVA	Digital Input Voice Answerback
	Digital Inquiry Voice Answerback
DIVOT	Digital-to-Voice Translator
DJ	Diffusion Junction
DK	Dark
	Deck
	Direct Kinematics
dkg	dekagram
DKI	Data Key Idle
dkL	dekaliter
dkm	dekameter
dks	dekastere
DL	Data Language
	Data Link
	Data List
	Deadload
	Delay Line
	Developed Length
	Diode Logic
	Double Layer
	Double Loop
	Drawing List
	Dynamic Loader
dL	deciliter
DLA	Data Link Adapter
	Data Link Address
	Defense Logistics Agency [US]
DLC	Data Link Control
	Direct Lift Control
	Duplex Line Control
	Dynamic Load Characteristics
DLCC	Digital Load Cell Comparison
DLCS	Data Line Concentration System
DLE	Data Link Escape (Character)
DLEA	Double Leg Elbow Amplifier
DLF	Direct Through-Connection Filter
DLI	Defense Language Institute [US]
DL/I	Data Language/I
DLIEC	Defense Language Institute, East Coast [US]
DLIR	Depot Level Inspection and Repair
	Downward Looking Infrared
DLK	Data Link
DLM	Dwight-Lloyd-McWane (Direct Iron Reduction)
DLO	Double Local Oscillator
DLP	Data Listing Program
	Deoxidized, Low-Phosphorus (Copper)
	Dynamic Low-Pass
DLS	Data Link Set
	Diffused Light Storage
	Direct Least Squares
DLSC	Defense Logistics Service Center [US]
DLT	Data Line Translator
	Data Link Terminal
	Data Link Translator
	Data-Loop Transceiver

	Decision Logic Table	**DMF**	Differential Matched Filter
	Depletion-Layer Transistor		Dimethylformamide
DLTDP	Dilaurythiodipropionate	**DMH**	Direct Manhours
DLTM	Data Link Test Message		Drop Manhole
DLTS	Deep Level Transient Spectroscopy	**DMI**	Desmethylimipramine
DLU	Data Line Unit		Dimethylimidazolidinone
DLVL	Diverted into Low Velocity Layers	**DMIRR**	Demand Module Integral Rocket Ramjet
DM	Data Management	**DMJTC**	Differential Multijunction Thermal
	Data Mark		Converter
	Decimal Multiply	**DML**	Data Manipulation Language
	Delta Modulation		Demolition
	Demand Meter	**DMLS**	Doppler Microwave Landing System
	Design Manual	**DMM**	Data Manipulation Mode
	Design Memorandum		Digital Multimeter
	Differential Mode		Dimethylmercury
	Digital Monolithic		Direct Metal Mastering
	Disconnected Mode	**DMN**	Dimension
D/M	Demodulation/Modulation		Dimethylnitrosamine
D&M	Dressed and Matched [lumber]	**DMO**	Data Management Office
dm	decimeter		Diode Microwave Oscillator
DMA	Dimethylacetamide	**DMOS**	Diffusive Mixing of Organic Solutions
	Dimethylaniline		Discrete Metal-Oxide Semiconductor
	Direct Memory Access	**DMP**	Dimethoxypropane
	Dynamic Mechanical Analyzer		Dimethylphenol
DMAB	Dimethylamineborane		Dimethylphtalate
	Dimethylaminobenzoic Acid		Dot Matrix Printer
DMAC	Defense Metals Information Center		Dump
	Direct Memory Access Controller	**DMPA**	Digitally-Modulated Power Amplifier
DMAE	Dimethylaminoethanol	**DMPD**	Dimethyl Phenylenediamine
DMAL	Dimethylacetal	**DMPE**	Dimethoxphenylethylamine
DMAP	Dimethylaminopyridine		Dimethylphosphinoethane
DMAPN	Dimethylaminoproionitrile	**DMPEA**	Dimethoxphenylethylamine
DMB	Data Management Block	**DMPO**	Dimethylpyrrolineoxide
DMBA	Dimethylbenzanthracene	**DMPP**	Dimethylphenylpiperazinium Iodide
DMC	Dichloromethylbenzhydrol	**DMPR**	Damper
	Digital Microcircuit	**DMPRT**	Dual-Mode Personal Rapid Transport
	Direct Multiplexed Control	**DMPS**	Dimercaptopropanesulfonic Acid
	Dough Molding Compound	**DMPU**	Dimethylpropyleneurea
	Duff Moisture Code	**DMR**	Digital Microwave Radio
DMCD	Dimethylcyclohexanedicarboxilate		Dynamic Modular Replacement
DMCS	Dimethyldichlorosilane	**DMS**	Database Management System
	Distributed Manufacturing Control System		Data Management Service
DMD	Dimethyloxozolidinedione		Data Management System
DMDCS	Dimethyldichlorosilane		Data Multiplex Switching
DMDS	Dimethyldisulfide		Defense Missile Systems
DME	Data Measuring Equipment		Differential Maneuvering Simulator
	Dimethylether		Digital Multimeter System
	Direct Measurements Explorer (Satellite)		Digital Multiplexing Synchronizer
	Distance Measuring Equipment		Dimethylsilyl
	Division of Mechanical Engineering [of		Dimethylsulfide
	NCR]		Display Management System
	Dropping Mercury Electrode		Documentation of Molecular Spectroscopy
	Dulbecco's Modified Eagle Medium		Duplex Microstructure
DMEA	Dimethylethylamine	**DMSCC**	Direct Microscopic Somatic Cell Count
DME/COTAR	Distance Measuring	**DMSO**	Dimethylsulfoxide
	Equipment/Correlation Tracking and	**DMSO$_2$**	Dimethylsulfone
	Ranging	**DMSP**	Defense Meteorological Satellite Program
DMED	Digital Message Entry Device		[US]
DMEP	Dimethoxyethylphtalate	**DMSR**	Director of Missile Safety Research
DMET	Distance Measuring Equipment TACAN	**DMSS**	Digital Multiplexing Subsystem
	(= Tactical Air Navigation)	**DMT**	Digital Message Terminal

	Dimethoxytrityl		Diesel Oil
	Dimethyl Terephthalate		Ditto
	Dimethyltriptamine		Draw Out
DMTC	Digital Message Terminal Computer	**D-O**	Decimal-to-Octal
DMTCNQ	Dimethyl Tetracyanquinonedimethane	**DOA**	Dead on Arrival
DMTR	Dornreay Materials Testing Reactor		Differential Operational Amplifier
DMTS	Department of Mines and Technical Survey [Canada]		Dioctyl Adipate
			Dominant Obstacle Allowance
DMU	Dictionary Management Utility	**DOC**	Data Optimizing Computer
	Diesel-Multiple Unit		Decimal-to-Octal Conversion
	Digital Message Unit		Deoxycorticosterone
	Dimethylolurea		Department of Commerce
	Dual Maneuvering Unit		Department of Communications [Canada]
DMV	Department of Motor Vehicles [US]		Depth of Cut
DMW	Dissimilar-Metal Weld		Direct Operating Costs
DMX	Data Multiplexer		Document
DMZ	Demilitarized Zone		Dynamic Overload Control
DN	Data Name	**DOCA**	Deoxycorticosterone Acetate
	Decimal Number	**DOCC**	DCA (= Defense Communications Agency) Operations Center Complex [US]
	Down		
	Nominal Diameter	**DOCS**	Disoriented Computer System
dn	delta amplitude [elliptic function]	**DOCSYS**	Display of Chromosome Statistics System
DNA	Deoxyribonucleic Acid	**DOCUS**	Display-Oriented Computer Usage System
	Digital Network Architecture	**DOD**	Department of Defense [US]
	Dinolylaniline		Development Operations Division
DNAF	Dinitrophenylazophenanthrol		Direct Outward Dialing
DNAG	Decade of North American Geology		Drop on Demand
DNAP	Dinitrophenylazophenol	**DODC**	Diethyloxadicarbocyanine
DNB	Departure from Nucleate Boiling	**DODCI**	Department of Defense Computer Institute [US]
DNC	Delayed Neutron Counting		
	Dinitrocellulose	**DODGE**	Department of Defense Gravity Experiment [US]
	Direct Numerical Control		
	Distributed Numerical Control	**DODGE-M**	DODGE, Multipurpose [US]
DNCCC	Defense National Communications Control Center [US]	**DODIS**	Distribution of Oceanographic Data at Isotropic Levels
DND	Department of National Defense [Canada]	**DODISS**	Department of Defense Index of Specifications and Standards [US]
dNDP	Desoxyribonucleoside Diphosphate		
DNE	Department of Nuclear Engineering [US]	**DOE**	Department of Electronics [India]
	Dinitroethane		Department of Energy [US]
DNI	Digital Noninterpolation		Department of Environment
DNIC	Digital Network Interface Circuit	**DOETS**	Dual-Object Electronic Tracking System
DNL	Differential Non-Linerarity	**DOF**	Degree of Freedom
	Dynamic Noise Limiter		Device Operating Failure
dNMP	Desoxyribonucleoside Monophosphate		Direction of Flight
DNOC	Dinitro-Ortho-Cresol	**DOFIC**	Domain-Originated Functional Integrated Circuit
DNP	Deoxyribonucleoprotein		
	Dinitrophenyl	**DOHC**	Double Overhead Camshaft
	Dynamic Nuclear Polarization	**DOI**	Department of Industry [UK]
DNPH	Dinitrophenylhydrazine		Department of the Interior [US]
DNPyr	Dinitropyridylamino Acid		Differential Orbit Improvement
DNR	Double Non-Return (Valve)		Distinctness of Image
	Dynamic Noise Reduction	**DO/IT**	Digital Output/Input Translator
DNS	Dimethylaminonaphthalenesulfonic Acid	**DOL**	Department of Labour [US]
	Dimethylnaphthylaminesulfonic Acid		Director of Laboratories
	Doppler Navigation System		Display-Oriented Language
DNSR	Director of Nuclear Safety Research		Dynamic Octal Load
DNT	Desmethylnortriplyline	**DOLARS**	Digital Off-Line Automatic Recording System
dNTP	Deoxyribonucleoside Triphosphate		
DO	Data Output		Doppler Location and Ranging System
	Digital Output	**DOLPHIN**	Deep Ocean Logging Profiler Hydrographic Instrumentation and Navigation
	Dissolved Oxygen		

DOM	Digital Ohmmeter		Double Paper (Wire)
DOMSAT	Domestic Satellite System		Double-Pole
DOP	Designated Overhaul Point		Drillpipe
	Developing-Out Paper		Drip-Proof
	Dioctylphthalate		Driving Power
DOPA	Dihydroxyphenylalanine		Drum Processor
DOPIC	Documentation of Program in Core		Dual-Port (Bus)
DOPLOC	Doppler Phase Lock		Dynamic Programming
DOPS	Digital Optical Protection System	**D&P**	Developing and Printing
	Dihydroxyphenylserine	**DPA**	Data Processing Activities
DOR	Digital Output Relay		Destructive Physical Analysis
DORAN	Doppler Range and Navigation [also: Doran]		Displacement per Atom
DORE	Defense Operational Research Establishment [of DRB]		Deoxidized Phosphorus (Copper), Arsenical
DORF	Diamond Ordinance Radiation Facility [US]	**DPAG**	Dangerous Pathogens Advisory Group
DORIS	Direct Order Recording and Invoicing System	**DPAGE**	Device Page
		D-PAT	Drum-Programmed Automatic Tester
DORV	Deep Ocean Research Vehicle	**DPB**	Data Processing Branch
DOS	Decision Outstanding		Defense Policy Board [US]
	Degree of Sensitization		Diphenylbutadiene
	Density of States	**DPBC**	Double-Pole, Back-Connected
	Digital Operation System	**D-PBS**	Dulbecco's Phosphate Buffered Saline
	Dioctyl Sebacate	**DPC**	Data Processing Center
	Disk Operating System		Data Processing Control
DOSV	Deep Ocean Survey Vehicle		Defense Planning Committee
DOS/VS	Disk Operating System/Virtual Storage		Direct Program Control
DOT	Deep Ocean Technology		Display Processor Code
	Deep Ocean Transponder	**DPCM**	Differential Pulse Code Modulation
	Department of Trade [UK]	**DPCT**	Differential Protection Current Transformer
	Department of Transportation [US]	**DPCX**	Distributed Processing Control Executive
	Department of Transport [Canada]	**DPD**	Data Processing Department
DOTP	Dioctyl Terephthalate		Data Processing Division
DOUG FIR	Douglas Fir		Diethyl-P-Phenylene Diamine
DOUG FIR-L	Douglas Fir - Lumber		Diffusion Pressure Deficit
DOUSER	Doppler Unbeamed Search Radar [also: Douser]		Digital Phrase Difference
		DPDC	Double Paper, Double Cotton (Wire)
DOVAP	Doppler Velocity and Position	**DPDT**	Double-Pole, Double-Throw
DOWB	Deep Operating Work Board	**DPDT DB**	Double-Pole, Double-Throw, Double-Break
DOXYL	Dimethyloxazolidinyoxyl	**DPDT SW**	Double-Pole, Double-Throw Switch
doz	dozen [unit]	**DPE**	Data Processing Equipment
DP	Dash Pot	**DPF**	Data Processing Facility
	Data Processing		Diesel Particulate Matter
	Data Processor	**DPFC**	Double-Pole, Front-Connected
	Depth	**DPFZ**	Destor-Porcupine Fault Zone [Canada]
	Deck Piercing	**DPG**	Digital Pattern Generator
	Deep Penetration		Diphosphoglycerate
	Deflection Plate		Diphosphoglyceric Acid
	Degree of Polymerization	**DPH**	Diamond Pyramid Hardness
	Dew Point		Diphenylhexatriene
	Dial Pulsing		Diphenylhydantoin
	Diametral Pitch	**DPI**	Differential Pressure Indicator
	Diamond Pin		Digital Pseudorandom Inspection
	Dichlorophenoxypropionate		Dots per Inch
	Differential Pressure	**DPI/O**	Data Processing Input/Output
	Differential Pulse	**DPL**	Diploma
	Digit Present	**DPLCS**	Digital Propellant Level Control System
	Digital Pulser	**DPM**	Data Processing Machine
	Disk Pack		Data Processing Management
	Dispersed Phase		Data Processing Manager
	Display Package		Diesel Particulate Matter
	Distribution Point		Digital Panel Meter
			Diphenylphosphinomethane

	Disintegrations per Minute
	Distributed Presentation Management
	Distributive Principle of Multiplication
DPMA	Data Processing Management Association
DPMS	Data Project Management System
DPN	Diamond Pyramid Hardness Number
	Diphosphopyridine Nucleotide
d(PN)	Deoxypolynucleotide
DPNH	Dihydrodiphosphopyridine Nucleotide, Reduced Form
	Diphosphopyridine Nucleotide, Reduced Form
DPO	Double Pulse Operation
DPP	Dextran Phosphate Precipitate
	Diphenyl Phthalate
DPPA	Diphenylphosphoryl Azide
DPPH	Diphenylpicrylhydrazyl
DPR	Digital Process Reporter
DPS	Data Processing Standards
	Data Processing Station
	Data Processing System
	Descent Power System
	Diphenylstilbene
	Disk Programming System
DPSA	Data Processing Supplies Association
	Deep Penetration Strike Aircraft
DPSK	Differential Phase-Shift Keying
DPSS	Data Processing Subsystem
	Data Processing System Simulator
DPST	Double-Pole, Single-Throw
DPST SW	Double-Pole, Single-Throw Switch
DP SW	Double Pole Switch
DPT	Department
	Dewpoint Temperature
	Differential Polarization Telegraphy
DPTE	Deoxidized Phosphorus (Copper), Tellurium Bearing
DPTH	Diphenylthiohydantoin
DPTT	Double-Pole, Triple-Throw
DPTT SW	Double-Pole, Triple-Throw Switch
DPU	Dip Pick-Up
DPV	Dry Pipe Valve
DPWM	Double Pulsewidth Modulation
DQ	Direct Quenching
	Drawing Quality
DQAK	Drawing Quality, Aluminum-Killed
DQC	Data Quality Control
DQE	Detective Quantum Efficiency
DQM	Data Quality Monitor
DQRS	Drawing Quality Rimmed Steel
DQSK	Drawing Quality, Special-Killed
DR	Data Receiver
	Data Recorder
	Data Reduction
	Dead Reckoning
	Defined Readout
	Differential Rate
	Digital Resolver
	Diploma in Radiology
	Direct Reduction
	Discrimination Radar

	Division of Research
	Divisor
	Doctor of Radiology
	Drafting Request
	Drag Reduction
	Drain
	Drill
	Drill Rod
	Drive
	Dynamic Range
Dr	Doctor
D/R	Direct/Reverse
	Downrange
D&R	Distiller and Rectifier
dr	dram [unit]
DRA	Dead Reckoning Analyzer
DR&A	Data Reduction and Analysis
dr ap	dram, apothecaries [unit]
DRADS	Degradation of Radar Defense System
DRAE	Defense Research Analysis Establishment
DRAI	Dead Reckoning Analog Indicator
DRAM	Dynamic Random-Access Memory [also: dRAM]
dr avdp	dram, avoirdupois [unit]
DRAW	Direct Read After Write
DRB	Defense Research Board [Canada]
	Dichlorobenzimidazole Riboside
DRC	Damage Risk Criterion
	Data Recording Control
	Data Reduction Compiler
	Defense Research Committee
	Denver Research Center
	Discontinuous Reinforced Composite
DRCL	Defense Research Chemical Laboratories [Canada]
DRD	Data Recording Device
	Director of Research and Development
	Draw-Redraw
	Drill Rig Duty
DRDTO	Detection Radar Data Takeoff
DRE	Data Recording Equipment
	Data Reduction Equipment
	Dead Reckoning Equipment
	Defense Research Establishment [Canada]
	Destruction and Removal Efficiency
	Director of Research and Engineering
DREA	Defense Research Establishment, Atlantic [of DND]
DREAC	Drum Experimental Automatic Computer
DRECT	Demonstration of Resource and Energy Conservation Technologies
	Development of Resource and Energy Conservation Technologies
DRED	Data Routing and Error Detecting
DREE	Department of Regional Economic Expansion [Canada]
DREO	Defense Research Establishment, Ottawa [of DND]
DREP	Defense Research Establishment, Pacific [of DND]
DRES	Defense Research Establishment, Suffield [of DND]

DRESS	Depth-Resolved Surface Coil Spectroscopy
DRET	Defense Research Establishment, Toronto [of DND]
	Direct Reentry Telemetry
DRETS	Direct Reentry Telemetry System
DREV	Defense Research Establishment, Valcartier [of DND]
DREWS	Direct Readout Equatorial Weather Satellite
dr fl	dram, fluid [unit]
DRFR	Division of Research Facilities and Resources [US]
DR FX	Drill Fixture
DRG	Division of Research Grants [NRC]
DR HD	Drill Head
DRI	Data Reduction Interpreter
	Dead Reckoning Indicator
	Defense Research Institute [US]
	Denver Research Institute [US]
	Descent Rate Indicator
	Direct-Reduced Iron
DRIC	Defense Research Information Center [US]
DRID	Direct Readout Image Dissector
DRIDAC	Drum Input to Digital Automatic Computer
DRIE	Department of Regional and Industrial Expansion [Canada]
DRIFT	Diffuse Reflectance Infrared Fourier Transform
	Diversity Receiving Instrumentation for Telemetry
	Diversity Reliability Instantaneous Forecasting Technique
DRIR	Direct Readout Infrared
DRL	Defense Research Laboratory [US]
	Drilling Research Laboratory
DRLG	Drilling
DRM	Digital Radiometer
	Direction of Relative Movement
	Drafting Room Manual
	Dynamic Recoil Mixing
DRML	Defense Research Medical Laboratories [Canada]
DRMO	District Records Management Officer
DRNL	Defense Research Northern Laboratories [Canada]
DRO	Destructive Readout
	Digital Readout
DROD	Delayed Readout Detector
DROS	Direct Readout Satellite
	Disk Resident Operating System
DRP	Data Reduction Program
	Dead Reckoning Plotter
	Dense Random Packed
	Digital Recording Process
	Distribution Requirements Planning
	Distribution Resource Planning
DRPHS	Dense Random Packing of Hard Spheres
DRQ	Data Ready Queue
DRR	Digital Radar Relay
DRRL	Digital Radar Relay Link
DRS	Data Recording Set
	Detection Ranging System

	Diffuse Reflectance Spectroscopy
	Digital Radar Simulator
	Direct Reception System
	Distributed Resource System
	Dress(ing)
	Dynamic Reflectance Spectroscopy
DRSS	Data Relay Satellite System
DRT	Data Reckoning Tracer
	Data Recording Terminal
	Dead Reckoning Tracer
	Diode Recovery Tester
dr t	dram, troy [unit]
DRTC	Documentation Research and Training Center
DRTE	Defense Research Telecommunications Establishment [of DND]
DRTL	Diode-Resistor-Transistor Logic
DRTR	Dead Reckoning Trainer
DRV	Data Recovery Vehicle
	Deep-Diving Research Vehicle
DRX	Dynamic Recrystallization
DS	Data Scanning
	Data Set
	Data Sheet
	Data Stream
	Data Structure
	Data System
	Decimal Subtract
	Descent Stage
	Device Selector
	Dial System
	Difference Spectrophotometry
	Digital Signal
	Diode Switch
	Dip Soldering
	Directional Solidification
	Disconnect Switch
	Disk Storage
	Dispersion Strengthening
	Double Silk (Wire)
	Double Space
	Double Stranded (Wire)
	Downspout
	Drum Storage
D&S	Display and Storage
ds	decistere [unit]
DSA	Deoxystreptamine
	Device Specific Adapter
	Dial Service Assistance
	Digital Signal Analyzer
	Digital Storage Architecture
	Digital Surface Analyzer
	Doppler Spectrum Analyzer
	Dynamic Spring Analysis
	Dynamic Strain Aging
DSAP	Data Systems Automation Program
	Defense Systems Application Program [US]
DSAR	Data Sampling Automatic Receiver
DSB	Defense Science Board [US]
	Double Sideband
DSBAM	Double Sideband Amplitude Modulation [also: DSB-AM]

DSBAMRC	Double Sideband Amplitude Modulation, Reduced Carrier [DSB-AM-RC]
DSBSC	Double Sideband Suppressed Carrier [DSB-SC]
DSBSCAM	Double Sideband Suppressed Carrier Amplitude Modulation [DSB-SC-AM]
DSB-SC-AM w/QM	DSB-SC-AM with Quadrature Multiplexing
DSBSCASK	Double Sideband Suppressed Carrier Amplitude Shift Keyed [DSB-SC-ASK]
DSC	Data Stream Compatibility
	Data Synchronizer Channel
	Differential Scanning Calorimeter
	Differential Scanning Calorimetry
	Discount
	Displacement Shift Complete Lattice
	Disuccinimidyl Carbonate
	Double Silk Covered (Wire)
DSc	Doctor of Science
DSCA	Default System Control Area
DScA	Doctor of Science in Agriculture
DSCB	Data Set Control Block
DSCC	Deep Space Communications Complex
DScC	Doctor of Commercial Science
DSCCD	Discount Code
DSCL	Displacement Shift Complete Lattice
DScMil	Doctor in Military Sciences
DScNat	Doctor of Natural Sciences
DSCRM	Discriminator
DSCS	Defense Satellite Communications System [US]
	Deskside Computer System
DSCT	Double Secondary Current Transformer
DSD	Data-Set Definition
	Defense Systems Division
	Digital System Design
DSDD	Double-Sided, Double-Density
DSDP	Deep Sea Drilling Project
DSDT	Data-Set Definition Table
DSE	Data Set Extension
	Data Storage Equipment
	Data-Switching Exchange
	Data Systems Engineering
	Distributed Systems Environment
DSEA	Data Storage Electronic Assembly
DSECT	Data Section
	Dummy Control Section
DSEG	Data Systems Engineering Group
DSEP	Distribution System Expansion Program
DSF	Data Secured File
	Directional Solidification Furnace
DSG	Double Strength Glass
DSGN	Design
DSI	Digital Speech Interpolation
	Direct Sample Insertion
	Dislocation-Solute Interaction
	Dynamic System Interchange
DSID	Direct Sample Insertion Device
DSIF	Deep Space Instrumentation Facility [of NASA]
DSIR	Department of Scientific and Industrial Research [UK]
DSIS	Department of Scientific Information Services [of DND]
	Directorate of Scientific Information Services
DSK	Disk
DSL	Data Set Label
	Data Simulation Language
	Data Structures Language
	Detroit Signal Laboratory [US Army]
	Dynamic Super Loudness
DSLC	Defense Logistics Services Center [USDOD]
DSM	Defense Suppression Missile
	Digital Simulation Model
	Dutch State Mines (Screen)
	Development of Substitute Materials
	Dynamic Scattering Mode
DSN	Data Set Name
	Deep Space Network [NASA]
DSNAME	Data Set Name
DSO	Data Systems Office
	Digital Storage Oscilloscope
DSORG	Data Set Organization
DSP	Defense Services Program
	Digital Signal Processing
	Digital Signal Processor
	Dynamic Support Program
DSPI	Digital Speckle-Pattern Interferometry
DSPL	Display
DSR	Data Scanning and Routing
	Data Set Ready
	Data Storage and Retrieval
	Digit Storage Relay
	Digital Stepping Recorder
	Discriminating Selector Repeater
DSRV	Deep Submergence Rescue Vehicle [US Navy]
DSRW	Dry-Sand, Rubber-Wheel
DSS	Decision Support Software
	Decision Support System
	Deep-Space Station
	Deep Submergence System
	Department of Supply and Services [Canada]
	Digital Subsystem
	Dimethylsilapentanesulfonate
	Director of Statistical Services
	Direct Station Selection
	Discrete Sync System
	Disk Support System
	Dynamic Support System
DSSB	Double Single-Sideband
DSSC	Double-Sideband Suppressed Carrier
DSSc	Diploma in Sanitary Sciences
DSSCS	Defense Special Secure Communications System
DSSD	Double-Sided, Single-Density
DSSM	Dynamic Sequencing and Segregation Model
DSSP	Deep Submergence Systems Project [US Navy]
	Defense Standardization and Specification Program [US]

DSSPTO	Deep Submergence Systems Project Technical Office [US Navy]
DSSRG	Deep Submergence Systems Research Group [US Navy]
DSSV	Deep Submergence Search Vehicle [US Navy]
DST	Data Summary Tape
	Data System Test
	Daylight Saving Time
	District
	Drill Stem Test
DSTCD	District Code
DSTE	Data Subcarrier Terminal Equipment
DSTLN	Distillation
DSU	Data Storage Unit
	Data Synchronization Unit
	Device Switching Unit
	Disk Storage Unit
	Drum Storage Unit
DSV	Deep Submergence Vehicle
	Double Silk Varnish (Wire)
DSW	Data Status Word
	Device Status Word
	Direct Step On Wafer
DT	Data Terminal
	Data Transfer
	Data Translator
	Data Transmission
	Date
	Decay Time
	Deflection Temperature
	Deformation Twin
	Destructive Testing
	Dial Tone
	Differential Time
	Digital Technique
	Digital Translator
	Digital Transmission
	Discrete Time
	Double-Throw
	Double Torsion
	Doubling Time
	Dravon Tube
	Drop Top
	Drop Tower
	Drop Tube
	Dummy Target
	Dynamic Tear
D/T	Deuterium/Tritium (Ratio)
	Disk/Tape
DTA	Development Test Article
	Differential Thermal Analysis
	Differential Thermal Analyzer
	Dispersion-Toughened Alumina
DTACK	Data Transfer Acknowledgement
DTAF	Dichlorotriazinylaminofluorescein
DTAN	Dithiobisaminonaphthalene
DTARS	Digital Transmission and Routing System
DTAS	Data Transmission and Switching System
DTB	Decimal-to-Binary
DTBP	Di-tert-Butyl Peroxide

DTC	Data Transmission Center
	Data Transmission Control
	Decision Threshold Computer
	Differential Thermal Coating
	Diode Transistor Compound
	Doppler Translation Channel
DTCP	Diode Transistor Compound Pair
DTCS	Digital Test Command System
DTCU	Data Transmission Control Unit
DTD	Data Transfer Done
	Dimethyl Tin Difluoride
DTDP	Ditridecyl Phthalate
DTDS	Digital Television Display System
DT/DT	Drop Tube/Drop Tower (Facility)
DTE	Data Terminal Emulator
	Data Terminal Equipment
	Data Transmission Equipment
	Destructive Testing Equipment
	Development Test and Evaluation
	Digital Test Executive
	Dithioerythritol
DTENT	Date of Entry
DTF	Default-the-File
	Define the File
	Definite Tape File
DTFT	Discrete Time Fourier Transform
DTG	Display Transmission Generator
DTI	Data Transfer Interface
	Department of Trade and Industry [US]
DTIE	Division of Technical Information Extension [of USAEC]
DTL	Diode-Transister Logic
DTL/TTL	Diode-Transistor Logic/Transistor-Transistor Logic
DTM	Delay Timer Multiplier
	Device Test Module
	Digital Terrain Model
DTMB	David Taylor Model Basin [US Navy]
DTMF	Dual-Tone Multifrequency
DTMS	Digital Test Monitoring System
DTN	Defense Telephone Network
DTNB	Dithiobisnitrobenzoic Acid
DTNSRDC	David Taylor Naval Ship Research and Development Center [US]
DTP	Directory Tape Processor
	Dithiophosphate
DTPA	Diethylenetriaminepentaacetic Acid
DTPL	Domain Tip Propagation Logic
DTPMT	Date of Payment
DTR	Daily Transaction Reporting
	Data Telemetering Register
	Definite Time Relay
	Demand Totalizing Relay
	Diffusion Transfer
	Digital Telemetering Register
	Disposable Tape Reel
	Distribution Tape Reel
DTRE	Defense Telecommunications Research Establishment
DT/RSS	Data Transmission/Recording Subsystem
DTS	Data Tape Service

	Data Transmission System
	Defense Telephone Service [US]
	Defense Telephone System
	Digital Termination System
	Double-Throw Switch
DTSS	Dartmouth Time-Sharing System
DTS-W	Defense Telephone Service, Washington [US]
DTT	Dithiothreitol
DTTF	Digital Tape and Tape Facility
dTTP	Deoxyribonucleoside Triphosphate
DTTT	Dynamic Time-Temperature Transformation
DTTU	Data Transmission Terminal Unit
DTU	Data Transfer Unit
	Data Transmission Unit
	Digital Tape Unit
	Digital Telemetry Unit
	Digital Transmission Unit
DTUL	Deflection Temperature under Load
DTV	Digital Television
D/TV	Digital-to-Television
DTVC	Digital Transmission and Verification Converter
DTVM	Differential Thermocouple Voltmeter
DTWX	Dial Teletypewriter Exchange
DU	Dalhousie University [Canada]
	Depleted Uranium
	Display Unit
DUAL	Dynamic Universal Assembly Language
DUF	Diffusion under Film
DUMAND	Deep Underwater Muon and Neutrino Detector
DUMD	Deep Underwater Measuring Device
DUMS	Deep Unmanned Submersible
DUNC	Deep Underwater Nuclear Counter
DUNS	Data Universal Numbering System
DUP	Disk Utility Program
	Diundecyl Phthalate
	Duplicate
DUT	Device under Test
DUV	Data under Voice
	Deep Ultraviolet
DV	Differential Voltage
DVA	Digital Voice Announcer
	Dynamic Visual Acuity
DVARS	Doppler Velocity Altimeter Radar Set
DVB	Divinylbenzene
DVC	Device
DVD	Detail Velocity Display
DVDA	Dollar Volume Discount Agreement
DVES	DOD (= Department of Defense) Value Engineering Services [US]
DVESO	DOD (= Department of Defense) Value Engineering Services Office [US]
DVF	Digital Variable Frequency
DVI	Digital Video Interactive [also: DV-I]
DVLPT	Development
DVM	Digital Voltmeter
	Displaced Virtual Machine
DVOM	Digital Volt-Ohmmeter
DVS	Digital Video System
DVST	Direct-View Storage Tube

DVTL	Dovetail
DVX	Digital Voice Exchange
DW	Daisy Wheel
	Deadweight
	Developed Width
	Die Welding
	Distilled Water
	Double Word
	Dried Weight
	Drop Wire
	Dumbwaiter
DWA	Double-Wire Armor
DWBA	Distorted-Wave Born Approximation
DWC	Deadweight Capacity
DWCM	Dried Weight of Cell Mass
DWG	Drawing
DWICA	Deep Water Isotopic Current Analyzer
DWL	Designed Water Line
	Dominant Wavelength
	Dowel
	Downwind Localizer
DWP	Daisy Wheel Printer
	Digital Waveform Pattern
DWS	Disaster Warning System
DWSMC	Defense Weapons Systems Management Center [USDOD]
DWT	Deadweight (Tonnage)
	Drop-Weight Test
dwt	pennyweight [unit]
DWTT	Drop-Weight Tear Test
DWV	Drain, Waste and Ventilation System
DX	Destroyer Experimental
	Distance
	Distance Reception
	Document Transfer
	Duplex (Repeater)
DXC	Data Exchange Control
DXF	Data Transfer Facility
Dy	Dysprosium
DYANA	Dynamic Analyzer
DYG	Dyeing
DYN	Dynamics
	Dynamo
DYNASAR	Dynamic System Analyzer
DYNAVIS	Dynamic Video Display System
DYNM	Dynamotor
DYSAC	Dynamic Storage Analog Computer
DYSTAC	Dynamic Storage Analog Computer
DYSTAL	Dynamic Storage Allocation Language
DZ	Depleted Zone
dz	dozen [unit]
DZ PR	Dozen Pairs
DZA	Doppler Zeeman Analyzer

E

E	East
	Einsteinium

Elbow
Electrode
Emitter
Enamel
Energy
English
Engineer(ing)
Error
Excellence
Execution
Exponent
Exposure
Expression
Extrinsic
Glutamin Acid

e	exponent(ial)
EA	Each
	Easy Axis
	Educational Age
	Effective Address
	Electrically-Alterable
	Enumeration Area
	Exhaust Air
EAA	Electric Auto Association
	Engineer in Aeronautics and Astronautics
	Essential Amino Acid
	Ethylene-Acrylic Acid
	Experimental Aircraft Association
EAAC	Experimental Aircraft Association of Canada
EAB	Engineering Activity Board [of SAE]
EABRD	Electrically-Activated Band Release Device
EAC	Error Alert Control
	Experiment Apparatus Container
	External Affairs Canada
EACA	Eta-Aminocaproic Acid
EACC	Error Adaptive Control Computer
EACS	Electronic Automatic Chart System
EACSO	East Africa Common Services Organization
EAD	Electroacoustic Dewatering
EADAS	Engineering and Administration Data Acquisition System
EADI	Electronic Altitude Director Indicator
EAEC	European Airlines Electronic Committee
	European Atomic Energy Commission
EAEG	European Association of Exploration Geophysicists
EAEP	Energy Action Educational Project
EAES	European Atomic Energy Society
EAF	Electric Arc Furnace
EAGE	Electrical Aerospace Ground Equipment
EAGLE	Elevation Angle Guidance Landing Equipment
EAI	Engineering Advance Information
EAIR	Extended Area Instrumentation Radar
EAL	Electromagnetic Amplifying Lens
EAM	Electric Accounting Machine
	Electronic Accounting Machine
	Electronic Automatic Machine
EAN	European Article Number
EANDC	European-American Nuclear Data Committee

EANDRO	Electrically-Alterable Nondestructive Readout
EAP	Employee Assistance Program
	Equivalent Air Pressure
EAPROM	Electrically-Alterable Programmable Read-Only Memory
EAR	Electron Affinity Rule
	Eroded Area Rate
EARB	European Airlines Research Bureau
EARC	Extraordinary Administrative Radio Conference
EAROM	Electrically-Alterable Read-Only Memory
EARP	Environmental Assessment Review Program
EAS	Equivalent Airspeed
	Extended Area Service
	Experimental Army Satellite
	Extensive Air Shower
EASCOMINT	Extended Air Surveillance Communications Intercept [US Air Force]
EASCON	Electronics and Aerospace Systems Conference
EASE	Electrical Automatic Support Equipment
	Engineering Automatic System for Solving Equations
EASI	Electrical Accounting for the Security Industry
EASINET	European Area Sales and Information Network
EASL	Engineering Analysis and Simulation Language
	Experimental Assembly and Sterilization Laboratory [NASA]
EASP	Electric Arc Spraying
EAST	Experimental Army Satellite, Tactical
EASTT	Experimental Army Satellite Tactical Terminal
EASY	Early Acquisition System
	Efficient Assembly System
	Engine Analyzer System
EAT	Estimated Arrival Time
	Expected Approach Time
EATCS	European Association for Theoretical Computer Science
EAU	Engineer Aviation Unit
EAUTC	Engineer Aviation Unit Training Center
EAVE	Experimental Autonomous Vehicle
EAVE-EAST	Experimental Autonomous Vehicle - East
EAW	Equivalent Average Words
EAX	Electronic Automated Exchange
EB	Ebony
	Electron Beam
	Elementary Body
	Equal Brake
	Exposure Back
	Eye Bolt
EBAM	Electron Beam Addressed Memory
EB ASB	Ebony Asbestos
EBB	Extra Best Best
EBC	Electron Beam Cutting
	Enamel Bonded (Single) Cotton (Wire)
EBCD	Extended Binary-Coded Decimal

EBCDIC	Extended Binary-Coded Decimal Interchange Code
EBCHR	Electron Beam Cold Hearth Refined (Electrode)
EBDC	Enamel Bonded Double Cotton (Wire)
EBDP	Enamel Bonded Double Paper (Wire)
EBDS	Enamel Bonded Double Silk (Wire)
EBF	Electron Bombardment Furnace
	Externally Blown Flap
EBHC	Equated Busy-Hour Call
EBHSS	Electron Beam High Speed Scan
EBI	Equivalent Background Input
EBIC	Electron Beam-Induced Conduction
	Electron Beam-Induced Current
EBL	Electronic Bearing Line
EBM	Electromagnetic Billetmaker
	Electron Beam Machining
	Electronic Bearing Marker
EBMA	Electron Beam Microprobe Analysis
	Engine, Booster Maintenance Area
EBMD	Electron Beam Mode Discharge
EBMF	Electron Beam Microfabricating System
EBOR	Experimental Beryllium Oxide Reactor [USAEC]
EBP	Enamel (Single) Bonded Paper (Wire)
EBPA	Electron Beam Parametric Amplifier
EB-PVD	Electron Beam Physical Vapor Deposition
EBR	Electron Beam Recorder
	Electron Beam Remelting
	Electronic Beam Recording
	Epoxy Bridge Rectifier
	Experimental Breeder Reactor
EBRD	Electron Beam Rotating Disk
EBS	Enamel Bonded (Single) Silk (Wire)
EBSS	Earle's Balanced Salt Solution
EBT	Eccentric Bottom Tapping
	Electroless Bath Treatment
EBU	European Broadcasting Union
EBW	Electron Beam Welding
	Exploding Bridgewire
EBW-HV	Electron Beam Welding - High Vacuum
EBW-MV	Electron Beam Welding - Medium Vacuum
EBW-NV	Electron Beam Welding - Nonvacuum
EBWR	Experimental Boiling Water Reactor
EC	Economy Cartridge
	Edge Clamp
	Edge Connector
	Electrical Conductivity
	Electrical Conductor
	Electrification Council [US]
	Electrochemistry
	Electron Capture
	Electronic Calculator
	Enamel Covering
	End Cell
	Endothelial Cell
	Engineering Change
	Engineering Construction
	Environment Canada
	Error Correction
	Error Counter

	Esterified Cholesterol
	Ethyl Cellulose
	Ethyl Centralite
	European Community
	Extended Control (Mode)
ECA	Economic Cooperation Administration
	Electrical Contractors Association
	Electronic Confusion Area
	Engineering and Computer Science Association
	Engineering Critical Assessment
ECAC	Electromagnetic Compatibility Analysis Center [USDOD]
	European Civil Aviation Conference
ECAP	Electronic Circuit Analysis Program
	Energy-Compensated Atom Probe
ECARS	Electronic Coordinator and Readout System
ECB	Event Control Block
ECC	Eccentricity
	Economic Council of Canada
	Electron Channeling Contrast
	Electronic Calibration Center [NBS]
	Electronic Components Conference
	Energy Conservation Coalition
	Environmental Control Council
	Error Checking and Correction
	Error Correction Code
	European Communities Commission
ECCA	European Coil Coating Association
ECCANE	East Coast Conference on Aerospace and Navigational Electronics [US]
ECCM	Electronic Counter-Countermeasures
ECCMF	European Council of Chemical Manufacturers' Federations
ECCP	Engineering Concepts Curriculum Project [US]
ECCS	Electrolytic Chromium-Coated Steel
	Emergency Core Cooling System
ECD	Electron Capture Detector
	Energy Conversion Device
	Enhanced Color Display
ECDC	Electrochemical Diffused Collector
ECDM	Electrochemical Discharge Machining
ECE	Economic Commission for Europe
	Engineering Capacity Exchange
ECF	Emission Contribution Fraction
	Extracellular Fluid
ECFA	European Committee for Future Accelerators
ECG	Electrocardiogram
	Electrocardiograph
	Electrochemical Grinding
	Electro-Epitaxial Crystal Growth
ECGS	Endothelial Cell Growth Supplement
ECH	Earth Coverage Horizon Measurement
	Echo Cancellation Hybrid
	Echelon
ECHO	Electronic Computing Hospital-Oriented
ECI	Cast Iron Electrode
	Eddy Current Inspection
	Electrochemical Interface
	Engine Component Improvement

ECIO	European Conference on Integrated Optics
ECL	East Coast Laboratory [ESSA]
	Eddy Current Loss
	Electronics Components Laboratory
	Emitter-Coupled Logic
	Equipment Component List
ECLAT	European Conference on Laser Treatment (of Materials)
ECM	Electric Coding Machine
	Electrochemical Machining
	Electrochemical Milling
	Electronic Control Module
	Electronically-Commutated Motor
	Electronic Countermeasures
	European Common Market
	Extended Core Memory
	Extracellular Matrix (Coating)
ECMA	Electronic Computer Manufacturers Association
	European Computer Manufacturers Association
ECMC	Electric Cable Makers Confederation
ECME	Electronic Countermeasures Environment
ECMP	Electronic Countermeasures Program
ECMSA	Electronics Command Meteorological Support Agency [US Army]
ECMT	European Conference of Ministers of Transport
ECN	Engineering Change Notice
	Equivalent Carbon Number
ECNE	Electric Council of New England [US]
ECNG	East Central Nuclear Group [US]
ECO	Electron-Coupled Oscillator
	Electronic Central Office [US]
	Electronic Checkout
	Electronic Contact Operate
	Engineering Change Order
	European Coal Organization
ECOC	European Conference on Optical Communication
ECOM	Electric Computer-Originated Mail
	Electronics Command [US Army]
ECOMS	Early Capability Orbital Manned Station
ECON	Economics
	Economizer
	Economy
ECORQ	Economic Order Quantity
ECOS	Energy Conservation and Substitution Branch
ECOSS	European Conference on Surface Science
ECP	Electromagnetic Containerless Processing
	Electromagnetic Capability Program [US Air Force]
	Electronic Circuit Protector
	Electron Channeling Pattern
	Engineering Change Proposal
	External Casing Packer
ECPD	Engineers Council for Professional Development [US]
ECPI	Electronic Computer Programming Institute
ECPS	Expanded Control Program Store

ECR	Electron-Cyclotron Resonance
	Electronic Cash Register
	Electronic Control Receiver
	Electronic Control Relay
	Engineering Change Request
	Engineering Change Requirement
	External Control Register
ECRC	Electricity Council Research Center
	Electronic Components Reliability Center
	Electronic Components Research Center [US]
ECRD	Eddy Current Resonance Digitizing
ECS	Electrochemical Society
	Electroconvulsive Shock
	Electronic Control Switch
	Emergency Coolant System
	Emission Control System
	Engine Control System
	Environmental Conservation Service
	Environmental Control System
	Etched Circuits Society [US]
	European Committee for Standardization
	European Communication Satellite
	Evaporation Control System
	Expanded Control Store
	Extended Core Storage
ECSA	European Communications Security Agency
ECSC	European Community for Steel and Coal
ECSG	Electronic Connector Study Group
ECSS	European Conference on Surface Science
EC SW	End Cell Switch
ECT	Eddy Current Testing
	Emission-Controlled Tomography
	Equicohesive Temperature
	Evans Clear Tunnel
ECTFE	Ethylenechlorotrifluoroethylene
ECTL	Emitter-Coupled Transistor Logic
ECTN	Eastern Canada Telemetered Network
ECU	East Carolina University [US]
	Electronic Conversion Unit
	Environmental Control Unit
	European Currency Unit
ECUT	Energy Conservation and Utilization Technology
ECV	Enamel (Single) Cotton Varnish (Wire)
ECWU	Energy and Chemical Workers Union
ED	Edition
	Editor
	Electrical Differential
	Electrodeposition
	Electrodialysis
	Electrodynamics
	Electronic Device
	Engine Drive
	Engineering Drawing
	Expanded Display
	External Deflector
	External Device
	Extra-Low Dispersion (Glass)
EDA	Economic Development Administration [US]
	Electrical Development Association

	Electronic Design Automation
	Electronic Differential Analyzer
	Electronic Digital Analyzer
	Energy-Dispersive Analysis
	Ethylenediamine
EDAC	Error Detection and Correction
	Ethyldimethylaminopropylcarbodiimide
EDANS	Aminoethylaminonaphthalenesulfonic Acid
EDB	Economic Development Board [US]
	Educational Data Bank
	Ethylene Dibromide
EDC	Economic Development Committee
	Education Development Center
	Effective Dielectric Constant
	Electronic Desk Calculator
	Electronic Digital Computer
	Enamel Double Cotton (Wire)
	Energy Distribution Curve
	Error Detection and Correction
	Error Detection Code
	Ethylene Dichloride
	European Defense Community
	European Documentation Center
EDCC	Electronic Data Council of Canada
EDCL	Electrical Discharge Correction Laser
EDCOM	Editor and Compiler
EDCPF	Environmental Data Collection and Processing Facility
EDCS	Engineering Data Control System
EDCV	Enamel Double Cotton Varnish (Wire)
EDCW	External Device Control Word
EDD	Electronic Data Display
	Envelope Delay Distortion
EDDA	Ethylenediaminediacetic Acid
EDDF	Error Detection and Decision Feedback
EDDHA	Ethylenediaminediohydroxyphenylacetic Acid
EDDQ	Extra Deep Drawing Quality
EDE	Emergency Decelerating
	Emitter Dip Effect
	External Document Exchange
EDF	Engineering Data Form
EDFR	Effective Date of Federal Recognition
EDG	Edge
	Electrodischarge Grinding
	Exploratory Development Goals
EDGF	Electrodynamic Gradient Freeze
EDGE	Electronic Data Gathering Equipment
EDI	Electromagnetic Discharge Imaging
	Electron Diffraction Instrument
	Electronic Data Interchange (Network)
EDIAC	Electronic Display of Indexing Association and Content
EDIC	Exploration Drilling Incentive Program [Canada]
EDICT	Engineering Departmental Interface Control Technique
	Engineering Document Information Collection Technique
EDIF	Electronic Design Interchange Format
EDIP	European Defense Improvement Program [NATO]

EDIS	Engineering Data Information System
	Environmental Data and Information Service
	Exploratory Drill Incentives System
EDIT	Edition
	Editor
	Engineering Development Integration Test
EDITAR	Electronic Digital Tracking and Ranging
EDL	Electroless Discharge Lamp
EDLCC	Electronic Data Local Communications Central
EDLIN	Line Editor (Program)
EDM	Electrical Discharge Machining
	Electrodischarge Machining
	Electrodischarge Machine
EDMS	Electronic Device and Materials Symposium
	Engineering Data Microreproduction System
EDO	Economic Development Office
	Effective Diameter of Objective
	Engineering Duties Only
EDOS	Extended Disk Operating System
EDP	Electronic Data Processing
	Enterprise Development Program
	Experimental Development Program
EDPA	Exhibition Designers and Producers Association [US]
EDPC	Electronic Data Processing Center
EDPE	Electronic Data Processing Equipment
EDPI	Electronic Data Processing Institute
EDPM	Electronic Data Processing Machine
EDPS	Electronic Data Processing System
EDR	Electrodermal Reaction
	Equivalent Direct Radiation
EDRCC	Electronic Data Remote Communications Complex
EDRI	Electronic Distributors Research Institute [US]
EDRL	Effective Damage Risk Level
EDRS	Engineering Data Retrieval System
EDS	Electronic Data System
	Enamel Double Silk (Wire)
	Emergency Detection System
	Energy-Dispersive Spectrometry
	Energy-Dispersive Spectroscopy
	Energy-Dispersive System
	Engineering Data System [USDOD]
	Engineering Data Sheet
	Environmental Data Service [of ESSA]
EDSAC	Electronic Data Storage Automatic Calculator
	Electronic Data Storage Automatic Computer
	Electronic Delay Storage Automatic Computer
	Electronic Discrete Sequential Automatic Computer
EDSC	Engineering Data Service Center [US Air Force]
EDST	Eastern Daylight Saving Time
	Elastic Diaphragm Switch Technology
	Electric Diaphragm Switch Technique

EDSV	Enamel Double Silk Varnish (Wire)
EDT	Early Decay Time
	Eastern Daylight Time
	Electric Discharge Tube
	Electronic Data Transmission
	Engineer Design Test
EDTA	Ethylenediaminetetraacetate
	Ethylenediaminetetraacetic Acid
EDTCC	Electronic Data Transmission Communications Central
EDTN	Ethoxydichlorotriazinylnaphthalene
EDU	Electronic Display Unit
	Experimental Diving Unit
EDUC	Education
	Educator
EDUCOM	Educational Communications
EDVAC	Electronic Digital-Vernier Analog Computer
	Electronic Discrete Variable Automatic Calculator
	Electronic Discrete Variable Automatic Computer
EDWC	Electrical Discharge Wire Cutting
EDX	Energy-Dispersive X-Ray Analysis
EDXA	Energy-Dispersive X-Ray Analysis
EDXRD	Energy-Dispersive X-Ray Diffraction
EDXRF	Energy-Dispersive X-Ray Fluorescence
EE	Electrical Engineer(ing)
	Electrically-Erasable
	Electronics Engineering
	Environmental Engineer(ing)
	Errors Excepted
	External Environment
EEA	Electronic Engineering Association
	Ethyleneethyl Acrylate
	Euonymus Europaeus Acetone Powder
EEB	European Environmental Bureau
EEC	Electronic Engine Control
	European Economic Commission
	European Economic Community
	Evaporative Emission Control
EECL	Emitter-Emitter Coupled Logic
EED	Electrical and Electronics Division
	Electroexplosive Device
EEDF	Electron Energy Distribution Function
EEDQ	Ethoxycarbonylethoxydihydroquinoline
EEEI	Energy, Economics and Environment Institute
EEG	Electroencephalogram
	Electroencephalograph
	Environmental Engineers Group [Canada]
EEI	Edison Electric Institute [US]
EEIB	Environmental Engineering Intersociety Board
EELS	Electron Energy Loss Spectrometry
EEM	Earth Entry Module
	Electronic Equipment Monitoring
	Emission Electron Microscopy
EEMA	Electrical and Electronic Manufacturers Association
EEMAC	Electrical and Electronic Manufacturers Association of Canada
EEMJEB	Electrical and Electronic Manufacturers Joint Education Board
EEMTIC	Electrical and Electronic Measurement and Test Instruments Conference
EEO	Electroendosmosis
	Equal Employment Opportunity
EEOC	Equal Employment Opportunity Commission [US]
EEPNL	Estimated Effective Perceived Noise Level
EEPROM	Electrically-Erasable Programmable Read-Only Memory
E²PROM	Electrically-Erasable Programmable Read-Only Memory
EER	Electrolyte Electroreflectance
	Explosive Echo Ranging
EERI	Earthquake Engineering Research Institute [US]
EERJ	External Expansion Ramjet
EERL	Earthquake Engineering Research Laboratory [US]
	Electrical Engineering Research Laboratory
EEROM	Electrically-Erasable Read-Only Memory
E²ROM	Electrically-Erasable Read-Only Memory
EES	Engineering Experiment Station
EESMB	Electrical and Electronics Standards Management Board
EET	Electrical Engineering Technology
	Electronic Excitation Transfer
	Electronics Engineering Technology
	Environmental Engineering Technology
EETF	Electronic Environmental Test Facility
EEUA	Engineering Equipment Users Association
EEVT	Electrophoresis Equipment Verification Test
EEWD	Enhanced Exchange Wide Dial
EF	Each Face
	Emitter Follower
	Engineering Foundation
	Error-Free
	Evaluation Finder
	Extra Fine (Thread)
	Extremely Fine
EFA	Esterified Fatty Acid
EFAS	Electronic Flash Approach System
EFATCA	European Federation of Air Traffic Controllers Associations
EFB	Electric Flash Butt (Welding)
EFC	Electronic Frequency Control
	European Federation of Corrosion
	Expect Further Clearance
EFD	Equivalent Full Discharge
EFD	Electrofluid Dynamic (Process)
EFDAS	Epsilon Flight Data Acquisition System
EFE	Early Fuel Evaporation
EFF	Efficiency
	Effective(ness)
EFFCY	Efficiency
EFFE	European Federation of Flight Engineers
EFFLOR	Efflorescence
EFG	Edge-Defined Film-Fed Growth
	Electric Field Gradient
EFI	Electronic Fuel Injection

	Engineered, Furnished and Installed
	Error-Free Interval
EFIS	Electronic Flight Instrument System
EFL	Effective Focal Length
	Emitter Follower Logic
	Equivalent Focal Length
EFM	Electric-Field Monitor
EFNS	Educational Foundation for Nuclear Science
EFP	European Federation of Purchasing
EFPD	Equivalent Full Power Days
EFPH	Equivalent Full Power Hours
EFPW	European Federation for the Protection of Waters
EFR	Emerging Flux Regions
EFS	Electronic Frequency Selection
EFT	Earliest Finish Time
	Electronic Funds Transfer
EFTA	European Free Trade Association
EFTR	Ethernet Frequency Translator
EFTS	Electronic Funds Transfer System
EFW	Energy from Waste
EG	Edge Grain
	Electron Gas
	Electron Gun
	Ethylene Glycol
	(Single) Enamel (Single) Glass (Wire)
e.g.	(exempli gratia) - for example
EGA	Enhanced Graphics Adapter
	Evolved Gas Analysis
EGBD	Extrinsic Grain Boundary Dislocation
E/GCR	Extended Group Coded Recording
EGD	Electrogasdynamics
EGDN	Ethylene Glycol Dinitrate
E GER	East Germany
EGF	Electrodynamic Gradient Freeze
	Epidermal Growth Factor
EGI	End of Group Indicator
EGIF	Equipment Group Interface
EGM	Electrogel Machining
EGMBE	Ethylene Glycol Monobutylether
EGME	Ethylene Glycol Monomethylether
EGO	Eccentric Geophysical Observatory
EGPS	Extended General-Purpose Simulator
EGR	Exhaust Gas Recirculation
EGRESS	Emergency Global Rescue, Escape and Survival System [NASA]
	Evaluation of Glide Reentry Structural Systems
EGSE	Electrical Ground Support Equipment
EGSMA	Electrical Generating Systems Marketing Association
EGT	Exhaust Gas Temperature
EGTA	Ethylene Glycol Bistetraacetic Acid
EGW	Electrogas Welding
EH	Electrohydraulics
	Electron Hole
EHA	Ethyl Hexyl Alcohol
EHC	Electrochemical Hydrogen Cracking
EHD	Elastohydrodynamics
	Electrohydrodynamics
	Electron-Hole Drop

EHDA	Electrohydrodynamic Atomization
EHE	External Heat Exchanger
EHF	Electrohydraulic Forming
	Extra-High Frequency
	Extremely-High Frequency
EHM	Extended Hueckel Method
EHP	Effective Horsepower
	Electric Horsepower
	Electron Hole Pair
EHPA	Ethylhexyl Phosphoric Acid
EHT	Extra-High Tension
EHV	Extra-High Voltage
	Extremely High Voltage
EHW	Extremely High Water
EI	Electrical Insulation
	Electromagnetic Interference
	Electronic Interference
	End Injection
	Error Indicator
	Exposure Index
EIA	Electronic Industries Association [US]
	Energy Information Administration
	Engineering Industries Association [UK]
	Enzyme Immunoassay
EIAA	Electronic Industry Association of Alberta [Canada]
EIAC	Electronic Industries Association of Canada
EIB	Export-Import Bank
EIC	Electrical Insulation Conference
	Employment and Immigration Canada
	Engineer-in-Charge
	Engineering Institute of Canada
	Equipment Identification Code
EICAS	Engine Indication and Crew Alerting System
EICR	Eastern Interior Coal Region
EID	Electronic Instrument Digest
	Electron-Induced Desorption
	Electron-Stimulated Ion Desorption
EIDLT	Emergency Identification Light
EIDS	Electronic Information Delivery System
EIED	Electrically-Initiated Explosive Device
EIES	Electron Impact Emission Spectroscopy
EIL	Electron Injection Laser
EIM	Excitability Inducing Material
EIMO	Electronic Interface Management Office [US Navy]
EIN	Education Information Network
EIR	Engineering Investigation Request
EIRMA	European Industrial Research Management Association
EIRP	Effective Isotropic Radiated Power
EIS	Economic Information System
	Electrical Induction Steel
	Electromagnetic Intelligence System [US Air Force]
	End-Interruption Sequence
	Executive Information System
EIT	Engineer-in-Training
EITB	Engineering Industry Training Board
EJ	Electronic Jamming
EJC	Engineers Joint Council [US]

EJCC	Eastern Joint Computer Conference [US]
EJECT	Ejector
EKG	Electrocardiogram
	Electrocardiography
EKS	Electrocardiogram Simulator
EKY	Electrokymogram
EL	Education Level
	Elasticity
	Elastic Limit
	Electroluminescence
	Electronics Laboratory
	Elevation
	Elongation
ELAH	Earle's Lactalbumin Hydrolysate
ELATS	Expanded Litton Automated Test Set
ELC	Extra-Low Carbon
ELCD	Evaporative Loss Control Device
ELCO	Electrostatic Coaxial
	Eliminate and Count Coding
ELD	Economic Load Dispatching
	Edge-Lighted Display
	Electroless Deposition
	Edge-Lit Display
	Extra-Long Distance
ELDO	European Launcher Development Organization
ELDOR	Electron-Electron Double Resonance
ELE	Equivalent Logic Element
ELEC	Electric(al)
	Electrician
	Electricity
	European League for Economic Cooperation
ELECOM	Electronic Computer
ELECT	Electrolyte
	Electrolytic
ELECTN	Electrician
ELECTR	Electric(al)
	Electrician
	Electricity
	Electronic
ELED	Entry-Level Employee Development
ELEED	Elastic Low-Energy Electron Diffraction
ELEM	Element
ELEPLTG	Electroplating
ELEV	Elevation
	Elevator
ELEX	Electronics
ELF	Electroluminescent Ferroelectricity
	Ellipsometry, Low-Field
	Extensible Language Facility
	Extremely Low Frequency
ELFA	Electric Light Fittings Association [UK]
ELG	Electrolytic Grinding
ELGMT	Ejector-Launcher Guided-Missile Transporter
ELGRA	European Low Gravity Research Association
ELH	Enol-Lactone Hydrolase
ELI	English Language Interpreter
	Equitable Life Interpreter
	Extended Lubrication Interval
	Extensible Language I

	Extra-Low Interstitial
ELINT	Electromagnetic Intelligence
	Electronic Intelligence
ELIP	Electrostatic Latent Image Photography
ELISA	Enzyme-Linked Immunosorbent Assay
ELK	External Link
ELL	Eccentric Leveling Lugs
ELLIPT	Ellipticity
ELME	Emitter Location Method
ELMS	Elastic Loop Mobility System
	Engineering Lunar Model Surface
ELNI	Ethernet Local Network Interconnect
ELOISE	European Large Orbiting Instrumentation for Solar Experiments
ELONG	Elongation
ELOP	Extended Logic Plan
ELP	Electropolishing
	English Language Program
ELPC	Electroluminescent - Photoconductive
ELPG	Electric Light and Power Group
ELPH	Elliptical Head
ELPO	Electrodeposition
ELR	Engineering Laboratory Report
ELRAC	Electronic Reconnaissance Accessory (Set)
ELRO	Electronics Logistics Research Office [US Army]
ELS	Earth Landing System
	Economic Lot Size
	Electroluminescence Screen
	Energy Loss Spectroscopy
	Equal Load Sharing
ELSA	Energy Loss Spectral Analysis
ELSB	Edge-Lighted Status Board
ELSIE	Electronic Letter Sorting and Indicating Equipment
	Electronic Signalling and Indicating Equipment
ELSS	Extravehicular Life Support System
ELT	Electronic Level Transducer
	Element
	Emergency Location Transmitter
ELTAD	Emergency Location Transmitter, Automatic Deployable
ELTAF	Emergency Location Transmitter, Automatic Fixed
ELTAP	Emergency Location Transmitter, Automatic Portable
ELTR	Emergency Location Transmitter/Receiver
ELU	Existing Carrier Line-Up
ELV	Earth Launch Vehicle
	Electrically-Operated Valve
	Expendable Launch Vehicle
ELVAC	Electric Furnace Melting, Ladle Refining, Vacuum Degassing and Continuous Casting
ELW	Extreme Low Water
ELWAR	Electronic Warfare
EM	Effective Modulus
	Efficiency Modulation
	Electromagnet(ism)
	Electromechanical

	Electromechanics	**EMCON**	Emission Control
	Electromigration	**EMCP**	Electromagnetic Compatibility Program
	Electron Microscopy	**EMCS**	Electromagnetic Compatibility
	Electrophoretic Mobility		Standardization
	End-of-Medium (Character)		Energy Management and Controls Society
	Engineering Manual		Energy Monitoring and Control System
	Engineering Material	**EMCTP**	Electromagnetic Compatibility Test Plan
	Engineering Memo(randum)	**EMD**	Electric-Motor Driven
	Engineering Model		Electronic Map Display
	Engineer of Mines		Extractive Metallurgy Division [of AIME]
	Enlisted Men	**EMDI**	Energy Management Display Indicator
	Epitaxial Mesa	**EME**	Electromagnetic Energy
	Ethoxylated Monoglyceride	**EMEC**	Electronic Maintenance Engineering Center
	Extensible Machine		[US Army]
	Extractive Metallurgy	**EMER**	Emergency
EMA	Effective Medium Approximation	**EMEX**	Equatorial Mesoscale Experiment
	Electronic Mail Association [US]	**EMF**	Effective Mass Filter
	Electronic Missile Acquisition		Electromagnetic Forming
	Electronics Manufacturers Association [US]		Electromotive Force
	Electronics Materiel Agency [US Army]		European Monetary Fund
	Emergency Minerals Administration [US]		Evolving Magnetic Features
	Engineered Materials Abstracts	**EMG**	Electromyography
	Ethylene Methyl Acrylate	**EMH**	Expedited Message Handling
	Excavator Makers Association	**EMHS**	Electronic Message Handling System
	Extended Mercury Autocoder	**EMI**	Early Manufacturing Involvement
EMAA	Engineering Materials Achievement Award		Electromagnetic Interference
	[of ASM]		End-of-Message Indicator
	Ethylene Methacrylic Acid	**EMIC**	Electronic Materials Information Center [of
EMABC	Electronics Manufacturers Association of		RRE]
	British Columbia [Canada]	**EMICE**	Electromagnetic Interference Control
EMAC	Electronics Manufacturers Association of		Engineer
	Canada	**EMINT**	Electromagnetic Intelligence
EMAR	Experimental Memory-Address Register	**EMI/RFI**	Electromagnetic Interference/Radio
EMAS	Electromagnetic Acoustic System		Frequency Interference
EMAT	Electromagnetic Acoustic (Wave) Transducer	**EMIRS**	Electrochemically-Modulated Infrared
EMATS	Emergency Mission Automatic Transmission		Reflectance Spectroscopy
	System	**EMIS**	Educational Management Information
EMB	Engineering in Medicine and Biology		System
	Eosine Methylene Blue		Electromagnetic Intelligence System
	Experimental Model Basin [US Navy]	**EMIT**	Engineering Management Information
EMBC	European Molecular Biology Conference		Technique
EMBERS	Emergency Bed Request System	**EMKO**	Ethyl Michler's Ketone Oxime
EMBET	Error Model Best Estimate of Trajectory	**EML**	Electromagnetic Levitator
EMBL	European Molecular Biology Laboratory		Engineering Mechanics Laboratory [of NBS]
	[FRG]		Equipment Modification List
EMBO	European Molecular Biology Organization		Exterior Metal Loss
EMBS	Eta-Maleimidocaproyloxysuccinimide	**EMM**	Ebers-Moll Model
EMC	Elastomeric-Molding Tooling Compound		Electromagnetic Measurement
	Electromagnetic Casting	**EMMA**	Electron Manual Metal-Arc
	Electromagnetic Compatibility		Electron Microscope Microanalyzer
	Electronic Materials Conference		Electron Microscopy and Microanalysis
	Engineered Military Circuit		Eye Movement Measuring Apparatus
	Engineering Manpower Commission	**EMO**	Emergency Measures Organization
	Ensemble Monte Carlo	**EMOS**	Earth's Mean Orbital Speed
	Equilibrium Moisture Content	**EMP**	Electromagnetic Pulse
	European Military Communication		Electromechanical Power
	Excess Minority Carrier		Emission Pattern
EMCAB	Electromagnetic Compatibility Advisory	**EMPF**	Electronic Manufacturing Productivity
	Board		Facility
EMCCC	European Military Communication	**EMPIRE**	Early Manned Planetary-Interplanetary
	Coordinating Committee		Roundtrip Expedition

EMPR	Ethernet Multiport Repeater
EMPRA	Emergency Multiple Person Rescue Apparatus
EMPS	Ethernet Multiport Station
EMQ	Economic Manufacturing Quantity
	Electromagnetic Quiet
EMR	Electrolytic Metal Recovery
	Electromagnetic Radiation
	Electromechanical Research
	(Department of) Energy, Mines and Resources [Canada]
	Engine Mature Ratio
EMRA	Electrical Manufacturers Representative Association
EMRIC	Educational Media Research Information Center [US]
EMRL	Engineering Mechanics Research Laboratory [US]
EMRS	European Materials Research Society
EMS	Earth and Mineral Sciences
	Electromagnetic Stirring
	Electromagnetic Submarine
	Electromagnetic Surveillance
	Electromagnetic Susceptibility
	Electronic Management System
	Electronic Message Service
	Electronic Monitoring System
	Electron Microscopy Society
	Energy Management System
	Engineering Management Society
	Engine Management System [US Army]
	Enhanced Memory Specification
	Environmental Mutagen Society
	Ethylmethane Sulfonate
	European Monetary System
	Expanded Memory Specification
EMSA	Electron Microscopy Society of America
	Electronic Materiel Support Agency [US Army]
EMSC	Electrical Manufacturers Standards Council [of NEMA]
EMT	Electrical Mechanical Tubing
	Electrical Metal Tubing
	Electromagnetic Tubing
	Electronic Maintenance Technician
EMTECH	Electromagnetic Technology
EMTTF	Equivalent Mean Time to Failure
EMU	Electrical-Multiple Unit
	Electromagnetic Unit
	European Monetary Unit
	Expanded Memory Unit
	Extravehicular Maneuvering Unit
	Extravehicular Mobility Unit
EMW	Elecromagnetic Warfare
ENAA	Epithermal Neutron Activation Analysis
ENAM	Enamel
ENC	Enclosure
	Equivalent Noise Charge
ENCL	Enclosure
ENCY	Encyclopedia
ENCYC	Encyclopedia

END	Equivalent Neutral Density
ENDOR	Electron-Nuclear Double Resonance
ENE	East-Northeast
	Estimated Net Energy
ENEA	European Nuclear Energy Agreement
	European Nuclear Energy Agency
ENET	Engineering Network
ENFIA	Exchange Network Facilities for Interstate Access
ENFOR	Energy from the Forest Program [Canada]
ENG	Engine
	Engineer(ing)
	England
	English
	Engraver
	Engraving
ENGL	England
	English
ENGR	Engineer
	Engraver
	Engraving
ENGRG	Engineering
ENIAC	Electronic Numerical Integrator and Calculator
	Electronic Numerical Integrator and Computer
ENL	Enamel
ENQ	Enqueue
	Enquiry (Character)
ENR	Equivalent Noise Resistance
ENS	Electrostatic Nonmetallic Separator
ENSI	Equivalent Noise Sideband Input
ENSIP	Engine Structural Integrity Program [US Air Force]
ENSO	El Nino Southern Oscillation
ENSP	Engineering Specification
ENT	Emergency Negative Thrust
	Entrance
	Equivalent Noise Temperature
ENTAC	Engine-Teleguide Anti-Char
ENTC	Engine Negative Torque Control
ENV	Envelope
EO	Engineering Order
	Errors and Omissions
	Ethylene Oxide
	Executive Organ
EOA	End of Address
EOAP	Earth Observation Aircraft Program [US]
EOAR	European Office of Aerospace Research
EOB	End of Block
EOC	Emergency Operating Center
EOD	End of Data
	Explosive Ordnance Disposal
EODD	Electro-Optical Digital Deflector
EOE	End of Extent
	Equal Opportunity Employer
	Error and Omission Excepted
E&OE	Error and Omission Excepted
EOF	End of File
	Energy Optimizing Furnace
EOG	Electrooculography

	Electroolfactogram	**EPAC**	Expanded Polystyrene Association of Canada
EOI	End of Inquiry		
EOJ	End of Job	**EPAM**	Elementary Perceiver and Memorizer
EOL	End of Life	**EPBM**	Enhanced Probability-Based Matching
	Expression-Oriented Language	**EPBS**	Earth-Pressure Balanced Shield
EOLM	Electron-Optical Light Modulator	**EPC**	Easy Processing Channel
EOLN	End of Line		Electric Propulsion Conference
EOLT	End of Logical Tape		Electronic Power Conditioner
EOM	End of Message		Electronic Program Control
	End of Month		Elementary Processing Center
EOP	End of Program		Engineering, Procurement and Construction
	End Output		Environmental Protection Control
	Even-Odd Predominance		Emergency Planning Canada
EOQ	Economic Order Quantity		European Patent Convention
EOQC	European Organization for Quality Control		Evaporative Pattern Casting
EOR	Electro-Optical Research	**EPCCS**	Emergency Positive Control Communications System
	End of Record		
	End of Reel	**EPCM**	Electropulse Chemical Machining
	Enhanced Oil Recovery		Engineering, Procurement and Construction Management
	Explosive Ordnance Reconnaissance		
EORS	Emergency Oil Spill Response System	**EPCO**	Engine Parts Coordinating Office
EOS	Electrical Overstress	**EPD**	Electric Power Database
	Electronic Office Service		Electric Power Distribution
	Electro-Optical System		Electronic Proximity Detector
	Electrophoresis Operations in Space	**EPDM**	Ethylene-Propylene-Dimonomer
	End of Selection		Ethylene-Propylene Diene Monomer
	Equation of State	**EPDT**	Estimated Project Duration Time
EOSAT	Earth Observation Satellite	**EPE**	Electron-Plastic Effect
EOT	End of Tape		Energetic Particles Explorer (Satellite) [NASA]
	End of Task		
	End of Transmission	**EPEC**	Emerson Programmer, Evaluator and Controller
	Engine Order Telegraph		
EOV	End of Volume	**EPFM**	Elastic-Plastic Fracture Mechanics
EOW	Engineering Order Wire	**EPG**	Electrostatic Particle Guide
	Engine over the Wing	**EPGA**	Emergency Petroleum and Gas Administration [US]
EP	Electrical Polarization		
	Electrical Properties	**EPH**	Electric Process Heating
	Electric Power	**EPI**	Earth Path Indicator
	Electron Probe		Electronic Position Indicator
	Electrophoresis		Electron Probe Instrument
	Electroplating		Emergency Position Indicator
	Electropulse		Emulsion Polymers Institute
	Elongated Punch		Environmental Policy Institute
	Emulsion Polymer	**EPIB**	Emergency Position Indicating Beacon
	End of Program	**EPIRB**	Emergency Position Indicator Radio Beacons
	Engineering Personnel		
	Epitaxial Planar	**EPIC**	Earth-Pointing Instrument Carrier
	Epoxy		Electrical Properties Information Center [US]
	Etched Plate		
	Ethylene Propylene		Electronic Photochromic Integrating CRT (= Cathode Ray Tube)
	Expanded Polystyrene		
	Explosionproof		Electronic Products Information Center
	Extended Performance		Evaluator Programmer Integrated Circuit
	Extended Play		Extended Performance and Increased Capability
	Extra Protection		
	Extra Pulse	**EPIRBS**	Emergency Position Indicating Radio Beacon System
	Extreme Pressure		
EPA	Enhanced Performance Architecture	**EPLA**	Electronics Precedence List Agency
	Environmental Protection Agency [US]	**EPLANS**	Engineering, Planning and Analysis System
	Expanded Polystyrene Association	**EPM**	Earth-Probe-Mars [NASA]
EPAA	Epithermal Neutron Activation Analysis		Economic Performance Monitoring

	External Polarization Modulation
EPMA	Electronic Parts Manufacturers Association
	Electron Probe Microanalysis
	Electron Probe Microanalyzer
EPMAU	Expected Present Multi-Attribute Utility
EPN	Epoxy Phenol Novolac
	External Priority Number
EPNdB	Equivalent Perceived Noise Level in Decibels
EPNL	Equivalent Perceived Noise Level
EPNS	Electroplated Nickel Silver
EPO	Erythropoietin
	European Patent Office
	European Patent Organization
	Examination Procedure Outline
EPOA	Eastcoast Petroleum Operators Association [US]
EPP	End Plate Potential
EPPS	Ethylpiperazinepropanesulfonic Acid
EPR	Electrochemical Potentiokinetic Reactivation
	Electronic Planning and Research
	Electron Paramagnetic Resonance
	Engine Pressure Ratio
	Ethylene Propylene Rubber
EPRI	Electric Power Research Institute [US]
EPROM	Electrically Programmable Read-Only Memory
	Erasable Programmable Read-Only Memory
EPS	Electrical Power Subsystem
	Electrical Power Supply
	Electromagnetic Position Sensor
	Emergency Power Supply
	Engineering Performance Standards
	Environmental Protection Service [Canada]
	Equilibrium Problem Solver
	Equivalent Prior Sample
	European Physical Society
	Expandable Polystyrene
	Expanded Polystyrene
EPSCS	Enhanced Private Switched Communications Service
EPSP	Excitatory Postsynaptic Potential
EPSS	Experimental Packet Switching System
EPT	Electromagnetic Propagation Tool
	Electrostatic Printing Tube
	Ethylene Propylene Terpolymer
EPTA	Expanded Program of Technical Assistance [of UN]
EPTE	Existed Prior to Entry
EPU	Electrical Power Unit
	Emergency Power Unit
EPUT	Events per Unit Time
EQ	End-Quench
	Equal(izer)
	Equation
	Equator
	Equivalent
EQCC	Entry Query Control Console
EQL	Equal
EQMT	Equipment
EQP	Equipment

EQPT	Equipment
EQSP	Equally Spaced
EQUALANT	Equatorial Atlantic
EQUATE	Electronic Quality Assurance Test Equipment
EQUIP	Equipment
	Equipment Usage Information Program
EQUIV	Equivalent
ER	Echo Ranging
	Effectiveness Report
	Electrical Resistance
	Electronic Ram
	Energy Resources
	Enhanced Radiation
	Error
	Error Recovery
	External Resistance
E&R	Engineering and Repair
Er	Erbium
ERA	Electrical Research Association [UK]
	Electronic Reading Automation
	Electronic Representatives Association
	Energy Reduction Analysis
	Engineering Research Associate
ERAP	Error Recording and Analysis Procedure
ERAS	Electronic Reconnaissance Access Set
ERASER	Elevated Radiation-Seeking Rocket
ERB	Equipment Review Board
ERBM	Extended Range Ballistic Missile
ERC	Electronics Research Center [of NASA]
	Engineering Research Center
	Equatorial Ring Current
	Error Retry Count
ERCB	Energy Resources Conservation Board
ERCEM	European Regional Conference on Electron Microscopy
ERCR	Electronics Retina Computing Reader
ERCS	Emergency Rocket Communications System
ERCSS	European Regional Communications Satellite System
ERD	Elastic Recoil Detection
	Electronic Research Directorate [US Air Force]
	Emergency Return Device
	Equivalent Residual Dose
ERDA	Economic Regional Development Agreements
	Electronics Research and Development Activity [of US Army]
	Energy Research and Development Adminstration [US]
ERDAF	Energy Research and Development in Agriculture and Food
ERDE	Explosive Research and Development Establishment [UK]
ERDL	Electronics Research and Development Laboratory [US Army]
ERDS	Earth Resource Data System
	Environmental Recording Data Set
ERE	Edison Responsive Environment
EREP	Earth Resources Experimental Package

	Environmental Recording, Editing and Printing	ERSOS	Earth Resources Survey Operational Satellite
ERF	Epoxy Resins Formulators Division [of SPI]	ERSP	European Remote Sensing Program
ERF	Error Function	ERSR	Equipment Reliability Status Report
ERFC	Error Function Complement	ERSS	Earth Resources Survey Satellite
ERFPI	Extended Range Floating Point Interpretative System		Earth Resources Satellite System
ERG	Electroretinography	ERTS	Earth Resources Technology Satellite
	Energy Research Group [Canada]	ERW	Electric Resistance Welding
ERGS	Electronic Route Guidance System	ERX	Electronic Remote Switching
	Enroute Guidance System	ES	Earth Station
ERI	Energy Research Institute		Echo Sounding
ERIC	Educational Research Information Center		Econometric Society
	Educational Resources Information Center [USOE]		Electrochemical Society [US]
	Electronic Remote and Independent Control		Electromagnetic Storage
	Energy Rate Input Controller		Electromagnetic Switch(ing)
ERIE	Environmental Resistance Inherent in Equipment		Electronic Standard
ERIS	Electrostatic Reflex Ion Source		Electronic Switch(ing)
	Engineering Resins Information System		Electron Synchrotron [USAEC]
	Exoatmospheric Reentry Vehicle Interceptor System		Electrostatics
ERL	Echo Return Loss		Electrostatic Storage
	Energy Research Laboratory [of CANMET]		Emission Spectroscopy
	Environmental Research Laboratory		Enamel (Single) Silk (Wire)
	ESSA Research Laboratories		Experiment(al) Station
ERM	Earth Reentry Module		Expert Systems
	Earth Return Module		Extruded Shape
	Elastic-Reservoir Molding	Es	Einsteinium
	Electronic Recording Machine	ESA	Ecological Society of America
ERMA	Electrical Reproduction Method of Accounting		Electric Spark Alloying
	Expansion Rate Measuring Apparatus		Electronic Surge Arrester
ERN	Engineering Release Notice		Electrostatic Analyzer
ERNIE	Electronic Random Numbering and Indicating Equipment		Entomological Society of America
			Ethernet Station Adapter
ERO	Emergency Repair Overseer		EURATOM Supply Agency
EROM	Erasable Read-Only Memory		European Space Agency
EROP	Extensions and Restrictions of Operators		Explosive Safe Area [NASA]
EROS	Earth Resources Observation Satellite	ESAIRA	Electronically-Scanned Airborne Intercept Radar
	Eliminate Range O (Zero) System	ESAPS	Experimental Strain Analysis Processing System
	Experimental Reflector Orbit Shot		
ERP	Effective Radiated Power	ESAR	Electronically-Scanned Array Radar
	Electronics Research Paper		Electronically-Steerable Array Radar
	Emergency Response Program	ESARS	Earth Surveillance and Rendezvous Simulator
	Environmental Research Paper		
	Environmental Research Project		Employment Service Automated Reporting System
	Equivalent Reduction Potential		
	Error Recovery Procedure	ESB	Electrical Stimulation of the Brain
ERPLD	Extended-Range Phase-Locked Demodulator		Electrical Standards Board [of USASI]
ERR	Error	ESC	Electronic Spark Control
ERS	Earth Resources Satellite		Electronic Systems Command [US Navy]
	Economic Research Service		Electroslag Casting
	Electric Resistant Strain		Electrostatic Compatibility
	Engineering Reprographic Society		Engineering and Scientific Computing
	Environmental Research Satellite [NASA]		Engineering Service Circuit
	Experimental Research Society [US]		Engineering Society of Cincinnati [US]
ER&S	Exploratory Research and Study		Equipment Serviceability Criteria
ERSA	Electronics Research Supply Agency		Escape (Character)
ERSO	Electronics Research and Service Organization [Taiwan]		Escutcheon
			European Space Conference
		ESCA	Electron Spectroscopy for Chemical Analysis
		ESCAP	Economic and Social Commission for Asia and the Pacific
		ESCAPE	Expansion Symbolic Compiling Assembly Program for Engineering

ESCC	External Stress Corrosion Cracking
ESCD	Extended System Contents Directory
ESCES	Experimental Satellite Communication Earth Station
ESCOE	Engineering Societies Commission on Energy
ESCR	Environmental Stress Crack Resistance
ESD	Echo Sounding Device
	Electronic Systems Division [US Air Force
	Electron-Simulated Desorption
	Electrostatic Discharge
	Electrostatic Dissipation
	Electrostatic Storage Deflection
	Emergency Shutdown
	Ending Sequence, Done
	Engineering Society of Detroit [US]
	Environmental Services Division
	Estimated Standard Deviation
	External Symbol Dictionary
ESDAC	European Space Data Center [of ESRO]
ESDIAD	Electron-Stimulated Desorption Ion Angular Distribution
ESDS	Entry-Sequenced Data Set
	Environmental Satellite Data System
ESE	East-Southeast
	Electrical Support Equipment
	Electronic System Evaluator
ESEM	Electron Spin Envelope Modulation
ESF	Ethynylphenoxysulfone
	European Science Foundation
	Extended Super Frame
ESFA	Emergency Solid Fuel Administration [US]
ESG	Electrically-Suspended Gyroscope
	Electronic Sweep Generator
	Electrostatic Gyroscope
	(Old) English Standard Gauge
ESH	Electric Strip Heater
	Equivalent Standard Hours
ESHAC	Electric Space Heating and Air Conditioning [of IEEE]
ESHT	Electroslag Hot Topping
ESHU	Emergency Ship Handling Unit
ESI	Electron Spectroscopic Imaging
	End of Segment Indicator
	Engineering and Scientific Interpreter
	Equivalent-Step Index
	Externally-Specified Index
ESID	Electron-Stimulated Ion Desorption
ESIP	Employee Savings Investment Plan
ESL	Electron Beam Switched Latch
	Expected Significance Level
	Evans Signal Laboratory [US Army]
ESLAB	European Space (Research) Laboratory [of ESRO]
ESLO	European Satellite Launching Organization
ESM	Elastomeric Shield Material
	Equivalent Standard Minute
ESMA	Electronic Sales and Marketing Association
E-SMC	Epoxy-Matrix Sheet Molding Compound
ESME	Excited State Mass Energy
ESMS	East Sullivan Monzonitic Stock

ESNE	Engineering Societies of New England [US]
ESO	European Southern Observatory
ESOC	European Space Operations Center [FRG]
ESONE	European Standards of Nuclear Electronics
ESP	Electrical Submersible Pump
	Electronic Short Pathfinder
	Electron Spin Polarization
	Electrosensitive Programming
	Electrosonic Profiler
	Experimental Solids Proposal
	Extrasensory Perception
	Extravehicular Support Pack [NASA]
E&SP	Equipment and Spare Parts
ESPAR	Electronically Steerable Phased Array Radar
ESPI	Electronic Speckle-Pattern Interferometry
ESPOD	Electronic Systems Precision Orbit Determination
ESPOL	Executive System Problem-Oriented Language
ESPRIT	European Strategic Program for Research in Information Technology [also: Esprit]
ESR	Effective Signal Radiated
	Electronic Scanning Radar
	Electron Spin Resonance
	Electroslag Refining
	Electroslag Remelting
	Equivalent Series Resistance
	Erythocyte Sedimentation Rate
	Exchangeable Sodium Ratio
	External Standard Ratio
ESRANGE	European Space Range [of ESRO]
ESRF	European Synchrotron Radiation Facility
ESRI	Engineering and Statistical Research Institute
ESRIN	European Space Research Institute [of ESRO]
ESRO	European Space Research Organization
ESRT	Electroslag Refining Technology
ESS	Earle's Salt Solution
	Earth System Sciences
	Electronic Sequence Switching
	Electronic Switching System
	Emplaced Scientific Station [NASA]
	Environmental Stress Screening
	Environmental Survey Satellite
	European Symposium for Stereology
ESSA	Electronic Scanning and Stabilizing Antenna
	Environmental Science Services Administration [US]
	Environmental Survey Satellite
ESSDERC	European Solid-State Device Research Conference
ESSG	Engineer Strategic Studies Group [US Army]
ESSU	Electronic Selective Switching Unit
EST	Earliest Start Time
	Eastern Standard Time
	Electrolytic Sewage Treatment
	Electronic Spark Timing
	Electrostatic Storage Tube

	Emerging Sciences and Technologies
	Enlistment Screening Test
	Establishment
	Estimation
ESTA	Ethernet Station Adapter
ESTC	European Space Tribology Center [UK]
ESTEC	European Space Agency Technical Center
	European Space Research Technology Center [of ESRO]
ESTRAC	European Space Satellite Tracking and Telemetry Network [of ESRO]
ESTU	Electronic System Test Unit
EST WT	Estimated Weight
ESU	Electrostatic Unit
ESV	Enamel (Single) Silk Varnish (Wire)
ESW	Electroslag Welding
	Error Status Word
ESWL	Equivalent Single Wheel Load
	Extracorporal Shock-Wave Lithotripsy
ESWS	Earth Satellite Weapon System
ET	Eddy Current Testing
	Edge-Triggered
	Electrical Time
	Electronic Transformer
	Electron Transfer
	Embryo Transfer
	EMF (= Electromotive Force) - Temperature
	Energy Technology
	Engineering Test
	Environmental Technology
	Ephemeris Time
	Equal Taper
	Erector Transporter
	Ethylenedithiotetrathiafulvalene
	Extraterrestrial
	Extruded Tube
ETA	Electrothermal Analyzer
	Estimated Time of Arrival
	European Tube Association
et al	(et alii; et aliae) - and others
ETAC	Environmental Technical Applications Center [US Air Force]
ETAC-1	Equilibrium Transfer Alkylating Cross-Link
ETAP	Extended Technical Assistance Program [UN]
ETB	End of Transmission Block
	Estimated Time of Berthing
ETC	Electronic Time Card
	Electrothermal (Integrated) Circuit
	Ethylene Carbonate
	European Translations Center
etc	(et cetera) - and so forth
ETCG	Elapsed-Time Code Generator
ETD	Equivalent Transmission Density
	Estimated Time of Departure
ETE	Electrothermal Excitation
	Estimated Time Enroute
ETECG	Electronics Test Equipment Coordination Group [US]
ETF	Engine Test Facility
ETFE	Ethylenetetrafluoroethylene

ETH	Ether
	Ethyl
ETH ACET	Ethyl Acetate
ETHEL	European Tritium Handling Experimental Laboratory
ETI	Extraterrestrial Intelligence
ETIM	Elapsed Time
ETL	Electrical Testing Laboratories [US]
	Electrotechnical Laboratories [Japan]
	Ending Tape Label
	Etching by Transmitted Light
ETLS	Extruded Tunnel Lining System
ETM	Electronic Test and Measurement
ETN	Equipment Table Nomenclature
ETO	Ethylene Oxide
E to E	End to End
ETOG	European Technical Operations Group
ETOS	Extended Tape Operating System
ETP	Effluent Treatment Plant
	Electrical Tough Pitch
	Electrolytic Tough Pitch (Copper)
ETPAE	Ethyl-Terminated Polyarylene Ether
ETR	Eastern Test Range [NASA]
	Electron Transfer Reaction
	Engineering Test Reactor
	Experimental Test Reactor
ETS	Electronic Telegraph System
	Electronic Test Set
	Engineering Test Satellite
	Environmental Technology Seminar
ETSA	Ethyltrimethylsilylacetate
ETSAL	Electronic Terms for Space Age Language
ETSD	Enhanced Thermionically Supported Discharge
et seq	(et sequentes; et sequentia) - and the following
ETSQ	Electrical Time, Superquick
ETT	Electrothermal Thrusters
ETTA	Ethanediylidenetetrakisacetic Acid
ETU	Enhanced Telephone Unit
ETV	Educational Television
	Electric Transfer Vehicle
ETVM	Electrostatic Transistorized Voltmeter
ETX	End of Text
EU	End User
	Electronic Unit
	Entropy Unit
Eu	Euronorm
	Europium
EUCAPA	European Capsules Association
EUFMC	Electric Utilities Fleet Managers Conference
EUM-AFTN	European Mediterranean Aeronautical Fixed Telecommunications Network
EUMC	Enameled Utensil Manufacturers Council
EUPC	Electric Utility Planning Council
EUR	Europe(an)
EURATOM	European Atomic Energy Community
EURECA	European Research Coordination Agency [also: Eureca]
	European Retrievable Carrier
EUREM	European Congress on Electron Microscopy

EUREX	Enriched Uranium Extraction
EUROBIT	European Association of Manufacturers of Business Machines and Data Processing Equipment
EUROCAE	European Organization of Civil Aviation Electronics
EUROCOMP	European Computing Congress
EUROCON	European Convention [of IEEE]
EUROCONTROL	European Organization for the Safety of Air Navigation
EURODOC	European Joint Documentation Service
EUROMICRO	European Association for Microprocessing and Microprogramming
EURONET	European Public (Data) Network
EuropEx	European Information Center for Explosion Protection [also: EUROPEX]
EUROSPACE	European Industrial Space Research Group
EUSA	Electrical Utilities Safety Organization
EUT	Equipment under Test
EUV	Extreme Ultraviolet
EV	Efficient Vulcanization
	Equalizer Valve
eV	Electron Volt
EVA	Electronic Velocity Analyzer
	Ethylene Vinyl Acetate
	Extravehicular Activity
EVAP	Evaporation
	Evaporator
EVATMI	European Vinyl Asbestos Tile Manufacturers Institute
EVC	Electric Vehicle Council
EVCS	Extravehicular Communication System [NASA]
EVDE	External Visual Display Equipment [NASA]
EVM	Electronic Voltmeter
EVOM	Electronic Voltohmmeter
EVOP	Evolutionary Operation
EVR	Electronic Video Recorder
	Electronic Video Recording
EVTCM	Expected Value Terminal Capacity Matrix
EW	Early Warning
	Electronic Warfare
	Electroslag Welding
	Electrowinning
	Erftwerk (Coating Process)
	Tungsten Electrode
EWASER	Electromagnetic Wave Amplification by Simulated Emission of Radiation [also: Ewaser]
EWC	Electric Water Cooler
EWCAS	Early Warning and Control Aircraft System
EWCS	European Wideband Communications System
EWES	Engineering Waterways Experiment Station [US Army]
EWF	Electrical Wholesalers Federation
	Equivalent-Weight Factor
EWI	Edison Welding Institute [US]
EWO	Engineering Work Order
EWP	Exploding Wire Phenomena
EWR	Early Warning Radar

	Electromagnetic Wave Resistivity
EWTAT	Early Warning Threat Analysis Display
EWTMIP	European Wideband Transmission Media Improvement Program
EWTR	Electronic Warfare Test Range [US Air Force]
EWWS	ESSA Weather Wire Service [US]
EX	Examination
	Example
	Exponent
	Excess
	Exchange
	Execution
	Exercise
EXAFS	Extended X-Ray Absorption Fine Structure (Spectroscopy)
EXAM	Examination
EXC	Excavation
	Exception
	Excitation
	Exciter
EXCELS	Expanded Communications Electronics System [USDOD]
EXCH	Exchange
EXCL	Exclusion
EXCO	Exfoliation Corrosion
EXCP	Execute Channel Program
EXCTR	Exciter
EXCVTG	Excavating
EXD	Exchange Degeneracy
EXDAMS	Extended Debugging and Monitoring System
EXEC	Execute (Statement)
	Execution
	Executor
	Executive
EXELFS	Extended Energy Loss Fine Structure
EXES	Electron-Induced X-Ray Emission Spectroscopy
EXF	Ex Factory
	External Function
EXH	Exhaust
EX-HY	Extra-Heavy
EXIST	Existence
EXLST	Exit List
EXP	Expansion
	Experiment
	Explosion
	Exponent(ial)
	Export
	Express
	Expression
	Expulsion
EXPL	Explosives
EXPT	Experiment
EXR	Execute and Repeat
exsec	exterior secant
EXSTA	Experimental Station
EXT	Extension
	Extinction
	Extinguisher
	Exterior

	External
EXTERRA	Extraterrestrial Research Agency [US Army]
EXTR	Extrusion
EXTRADOP	Extended Range Doppler
EXTRN	External Reference
EXW	Explosion Welding
EZ	Electrical Zero

F

F	Facsimile
	Fahrenheit
	Fairing
	Farad
	Faraday
	February
	Filament
	File
	Filter
	Fire
	Flag
	Fluorescence
	Fluorine
	Force
	Formula
	French
	Frequency
	Friday
	Fuel
	Function
	Phenylalanine
f	fair
	fetch
	fixed
	flat
FA	Factory Automation
	Fatty Acid
	Fluorescent Antibody
	Forced-Air
	Free Air
	Full Adder
	Fully-Automatic
F/A	Fuel/Air (Ratio)
FAA	Federal Aviation Administration [US]
FAAAS	Fellow of the American Association for the Advancement of Science
	Fellow of the American Academy of Arts and Sciences
FAAB	Frequency Allocation Advisory Board
FAAD	Forward Area Air Defense
FAAD-LOS	Forward Area Air Defense - Line-of-Sight
FAAD-LOS(H)	Forward Area Air Defense - Line-of-Sight (Heavy)
FAAP	Federal Aid Airport Program [US]
FAAR	Forward Area Alerting Radar
FAAS	Fellow of the American Academy of Arts and Sciences
FAB	Fabricate

	Fabrication
	Fast Atom Bombardment
FABG	Fabricating
FABMDS	Field Army Ballistic Missile Defense System [US Army]
FABMS	Fast Atom Bombardment Mass Spectroscopy
FABS	Formulated Abstracting Service
FABU	Fuel Additive Blender Unit
FAC	Facsimile
	Fast Affinity Chromatography
	Field Accelerator
	Forward Air Control [US Air Force]
	Forward Air Controller
	Frequency Allocation Committee
FACE	Federation of Associations on the Canadian Environment
	Field-Alterable Control Element
	Field Artillery Computer Equipment
FACET	Fluid Amplifier Control Engine Test
FACISCOM	Finance and Controller Information System Command
FACR	Fellow of the American College of Radiology
FACS	Fine Attitude Control System
	Floating-Decimal Abstact Coding System
	Frequency Allocation Coordinating Subcommittee
	Fully Automatic Compiling System
FACSC	Frequency Allocation Coordinating Subcommittee, Canada
FACSI	Fast Access Coded Small Images
FACSIM	Facsimile
FACSS	Federation of Analytical Chemistry and Spectroscopy Societies
FACT	Facility for the Analysis of Chemical Thermodynamics
	Flexible Automatic Circuit Tester
	Flight Acceptance Composite Test
	Ford Anodized-Aluminum Corrosion Test
	Foundation for Advanced Computer Technology
	Fully Automatic Cataloguing Technique
	Fully Automatic Compiler Translator
	Fully Automatic Compiling Technique
FACTS	Facsimile Transmission System
	Financing Alternative Computer Terminal System
FACTY	Factory
FAD	Flavin Adenine Dinucleotide
	Floating Add
FADA	Federation of Automobile Dealers Associations
FADAC	Field Artillery Digital Automatic Computer [US Army]
FADEC	Full Authority Digital Electronic Control
FADES	Fuselage Analysis and Design Synthesis
FADIC	Field Artillery Digital Computer
FADP	Finnish Association for Data Processing
FADS	Force Administration Data System
FAE	Final Approach Equipment
FAES	Federated American Engineering Societies

FAETUA	Fleet Airborne Electronics Unit, Atlantic
FAETUP	Fleet Airborne Electronics Unit, Pacific
FAF	Final Approach Fix
	Fly Away Field
FAFPS	Five-Axis Fiber Placement System
FAGC	Fast Automatic Gain Control
FAGS	Federation of Astronomical and Geophysical Services [of IAPO]
	Fellow of the American Geographical Society
FAHQMT	Fully-Automatic High Quality Machine Translation
FAHR	Fahrenheit
FAI	Fresh Air Intake
FAIA	Fellow of the American Institute of Architects
FAIC	Fellow of the Architectural Institute of Canada
FAICE	Fellow of the Institute of Civil Engineers
FAIEE	Fellow of the American Institute of Electrical Engineers [now: FIEEE]
FAIMME	Fellow of the American Institute of Mining and Metallurgical Engineers
FAIR	Fairing
	Fast Access Information Retrieval
FAIRS	Federal Aviation Information Retrieval System
FAK	File Access Key
FAL	Frequency Allocation List
FALT	FADAC Automatic Logic Tester
FALTRAN	FORTRAN-to-ALGOL Translator
FAM	Fast Auxiliary Memory
FAME	Fatty Acid Methyl Ester
	Florida Association of Marine Explorers [US]
FAMECE	Family of Military Engineer Construction Equipment
FAMEM	Federation of Associations of Mining Equipment Manufacturers
FAMOS	Fleet Application of Meteorological Observations for Satellites
	Floating-Gate Avalanche-Injection Metal-Oxide Semiconductor
FAMSO	Methyl Methylsulfinylmethyl Sulfide
FAMU	Fuel Additive Mixture Unit
FANTAC	Fighter Analysis Tactical Air Combat
FAO	Finish All Over
	Food and Agricultural Organization [UN]
FAP	Floating-Point Arithmetic Package
	FORTRAN Assembly Program
	Frequency Allocation Panel
FAPIG	First Atomic Power Industry Group [Japan]
FAPS	Fellow of the American Physical Society
FAPUS	Frequency Allocation Panel, United States
FAPUS-MCEB	Frequency Allocation Panel, United States - Military Communications Electronics Board
FAQ	Fair Average Quality
FAR	Failure Analysis Report
	Flight Aptitude Rating
	Federal Airworthiness Regulations [US]
FARADA	Failure Rate Data

F/A	Fuel-Air (Ratio)
FARET	Fast Reactor Experiment Test
FARR	Forward Area Refuelling and Rearming
FARS	Fatal Accident Reporting System
FAS	Faculty of Administrative Studies
	Fastener
	Ferrous Aluminum Sulfate
	Field Advisory Service
	Free Alongside Ship
	Frequency Allocations Subcommittee
FASCE	Fellow of the American Society of Civil Engineers
FASE	Fundamentally Analyzable Simplified English
FASEB	Federation of American Societies for Experimental Biology
FASM	Fellow of the American Society for Metals
FASME	Fellow of the American Society of Mechanical Engineers
FAST	Facility for Automatic Sorting and Testing
	Fairchild Advanced Schottky Technology
	Fast Automatic Shuttle Transfer
	Field-Data Applications, Systems and Techniques
	Flexible Algebraic Scientific Translator
	Fluorescent Antibody Staining Technique
	Formal Auto-Indexing of Scientific Texts
	Formula and Statement Translator
	Forward Air Strike
FASTAC	Furnace Aerosol Sampling Technique with Autocalibration
FASTAR	Frequency Angle Scanning, Tracking and Ranging
FASTI	Fast Access to Systems Technical Information
FASWC	Fleet Antisubmarine Warfare Command
FAT	Factory Acceptance Test
	Fixed Analyzer Transmission
	Fluorescent Antibody Test
	Formula Assembler Language
FATAL	FADAC Automatic Test Analysis Language
FATCAT	Film and Television Correlation Assessment Technique
FATDL	Frequency and Time-Division Data Link
FATE	Fusing and Arming Test and Evaluation
	Fusing and Arming Test Experiment
fath	fathom [unit]
FATIMA	Fatigue Indicating Meter Attachment
FATR	Fixed Autotransformer
FATT	Fracture Appearance Transition Temperature
FAU	Florida Atlantic University [US]
	Frequency Allocation and Uses
FAUSTUS	Frame-Activated Unified Story Understanding System
FAV	Fast-Acting Valve
FAW	Fiber Areal Weight
FAWS	Flight Advisory Weather Service
FAX	Facsimile
FB	Fixed Block
	Flat Bottom
	Furnace Brazing

	Fuse Block
F&B	Fire and Bilge
FBA	Fixed-Block Architecture
	Furnace Bottom Ash
FBB	Fusion Breeder Blanket
FBC	Fluidized Bed Combustion
	Fully-Buffered Channel
FBCR	Fluidized Bed Combustion Residue
FBCR	Fluidized Bed Control Rod
FBDB	Federal Business Development Bank [Canada]
FBFM	Feedback Frequency Modulation
FBI	Federal Bureau of Investigation [US]
	Federation of British Industry [now: CBI]
FBLO	Foreign Branch Liaison Office
FBM	Feet Board Measure
	Fleet Ballistic Missile
FBMP	Fleet Ballistic Missile Program
FBMWS	Fleet Ballistic Missile Weapon System
FBO	Fixed Base Operator
FBOA	Fellow of the British Optical Association
FBP	Final Boiling Point
FBR	Fast Breeder Reactor
	Fast Burst Reactor
	Feedback Resistance
	Fiber
FBRL	Final Bomb Release Line
FBS	Fetal Bovine Serum
FBT	Functional Board Tester
FBW	Fly-by-Wire
FC	Faraday Cup
	Ferrite Core
	Fiber Glass Cover
	File Code
	File Conversion
	Fire Control
	First Class
	Flexible Connection
	Foam Cell
	Font-Change (Character)
	Foot-Candle [unit]
	Free Cholesterol
	Front Connected
	Fuel Cell
	Function Code
	Furnace-Cooled
4/C	Four Conductor
FCA	Faraday Cup Array
	Fire Control Area
	Frequency Control and Analysis
FCASI	Fellow of the Canadian Aeronautics and Space Institute
FCAW	Flux Cored Arc Welding
FCAW-EG	Flux Cored Arc Welding - Electrogas
FCB	File Control Block
	Forms Control Buffer
FCC	Face-Centered Cubic
	Federal Communications Commission [US]
	Flat Conductor Cable
	Flight Control Center
	Flight Control Computer

	Fluid Catalytic Cracking
	Food Contaminants Commission [US]
FCCTS	Federal COBOL Compiler Testing Service [US]
FCD	Failure-Correction Decoding
	Frequency Compression Demodulator
FCDR	Failure Cause Data Report
FCEA	Federal Capital Equipment Authority
FCF	Fuel Cycle Facility [USAEC]
FCFS	First-Come, First-Served
FCFT	Fixed Cost, Fixed Time (Estimate)
FCG	Fatigue Crack Growth
	Federal Coordination Group [US]
FCI	Functional Configuration Identification
FCIC	Fellow of the Chemical Institute of Canada
FCIM	Farm, Construction and Industrial Machinery
FCIR	Fatigue Crack Initiation Resistance
FCL	Feedback Control Loop
	Full Container Load
FCLTY	Facility
FCM	First Class Mail
FCMD	Fire Command
FCMJ	Federation of Canadian Manufacturers in Japan
FCMV	Fuel Consuming Motor Vehicle
FCO	Field Change Order
	Flight Clearance Office
	Functional Checkout
FCOH	Flight Controllers Operations Handbook
FCP	Fatigue Crack Propagation
	File Control Procedure
	File Control Processor
FCPC	Fleet Computer Programming Center
FCPP	Fuel-Cell Power Plant
FCPR	Fatigue Crack Propagation Resistance
FCR	Fast Cycling Resin
	Final Configuration Review
	Fire Control Radar
	Flight Control System
	Fuse Current Rating
FCS	Fellow of the Chemical Society
	Fellow of the College of Sciences
	Field Control Strain
	Fire Control System
	Flexible Clamping System
	Frame Checking Sequence
FCSLE	Forecastle
FCSRT	Fellow of the Canadian Society of Radiological Technicians
FCSS	Fire Control Sight System
FCST	Federal Council for Science and Technology [US]
FCT	Face-Centered Tetragonal
	Filament Center Tap
	File Control Table
	Function
FCTC	Fuel Centerline Thermocouple
FCTN	Function
FCTRY	Factory
FCWG	Frequency Coordination Working Group [US]

FD	Feed
	File Description
	File Directory
	Finite Difference
	Fire Damper
	Fire Department
	Flexible Disk
	Flight Director
	Floor Drain
	Floppy Disk
	Focal Distance
	Forced Draft
	Forging Direction
	Free Discharge
	Frequency Demodulator
	Frequency Diversity
	Frequency Division
	Frequency Doubler
	Full Duplex
	Fund
F/D	Focal Length to Diameter (Ratio)
F&D	Facilities and Design
	Freight and Demurrage
Fd	Ferredoxin
FDA	Food and Drug Administration [US]
FDAA	Fluorenyldiacetamide
FDAI	Flight Director Attitude Indicator
FDAS	Frequency Distribution Analysis Sheet
FDAU	Flight Data Acquisition Unit
FDB	Field Dynamic Braking
	Forced Draft Blower
FDC	Fire Department Connection
	Flight Director Computer
FDD	Formatted Data Disk
FDDL	Frequency Division Data Link
FDDS	Flight Data Distribution System
FDE	Field Decelerator
FDEP	Flight Data Entry Panel
FDFM	Frequency Division/Frequency Modulation
FDG	Fractional Doppler Gate
	Funding
FDI	Field Discharge
	Flight Detector Indicator
	Frequency Domain Instrument
FDL	Fast Deployment Logistics
	Flight Dynamic Laboratory [US Air Force]
FDLS	Fast Deployment Logistics Ship [US Navy]
FDM	Form Description Macro
	Frequency Division Multiplex(er)
FDMA	Frequency Division Multiple Access
FDM/FM	Frequency-Division-Multiplexed/Frequency Modulated
FDMS	Factory Data-Management System
FDN	Foundation
FDNR	Frequency-Dependent Negative Resistance
	Frequency-Dependent Negative Resistor
FDOS	Floppy Disk Operating System
FDP	Fast Digital Processor
	Fibrin Degradation Product
	Field-Developed Program
	Filter Drainage Protection
	Form Description Program
	Fructose Diphosphate
	Future Data Processor
FDPC	Federal Data Processing Center
FDR	Facility Development Research
	Feeder
	Final Design Review
	Flight Data Recorder
	Frequency Diversity Radar
	Frequency Domain Reflectometry
F DR	Fire Door
FDRY	Foundry
FDS	Fallout Decay Simulation
	Fluid Distribution System
	Frame Difference Signal
	Frequency Division Separator
FDSS	Fine Digital Sun Sensor
FDSSA	Fine Digital Sun Sensor Assembly
FDSU	Flight Data Storage Unit
FDT	Flowing-Gas Detonation Tube
	Formatted Data Tape
	Full Duplex Teletype
FDU	Form Description Utility
FDV	Fault Detect Verification
FDX	Full Duplex
FDY	Foundry
FE	Field Engineer
	Finite Element
	Flash Evaporation
	Forest Engineer(ing)
	Format Effector (Character)
	Free Electron
	Further Education
Fe	(ferrum) - Iron
FEA	Failure Effect Analysis
	Federal Energy Administration [US]
	Finite Element Analysis
	Finite Element Analyzer
FeAA	Ferric Acetylacetonate
FEAO	Federation of European and American Organizations
FEARO	Federal Environmental Assessment Review Office
FEAT	Frequency of Every Allowable Term
FEATH	Feathery
FEB	February
	Functional Electronic Block
FEBS	Federation of European Biochemical Societies
FEC	Forward Error Correction
	Front-End Computer
FECO	Fringes of Equal Chromatic Order
FED	Federal
	Field Emission Deposition
	Freeze Etching Device
FEDA	Farm Equipment Dealers Association
FEDC	Federation of Engineering Design Consultants
	Fusion Engineering Design Center
FEDSTRIP	Federal Standard Requisitioning and Issue Procedure

FEED	Field Electron Energy Distribution
	Floating Electrode Effect Development
FEEP	Field Emission Electric Propulsion System
FEF	Foundry Educational Foundation [US]
	Fusion Energy Foundation
FEFO	First-Ended, First-Out
FEG	Field Emission Gun
FEI	Financial Executive Institute [US]
FEIC	Fellow of the Engineering Institute of Canada
FEL	Free-Electron Laser
FEM	Field (Electron) Emission Microscopy
	Final Effluent Monitor
	Finite Element Method
FEMA	Farm Equipment Manufacturers Association
	Federal Emergency Management Agency
	Foundry Equipment and Materials Association
FEMF	Floating Electronic Maintenance Facility
FEMP	Fusion Engineering Materials Program
FEO	Field Engineering Order
FEP	Financial Evaluation Program
	Fluorinated Ethylene Propylene
	Front-End Processor
FEPA	Federation of European Producers of Abrasives
FEPE	Full Energy Peak Efficiency
FEPEM	Federation of European Petroleum Equipment Manufacturers
FER	Forward Engine Room
	Fusion Engineering Reactor
	Fusion Experimental Reactor
FERD	Fuel Element Rupture Detector
FERIC	Forest Engineering Research Institute of Canada
FERN	Forest Ecosystem Research Network
FERPIC	Ferroelectric Ceramic Picture Device
FERSI	Flat Earth Research Society International
FES	Field Emission Spectroscopy
	Finite Element Solver
	Florida Engineering Society [US]
	Fundamental Electrical Standards
FESA	Federation of Engineering and Scientific Associations
FESE	Field-Enhanced Secondary Emission
FESEM	Field Emission Scanning Electron Microscopy
FET	Federal Excise Tax
	Field-Effect Transistor
FETO	Factory Equipment Transfer Order
FETT	Field-Effect Tetrode Transistor
FEUS	French Engineers in the United States
FF	Fast-Fast (Wave)
	Fast Forward
	Filtration Fraction
	Fixed Focus
	Flip-Flop
	Form Feed (Character)
	Fuel Flow
	Full Field
F&F	Fire and Flushing

FFA	Free Fatty Acids
	Free from Alongside
FFAG	Fixed-Field Alternating Gradient
FFAR	Folding Fin Air Rocket
FFB	French Forces Broadcasting
FFC	Flexible Flatness Control
	Flip-Flop Complimentary
FFD	Failure Flux Density
FFEC	Field-Free Emission Current
FFI	Fuel Flow Indicator
FFL	Finished Floor
	First Financial Language
	Front Focal Length
FFMC	Fine Fuel Moisture Code
FFMS	Fast Fourier Mass Spectrometry
	Free Flight Melt Spinning
FFP	Fixed Fee Procurement
FFR	Fellow of the Faculty of Radiology
	Flat Face Rolling System
FFSA	Field Functional System Assembly
FFSF	Full Fat Soy Flour
FFT	Fast Fourier Transform
	Flicker Fusion Threshold
FFTA	Fast Fourier Transform Analyzer
FFTF	Fast Flux Test Facility [of USAEC]
FFTR	Fast Flux Test Reactor
FFTV	Free Flight Test Vehicle
FG	Fiberglass
	Filament Ground
	Fine Grain
	Flat Grain
	Foreground
	Forging
	Friction Glaze
	Fuel Gas
	Function Generator
FGAN	Fertilizer Grade Ammonium Nitrate
FGCS	Flight Guidance and Control System
FGD	Flue Gas Desulfurization
4GL	Fourth Generation Language
FGR	Fertility and Genetics Research
	Finger
FGS	Fellow of the Geographical Society
	Fellow of the Geological Society
FGSA	Fellow of the Geographical Society of America
	Fellow of the Geological Society of America
FH	Fire Hose
	Flat Head
	Full-Hole (Mining)
FHA	Federal Highway Administration
	Federal Housing Administration
FHC	Fire Hose Cabinet
FHD	Fixed Head Disk
	Flat Head
FHDA	Fir and Hemlock Door Association
FHP	Fractional Horsepower
	Friction Horsepower
FHR	Fire Hose Rack
FHS	Forward Head Shield
FHSA	Federal Hazardous Substances Act [US]

FHT	Fully Heat-Treated
FHWA	Federal Highway Administration [US]
FHY	Fire Hydrant
FI	Fault Identification
	Field Intensity
	Fixed Interval
	Flow Indicator
FIA	Field Information Agency [US]
	Flame Ionization Analysis
	Flow Injection Analysis
	Fluoroimmunoassay
	Forging Industry Association [US]
FIAeS	Fellow of the Institute of Aeronautical Sciences
FIAS	Fellow of the Institute of Aeronautical Sciences
FIAT	Field Information Agency, Technical [US]
FIB	Fellow of the Institute of Biology
	Focused Ion Beam
Fib	Fibrin
FIBR	Fiber
FIC	Federal Information Center [US]
	Fellow of the Institute of Chemists
	First-in-Chain
	Flight Information Center
	Frequency Interference Control
FICC	Frequency Interference Control Center [US Air Force]
FICE	Fellow of the Institution of Civil Engineers
FICON	File Conversion
FICS	Forecasting and Inventory Control System
FID	Flame Ionization Detector
	Forecasts-In-Depth
FIDAC	Film Input to Digital Automatic Computer
FIDO	Fog Investigation and Dispersal Operation
FIED	Field Ionization Energy Distribution
FIEE	Fellow of the Institute of Electrical Engineers
FIEEE	Fellow of the Institute of Electrical and Electronic Engineers
FIER	Foundation for Instrumentation Education and Research [of ISA]
FIERF	Forging Industry Educational and Research Foundation [US]
FIFO	First-In, First-Out
	Floating Input - Floating Output
FIFRA	Federal Insecticide, Fungicide and Rodenticide Act
FIG	Figure
	FORTH Interest Group [FORTH is a programming language]
FIGE	Field Inversion Gel Electrophoresis
FIGLU	Formiminoglutamic Acid
FIGS	Figures
	Figure Shift
FIIA	Fellow of the Institute of Industrial Administration
FIIG	Federal Item Identification Guide
FIILS	Full Integrity Instrument Landing System
FIL	Filament
	Fillet

	Fillister
FILA	Filament
FILH	Fillister Head
FILL	Filling
FILO	First-In, Last-Out
FILS	Flare-Scan Instrument Landing System
FIM	Fault Isolation Meter
	Field Intensity Meter
	Field Ion Microscopy
FIMATE	Factory-Installed Maintenance Automatic Test Equipment
FIME	Fellow of the Institute of Mechanical Engineers
FIMechE	Fellow of the Institute of Mechanical Engineers
FIML	Full-Information Maximum Likelihood
FIMM	Fellow of the Institute of Mining and Metallurgy
FIN	Finish
	Finland
	Finnish
FINAC	Fast Interline Non-Active Automatic Control
FINAL	Financial Analysis Language
FINCO	Finance Committee [of ISO]
FIND	File Interrogation of Nineteen Hundred Data
FINE	Fighter Inertial Navigation System
FINESSE	Fusion Integrated Nuclear Experiment Strategy Study Effort
FINSHG	Finishing
FInstP	Fellow of the Institute of Physics
FInstPet	Fellow of the Institute of Petroleum
FIO	Florida Institute of Oceanography [US]
	Free In and Out
FIOA	File Input/Output Area
FIOP	FORTRAN Input-Output Package
FIOR	Fluid Iron Ore Reduction
FIP	Fairly Important Person
	Finance Image Processor
	Fully-Ionized Plasma
FIPS	Federal Information Processing Standards
FIQS	Fellow of the Institute of Quantity Surveyors
FIR	Far Infrared
	Finite Impulse Response
	Firkin
	Flight Information Region
	Fuel Indicator Reading
FIRA	Foreign Investment Review Act
	Foreign Investment Review Agency
	Furniture Industry Research Association [UK]
FIRD	Fast-Induced Radioactivity Decay
FIRE	Fellow of the Institute of Radio Engineers
	Flight Investigation Reentry Environment
FIRETRAC	Firing Error Trajectory Recorder and Computer
FIRL	Franklin Institute Research Laboratories [US]
FIRM	Financial Information for Resource Management
FIRMS	Forecasting Information Retrieval of Management System

FIRST	Fabrication of Inflatable Reentry Structures for Test		Follow-the-Leader Feedback
	Financial Information Reporting System	**FLG**	Flag
	Fragment Information Retrieval of Structures		Flange
			Flooring
FIS	Field Information System	**FLGD**	Flanged
	Flight Information Service	**FLH**	Flat Head
FISH	Fully-Instrumented Submersible Housing	**FLHLS**	Flashless
FIST	Fault Isolation by Semi-Automatic Techniques	**FLHP**	Filter High Pass
		FLI	Field Length Indicator
FIT	Failure in Time	**FLIH**	First-Level Interrupt Handler
	Fast Installation Technique	**FLIMBAL**	Floated Inertial Measurement Ball
	Fault Isolation Test	**FLIP**	Film Library Instantaneous Presentation
	Federal Income Tax		Flight-Launched Infrared Probe
	File Information Table		Floated Instrument Platform
FITC	Fluorescein Isothiocyanate		Floated Lightweight Inertial Platform
FIX	Fixture		Floating Indexed Point
FIXS	Fixtures		Floating Point Interpretative Program
FJCC	Fall Joint Computer Conference [US]	**FLIR**	Forward-Looking Infrared
FJI	Federal Job Information	**FLITE**	Federal Legal Information through Electronics
FJSRL	Frank J. Seiler Research Laboratory [US Air Force]	**FLO**	Functional Line Organization
		FLO CON	Floating Container
FK	Fixture Key	**FLODAC**	Fluid-Operated Digital Automatic Computer
FKI	Federation of Korean Industry	**FLOLS**	Fresnel Lens Optical Landing System
FL	Field Length	**FLOOD**	Fleet Observation of Oceanographic Data [US Navy]
	Field Lens		
	Flashing	**FLOP**	Floating Octal Point
	Flight Level		Floating-Point Operation
	Floor	**FLOSOST**	Fluorine One-Stage Orbital Space Truck
	Floor Line	**FLOTRAN**	Flowcharting FORTRAN
	Florida [US]	**FLOW**	Flow Welding
	Fluid	**FLOX**	Fluorine-Liquid Oxygen
	Fluoroleucine	**fl oz**	fluid ounce
	Focal Length	**FLP**	Filter Low Pass
	Foot-Lambert [unit]		Foreign Language Program
	Free Length	**FLPL**	FORTRAN List Processing Language
FLA	Fabric Laminators Association [US]	**FLR**	Fluoroleucine Resistance
Fla	Florida [US]		Forward-Looking Radar
FLAC	Florida Automatic Computer	**FLSC**	Flexible Linear-Shaped Charge
FLAM	Forward-Launched Aerodynamic Missile	**FLSP**	Flame Spraying
FLAMR	Forward-Looking Advanced Multilobe Radar		Fluorescein Labeled Serum Protein
FLAP	Flores Assembly Program	**FLT**	Fault Location Technology
FLAR	Forward-Looking Airborne Radar		Filter
FLB	Flow Brazing		Fleet
FLBE	Filter Band Elimination		Flight
FLBIN	Floating-Point Binary		Float
FLBP	Filter Bandpass	**FLTR**	Fusible Link Top Register
FLBR	Fusible Link Bottom Register	**FLTR**	Filter
FLC	Forming Limit Curve	**FLTSATCOM**	Fleet Satellite Communication
FLCS	Fiberoptics Low-Cost System	**FLU**	Full Line-Up
FLD	Field	**FLUOR**	Fluorescence
	Formability Limit Diagram	**FLUORES**	Fluorescence
	Forming Limit Diagram	**FLW**	Feetlot Waste
FLDEC	Floating Point Decimal	**FLWF**	Feetlot Waste Fiber
fl dr	fluid dram [unit]	**FM**	Face Measurement
FLEA	Flux Logic Element Array		Feedback Mechanism
FLECHT	Full-Length Emergency Cooling Heat Transfer		Ferromagnet(ism)
			Field Manual
FLEEP	Feeble Beep		File Maintenance
FLEX	Flexible		Fine Measurement
FLF	Fixed Length Field		Fire Main

	Fracture Mechanics
	Frequency Modulation
	Frequency Multiplex
Fm	Fermium
fm	fathom [unit]
FMA	Fabricators and Manufacturers Association
FMAC	Frequency Management Advisory Council
FMAR	Ferromagnetic Antiresonance
FMAS	Florida Marine Aquarium Society [US]
FMB	Foundation for Microbiology
FMC	Federal Maritime Commission
	Flatness Measuring and Control
	Flow Microcalorimeter
	Flutter Mode Control
	Forward Motion Compensation
	Frequency-Modulated Cyclotron [USAEC]
FMCE	Federation of Manufacturers of Construction Equipment
FMCW	Frequency-Modulated Continuous Wave
FMD	Function Management Data
FME	Frequency Measuring Equipment
FMEA	Failure Mode and Effect Analysis
FMECA	Failure Modes, Effects and Criticality Analysis
FMEP	Friction Mean Effective Pressure
FMES	Full Mission Engineering Simulator
F-MET-PHE	Formyl-Methionyl-Phenylalanine
FMEVA	Floating Point Means and Variance
FMFB	Frequency Modulation with Feedback
FM-FM	Frequency Modulation - Frequency Modulation
FMHS	Flexible Materials Handling System
FMIC	Frequency Monitoring and Interference Control
FMICW	Frequency-Modulated Intermittent Continuous Wave
FML	Feedback, Multiple Loop
FMLS	Full-Matrix Least Squares
FMN	Flavin Mononucleotide
FMO	Fleet Mail Office [military]
FMOI	First Moment of Inertia
FMOC	Fluorenylmethoxycarbonyl
FMPS	FORTRAN Mathematical Programming System
	Functional Mathematical Programming System
FMPT	First Material Processing Test
FMR	Ferromagnetic Resonance
	Frequency-Modulated Radar
	Frequency-Modulated Receiver
FMS	Farm Management System
	Federation of Materials Societies
	First Melt Sample
	Flexible Machining System
	Flexible Manufacturing System
	Forms Management System
	FORTRAN Monitor System
FMSA	Fellow of the Mineralogical Society of America
FMSC	Flexible Manufacturing System Complex
FMSL	Fort Monmouth Signal Laboratory [US Army]

FMSO	Fleet Material Supply Office [military]
FMSR	Finite Mass Sum Rule
FMSWR	Flexible Mild Steel Wire Rope
FMT	Flour Milling Technology
	Flush Metal Threshold
	Frequency-Modulated Transmitter
FMV	Fair Market Value
FMWA	Fixed Momentum Wheel Assembly
FMX	FM (= Frequency Modulation) Transmitter
FN	Ferrite Number
	Flange Nut
	Flat Nose (Projectile)
	Function
FNAA	Fast Neutron Activation Analysis
FNC	Ferritic Nitrocarburizing
FNCC	Foreign Claims Commission
FNDRY	Foundry
FNP	Fusion Point
FNPA	Foreign Numbering Plan Area
FNPS	Fluoronitrophenylsulfone
FNS	File Nesting Store
	Fusion Neutron Source
FNWF	Fleet Numerical Weather Facility [US Navy]
FNSH	Finish
FO	Factory Order
	Fast Operate
	Free Overside
	Forced Oil (Transformer)
	Fuel Oil
FOA	Forced Oil to Air-Cooled (Transformer)
	Free on Aircraft
FOB	Forward Operating Base
	Free on Board
FOBS	Fractional Orbital Bombardment System
FOC	Chemical Flux Cutting
	Fiber Optics Converter
	Focus
	Free on Car
FOCAL	Formula Calculator
FOCH	Forward Channel
FOG-M	Fiber-Optic Guided Missile
FOHMD	Fiber-Optic Helmet-Mounted Display
FOI	Follow-On Intercepter
FOIF	Free Oceanographic Instrument Float
FOIL	File-Oriented Interpretative Language
FOL	Following
FOLZ	First-Order Laue Zone
FOM	Figure of Merit
FOMC	Federal Open Market Commission
FOMCAT	Foreign Material Catalogue
FOPS	Falling Object Protective Structure
FOPT	Fiber-Optics Photon Transfer
FOQ	Free on Quay
FOR	Foreign
	Forestry
	Free on Rail
FORAC	Fisheries and Oceans Research and Advisory Council
FORACS	Fleet Operational Readiness Accuracy Check Site [US Navy]
FORAST	Formula Assembler Translator

FORC	Formula Coder
FORD	Floating Ocean Research Development Station
FORDACS	Fuel-Oil Route Delivery and Control System
FORDS	Floating Ocean Research and Development Station
FORESDAT	Formerly Restricted Data
FORG	Forging
FORMAC	Formula Manipulation Compiler
FORMAT	FORTRAN Matrix Abstraction Technique
FORTRAN	Formula Translation
FORTRANSIT	FORTRAN and Internal Translator (System)
FOS	Fiber Optic Sensor
	Free on Ship
	Fuel-Oxygen Scrap
FOSDIC	Film Optical Sensing Device for Input to Computers
FOSE	Federal Office Systems Exposition
FOT	Free on Truck
	Frequency Optimum Traffic
FOTP	Fiber-Optic Test Procedure
FOTS	Fiber-Optic Transmission System
FOV	Field of View
FOW	First Open Water
	Forced Oil to Water (Transformer)
	Forge Welding
	Free on Wagon
FP	Faceplate
	Feedback Positive
	Feedback Potentiometer
	Fine Paper
	Finger Pin
	Fire Place
	Fire Point
	Fireproof
	Flameproof
	Flash Point
	Focal Plane
	Forward Perpendicular
	Four-Pole
	Free Piston
	Freezing Point
	Fructose Phosphate
	Full Period
	Fusion Point
F/P	Flat Pattern
fp	foot-pound
4P	Four-Pole
FPA	Final Power Amplifier
	Fire Protection Association
	Floating Point Accelerator
	Focal Plane Array
	Free of Particular Average
FPAC	Flight Path Analysis and Command
FPAP	Floating Point Array Processor
FPAPA	Forest Products Accident Prevention Association
FPC	Federal Petroleum Commission
	Federal Power Commission [US]
	Final Processing Center

	Flexible Program Control
	Free-Programmable Controller
FPCE	Fission Products Conversion and Encapsulation
FPCS	Free Polar Corticosteroid
FPD	Focal Plane Deviation
FPDT	Four-Pole, Double-Throw
4PDT	Four-Pole, Double-Throw
FPDT SW	Four-Pole, Double-Throw Switch
4PDT SW	Four-Pole, Double-Throw Switch
FPH	Floating Point Hardware
FPhysS	Fellow of the Physical Society
FPIS	Forward Propagation by Ionospheric Scatter(ing)
FPL	Final Protective Line
	Forest Products Laboratory [US]
	Frequency Phase Lock
	Functional Problem Log(ging)
FPLA	Fair Packaging and Labeling Act [US]
	Field-Programmable Logic Array
FPM	Feet per Minute [also: fpm]
	File Protect Mode
	Flashes per Minute
	Functional Planning Matrices
FPN	Fixed Pattern Noise
FPP	Floating Point Processor
FPPS	Flight Plan Processing System
FPQA	Fixed Portion Queue Area
FPRF	Fireproof
FPRL	Forest Products Research Laboratory [UK]
FPRS	Forest Products Research Society
FPS	Feet per Second [also: fps]
	Fiber Placement System
	Film Performance Score
	Floating Point Systems
	Fluid Power Society [US]
	Focus Projection and Scanning
	Foot-Pound-Second System
	Frames per Second
FPSE	Foot-Pound-Second Electrostatic System
	Free Piston Stirling Engine
FPSK	Frequency and Phase-Shift Keying
FPSM	Foot-Pound-Second Electromagnetic System
FPST	Four-Pole, Single-Throw
4PST	Four-Pole, Single-Throw
FPST SW	Four-Pole, Single-Throw Switch
4PST SW	Four-Pole, Single-Throw Switch
FP SW	Four-Pole Switch
4P SW	Four-Pole Switch
FPT	Female Pipe Thread
	Frame Paperfeed Transport
	Full Power Trial
FPTS	Forward Propagation Tropospheric Scatter
FPY	First-Pass Yield
FQHE	Fractional Quantum Hall Effect
FQPR	Frequency Programmer
FR	Faculty of Radiology
	Failure Rate
	Fast Release
	Federal Reserve
	Fiber Reinforced

	Field Reversing
	Fineness Ratio
	Fire Resistance
	Fire-Retardant
	Flash Ranging
	Fragment
	Frame
	France
	French
	Friday
	Fusion Reactor
Fr	Francium
FRA	Federal Railway Administration [US]
	Frequency Response Analyzer
FRAeS	Fellow of the Royal Aeronautical Society
FRAC	Fraction
FRAG	Fragment(ation)
FRAIC	Fellow of the Royal Architectural Institute of Canada
FRAS	Fellow of the Royal Astronomical Society
FRAT	Fiber-Reinforced Advanced Titanium
FRB	Federal Reserve Bank
FR BEL	From Below
FRC	Fast Reaction Concept
	Federal Radio Commission [US]
	Fiber-Reinforced Composite
	File Research Council
	Flight Research Center [NASA]
	Functional Residue Capacity
FRCI	Fibrous Refractory Composite Insulation
FRCS	Forged Radius Clamp Straps
FRCTF	Fast Reactor Core Test Facility [USAEC]
FRD	Fiber Resin Development
	Functional Reference Device
FRED	Figure Reading Electronic Device
	Fractionally Rapid Electronic Device
FREDI	Flight Range and Endurance Data Indicator
FREEBD	Freeboard
FREL	Feltman Research and Engineering Laboratory [US Army]
FREQ	Frequency
FREQMULT	Frequency Multiplier
FRESCAN	Frequency Scanning
FRESCANNAR	Frequency Scanning Radar
FRG	Federal Republic of Germany
FRGS	Fellow of the Royal Geographical Society
FRI	Friday
FRIBA	Fellow of the Royal Institute of British Architects
FRIC	Fellow of the Royal Institute of Chemistry
FRICS	Fellow of the Royal Institute of Chartered Surveyors
FRINGE	File and Report Information Processing Generator
FRL	Feltman Research Laboratory [US Army]
	Filter, Regulator and Lubricator Unit
FRM	Fiber-Reinforced Metal
FRMetS	Fellow of the Royal Meteorological Society
FRMS	Fellow of the Royal Microscopical Society
FROM	Fusable Read-Only Memory
FRP	Fiber-Reinforced Plastic

	Fiber-Reinforced Polyester
FRR	Functional Recovery Routine
FRRS	Full Remaining Radiation Service [US]
FRS	Fellow of the Royal Society
	Federal Reserve System
	Fiber-Reinforced Superalloy
	Fire Research Station
	Fragility Response Spectrum
FRSB	Frequency-Referenced Scanning Beam
FRSM	Fellow of the Royal School of Mines
FRSNA	Fellow of the Royal School of Naval Architecture
FRSS	Fellow of the Royal Statistical Society
FRSTP	Fire-Refined Tough Pitch (Copper) with Silver
FRT	Finish-Rolling Temperature
	Freight
FRTP	Fiber(glass)-Reinforced Thermoplastic
	Fiber-Reinforced Thermosetting Plastic
FRTP	Fire-Refined Tough Pitch (Copper)
FRU	Field Replaceable Unit
FRUGAL	FORTRAN Rules Used as a General Applications Language
FRUSTUM	Frame-Based Unified Story-Understanding Model
FRW	Friction Welding
FRWI	Framingham Relative Weight Index
FRWK	Framework
FS	Factor of Safety
	Far Side
	Fast-Slow (Wave)
	Federal Specification
	Federal Standard
	Feedback, Stabilized
	Female Soldered
	Fiber Society
	Field Separator
	Field Service
	File Separator (Character)
	Flat Slip
	Floating Sign
	Foresight
	Forged Steel
	Free Sterols
	Freeze Substitution
	Frequency Synthesizer
	Full Scale
	Full Size
	Functional Schematic
	Furnace Soldering
fs	foot-second
FSA	Field Search Argument
	Fracture Surface Analysis
FSAA	Flat Slips All Around
	Flight Simulator for Advanced Aircraft
FSB	Federal Specifications Board [US]
	Flat Slip on Bottom
FSBL	Fusible
FSC	Federal Supply Classification [US]
	Flame-Sprayed Coating
	Foundation for the Study of Cycles

	Free Secretory Component
FSCI	Frequency Space Characteristic Impedance
FSCS	Forged Straight Clamp Strap
FSCT	Federation of Societies for Coating Technology
FSD	Flying Spot Digitizer
	Full Scale Deflection
	Full-Size Detail
FSDC	Federal Statistical Data Center
FSE	Fellow of the Society of Engineers
FSF	Fibrin-Stabilizing Factor
	Flash Smelting Furnace
FSFE	Flash Smelting Furnace with Furnace Electrodes
FSG	First Stage Graphitization
FSH	Follicle-Stimulating Hormone
FSI	Fellow of the Surveyors Institute
FSLP	First Spacelab Program
FSIS	Fast Sample Insertion System
FSK	Frequency-Shift Keying
FSL	Flexible Satellite Link
	Flexible System Link
	Formal Semantic Language
FSM	Field Strength Meter
	Finite Strip Method
	Finite State Machine
FSMWO	Field Service Modification Work Order
FSN	Federal Stock Number
FSP	Ford Satellite Plan
	Frequency Shift Pulsing
	Frequency Standard, Primary
	Full-Screen Processing
FSPT	Federation of Societies for Paint Technology
FSR	Farming Systems Research
	Fast Sodium-Cooled Reactor
	Feedback Shift Register
	Field Strength Ratio
	Force Sensing Resistor
	Foundation for Scientific Relaxation
FSS	Fatigue Striation Spacing
	Federal Supply Services
	Fixed Satellite Service
	Flight Service Station
	Flight Standards Service
	Flying Spot Scanner
	Formatted System Services
FST	Federal Sales Tax
	Forged Steel
	Flat Slip on Top
	Frequency Shift Transmission
	Functional Subassembly Tester
	Functional Systems Test
FSTC	Foreign Science and Technology Center [US Army]
FSTNR	Fastener
FSTU	Fluid Sealing Technology Unit
FSU	Field Select Unit
	Florida State University [US]
FSUC	Federal Statistics User Conference [US]
FSV	Floating-Point Status Vector
FSVM	Frequency Selective Voltmeter

FSW	Flexible Steel Wire
FSWR	Flexible Steel Wire Rope
FT	Film Thickness
	Firing Tables
	Flat Top
	Flame Tight
	Flush Threshold
	Fourier Transform
	Frequency and Time
	Frequency Tracker
	Full Time
	Fume Tight
	Functional Test
F&T	Fuel and Transportation
ft	foot (or feet)
FTA	Fault Tree Analysis
FTAM	File Transfer and Access Method
FTAB	Field Tab(ulator)
ft-c	foot-candle
FTC	Fast Time Constant
	Federal Trade Commission [US]
	Frequency Time Control
FTD	Folded Triangular Dipole
	Foreign Technology Division [of USAFSC]
	Frequency Translation Distortion
FTE	Fracture Transition Elastic
	Full Time Equivalent
FTESA	Foundry Trades Equipment and Supplies Association
FTF	Flared Tube Fitting
FTFET	Four-Terminal Field-Effect Transistor
FTG	Footing
	Fitting
fth	fathom [unit]
fthm	fathom [unit]
FTI	Federal Tax Included
	Fellow of the Textile Institute
FTIR	Fourier Transform Infrared Spectrometer
	Fourier Transform Infrared Spectroscopy
FTIRRS	Fourier Transform Infrared Reflection Spectroscopy
FTL	Federal Telecommunications Laboratory [US]
ft-L	foot-Lambert
ft-lb	foot-pound
FTM	Flight Test Missile
	Folded Triangular Monopole
	Functional Test Manager
FTMC	Frequency and Time Measurement Counter
FTMP	Fault Tolerant Multiprocessor System
FTMS	Federal Test Method Standards
	Fourier Transform Mass Spectrometry
FTMT	Final Thermomechanical Treatment
FTNMR	Fourier Transform Nuclear Magnetic Resonance
F to F	Face to Face
FTP	Federal Test Procedure
	Fuel Transfer Pump
FTR	Feed per Tooth per Revolution
	Fixed Transom
	Flat Tile Roof

	Functional Test Requirement
FTS	Facsimile Text Society [US]
	Federal Telecommunications System [US]
	Fourier Transform Spectrometer
	Free-Time System
FTSC	Fault Tolerant Spaceborne Computer
FTTU	Field Technical Training Unit
FTU	Flight Test Unit
	Formazin Turbidity Unit
	Frascati Tokamak Upgrade [Italian Reactor]
FTZ	Free Trade Zone
FU	Fume Concentration
	Functional Unit
FUDR	Failure and Usage Data Report
	Fluorodeoxyuridine
FUIF	Fire Unit Integration Facility
FUNC	Function
FUNCTLINE	Functional Line Diagram
FUNOP	Full Normal Plot
FUNY	Free University of New York [US]
FUR	Furring
	Furnish
fur	furlong [unit]
FURN	Furnish
FUS	Fuselage
FUSE	Federation for Unified Science and Engineering
FV	Fiber Volume
	Flux Value
	Front View
	Full Voltage
F/V	Frequency-to-Voltage
fV	femtovolt
FVN	Failed Vector Numbers
FVPRA	Fruit and Vegetable Preservation Research Association
FVR	Fuse Voltage Rating
FVRDE	Fighting Vehicles Research and Development Establishment [of MOD]
FVS	Flight Vehicle Systems
FVT	Flash Vacuum Thermolysis
FW	Face Width
	Feed Water
	Filament Wound
	Firmware
	Flash Welding
	Flat Washer
	Formula Weight
	Fresh Water
	Full-Wave
FWA	File Work Area
	First Word Address
	Fluorescent Whitening Agent
	Forward Wave Amplifier
FWB	Four Wheel Brake
FWC	Fourdrinier Wire Council
	Fully Loaded Weight and Capacity
FWD	Forward
	Four Wheel Drive
	Free Working Distance
	Fresh Water Damage

FWE	Finished with Engines
FWHM	Full Width at Half Maximum
FWI	Fire Weather Index
FWL	Fixed Word Length
FWM	Fourier Wave Mixing
FWPCA	Federal Water Pollution Control Administration [US]
FWQA	Federal Water Quality Administration
FWR	Full-Wave Rectifier
FWRC	Federal Water Resources Council [US]
FWS	Filter Wedge Spectrometer
FWTM	Full Width Tenth Maximum
FWTT	Fixed Wing Tactical Transport
FXD	Fixed
FXS	Fein-Focus X-Ray Series
FY	Fiscal Year
FYDP	Five Year Defense Program [US]
FZ	Fusion Zone
FZES	Float Zone Experiment System
FZP	Fresnel Zone Plate

G

G	Gain
	Gap
	Gas
	Gate
	Gauge
	Gauss [unit]
	German(y)
	Gilbert [unit]
	Girder
	Glass
	Glycine
	Graph
	Grid
	Guanine
g	gram
	gravity
GA	Gauge
	Gain of Antenna
	Gas Amplification
	General Assembly
	Georgia [US]
	Gibberellic Acid
	Glide Angle
	Glyoxylic Acid
	Graphic Ammeter
G&A	General and Administrative
G-A	Ground-to-Air
Ga	Gallium
	Georgia [US]
GAAP	General Adjustment Assistance Program
	Generally Accepted Accounting Principles
GAATS	Gander Automatic Air Traffic System [Canada]
GAB	Gusseted Angle Bracket
GABA	Gamma-Aminobutyric Acid

GAC	Geological Association of Canada	GASCan	Getaway Special Canister
	Granular Activated Carbon	GASL	General Applied Science Laboratory
GACS	Gun Alignment and Control System	GASO	Gasoline
GAD	Glutamic Acid Decarboxylase	GASP	General Academic Simulation Program
GADL	Ground-to-Air Data Link		General Activity Simulation Program
GAELIC	Grumman Aerospace Engineering Language for Instructional Checkout		Grand Accelerated Space Platform
			Graphic Applications Subroutine Package
GAGR	Group Automatic Gain Regulator	GASS	Generalized Assembly System
GAI	Generalization Area of Intersection	GAT	General Air Traffic
GAIBA	Gamma-Aminoisobutyric Acid		Generalized Algebraic Translator
GAINS	Gimballess Analytic Inertial Navigation System		Georgetown Automatic Translator [US]
		GATB	General Aptitude Test Battery
GAIBA	Gamma-Aminoisobutyric Acid	GATD	Graphic Analysis of Three-Dimensional Data
gal	gallon [unit]		
GALCIT	Guggenheim Aeronautical Laboratory of California Institute of Technology [US]	GATE	GARP Atlantic Tropical Experiment
			Generalized Algebraic Translator Extended
GALL	Gallery	GATF	Graphic Arts Technical Foundation [US]
gal per min	gallons per minute	GATNIP	Graphic Approach to Numerical Information Processing
GALV	Galvanize		
	Galvanometer	GATR	Ground-to-Air Transmitter Receiver
GAM	Graduate in Aerospace Mechanical Engineering	GATT	General Agreement on Tariffs and Trade [of UN]
	Graphic Access Method		Ground-to-Air Transmitter Terminal
	Ground-to-Air Missile	GATU	Geophysical Automatic Tracker Unit
	Guided Aircraft Missile	GATV	Gemini Agena Target Vehicle [NASA]
GAMA	Gas Appliances Manufacturers Association	GAWR	Gross Axle Weight Rating
	General Aviation Manufacturers Association	GB	General Background
	Graphics-Assisted Management Application		Grain Boundary
GAMIS	Graphic Arts Marketing and Information Service [of PIA]		Great Britain
			Greenish Blue
GAMLOGS	Gamma-Ray Logs		Grid Bias
GAMM	German Association for Applied Mathematics and Mechanics		Guide Block
			Gun Branch
GAN	Generalized Activity Network	Gb	Gigabit
	Generating and Analyzing Network		Gilbert [unit]
GAO	General Accounting Office [US]	GBA	Grain Boundary Allotriomorph
GAP	General Assembly Program	GBBA	Glass Bottle Blowers Association
	Glyceraldehydephosphate	GBD	Grain Boundary Dislocation
GAPA	Ground-to-Air Pilotless Aircraft	GBM	Glomerular Basement Membrane
GAPD	Glyceraldehydephosphate Dehydrogenase		Grain Boundary Migration
GAPDH	Glyceraldehydephosphate Dehydrogenase	GBP	Gain Bandwidth Product
GAPSFAS	Graduate and Professional School Financial Aid Service	GBQ	Generating Bearing Quality
		GC	Gas Chromatography
GAPT	Graphical Automatically Programmed Tools		General Circular
GAR	Growth Analysis and Review		Glassy Carbon
	Guided Aircraft Rocket		Government of Canada
GARBD	Garboard		Grain Cube
GARC	Graphic Arts Research Center [US]		Guanine-Plus-Cytosine
GARD	General Address Reading Device		Guidance Computer
GARDAE	Gather, Alarm, Report, Display and Evaluate		Gyro Compass
		Gc	gigacycles
GAREX	Ground Aviation Radio Exchange System	G&C	Guidance and Control
GARF	Graphic Arts Research Foundation	GCA	Gain Control Driver
GARMI	General Aviation Radio Magnetic Indicator		General Control Approach
GARP	Global Atmospheric Research Program [of NAS]		Glazing Contractors Association
			Ground-Controlled Approach
GAS	Gasoline	g-cal	gram-calorie
	General Aviation Services	GCAP	Generalized Circuit Analysis Program
	Getaway Special	GCAW	Gas Carbon-Arc Welding
	Goods Acquisition System	GCB	General Circuit Breaker
	Graphics Attachment Support	GCC	Gas Chromatograph Column

	General Channel Coordinator
	Global Competitiveness Council
	Ground Control Center
GCD	Greatest Common Divisor
GCE	Glassy Carbon Electrode
	Ground Communication Equipment
	Ground Control Equipment
GC ENV	Government of Canada, Environment Canada
GCF	Greatest Common Factor
GCFAP	Guidance and Control Flight Analysis Program
GCFR	Gas-Cooled Fast Reactor
GCFRE	Gas-Cooled Fast Reactor Experiment
GCHQ	Government Communications Headquarters
GCI	Ground-Controlled Interception
GC-IR	Gas Chromatography - Infrared Spectroscopy
GCIS	Ground Control Intercept Squadron
GCIU	Graphic Communications International Union
GCM	Greastest Common Measure
	Ground Check Monitor
GCMA	Government Contract Management Association of America
GC/MS	Gas Chromatography - Mass Spectroscopy
GCMSC	George C. Marshall Spaceflight Center [NASA]
GCN	Gauge Code Number
GCOS	Great Canadian Oil Sands
GCP	Geometrically Close-Packed
GCR	Galvanocutaneous Reaction
	Gas-Cooled Reactor
	General Control Relay
	General Component Reference
	Group-Coded Recording
GCS	Gate-Controlled Switch
	Ground Control System
GCT	General Classification Test
	Greenwich Civil Time [UK]
	Guard Control System
GCTS	Ground Communication Tracking System
GCU	Generator Control Unit
GCVS	General Catalogue of Variable Stars
GCW	Global Chart of the World
GD	Gate Driver
	Grade
	Ground Detector
	Grown Diffused
	Guard
Gd	Gadolinium
GDC	Generalized Dynamic Charge
	Glow Discharge Condition
	Graphic Designers of Canada
GDCI	Gypsum Drywall Contractors Association
GDE	Ground Data Equipment
GDF	Gas Dynamic Facility [US Air Force]
	Group Distribution Frame
GDG	Generation Data Group
GDH	Glutamate Dehydrogenase
	Glycerophosphate Dehydrogenase
GDIFS	Gray and Ductile Ironfounders Society [US]
GDL	Gas Dynamic Laser
	Glow Discharge Lamp
GDMS	Generalized Data Management System
	Glow Discharge Mass Spectroscopy
GDN	Government Data Network
GDO	Grid-Dip Oscillator
	Gross Domestic Output
GDOES	Glow Discharge Optical Emission Spectroscopy
GDOP	Geometric Dilution of Precision
GDOS	Glow Discharge Optical Spectroscopy
GDP	Glow Discharge Polymer
	Graphic Data Processing
	Gross Domestic Product
	Guanosine Diphosphate
GDPG	Guanosine Diphosphoglucose
GDPS	Global Data Processing System
GDR	General Design Requirement
	German Democratic Republic
	Ground Delay Response
GDRS	Geoscience Data Referral System
GDS	Glow Discharge (Optical) Spectroscopy
GDT	Generator Development Tool
GE	Gauge
	Gas Ejection
	Gaussian Elimination
	Genetic Engineer(ing)
	Greater than or Equal to
	Guanidoethyl
Ge	Germanium
GEAP	General Electric Atomic Power
GEBCO	General Bathymetric Chart of the Oceans
GEC	Generalized Equivalent Cylinder
GECAP	General Electric Computer Analysis Program
GECOM	Generalized Compiler
GECOS	General Comprehensive Operating System
GED	General Educational Development
GEEI	General Electric Electronic Installation
GEEP	General Electric Electronic Processor
GEESE	General Electric Electronic Systems Evaluator
GEG	General Euclidian Geometry
GEGB	General Electricity Generating Board
GEIS	General Electric Information Service
GEISHA	Geodetic Inertial Survey and Horizontal Alignment
GEK	Geomagnetic Electrokinetograph
GEL	Gelatine
	Gelatinous
GEM	General Electronics Module
	General Epitaxial Monolith
	Generic Experiment Module
	Geostatistical Evaluation of Mines
	Governmental Energy and Minerals Committee [of TMS, AIME]
	Ground Effect Machine
	Graphics Environment Manager
GEMAC	General Electric Measurement and Control [also: GE/MAC]

GEMM	Generalized Electronics Maintenance Model
GEMS	General Education Management System
	General Electric Manufacturing Simulator
	Global Environmental Monitoring Service
GEMSIP	Gemini Stability Improvement Program [NASA]
GEN	General
	Generator
GENDA	General Data Analysis
GENDARME	Generalized Data Reduction, Manipulation and Evaluation
GENDAS	General Data Analysis and Simulation
GENERIC	Generation of Integrated Circuits
GENESYS	General Engineering System
	Graduate Engineering Education System
GENL	General
GEO	Geostationary Orbit
GEOCHEM	Geochemical
	Geochemist
	Geochemistry
GEOG	Geographer
	Geographical
	Geography
GEO-IRS	Geostationary Orbit - Infrared Sensor
GEOIS	Geographic Information System
GEOL	Geologic(al)
	Geologist
	Geology
GEOM	Geometric(al)
	Geometrician
	Geometry
GEON	Gyro-Erected Optical Navigation
GEOPHYS	Geophysical
	Geophysicist
	Geophysics
GEOPS	Geodetic Estimates from Orbital Perturbations of Satellites
GEOS	Geodetic Earth Orbiting Satellite [NASA]
GEOSCAN	Geological Survey of Canada
	Ground-Based Electronic Omnidirectional Satellite Communications Antenna
GEP	Goddard Experiment Package [NASA]
GEPAC	General Electric Programmable Automatic Comparator
	General Electric Process Automation Computer
GER	German(y)
GERD	Gross Expenditure on Research and Development
GEREP	Generalized Equipment Reliability Evaluation Procedure
GERSIS	General Electric Range Safety Instrumentation System
GERT	Graphical Evaluation and Review Technique
GERTS	General Electric Range Tracking System
	General Electric Remote Terminal System
GESAC	General Electric Self-Adaptive Control System
GESO	Geodetic Earth-Orbiting Satellite
GESOC	General Electric Satellite Orbit Control
GESS	Grinding Energy Saving System

GET	Gas Evaporation Technique
	General Equivalence-Point Titration
	Genetic Engineering Technology
	Geological Engineering Technology
	Ground Elapsed Time
GE/TAC	General Electric Telemetering and Control
GETEL	General Electric Test Engineering Language
GETIS	Ground Environment Team of the International Staff [NATO]
GETOL	General Electric Training Operational Logic
	Ground Effect Takeoff and Landing
GETS	Ground Equipment Test Set
GEV	Ground Effect Vehicle
GeV	Gigaelectronvolt
GEVIC	General Electric Variable Increment Computer
GEWS	Group on Engineering Writing and Speech [of IEEE]
GF	Gold Filling
	Ground Fault
gf	gram force
GFA	Glass Formation Ability
GFAAS	Graphite Furnace Atomic Absorption Spectroscopy
GFAE	Government-Furnished Aeronautical Equipment
GFC	Gel Filtration Chromatography
GFCI	Ground Fault Circuit Interrupter
GFE	Government-Furnished Equipment
GFF	Glass Fiber Filter
GFFAR	Guided Folding Fin Aircraft Rocket
GFI	Ground Fault Interrupter
GFLOPS	Billion Floating Point Operations per Second [G stands for "Giga" = 10^9 = 1 Billion]
GFM	Government Furnished Materials
GFO	Gap-Filler Output
GFP	Glass Fiber Pulling
GFQ	Gradient Furnace with Quenching Device
GFR	Gas-Filled Rectifier
	Glomerular Filtration Rate
	Ground Fault Monitor
GFRP	Glass Fiber Reinforced Plastic
	Graphite Fiber Reinforced Plastic
GFW	Glass Filament Wound
	Ground-Fault Warning
G-G	Ground-to-Ground
GGG	Gadolinium Gallium Garnet
GGR	Gas Graphite Reactor
G Gr	Great Gross [unit]
GGSE	Gravity-Gradient Stabilization Experiment [NASA]
GGTP	Glutamyltranspeptidase
GGTS	Gravity-Gradient Test Satellite [NASA]
GH	Growth Hormone
GHA	Greenwich Hour Angle
GHCP	Georgia Hospital Computer Group [US]
GHF	Gradient Heating Facility
GHIUD	Global Human Information Use per Decade
GHOST	Global Horizontal Sounding Technique
GHz	Gigahertz
GI	Galvanized Iron

	Geodesic Isotensoid
	Governmental and Industrial
	Government-Initiated
	Gross Investment
gi	gill [unit]
GIA	General Industrial Application
GIANT	Genealogical Information and Name Tabulating System
GIC	General Impedance Converter
GIDEP	Government-Industry Data Exchange Program [US]
GIF	Gravito-Inertial Force
GIFS	Generalized Interrelated Flow Simulation
GIFT	General Internal FORTRAN Translator
GIGO	Garbage-In, Garbage-Out
GILSP	Good Industrial Large-Scale Practice
GIM	General Information Management
GIMRADA	Geodesy Intelligence and Mapping Research and Development Agency [US Army]
GIMU	Gimballess Inertial Measuring Unit
GIOC	Generalized Input/Output Controller
GIP	Gastric Inhibitory Polypeptide
	Group Interface Processor
GIPS	Ground Information Processing System
GIPSE	Gravity-Independent Photosynthetic Exchange
GIPSY	General Information Processing System
GIRD	General Incentive for Research and Development [Canada]
GIRI	Government Industrial Research Institute
GIRIN	Government Industrial Research Institute at Nagoya [Japan]
GIRL	German Infrared Laboratory [FRG]
	Graph Information Retrieval Language
GIRLS	Generalized Information Retrieval and Listing System
GIS	Generalized Information System
	Geophysical Incentive System
	Geoscience Information Society [US]
GIT	General Industrial Training
	Georgia Institute of Technology [US]
	Graph Isomorphism Tester
GITC	Glucopyranosyl Isothiocyanate
GJ	Gigajoule
	Grown Junction
GJE	Gauss-Jordan Elimination
GJP	Graphic Job Processor
GK	Greek
GKS	Graphical Kernel System
GL	Gate Leads
	Ginsburg-Landau (Theory)
	Glass
	Glaze
	Gothic Letter
	Grade Line
	Grid Leak
	Gross Leak
	Ground Level
gl	gill [unit]
GLAG	Ginzburg-Landau-Abrikosov-Gorkov (Theory)

GLANCE	Global Lightweight Airborne Navigation Computer Equipment
GLBC	Great Lakes Basin Commission
GLC	Gas Liquid Chromatography
GLCM	Ground-Launched Cruise Missile
GLDC	Glutamic Decarboxylase
GLDH	Glutamic Dehydrogenase
GLDP	Glutamic Dephosphatase
GLEAM	Graphic Layout and Engineering Aid Method
GLEEP	Graphite Low-Energy Experimental Pile
GLGST	Geologist
GLHK	Granato-Luecke Theory for High Kelvin (Temperatures)
GLINT	Global Intelligence
GLIT	Glitter(ing)
GLLD	Ground Laser Locator Designator
Gln	Glutamine
GLO	Goddard Launch Operations [NASA]
GLOC	Gravity-Induced Loss of Consciousness
GLOK	Granato-Luecke Theory for 0 (= Zero) Kelvin
GLOMEX	Global Meteorological Experiment
GLOPAC	Gyroscopic Lower Power Attitude Controller
GLORIA	Geological Long-Range Inclined Asdic
GLOSS	Global Ocean Surveillance System
	Glossary
GLOTRAC	Global Tracking Network
GLR	Gas/Liquid Ratio
GLS	Glass
Glu	Glutamin Acid
GLV	Gemini Launch Vehicle
	Glove Valve
Gly	Glycine
Glyc	Glycerine
GM	Gaseous Mixture
	Geiger-Mueller (Tube) [also: G-M]
	Geometric Scan
	Greenwich Meridan
	Grid Modulation
	Group Mark
	Guided Missile
	Gun Metal
gm	gram
GMA	Gas Metal-Arc
	Glycol Methacrylate
GMAC	Gas Metal-Arc Cutting
GMAG	Genetic Manipulation Advisory Group
GMAP	Geometric Modeling Application Program
GMAT	Graduate Management Admission Test
	Greenwich Mean Astronomical Time
GMAW	Gas Metal-Arc Welding
GMAW-EG	Gas Metal-Arc Welding, Electrogas
GMAW-P	Pulsed Gas Metal-Arc Welding
GMAW-S	Short Circuiting Gas Metal-Arc Welding
GMBS	Gamma-Maleimidobutyryloxysuccinimide
GMC	Ground Movement Controller
GMCM	Guided Missile Countermeasures
GMD	Guided Missiles Division [US Air Force]
GMDA	Groundwater Management Districts Association

GMDEP	Guided Missile Data Exchange Program [US Navy]
GMFB	Gang Mill Fixture Base
GMFC	Gem and Mineral Federation of Canada
GMFCS	Guided Missile Fire Control System
GMIS	Government Management Information Sciences
GML	Generalized Mark-Up Language
GMM	General Matrix Manipulator
GMP	Good Manufacturing Practices
	Ground Movement Planner
	Guanosine Monophosphate
GMR	Greatest Meridional Radius
	Ground Mapping Radar
	Ground Movement Radar
GMRWG	Guided Missile Relay Working Group [US Navy]
GMS	Gemini Mission Simulator [NASA]
	Generalized Main Scheduling
	Geometric Modeling System
	Geostationary Meteorological Satellite
	Ground Maintainance Support
GMSFC	George Marshall Spaceflight Center [NASA]
GMT	Global Money Transfer
	Greenwich Mean Time
GMV	Guaranteed Minimum Value
GM/WM	Group Mark/Word Mark
GN	Green
G&N	Guidance and Navigation
GNC	Global Navigation Chart
	Graphic Numerical Control
GND	Ground
GNE	Gross National Expenditure
GNI	Generation of New Ideas
GNP	Gross National Product
	Grain Neutral Spirits
GNSI	Guild of Natural Science Illustrators
GO	Graphitic Oxide
G/O	Gas/Oil (Ratio)
GOAL	Ground Operations Aerospace Language
GOC	Gas/Oil Contact
GOCI	General Operator-Computer Interaction
GOCR	Gated-Off Controlled Rectifier
GOD	Glucose Oxidase
GOE	Ground Operating Equipment
GOES	Geostationary Operational Environmental Satellite
GOI	Gate Oxide Integrity
GOIC	(Persian) Gulf Organization for Industrial Consulting
GOL	General Operating Language
	Graphic On-Line Language
GOP	General Operational Plot
GOR	Gained Output Ratio
	Gas/Oil Ratio
	General Operational Requirement
GORID	Ground Optical Recorder for Intercept Determination
GOSIPS	Government Open Systems Interconnect(ion) Procurement Specification [US]

GOSS	Ground Operational Support System
GOT	Glutamic-Oxalacetic Transaminase
GOTS	Gravity-Oriented Test Satellite [NASA]
GOV	Governor
	Government
GOVT	Government
GP	Generalized Programming
	General Purpose
	Geometric Progression
	Glide Path
	Glycerophosphate
	Ground Protection
	Group
	Guaranteed Performance
	Guinier-Preston (Zone)
GPA	General-Purpose Amplifier
	General-Purpose Analysis
	Grade Point Average
	Graphical PERT Analog
GPa	Gigapascal
GpA	Guanylyladenosine
GPAC	Government and Public Affairs Committee
GPAD	Gallons per Acre per Day [also: gpad]
GPAS	General-Purpose Airborne Simulator
GPATS	General-Purpose Automatic Test System
GPB	Ground Power Breeder
GPC	Gel Permeation Chromatography
	General-Purpose Computer
	Gross Profit Contribution
GpC	Guanylylcytidine
GPCP	Generalized Process Control Programming
GpCpC	Guanylylcytidylylcytidine
GPD	Gallons per Day [also: gpd]
GPDC	General-Purpose Digital Computer
GPDS	General-Purpose Display System
GPES	Ground Proximity Extraction System
GPF	Generalized Production Function
	General-Purpose Furnace
GPG	Grains per Gallon
GpG	Guanylylguanosine
GPGAP	Great Plains Gasification Associates' Project [US]
GPGL	General-Purpose Graphic Language
GPH	Gallons per Hour [also: gph]
	Graphite
GPI	Glucose Phosphate Isomerase
	Ground Position Indicator
GPIB	General-Purpose Instrumentation Bus
	General-Purpose Interface Bus
GPIS	Gemini Problem Investigation Status
GPL	Generalized Programming Language
	General-Purpose Language
GPLP	General-Purpose Linear Programming
GPM	Gallons per Minute [also: gpm]
	Generalized Perturbation Method
	General-Purpose Macrogenerator
	Geopotential Meter
	Groups per Message
GPMS	General-Purpose Microprogram Simulator
GPO	General Post Office
	Government Printing Office [US]

GPOS	General-Purpose Operating System
GPP	Gross Provincial Product [Canada]
GPR	General-Purpose Radar
	Ground-Penetrating Radar
GPRF	General-Purpose Rocket Furnace
GPRF-G	General-Purpose Rocket Furnace - Gradient
GPRF-I	General Purpose Rocket Furnace - Isothermal
GPS	Gallons per Second [also: gps]
	Gas-Pressure Sintering (System)
	General Problem Solver
	Global Positioning System
	Graphic Programming Services
GPSS	General-Purpose Systems Simulator
	General-Purpose Simulation System
GPSSN	Gas Pressure Sintered Silicon Nitride
GPT	Gas Power Transfer
	Gemini Pad Test [NASA]
	Glutamic-Pyruvic Transaminase
GPT-C	Glutamic-Pyruvic Transaminase-C
GPTM	Gross Profit this Month
GPTY	Gross Profit this Year
GpU	Guanylyluridine
GpUpG	Guanylyluridylylguanosine
GPWS	Ground Proximity Warning System
GPX	Generalized Programming Extended
GPZ	Guinier-Preston Zone
GQE	Generalized Queue Entry
GR	Gear
	General Register
	General Relativity
	General Reconnaissance
	General Reserve
	Germanium Rectifier
	Glass-Reinforced
	Government Reserve
	Grade
	Grain
	Gray
	Greece
	Greek
	Grid Resistor
	Grid Return
	Group
gr	gram
	gross [unit]
GRACE	Graphic Arts Composing Equipment
GRAD	General Recursive Algebra and Differentiation
	Gradient
	Graduate
	Graduate Resume Accumulation and Distribution
	Graduation
GRADB	Generalized Remote Access Database
GRADS	Generalized Remote Access Database System
GRAF	Graphic Addition to FORTRAN
GRAMPA	General Analytical Model for Process Analysis
GRAN	Global Rescue Alarm Network
GRAPE	Gamma-Ray Attenuation Porosity Evaluator
GRAPH	Graphic(s)
GRAPHDEN	Graphic Data Entry
GRARR	Goddard Range and Range Rate [NASA]
GRASER	Gamma Ray Amplification by Stimulated Emission of Radiation [also: Graser]
GRASP	Generalized Retrieval and Storage Program
	Graphical Robot Applications Stimulation Package
	Graphic Service Program
GRATIS	Generation, Reduction, and Training Input System
GRB	Geophysical Research Board [of NRC, US]
	Government Reservation Bureau
GR BR	Great Britain
GR BRIT	Great Britain
GRC	Geothermal Resources Council
	Graduate Research Center
GRCSW	Graduate Research Center of the Southwest [US]
GRD	Grind
	Ground
	Ground Detector
	Ground-Resolved Distance
GRE	Graduate Record Examination
	Graduate Reliability Engineering
GREAT	Graduate Research in Engineering and Technology
GREB	Galactic Radiation Experiment Background
GRED	Generalized Random Extract Device
GREMEX	Goddard Research Engineering Management Exercise [NASA]
GRETA	Ground Radar Emissions Training Aviator
GRF	Gravity Research Foundation
	Ground Repetition Frequency
	Growth-Hormone Releasing Factor
GRG	Glass-Fiber Reinforced Gypsum
	Geodesy Research Group
GRI	Gas Research Institute
	Group Repetition Interval
GRID	Graphical Interactive Display
GRIN	Graphical Input
GRINS	General Retrieval Inquiry Negotiation Structure
GRIPHOS	General Retrieval and Information Processing for Humanities-Oriented Studies
GRIT	Graduated Reduction in Tensions
GRL	Gamma Ray Laboratory [FRG]
GRM	Generalized Reed-Muller (Code)
	Global Range Missile
GRN	Green
GROM	Grommet
GROUT	Graphical Output
GRP	Glass-Reinforced Plastic
	Group
	Group Reference Point
GRR	Geneva Radio Regulations
	Guidance Reference Release
GR&R	Gauge Repeatablilty and Reproducibility
GRS	Gears

	Generalized Retrieval System	**GSS**	Global Surveillance System
	General Radio Service [Canada]		Graphic Service System
GRT	Gross Register Tonnage	**GSSC**	Ground Support Simulation Computer
GRTG	Grating	**GSSW**	Gas-Shielded Stud Welding
GRU	Gyro Reference Unit	**GSTA**	Ground Surveillance and Target Acquisition
GRV	Groove	**GSU**	General Service Unit
GR WT	Grain Weight		Guaranteed Supply Unit
	Gross Weight	**GSV**	Guided Space Vehicle
GS	Galvanized Steel	**GSWR**	Galvanized Steel Wire Rope
	General Semantics	**GT**	Game Theory
	General Solution		Gemini-Titan [NASA]
	Geochemical Society		Glutamyltransferase
	Geographical Society		Grease Trap
	Geological Society		Great
	Glaciological Society		Greater than
	Gold Standards		Gross Tons
	Grain Size		Ground Transmit
	Ground Speed		Group Technology
	Ground Support	**G/T**	Gain-to-Noise Temperature (Ratio)
	Group Separator	**GTA**	Gas Tungsten-Arc
GSA	General Services Administration [US]		Gemini-Titan-Agena [NASA]
	General Services Agencies		Glycerol Triacetate
	Genetics Society of America		Government Telecommunications Agency [US]
	Geographical Society of America		
	Geological Society of America	**GTAC**	Gas Tungsten-Arc Cutting
GSAM	Generalized Sequential Access Method	**GTAW**	Gas Tungsten-Arc Welding
GSC	Gas-Solid Chromatography	**GTAW-P**	Pulsed Gas Tungsten-Arc Welding
	Genetics Society of Canada	**GTBA**	Gasoline Grade Tertiary Butylacetate
	Geological Survey of Canada	**GT BR**	Great Britain
	Grant Selection Committee	**GT BRIT**	Great Britain
GSCG	Ground Systems Coordination Group	**GTC**	Gain Time Constant
GSCU	Ground Support Cooling Unit		Gain Time Control
GSD	General Supply Depot		Gas Turbine Compressor
	General Systems Division		Geological Testing Consultant
	Generator Starter Drive	**GTD**	Geometrical Theory of Diffraction
GSDS	Goldstone Duplicate Standard		Graphic Tablet Display
GSE	Ground Support Equipment		Guaranteed
GSFC	Goddard Spaceflight Center [of NASA]	**GTE**	Ground Transport Equipment
GSG	Galvanized Sheet Gauge		Group Translating Equipment
GSI	Government Source Inspection	**GTF**	Generalized Trace Facility
	Grand Scale Integration		Generalized Transformation Function
GSKT	Gasket	**GTG**	Gas Turbine Generator
GSL	Generation Strategy Language	**GTM**	Ground Test Missile
GSM	Generalized Sequential Machine		Gyratory Testing Machine
	General Syntactic Processor	**GTMA**	Gauge and Tool Makers Association [US]
	Greenough Stereomicroscope	**GTO**	Gate Turn-Off (Switch)
GSMB	Graphics Standards Management Board		Guide to Operations
GSO	Geostationary Orbit	**GTOW**	Gross Takeoff Weight
	Ground Safety Office	**GTP**	General Test Plan
	Ground Speed Oscillator		Group Transfer Polymerization
GSOC	German Space Operations Center [FRG]		Guanosine Triphosphate
GSOP	Guidance Systems Operations Plan	**GTR**	Gas Transmission Rate
GSP	General Simulation Program	**GTS**	General Technical Service
	Graphic Subroutine Package		Global Telecommunication System
	Guidance Signal Processor		Geostationary Technology Satellite
GSPR	Guidance Signal Processor Repeater		Ground Telemetry Subsystem
GSPO	Gemini Spacecraft Project Office [NASA]	**GTV**	Gas Turbine Vessel
GSR	Galvanic Skin Response		Gate Valve
	Glide Slope Receiver		Ground Test Vehicle
	Global Shared Resources	**GTW**	Gross Takeoff Weight
	Gray Scale Recording	**GU**	Guam

GUAR	Guarantee(d)
GUIDE	Guidance of Users of Integrated Data Processing Equipment
GULP	General Utility Library Program
GUN	Gunnery
GUSTO	Guidance using Stable Tuning Oscillations
GUY	French Guyana Space Center [ESA]
GVL	Gravel
GVMR	Gross Vehicle Mass Rating
GVT	Gravity Vacuum Tube System
GVW	Gross Vehicle Weight
GVWR	Gross Vehicle Weight Rating
GW	Gas Welding
	General Warning
	Gigawatt
	Glass Wool
GWC	Global Weather Center [US Air Force]
	Ground Water Council
GWe	Gigawatt, electrical
GWh	Gigawatthour
GWI	Ground Water Institute
GWP	Guelph-Waterloo Program (for Graduate Work in Physics) [Canada]
GWRI	Ground Water Research Institute
GWth	Gigawatt, thermal
GWU	George Washington University [US]
GY	Greenish Yellow
	Grey
GYFM	General Yielding Fracture Mechanics
GYP	Gypsum
GYR	Gyration
GYRO	Gyroscope
GZ	Ground Zero
GZT	Greenwich Zone Time

H

H	Hardness
	Hardware
	Hatch
	Head
	Heater
	Henry [unit]
	Hexapole
	Hexode
	Histidine
	Host
	Humidity
	Hydrant
	Hydrogen
h	hard
	heavy
	height
	high
	hour
HA	Half Add(er)
	Hard Axis
	Hemagglutinin

	High Altitude
	Home Address
	Hot Air
	Hydraulics Association
	Hydroxyapatite
Ha	Hahnium
ha	hectare
HAA	Height Above Airport
	Helix Aspersa Agglutinin
HAATC	High-Altitude Air Traffic Control
HABA	Hydroxyphenylazobenzoic Acid
HAC	High Alumina Cement
HAD	Half-Amplitude Duration
	High-Altitude Density
HADES	Hypersonic Air Data Entry System
HADR	Hughes Air Defense Radar
HADS	Hypersonic Air Data Sensor
	Hypersonic Air Data System
HAES	High Altitude Effects Simulation
HAF	High Abrasion Furnace
	High Altitude Fluorescence
HAFB	Holloman Air Force Base [US]
HAIC	Hearing Aid Industry Conference
HAIRS	High-Altitude Infrared Source
HAL	Highly-Automated Logic
HALO	High-Altitude, Low-Opening (Parachute)
HALSIM	Hardware Logic Simulator
HALT	Hydrate Addition at Low Temperatures
HAM	Hardware Associative Memory
	Hold-and-Modify (Image)
HAMS	High Altitude Mapping System
	Hour Angle of the Mean Sun
HAMT	Human-Aided Machine Translation
HANDS	High-Altitude Nuclear Detection Studies [NBS]
HANE	High-Altitude Nuclear Effects
HAOSS	High-Altitude Orbital Space Station
HAP	High-Altitude Platform
	Hydroxyapatite
HAPDAR	Hard Point Demonstration Array Radar
HAQO	Hydroxyaminoquinoline Oxide
HARA	High-Altitude Radar Altimeter
	High-Altitude Resonance Absorption
HARAC	High-Altitude Resonance Absorption Calculation
HARC	Houston Area Research Center [US]
HARCO	Hyperbolic Area Coverage
HARDTS	High-Accuracy Radar Transmission System
HARDWR	Hardware
HARM	High-Speed Anti-Radar Missile
HARP	Halpern Antiradar Point
	High Altitude Research Probe
	High Altitude Research Project
	High Altitude Rocket Probe
	Hitachi Arithmetic Processor
HARTRAN	Hardwell FORTRAN
HAS	Heading Altitude System
	Helicopter Avionics System [US Air Force]
HASCI	Human Applications Standard Computer Interface
HASL	Health and Safety Laboratory [USAEC]

HASP	High Altitude Sampling Program
	High Altitude Sampling Plane
	High Altitude Sounding Project
	High-Level Automatic Scheduling Program
	Houston Automatic Spooling Program
HAST	Highly-Accelerated Stress Test
HAT	High-Altitude Testing
HATAPH	Hexaalkyltriamidophosphazohydride
HATS	Helicopter Advanced Tactical System
	High-Accuracy Targeting Subsystem
	High-Altitude Terrain Sensor
	Hour Angle of the True Sun
hav	haversine
HAW	Heavy Anti-Armor Weapon
HAWS	Heavy Anti-Armor Weapon System
HAZ	Heat-Affected Zone
HB	Hard Black
	Brinell Hardness
	Horizontal Bridgman (Growth)
	Hose Bib
H/B	Hexadecimal-to-Binary
Hb	Hemoglobin
HBA	Hydroxybenzoic Acid
	Hydroxybutyrate
HBABA	Hydroxybenzeneazobenzoic Acid
HBD	Hydroxybutyrate Dehydrogenase
HBFP	Hematoxylin Basic Fuchsin Pecric
HBGM	Hypersonic Boost-Glide Missile
HBO	Heavy Batch Oven
HBSS	Hank's Balanced Salt Solution
HBT	Heterojunction Bipolar Transistor
HBWR	Halden Boiling Water Reactor
HC	Hand Control
	Hanging Ceiling
	Heating Coil
	Heuristic Concepts
	High Capacity
	High Carbon
	High Compression
	Holding Coil
	Hose Connector
	Host Computer
	Hot Cathode
	Hybrid Computer
	Hydrocarbon
	Hydrogen Chemisorption
H/C	Hand Carry
Hc	Hermitian conjugate
hc	(honoris causa) - honorary
HCB	Hexachlorobenzene
HCD	High Current Density
	Hollow Cathode Discharge
	Hot-Carrier Diode
HCE	Heater Control Electronics
	Hollow-Cathode Effect
HCEX	Hypercharge Exchange
HCF	High-Cycle Fatigue
	Highest Common Factor
	Host Command Facility
HCG	Hardware Character Generator
	Hexagonal Coupling

	Horizontal Location of Center of Gravity
hCG	Human Chorionic Gonadotropin
HCL	Hollow Cathode Lamp
	High, Common, Low (Relay)
	High Cost of Living
	Horizontal Centerline
HCMOS	High-Density Complimentary Metal-Oxide Semiconductor
	High-Speed Complimentary Metal-Oxide Semiconductor
HCMTS	High-Capacity Mobile Telecommunications System
HCP	Hepatocatalase Peroxidase
	Hexagonal Close-Packed
HCS	High-Carbon Steel
	Hundred Call Seconds
hCS	Human Chorionic Somatotropin
HCSS	Hospital Computer Sharing System
HCT	Heater Center Tap
HCU	Homing Comparator Unit
HCZ	Hydrogen Convection Zone
HD	Half Duplex
	Harbor Defense
	Hard Disk
	Hard-Drawn
	Harmonic Distortion
	Head
	Head Diameter
	Heavy Duty
	High Density
	Horizontal Distance
	Hot Drawing
	Hydrodynamics
H/D	Hexadecimal-to-Decimal
HDA	Head/Disk Assembly
	Hexadecenylacetate
	Horizontal Danger Angle
	Hydrodealkylation
HDAL	Hexadecenal
HDAM	Hierarchical Direct Access Method
HDAOS	Hydroxysulfopropyldimethooxyaniline
HDBK	Handbook
HDC	Heavy Duty Clamp
	Helium Direct-Current
HDDA	Hexadecadienacetate
HDDR	High-Density Digital Recording
HDDS	High-Density Data System
HDEP	High-Density Electronic Packaging
HDL	Handle
	Hardware Description Language
	Harry Diamond Laboratories [US Army]
	High-Density Lipoprotein
	High-Level Data Link
HDLC	High-Level Data Link Control
	Hierarchical Data Link Control
HDLG	Handling
HDLS	Headless
HDM	Hydrodensimeter
	Hydrodynamic Machining
HDME	Hanging Drop Mercury Electrode

HDMR	High-Density Moderated Reactor	**HECTOR**	Heated Experimental Carbon Thermal Oscillator Reactor
HDMS	Honeywell Distributed Manufacturing System	**HECV**	Heavy Enamel (Single) Cotton Varnish (Wire)
HDN	Harden		
	Hydrodenitrogenation	**HED**	Horizontal Electric Dipole
HDNP	High-Density Nickel Powder	**HEDC**	Heavy Enamel Double Cotton (Wire)
HDOC	Handy Dandy Orbital Computer		Houston Economic Development Council [US]
HDODA	Hexanediol Diacrylate		
HDOL	Hexadecanol	**HEDCV**	Heavy Enamel Double Cotton Varnish (Wire)
HDOS	Heath Disk Operating System		
HDP	High Detonation Pressure	**HEDI**	High Endoatmospheric Defense Interceptor
	Horizontal Data Processing	**HEDS**	Heavy Enamel Double Silk (Wire)
HDPE	High-Density Polyethylene	**HEDSV**	Heavy Enamel Double Silk Varnish
HDQRS	Headquarters	**HEDTA**	Hydroxyethylethylenediaminetriacetic Acid
HDR	Header	**HEE**	Hydrogen Environment Embrittlement
HDS	Huang Diffuse Scattering	**HEED**	High-Energy Electron Diffraction
	Hydrodesulfurization	**HEEDTA**	Hydroxyethylethylenediaminetriacetic Acid
HD/SCSI	Hard Disk/Small Computer System Interface	**HEEP**	Highway Engineering Exchange Program
		HEF	High-Energy Forging
HDSS	Hospital Decision Support System		High-Energy Forming
HDST	High-Density Shock Tube		High-Energy Fuel
HDT	Heat Deflection Temperature		High-Expansion Foam
	Heat Deflection Test	**HEG**	Heavy Enamel (Single) Glass (Wire)
	Heat Distortion Temperature	**HEI**	High-Energy Ignition
H/D/T	Hydrogen/Deuterium/Tritium (Ratio)	**HE-I**	High-Explosive Incendiary
HDTV	High Dissolving Television	**HEIS**	High-Energy Ion Scattering
HDW	Hardware	**HEI-T**	High-Explosive Incendiary with Tracer
HDWC	Hawaii Deep Water Cable [US]	**HELEX**	Hydrogenous Exponential Liquid Experiment
HDWD	Hardwood		
HD WHL	Hand Wheel	**HELMS**	Helicopter Multifunction System
HDX	Half Duplex	**HELP**	Hazardous Emergency Leaks Procedure
HE	Hall Effect		Helicopter Electronic Landing Path
	Handling Duplex		Highly-Extendable Language Processor
	Heat Exchanger	**HEM**	Hazardous Environment Machine
	Heptachlorine Epoxide		Hybrid Electromagnetic
	Heavy Enamel (Wire)		Hydrogen Embrittlement
	High Efficiency	**HEMA**	Hydroxyethyl Methacrylate
	High Explosive	**HEMAC**	Hybrid Electromagnetic Antenna Coupler
	Horizontal Equivalent	**HEM FIR**	Hemlock Fir
	Hydrogen Embrittlement	**HEMT**	High-Electron Mobility Transistor
He	Helium	**HENRE**	High-Energy Neutron Reactions Experiment [USAEC]
HEA	High-Efficiency Antireflection (Coating)		
HEAO	High-Energy Astronomy Observatory	**HEOS**	Highly-Eccentric Orbit Satellite
HEAP	Helicopter Extended Area Platform	**HEP**	Heterogeneous Element Processor
	High-Explosive Armor Piercing		High-Explosive Penetrating
HEAT	High-Explosive Antitank		Hydroelectric Power
	Hydroxyphenyl Ethyl Aminoethyl Tetralone	**HEPA**	High-Efficiency Particulate Air (Filter)
HEBC	Heavy Enamel Bonded (Single) Cotton (Wire)	**HEPC**	Hydro-Electric Power Commission [Canada]
		HEPCAT	Helicopter Pilot Control and Training
HEBDC	Heavy Enamel Bonded Double Cotton (Wire)	**HEPES**	Hydroxylethylpiperazine Ethanesulfonic Acid
HEBDP	Heavy Enamel Bonded Double Paper (Wire)	**HEPL**	High Energy Physics Laboratory [US]
HEBDS	Heavy Enamel Bonded Double Silk (Wire)	**HEPP**	Hoffman Evaluation Program and Procedure
HEBP	Heavy Enamel Bonded (Single) Paper (Wire)	**HEPPS**	Hydroxyethylpiperazine Propanesulfonic Acid
HEBS	Heavy Enamel Bonded (Single) Silk (Wire)	**HEPPSO**	Hydroxyethylpiperazine Hydroxypropane-sulfonic Acid
HEC	Heavy Enamel (Single) Cotton (Wire)		
	Hollerith Electronic Computer	**HEPTA**	Hydroxyethylethylenediaminetriacetic Acid
	Hydrogen Embrittlement Cracking	**HERA**	High-Explosive Rocket-Assisted
	Hydroxyethylcellulose	**HERALD**	Highly Enriched Reactor Aldermaston
hectol	hectoliter	**HERF**	High Energy Rate Forging

	High Energy Rate Forming
HERMES	Heavy Element and Radioactive Material Electromagnetic Separator
HERO	Hazards of Electromagnetic Radiation to Ordnance
	Heath Robot
	Hot Experimental Reactor of 0 (= Zero) Power
HES	Heavy Enamel (Single) Silk (Wire)
HESS	History of Earth Sciences Society
HET	Heavy Equipment Transporter
HE-T	High Explosive with Tracer
HETE	Hydroxyeicosatetraenoic Acid
HETP	Height Equivalent to a Theoretical Plane
	Hexaethyltetraphosphate
HETS	Height Equivalent to a Theoretical Stage
HEU	High-Enriched Uranium
	Hydroelectric Unit
HEVAC	Heating, Ventilating and Air Conditioning
HEW	Health, Education, and Welfare
HEX	Hexagon
	Hexadecimal
	Hexagonal
HEX HD	Hexagonal Head
HF	Half
	Hardenability Factor
	Hartree-Fock (Field)
	Haze Filter
	Height Finder
	High Frequency
	History File
	Hydraulic Fluid
	Hyperfine
Hf	Hafnium
HFB	Hopper-Feeder-Bolter
HFBR	High Flux Beam Reactor
HFC	High-Frequency Current
HFDF	High-Frequency Direction Finder
	High-Frequency Distribution Frame
HFE	Human Factors in Electronics
HFG	Heavy Free Gas
HFI	High-Frequency Inductance
HFIB	Hexafluoroisobutylene
HFIM	High-Frequency Instruments and Measurements [IEEE Committee]
HFIP	Hexafluoroisopropanol
HFIR	High Flux Isotope Reactor
HFIW	High-Frequency Induction Welding
HFL	High Free Lift
HFM	High-Frequency Microphone
HFO	High-Frequency Oscillator
HFORL	Human Factors Operations Research Laboratory
HFPS	High-Frequency Phase Shifter
HFRSc	Forged Roll Scleroscope Hardness Number, Model c
HFRSd	Forged Roll Scleroscope Hardness Number, Model d
HFRT	High-Frequency Resonance Technique
HFRW	High-Frequency Resistance Welding
HFS	Hyperfine Structure

	Hypothetical Future Samples
HF-SCF	Hartree-Fock Self-Consistent Field
hFSH	Human Follicle Stimulating Hormone
HFW	Horizontal Full Width
HFX	High Frequency Transceiver
HG	Hand Generator
	High Grade
	Horizon Grow
Hg	(Hydrargyrum) - Mercury
hg	hectogram
HGA	High Gain Antenna
HGC	Hercules Graphics Card
HGM	Human Gene Mapping
HGMS	High Gradient Magnetic Separation
HGR	Hanger
HGPRT	Hypoxanthineguanine Phosphoribosyl Transferase
HGRF	Hot Gas Radiating Facility
HGT	Height
	High Group Transmit
	Hypergeometric Group Testing
HGV	Heavy Goods Vehicle
HH	Double Hard
	Half Hard
	Handhole
	Hitchhiker [NASA]
1/2 H	Half-Hard
HHC	Hand-Held Computer
HHD	Hogshead
HH-G	Hitchhiker, Goddard [NASA]
HHHMU	Hydrazine Hand-Held Maneuvering Unit
HH-M	Hitchhiker, MSFC [NASA]
HHMU	Hand-Held Maneuvering Unit
HHTU	Hand-Held Teaching Unit
HHV	High Heat Value
HI	Hawaii [US]
	High Intensity
	Hydraulic Intensifier
	Hydraulics Institute [US]
	Hydriodic Acid
	Hydronics Institute
HIAA	Hydroxyindole Acetic Acid
HIAC	High Accuracy
HIB	High Iron Briquetting
HIBA	Hydroxyisobutyric Acid
HIBCC	Health Industry Bar Code Council
HIBEX	High-Acceleration Boost Experiment
	High Impulse Booster Experiment
HIC	Hybrid Integrated Circuit
	Hydrogen-Induced Cracking
HICAT	High-Altitude Clear Air Turbulence
HID	High-Intensity Discharge
HIDC	Hexamethylindodicarbocyanine
HIDE	Hydrogen-Induced Deformation Experiment
HIDEC	Highly-Integrated Digital Electronic Control
HIDES	High Absorption Integrated Defense Electromagnetic System
HIDM	High Information Delta Modulation
HIF	Hot Isostatic Forging
HIFAR	High Flux Australian Reactor
HiFi	High Fidelity

HIG	Hermetically-Sealed Integrated Gyro
HIGFET	Heterostructure-Insulated Gate Field-Effect Transistor
HIGHCOM	High Fidelity Compander
HIGH GASSER	High Geographic Aerospace Search Radar
HIGH TECH	High Technology
HIHAT	High-Resolution Hemispherical Reflector Antenna Technique
HI-HICAT	High High Altitude Clear Air Turbulence
HIIS	Honeywell Institute for Information Sciences [US]
HIL	High-Intensity Lighting
HILAC	Heavy Ion Linear Accelerator
HILAN	High-Level Language
HIMES	Highly-Maneuverable Experimental Spacecraft
HIOMT	Hydroxyindole-O-Methyltransferase
HIOS	Heath/Zenith Instrument Operating Software
HIP	High-Impact Polystyrene
	Hot Isostatic Pressing
	Hitachi Parametron Automatic Computer
HIPAR	High-Power Acquisition Radar
HIPERNAS	High-Performance Navigation System
HIPO	Hierarchy plus Input-Process-Output
HIPOT	High-Potential (Test)
HIPS	High-Impact Polystyrene
HIPSN	Hot Isostatically Pressed Silicon Nitride
HIQSA	Hydroxyiodoquinolinesulfonic Acid
	Horizontal Impulse Reaction
HIRAC	High Random Access
HIRD	High-Intensity Radiation Device
HIRES	Hypersonic In-Flight Refueling System
hi-res	high-resolution graphics
HIRIS	High-Resolution Interferometer/Spectrometer
HIRF	High-Intensity Reciprocity Failure
HIRL	High-Intensity Runway Lights
HIRNS	Helicopter Infrared Navigation System
HIRS	High-Impulse Retro-Rocket System
	High-Resolution Infrared Radiation Sounder
HIS	Helicopter Integrated System
	Honeywell Information System
	Hospital Information System
His	Histidine
HISAM	Hierarchical Indexed-Sequential Access Method
HIT	Houston International Teleport [US]
	Hypersonic Interference Technique
HITC	Hexamethylindotricarbocyanine
HITEC	High-Temperature Emission Control System
HITRAC	High Technology Training Access
HITS	Holloman Infrared Target Simulator
HIU	Hydrostatic Interface Unit
HIVOS	High Vacuum Orbital Simulation
HJ	Hose Jacket
HK	Hand Knob
	Hexakinase
	Knoop Hardness (Number)
HKSM	Henry Krumb School of Mines [US]

HKTDC	Hong Kong Trade Development Council
HL	Hearing Level
	Height-Length
	High-Level
	Hinge Line
	Hirth Lock
	Horizontal Line
	Hot Line
hL	hectoliter
hLA	Human Lymphocyte Antigen
HLAD	Hearing-Lookout Assist Device
HLAH	Hank's Lactalbuminhydrolysate
HLAS	Hot Line Alert System
HLB	Hydrophilic Lipophilic Balance
HLD	Helium Leak Detector
HLH	Heavy Lift Helicopter
	High-Level Heating
hLH	Human Luteinizing Hormone
HLL	High-Level Language
	High-Level Logic
HLMI	High Load Melt Index
HLN	Hexagonal Long Nipple
HLR	Holder
HLS	Heavy Logistic Support [US Air Force]
HLSE	High-Level, Single-Ended
HLSM	Homopolar Linear Synchronous Motor
HLT	Halt
HLTL	High-Level Transistor Logic
HLTTL	High-Level Transistor-Transistor Logic
HM	Hand-Made
	Heavy Mobile
	High-Melting
	High-Modulus
	Hydrometallurgy
hm	hectometer
HMB	Hydroxymethoxybenzaldehyde
HMBA	Hydroxymercurbenzoate
HMC	High-Strength Molding Compound
	Horizontal Motion Carriage
HMCS	Her (or His) Majesty's Canadian Ship
HMD	Hexamethylenediamine
	Hydraulic Mean Depth
HMDE	Hanging Mercury Drop Electrode
HMDS	Hexamethyldisilazane
HMDSO	Hexamethyldisiloxane
HME	High Vinyl Modified Epoxy
HMF	Heavy-Metal Fluoride
HMG	Heavy Machine Gun
hMP	Human Menopausal Gonadotropin
HMI	Hoisting Machinery Institute
HML	Hardware Modelling Library
HMM	Heavy Meromyosin
HMMS	Highway Maintenance Management System
HMMWV	High Mobility Multipurpose Wheeled Vehicle
HMO	Hueckel Approximation for Molecular Orbitals
HMOS	High-Density Metal-Oxide Semiconductor
HMP	Hexose Monophosphate
HMPA	Hexamethylphosphoramide
HMPT	Hexamethylphosphorous Triamide

HMRB	Hazardous Materials Regulation Board		Horsepower
HMS	Her (or His) Majesty's Ship		Host Processor
HMSF	Hexamethylenetetraselenafulvalene		Hot Pressing
HMSS	Hospital Management Systems Society		Hydroxyproline
HMSLD	Helium Mass Spectrometer Leak Detector	**HPA**	Helix Pomatia Agglutinin
HMTA	Hazardous Materials Transportation Act		High-Power Amplifier
HMTTeF	Hexamethylenetetratellurafulvalene	**HPBC**	Homopolar Pulse Billet Heating
HMW	High Molecular Weight	**HPBW**	Half Power Beamwidth
HMWPE	High Molecular Weight Polyethylene	**HPC**	High Pin Count
HN	Hexagonal Nipple		Hydroxypropylcellulose
	Hexagon Nut	**HPCS**	High-Pressure Core Spraying System
HNA	Hydroxynaphthoic Acid	**HPD**	Hearing Protective Device
HNED	Horizontal Null External Distance		Highest Posterior Density
HNIL	High Noise Immunity Logic		Horizontal Polar Diagram
HNL	Helium Neon Laser		Hough-Powell Device
HNPA	Home Numbering Plan Area	**HPF**	Heat Pipe Furnace
HNS	Hexanitrostilbene		Highest Probable Frequency
HNTD	Highest Non-Toxic Dose		High-Pass Filter
HNVS	Helicopter Night Vision System		Hot-Pressed Ferrite
HO	Head Office	**HPFD**	Hybrid Personal Floating Device
	High Order	**HPG**	Homopolar Generator
	Home Office		Hydroxypropyl Guar
	House	**HPhr**	Horsepower-Hour
	Hybrid Orbital	**HPIA**	Hydroxyphenlisopropyladenosine
Ho	Holmium	**HPIT**	High-Performance Infiltrating Technology
HOB	Homing on Offset Beacon	**HPL**	Human Placenta Lactogen
	Hot Ore Briquetting	**HPLC**	High-Performance Liquid Chromatography
HOBO	Homing Official Bomb	**HPLC**	High-Pressure Liquid Chromatography
HOBOS	Homing Bomb System	**HP-LEC**	High-Pressure Liquid Encapsulated
HODRAL	Hokushin Data Reduction Algorithm		Czochralski
	Language	**HPM**	Hybrid Phase Modulation
HOE	Height-of-Eye	**HPMA**	Hardwood and Plywood Manufacturers
HOI	Hypoiodous Acid		Association [US]
HOL	Hollow		Hydroxylpropylmethacrylate
HOLZ	Higher Order Laue Zone	**HPMC**	Hydroxypropylmethylcellulose
Hon	Honorable	**HPN**	High-Pass Notch
	Honorary		Horsepower Nominal
	Honors	**HPOT**	Helipotentiometer
HOP	Hybrid Operating Program	**HPP**	Harward Project Physics [US]
HOPE	Hydrogen-Oxygen Primary Extraterrestrial		Hydroxypyrazolopyrimidine
	[NASA]	**HPPH**	Hydroxylphenylphenylhydantoin
HOPG	Highly-Oriented Pyrolytic Graphite	**H-PRESS**	High-Pressure
HOPL	History of Programming Languages	**HPS**	Hazardous Polluting Substance
HOR	Heliocentric Orbit Rendezvous		Health Physics Society
	Horizontal		High Pressure Sintering
HORACE	H$_2$O Reactor Aldermaston Critical		Horizontal Pull Slipmeter
	Experiment	**HPSC**	High-Pressure Self-Combustion Sintering
HORIZ	Horizontal		Hot-Pressed Silicon Carbide
HOST	Hot Section Technology	**HPSN**	Hot-Pressed Silicon Nitride
HOT	Hand Over Transmitter	**HPT**	High-Precision Thermostat
HOTOL	Horizontal Takeoff and Landing		High-Pressure Test
HOTRAN	Hover and Transition Simulator		High-Pressure Turbine
HOW	Howitzer		Horizontal Plot Table
HOWAQ	Hot Water Quenching		Hydroxypyrenetrisulfonic Acid
HP	Hand Pump	**H PT**	High Point
	High-Pass (Filter)	**HPTE**	High-Performance Turbine Engine
	High Performance	**HPTET**	High-Performance Turbine Engine
	High Power		Technology
	High Pressure	**HPTLC**	High-Performance Thin Layer
	High Purity		Chromatography
	Horizontal Polarization	**HPW**	Homopolar Pulse Welding

	Hot Pressure Welding
HQ	Headquarters
	High Quality
	Hydroquinone
HQDA	Headquarters Department of the Army [US]
HQO	Hydroxyquinoline Oxide
HQIR	Hydro-Quebec Institute of Research [Canada]
HR	Handling Routine
	Hand Reset
	Harbour
	High Resilience
	High Resistance
	High Resolution
	Hoist Ring
	Hot Rolling
	Rockwell Hardness
hr	hour
HRA	Human Reliability Analysis
	Rockwell 'A' Hardness
HRAES	High-Resolution Auger Electron Spectroscopy
HRAI	Heating, Refrigerating and Air Conditioning Institute
HRB	Highways Research Board [US]
	Rockwell 'B' Hardness
HRC	Hardwood Research Council [US]
	High Rupturing Capacity
	Harmonically Related Carrier Frequency
	Horizontal Redundancy Check
	Rockwell 'C' Hardness
HRCG	Hexagonal Reducing Coupling
HRD	Human Resources Department
	Human Resources Development
H RD	Half Round
1/2 RD	Half Round
HRE	Rockwell 'E' Hardness
	Hypersonic Research Engine
HREELS	High-Resolution Electron Energy Loss Spectroscopy
HREM	High-Resolution Electron Microscopy
HRES	High-Resolution Electron Spectroscopy
HRF	Rockwell 'F' Hardness
HR15T	Rockwell '15T' Superficial Hardness
HRH	Rockwell 'H' Hardness
HRIO	Height-Range Indicator Operator
HRIR	High-Resolution Infrared Radiometer
HRIS	Highway Research Information Service [of HRB]
HRL	Horizontal Reference Line
	Human Resources Laboratory [US Air Force]
HRLC	High-Resolution Liquid Chromatography
HRLEELS	High-Resolution Low-Energy Electron Loss Spectroscopy
HRM	Hard-Rock Mining
HRN	Hexagonal Reducing Nipple
HRP	Hand Retractable Plunger
	Heat-Resistant Phenolic
	Horseradish Peroxidase
H&RP	Holding and Reconsignment Point
HRRC	Human Resources Research Center

HRRTS	High-Resolution Remote Tracking Sonar
HRS	High-Resolution Spectrography
	High-Resolution Spectometry
	High-Speed Rail
	Hot-Rolled Steel
	Hovering Rocket System
	Hydraulics Research Station
HRSEM	High-Resolution Scanning Electron Microscopy
HRSI	High-Temperature Reusable Surface Insulator
HRSIM	High-Resolution Selected Ion Monitoring
HRT	High-Resolution Tracker
HRU	Hydrological Research Unit
HRV	Heat Recovery Ventilator
	Hypersonic Research Vehicle
HRZ	Hazard Ranking System
HS	Half-Subtract(or)
	Heating Surface
	High Speed (Steel)
	High Stability
	High Strength
	Holographic Stereogram
	Horizon Scanner
	Horizon Sensor
	Hypersonics
HSA	Hemispherical Analyzer
HSAC	Health Safety and Analysis Center
HSAM	Hierarchical Sequential Access Method
HSC	Hydrocarbon Subcommittee
HSCC	Hydrogen-Induced Stress Corrosion Cracking
HSc	Scleroscope Hardness Number, Model c
HSD	High-Speed Data
	High-Speed Displacement
	Horizontal Situation Display
HSd	Scleroscope Hardness Number, Model d
HSDA	High-Speed Data Acquisition
HSDC	Hybrid Synchro-to-Digital Converter
HSE	House
HSG	Horizontal Sweep Generator
	Housing
HSGT	High-Speed Ground Transport
HSGTC	High-Speed Ground Test Center
HSI	Horizontal Situation Indicator
HSIP	Hsinchu Science-Based Industrial Park [Taiwan]
HSL	Hazardous Substances List
HSLA	High-Strength Low-Alloy (Steel)
HSM	High-Speed Memory
HSP	High-Speed Photometer
	High-Speed Printer
HSR	High-Speed Reader
HSRI	Highway Safety Research Institute [US]
HSRO	High-Speed Repetitive Operation
HSRTM	High-Speed Resin-Transfer Molding
HSS	High-Speed Steel
	High-Speed Storage
	History of Science Society
HSST	Heavy Section Steel Technology
HST	Hawaiian Standard Time [US]

	High Speed Technology		High-Temperature Reactor
	Hubble Space Telescope	HTRB	High-Temperature Reverse Bias
	Hypersonic Transport	HTS	Height Telling Surveillance
	Hypervelocity Shock Tunnel		High-Tensile Steel
HSTCO	High-Stability Temperature-Compensated		High-Tensile Strength
	Crystal Oscillator	HT/SPC	Heat Treating/Statistical Process Control
HSVP	High-Speed Vector Processor	HTSS	Hamilton Test Simulation System
HSWA	Hazardous and Solid Waste Amendment		Honeywell Time-Sharing System
HSZD	Hermetically-Sealed Zener Diode	HTTL	High-Power Transistor-Transistor Logic
HT	Heat Treat	HTTMT	High-Temperature Thermomechanical
	Heat Treatment		Treatment
	Heavy Truck	HT TR	Heat Treat(ment)
	Height	HTTT	High-Temperature Turbine Technology
	Height of Target	HTTVMT	High-Temperature Thermovibrational
	Hematoxylin		Mechanical Treatment
	High Temperature	HTU	Heat Transfer Unit
	High Tension	HTV	High Temperature and Velocity
	High Tide	HTW	High-Temperature Water
	Homing Transponder	HTX	High-Temperature Crystalline
	Horizontal Tab(ulation Character)	HU	Harvard University [US]
	Hydroxytryplamine	HUC	High Usage Circuit
HTA	Heavier than Air	HUD	Head-Up Display
	High-Temperature Amorphous	HUGO	Highly Unusual Geophysical Operation
	Hypophysiotropic Area	HUMRRO	Human Resources Research Office
HTC	High-Temperature Crystallizable	HV	Hardware Virtualizer
	High-Temperature Crystallization		Heating and Ventilation
	Horizontal Toggle Clamp		Heavy
HTCI	High-Tensile Cast Iron		High Vacuum
HT-CVD	High Temperature - Chemical Vapor		High Velocity
	Deposition		High Voltage
HTD	Hand Target Designator		High Volume
	Heterojunction Tunneling Diode		Hypervelocity
	High Torque Drive		Vickers Hardness
HTDA	High-Temperature Dilute Acid	H/V	Horizontal/Vertical
HTDS	High-Temperature Drawing Salt	HVA	High Voltage Apparatus
HTFFR	High-Temperature Fast-Flow Reactor		Homovanillic Acid
HTG	Hydrostatic Tank Gauging	HVAC	Heating, Ventilation and Air Conditioning
HTGCR	High-Temperature Gas-Cooled Reactor		High-Voltage Alternating Current
HTGPF	High-Temperature General Purpose Furnace	HVACC	High Voltage Apparatus Coordinating
HTGR	High-Temperature Gas-Cooled Reactor		Committee [ANSI]
HTH	High-Test Hypochlorite	HVAP	Hypervelocity, Armor Piercing
HTHP	High-Temperature/High-Pressure	HVAR	High-Velocity Airborne Rocket
HTI	High-Technology Intensive		High-Velocity Aircraft Rocket
	High-Temperature Impact	HVAT	High-Velocity Antitank
HTL	High-Threshold Logic	HVBO	Heterojunction Valence-Band Offset
	Hydroxyl Terminated Liquid	HVC	High-Velocity Clouds
HTLA	High-Temperature Low-Activity		Horizontal-Vertical Control
HTLV	Human T-Cell Lymphotrophic Virus	HVCA	Heating and Ventilating Contractors
HTM	High Temperature		Association
	Hypothesis Testing Model	HVCMOS	High-Voltage Complimentary Metal-Oxide
HTMP	Hydrooxytetramethylpiperidine		Semiconductor
H-TMS	Honeywell Test Management System	HVD	High-Velocity Detonation
HTN	Heat Treatable Nodular	HVDC	High-Voltage Direct Current
HTO	High-Temperature Oxidation	HVDF	High and Very-High Frequency Direction
	Horizontal Takeoff		Finding
HTP	High-Temperature Performance	HVE	Horizontal Vertex Error
	High-Temperature Pretreatment	HVEM	High-Voltage Electron Microscopy
	High Test Peroxide	HVF	High Viscosity Fuel
	Hydroxytryptophane	HVG	High-Voltage Generator
HTPB	Hydroxy Terminated Polybutadiene	HVHMD	Holographic Visor Helmet-Mounted Display
HTR	Heater	HVI	Human Visual Inspection

HVIC	High-Voltage Integrated Circuit
HVIS	Hypervelocity Impact System
HVL	Half-Value Layer
HVMS	High Voltage Mass Separator
HVOSM	Highway Vehicle Object Simulation Model
HVP	High Video Pass
HVPS	High-Voltage Power Supply
HVR	High-Vacuum Rectifier
	High-Voltage Regulator
HVRA	Heating and Ventilating Research Association [UK]
HVS	High Vacuum Seal
HVSCR	High-Voltage Selenium Cartridge Rectifier
HVT	Hydraulic Variable Timing
HVTP	High-Velocity Target Practice
	Hypervelocity Target Practice
HVTS	Hypervelocity Techniques Symposium
HVY	Heavy
HW	Half Wave
	Hammer Welding
	Hardware
	High Water
	Hollow
	Hot Water
H/W	Hardware
HWC	Hot Water Circulation
HWCTR	Heavy Water Components Test Reactor
HWE	Hot Wall Epitaxy
HWERL	Hazardous Waste Engineering Research Laboratory [EPA, US]
HWL	High Water Line
HWM	High Water Mark
HWOCR	Heavy Water Organic-Cooled Reactor
HWOR	Heavy Water Organic-Cooled Reactor
HWOST	High Water Ordinary Spring Tide
HWR	Heavy Water Reactor
HWTC	Hazardous Waste Treatment Council
HWW	Hot and Warm Working
HWY	Highway
HXSA	Hexenylsuccinic Anhydride
HYACS	Hybrid Analog-Switching Attitude Control System
HY BALL	Hydraulic Ball
HYBLOC	Hybrid Computer Block-Oriented Compiler
HYCOL	Hybrid Computer Link
HYCOTRAN	Hybrid Computer Translator
HYD	Hydraulic(s)
HYDAC	Hybrid Digital Analog Computer
HYDAPT	Hybrid Digital-Analog Pulse Timer
HYDR	Hydraulic(s)
HYDRO	Hydrodynamic(s)
	Hydrostatic(s)
HYDROLANT	Hydrographic Warning - Atlantic [NOO]
HYDROPAC	Hydrographic Warning - Pacific [NOO]
HYFES	Hypersonic Flight Environmental Simulator
HYG	Hygroscopic
HYL	Hojalata Y Lamina (Process)
HYLA	Hybrid Language Assembler
HYLO	Hybrid LORAN
HYP	Harvard, Yale, and Princeton Universities [US]

	Hyphen (Character)
	Hypothesis
	Hypotenuse
HYPERDOP	Hyperbolic Doppler
hyp log	hyperbolic logarithm
HYPSES	Hydrographic Precision Scanning Echo Sounder
HYSTAD	Hydrofoil Stabilization Device
HYTRESS	High Test Recorder and Simulator System
Hz	Hertz
HZIMP	Horizontal Impulse

I

I	Inertia
	Indicator
	Information
	Input
	Instruction
	Intensity
	Interruption
	Intrinsic
	Iodine
	Iron
	Island
	Isle
	Isoleucine
i	insoluble
IA	Incremental Analysis
	Indirect Addressing
	Industrial Application
	Instruction Address
	Instrumentation Amplifier
	Intermediate Amplifier
	International Ångstrom
	Iowa [US]
	Itaconic Acid
Ia	Iowa [US]
IAA	Indole Acetic Acid
	International Academy of Astronautics
IAAB	Inter-American Association of Broadcasting
IAAC	International Agriculture Aviation Center
IAAEES	International Association for the Advancement of Earth and Environmental Sciences
IAAS	Incorporated Association of Architects and Surveyors
IABSE	International Association for Bridge and Structural Engineering
IABTI	International Association of Bomb Technicians and Investigators
IAC	Industry Advisory Committee [US]
	Information Analysis Center
	Interference Absorption Circuit
	International Academy of Ceramics
	International Algebraic Compiler
	International Apple Core
	Ion-Assisted Coating

IACA	International Air Carriers Association
IACC	IIM-ASM Cooperation Committee
IACP	International Association of Computer Programmers
IACS	Inertial Attitude Control System
	Integrated Armament Control System
	International Annealed Copper Standard
	International Association of Classification Society
IAD	Integrated Automatic Documentation
IADES	Integrated Attitude Detection and Estimation System
IADIC	Integration Analog-to-Digital Converter
IADP	INTELSAT Assistance and Development Program
IAE	Institute for the Advancement of Engineering
	Integral Absolute Error
IAEA	International Atomic Energy Agency [of UN]
	International Atomic Energy Authority
I-AEDANS	Iodoacetamidoethylaminonaphthalenesulfonic Acid
IAEE	International Association of Earthquake Engineers
	International Association of Energy Economists
IAEG	International Association of Engineering Geology
IAEO	International Atomic Energy Organization [of UN]
IAES	Ion-Induced Auger Electron Spectroscopy
IAESTE	International Association for the Exchange of Students for Technical Experience
IAF	Initial Approach Fix
	International Astronautical Federation
IAFF	International Association of Fire Fighters
IAG	IFIP Administrative Group
	International Applications Group [of IFIP]
	International Association of Geodesy
IAGA	International Association of Geomagnetism and Aeronomy
IAGC	Instantaneous Automatic Gain Control
IAGOD	International Association on the Genesis of Ore Deposits
IAGP	International Antarctic Glaciological Project
IAGS	Inter-American Geodetic Survey [US]
IAH	International Association of Hydrogeologists
IAHE	International Association for Hydrogen Energy
IAHR	International Association for Hydraulic Research
IAI	Informational Acquisition and Interpretation
IAIA	International Association for Impact Assessment
IAIAS	Inter-American Institute of Agricultural Sciences
IAIN	International Association of Institutes of Navigation
IAL	International Algebraic Language
	Investment Analysis Language
IALA	International Association of Lighthouse Authorities
IAM	Initial Address Message
	Interactive Algebraic Manipulation
	International Association of Machinists
IAMAP	International Association of Meteorology and Atmospheric Physics
IAMAW	International Association of Machinists and Aerospace Workers
IAMC	Institute for the Advancement of Medical Communication
IAMCS	International Association for Mathematics and Computers in Simulation
IAMG	International Association for Mathematical Geology
IAMR	Institute of Arctic Mineral Resources [UA, US]
IAMS	Institute of Advanced Manufacturing Sciences
	International Advanced Microlithography Society
	International Association of Microbiological Societies
IAMTCT	Institute of Advanced Machine Tool and Control Technology [US]
IAMTEC	Institute of Advanced Machine Tool and Control Technology [US]
IAN	Integrated Acoustic Network
IANAP	Interagency Noise Abatement Program
IANEC	Inter-American Nuclear Energy Commission [US]
IAO	Internal Automation Operation
IAOPA	International Aircraft Owners and Pilots Association
IAP	Imaging Atom Probe
	Institutional Assistance Program
IAPA	Industrial Accident Prevention Association
IAPIP	International Association for the Protection of Industrial Property
IAPMO	International Association of Plumbing and Mechanical Officials
IAPO	International Association of Physical Oceanography
IAPR	International Association for Pattern Recognition
IAPS	International Association for the Production of Steam
IAQC	International Association for Quality Circles
IAR	Instruction Address Register
IARC	International Agricultural Research Center [of CGIAR, US]
IARIGAI	International Association of Research Institutes for the Graphic Arts Industry
IARU	International Amateur Radio Union
IAS	Immediate Access Storage
	Indicated Air Speed
	Industrial Applications Society
	Institute of Advanced Studies [US Army]
	Institute of Aeronautical Sciences
	Institute of Aerospace Sciences [US]

	Institute for Atmospheric Sciences [of ESSA]
	Instrument Approach System
	Integrated Analytical System
	International Association of Sedimentology
IASA	Insurance Accounting and Statistical Association
	International Air Safety Association
IASC	International Association for Statistical Computing
IASF	Instrumentation in Aerospace Simulation Facilities [IEEE Committee]
IASH	International Association of Scientific Hydrology
IASI	Inter-American Statistical Institute
IASPEI	International Association of Seismology and Physics of the Earth's Interior
IASY	International Active Sun Years
IAT	Internal Average Temperature
	Institute for Advanced Technology [of NBS]
	Institute for Applied Technology
	Institute of Automatics and Telemechanics
	International Automatic Time
IATA	International Air Transport Association
IATAL	International Association of Theoretical and Applied Limnology
IATC	International Air Traffic Communications
IATM	International Association for Testing Materials
IATUL	International Association of Technical University Libraries
IAU	International Association of Universities
	International Astronomical Union
IAV	International Association of Volcanology
IAVC	Instantaneous Automatic Video Control
	Instantaneous Automatic Volume Control
IAWPR	International Association of Water Pollution Research
IB	Induction Brazing
	Institute of Biology
	Instruction Book
	Interface Bus
ib	(ibidem) - in the same place
IBA	Indolebutyric Acid
	Institute of British Architects
	International Bauxite Association
	International Broadcasting Authority
IBAE	Ion Beam Assisted Etching
IBAM	Institute of Business Administration and Management [Japan]
IBBM	Iron Body Brass Mounted
	Iron Body Bronze Mounted
IBC	International Broadcasting Convention
IBCC	International Building Classification Committee
IBDH	Inboard Chromatography Data Handler
IBE	Ion Beam Epiplantation
IBED	Ion Beam Enhanced Deposition
IBEF	International Bio-Environmental Foundation
IBEW	International Brotherhood of Electrical Workers

IBF	Incident Bright-Field
	Institute of British Foundrymen
IBG	Interblock Gap
IBI	Intergovernmental Bureau for Informatics
	International Broadcasting Institute
ibid	(ibidem) - in the same place
IBIS	Intense-Bunched Ion Source
IBMM	Ion Beam Modification of Materials
IBMX	Isobutylmethylxanthine
IBN	Identification Beacon
IBP	Initial Boiling Point
	International Biological Program [NRC, US]
IBR	Institute for Basic Research [of NBS]
	Integral Boiling Reactor
	Integrated Bridge Rectifier
IBRD	International Bank for Reconstruction and Development [of UN]
IBRL	Initial Bomb Release Line
IBS	Institute for Basic Standards [of NBS]
	INTELSAT Business Services
IBSAC	Industrialized Building Systems and Components
IBSE	Ion Beam Sputter Etching
IBT	Inclined Bottom Tank
IBTO	International Broadcasting and Television Organization
IBTTA	International Bridge, Tunnel and Turnpike Association
IBW	Impulse Bandwidth
IBWM	International Bureau of Weights and Measures
IBY	International Biological Year
IC	Infrared Cell
	Immediate Constituent
	Impulse Conductor
	Inductance-Capacitance
	Inertial Component
	Information Circular
	Input Circuit
	Institute of Chemistry
	Instruction Counter
	Instrument Correction
	Insulated Conductor
	Integrated Circuit
	Intercommunications
	Intercrystalline Corrosion
	Interface Control
	Intergranular Corrosion
	Interior Communication
	Internal Connection
	Internal Conversion
	Ion Chamber
	Ion Chromatography
I&C	Installation and Checkout
i/c	in charge of
ICA	In-Circuit Analyzer
	Industrial Communications Association
	Initial Cruise Altitude
	Intercompany Agreement
	International Cartographic Association
	International Communications Association

	International Council on Archives
	Inventors Club of America
ICAD	Integrated Computer-Aided Design
	Integrated Control and Display
ICAF	Industrial College of the Armed Forces [US]
	International Committee on Aeronautical Fatigue
ICAI	International Commission for Agricultural Industries
ICALEO	International Congress on Applications of Lasers in Electrooptics
ICAM	Integrated Computer-Aided Manufacturing
ICAMRS	International Civil Aviation Message Routing System
ICAMS	Industrial Central Atmosphere Monitoring System
ICAN	International Committee of Air Navigation
ICAO	International Civil Aviation Organization [of UN]
ICAP	Inductively-Coupled Argon Plasma
	Industrial Conversion Assistance Program
ICAR	Indian Council of Agricultural Research [India]
	Inventory of Canadian Agricultural Research
ICARVS	Interplanetary Craft for Advanced Research in the Vicinity of the Sun
ICAS	Independent Collision Avoidance System
	Intermittent Commercial and Amateur Service
	International Council of the Aeronautical Sciences
ICB	Ionized Cluster Beam
ICBM	Intercontinental Ballistic Missile
ICBO	International Conference of Building Officials
ICBT	Intercontinental Ballistic Transport
ICC	Independent Channel Controller
	Industrial Communication Council [US]
	Information Commissioner of Canada
	Instrument Center Correction
	Internal Conversion Coefficient
	International Chamber of Commerce
	International Computation Center
	International Computer Center
	Interstate Commerce Commission [US]
ICCA	Independent Computer Consultants Association
ICCAD	IEEE Conference on Computer-Aided Design
ICCAIA	International Coordinating Council of Aerospace Industries Associations
ICCC	International Color Computer Club
	International Council for Computer Communication
ICC-CAPA	International Chamber of Commerce Commission on Asian and Pacific Affairs
ICCG	International Conference on Crystal Growth
ICCGR	International Cyclic Crack Growth Rate
ICCI	International Conference on Composite Interfaces
ICCM	International Conference on Composite Materials

	Isoconcentration Contour Migration
ICCP	Institute for Certification of Computer Professionals
	International Conference on Cataloguing Principles
ICCS	Ice Center Communications System
ICC-TM	Interstate Commerce Commission - Transport Mobilization [US]
ICD	Isocitrate Dehydrogenase
ICDA	International Cooperative Development Association
ICDD	International Center for Diffraction Data
ICDES	Item Class Description
ICDH	Isocitrate Dehydrogenase
ICE	In-Circuit Emulator
	Input-Checking Equipment
	Institute of Chartered Engineers
	Institute of Civil Engineers
	Institution of Civil Engineers [UK]
	Integrated Circuit Engineering
	Integrated Cooling for Electronics
	Intercity Experimental (Train)
	Intermediate Cable Equalizer
	Internal Combustion Engine
ICEA	Insulated Cable Engineers' Association
ICEC	Institute of Chartered Engineers of Canada
	International Cryogenic Engineering Conference
ICED	International Conference on Engineering Design
	Interprofessional Council on Environmental Design
ICEF	International Congress on Engineering and Food
ICEI	International Combustion Engine Institute
ICEM	Integrated Computer-Aided Engineering and Manufacturing
	International Conference on Electron Microscopy
	International Congress on Electron Microscopy
	International Council for Education Media
ICEPAK	Intelligent Classifier Engineering Package
ICER	Infrared Cell, Electronically-Refrigerated
ICES	Integrated Civil Engineering System
	International Conference of Engineering Societies
	International Council for the Exploration of the Sea
ICET	Institute for the Certification of Engineering Technicians
ICETK	International Council of Electrochemical Thermodynamics and Kinetics
ICETT	Industrial Council for Educational and Training Technology
ICF	Incommunication Flip-Flop
	Incremental Cost per Foot
	Inertial Confinement Fusion
	International Casting Federation
ICFM	Inlet Cubic Feet per Minute
ICFRM	International Conference on Fusion Reactor Materials

ICG	Interactive Computer Graphics
	International Commission on Glass
ICH	Ion Channeling
ICHCA	International Cargo Handling Coordination Association
IChemE	Institute of Chemical Engineers
	Institution of Chemical Engineers [UK]
ICHTM	International Congress on the Heat Treatment of Materials
ICI	Imperial Chemical Industries [UK]
	International Commission of Illumination
	Investment Casting Institute [US]
ICID	International Commission on Irrigation and Drainage
ICIP	International Conference on Information Processing
ICIREPAT	International Cooperation in Information Retrieval among Examining Patent Offices
ICIS	International Conference on Ion Sources
ICL	Incoming Line
	Interface Control Layer
ICM	Improved Capability Missile
	Incoming Message
	Instrumentation and Communications Monitor
	Integral Charge-Control Model
	International Congress of Mathematicians
	Ion Chromatography Module
ICMA	International Congress on Metalworking and Automation
ICMC	International Cokemaking Congress
ICMF	International Conference on (Advances in) Metal Forming Techniques
ICMI	International Commission on Mathematical Instruction
ICMP	Interchannel Master Pulse
ICMSF	International Commission on Microbiological Specifications for Food
ICN	Idle Channel Noise
	Instrument Communication Network
ICNDT	International Committee on Nondestructive Testing
	International Conference on Nondestructive Testing
ICNV	International Committee for the Nomenclature of Viruses
ICO	Inter-Agency Committee on Oceanography [US]
	International Commission for Optics
ICOGRADA	International Council of Graphic Design Associations
ICOLD	International Congress on Large Dams
ICOMAT	International Conference on Martensitic Transformation
ICON	Integrated Control
ICOR	Intergovernmental Conference on Oceanic Research [of UNESCO]
ICOSS	Inertial-Command Offset System
ICOT	Institute for (New Generation) Computer Technology [Japan]

ICOTOM	International Conference on Textures of Materials
ICP	Indicator Control Panel
	Inductively-Coupled Plasma
	Industrial Cooperation Program
	Interconnected Processing
	International Communication Planning
	International Computer Programs
	International Control Plan
ICPA	International Cooperative Petroleum Association
ICP-AES	Inductively-Coupled Plasma - Atomic Emission Spectroscopy
ICPC	Interrange Communications Planning Committee
ICPCSH	International Conference on Physics and Chemistry of Semiconductor Heterostructures
ICP-ES	Inductively-Coupled Plasma - Emission Spectroscopy
ICPIG	International Conference on Phenomena in Ionized Gases
ICP-MS	Inductively-Coupled Plasma - Mass Spectroscopy
ICPP	Idaho Chemical Processing Plant [of USAEC]
ICPR	International Conference on Production Research
ICPS	Industrial and Commercial Power System
	International Conference on the Properties of Steam
	International Congress of Photographic Sciences
ICPVT	International Council for Pressure Vessel Technology
ICR	Inductance-Capacitance-Resistance
	Institute for Cooperative Research
	Ion Cyclotron Resonance
	Iron-Core Reactor
ICRH	Institute for Computer Research in the Humanities [US]
ICRP	International Commission on Radiological Protection
ICRS	International Conference on Residual Stresses
ICRUM	International Commission on Radiation Units and Measurements
ICS	Infinity Color-Corrected System
	Inland Computer Service [US]
	Input Control System
	Institite of Chartered Surveyors
	Institute of Computer Science
	Integrated Communication System
	Integration Control System
	Intercommunications System
	Intermittent Control System
	International Chamber of Shipping
	International Cogeneration Society
	Interphone Control Station
	Iron Casting Society
ICSAB	International Civil Service Advisory Board

ICSC	Interim Communications Satellite Committee
	Inter-Ocean Canal Study Commission [US]
ICSEMS	International Commission for the Scientific Exploration of the Mediterranean Sea
ICSEP	International Center for the Solution of Environmental Problems
ICSFS	International Conference on Solid Films and Surfaces
ICSI	International Commission on Snow and Ice [of IASH]
	International Conference on Scientific Information
ICSID	International Council of Societies of Industrial Design
ICSMA	International Conference on the Strength of Metals and Alloys
ICSMP	Interactive Continuous Systems Modelling Program
ICSOS	Internal Conference on Structure of Surfaces
ICSP	International Conference on Shot Peening
ICSPS	International Council for Science Policy Studies
ICSPTF	International Conference on Structure and Properties of Thin Films
ICSS	International Conference on Solid Surfaces
ICST	Institute for Computer Sciences and Technologies
	Institute for Chemical Science and Technology
ICSU	International Council of Scientific Unions
ICT	In-Circuit Tester
	In-Circuit Testing
	Incoming Trunk
	Institute of Computer Technology
	Insulating Core Transformer
	Integrated Computer Telemetry
	Interaction Control Table
	Interactive Command Test
	International Computers and Tabulators
	International Critical Tables
ICTE	Inertial Component Test Equipment
ICTF	International Conference on Thin Films
ICTP	International Center for Theoretical Physics [US]
	International Conference on the Technology of Plasticity
ICTS	Isothermal Capacitance Transient Spectroscopy
ICU	Indicator Control Unit
	Instruction Control Unit
ICV	Initial Chaining Value
	Internal Correction Voltage
ICVM	International Conference on Vacuum Metallurgy
ICW	Interface Control Word
	Interrupted Continuous Wave
ICWM	International Committee of Weights and Measures
ICXOM	International Congress on X-Ray Optics and Microanalysis

ICY	International Cooperation Year
ID	Idaho [US]
	Identification
	Identifier
	Identity
	Indication Device
	Induced Draft
	Industrial Development
	Information Distributor
	Inferometer and Doppler
	Inner Diameter
	Inside Diameter
	Inside Dimension
	Instruction Decoder
	Interior Department
	Intermediate Description
	Internal Diameter
	Isotope Dilution
I/D	Instruction Data
id	(idem) - the same
IDA	Input Data Assembler
	Institute for Defense Analysis
	Integrated Digital Avionics
	Interactive Debugging Aid
	International Development Association [of UN]
	Ionospheric Dispersion Analysis
Ida	Idaho [US]
IDAC	Industrial Developers Association of Canada
IDAL	Indirect Data Access List
IDAP	Industrial Design Assistance Program
	Intelligent Data Acquisition System
	Interactive Data Access System
IDAST	Interpolated Data and Speech Transmission
IDAW	Indirect Data Address Word
IDB	Integrated Database
	Inter-American Development Bank
IDC	Image Dissector Camera
IDCC	International Data Coordinating Centers
IDCNS	Interdivisional Committee on Nomenclature and Symbols [of IUPAC]
IDCSP	Initial Defense Communication Satellite Project [US]
IDDD	International Direct Distance Dialing
IDDRG	International Deep-Drawing Research Group
IDEA	Industrial Designers Excellence Award
	Integrated Digital Electric Aircraft
IDEALS	Ideal Design of Effective and Logical Systems
IDEAS	Integrated Design and Engineering Automated System
IDEEA	Information and Data Exchange Experimental Activities
IDENT	Identification
IDEP	Interservice Data Exchange Program [USDOD]
IDES	Integrated Defense System
	Integrated Design and Engineering System
IDEX	Initial Defense Experiment
IDF	Incident Dark-Field
	Integrated Data File

	Intermediate Distribution Frame
IDFT	Inverse Discrete Fourier Transform
IDG	Individual Drop Glider
	Integrated Drive Generator
IDI	Improved Data Interchange
IDIAS	Ice Data Integration Analysis System
IDIIOM	Information Display Incorporated Input-Output Machine
IDIOT	Instrumentation Digital On-Line Transcriber
IDL	Interdisciplinary Research Laboratory
IDL&RS	International Data Library and Reference Service
IDM	Integral and Differential Monitoring
	Intelligent Data Mapper
	Interdiction Mission
IDMS	Integrated Database Management System
	Integrated Data Management System
ID-MS	Isotope Dilution - Mass Spectrometry
IDMS/R	Integrated Data Management System/Relation
IDN	Integrated Digital Network
IDOC	Inner Diameter of Outer Conductor
IDP	Industrial Data Processing
	Inosine Diphosphate
	Integrated Data Processing
	Intermodulation Distortion Percentage
IDOE	International Decade of Ocean Exploration
IDPC	Integrated Data Processing Center
IDPI	International Data Processing Institute
IDQ	Industrial Development Quotient
IDR	Industrial Data Reduction
	Intermediate Design Review
IDRC	International Development Research Center [Canada]
IDS	Identification Section
	Image-Dissector Scanner
	Instrument Development Section
	Integrated Data Store
	Interim Decay Storage
IDSA	Industrial Designers Society of America
IDSCS	Initial Defense Satellite Communications System
IDT	Instrumented Drop Tube
	Isodensiotracer
IDTS	Instrumentation Data Transmission System
IDTSC	Instrumentation Data Transmission System Controller
IDU	Industrial Development Unit
	Iododeoxyuridine
IE	Index Error
	Industrial Engineer(ing)
	Infrared Emission
	Initial Equipment
	Intermediate Electrode
	Isoelectric Point
I&E	Information and Education
ie	(id est) - that is
IEA	Institute of Environmental Action
	International Energy Agency
	International Ergonomics Association
IEB	International Executive Board

IEC	Information Exchange Center
	Integrated Electronic Component
	International Electrotechnical Commission
	Ion Exchange Chromatography
I&EC	Industrial and Engineering Chemistry [of ACS]
IECEJ	Institute of Electronic Communications Engineers of Japan
IECI	Industrial Electronics and Control Instrumentation [IEEE Group]
IECPS	International Electronic Packaging Symposium
IED	Individual Effective Dose
IEE	Institution of Electrical Engineers [UK]
IEEE	Institute of Electrical and Electronic Engineers [US]
IEEE-CS	Institute of Electrical and Electronic Engineers - Computer Society
IEES	Imaging Electron Energy Spectrometer
IEETE	Institution of Electrical and Electronics Technician Engineers [UK]
IEF	Isoelectric Focusing
IEG	Information Exchange Group
IEI	Implantation-Enhanced Interdiffusion
	Industrial Education Institute [US]
	Institution of Engineering Inspection [UK]
IEIS	Integrated Engine Instrument System
IEMT	International Electronic Manufacturing Technology
IEP	Image Edge Profile
	Immunoelectrophoresis
	Institute for Ecological Policies
	Instrumentation for the Evaluation of Pictures
	International Energy Program
IEPS	International Electronics Packaging Society
	International Electronic Packaging Symposium
IER	Institute for Econometric Research
	Institute for Environmental Research [of ESSA]
IERD	Industry Energy Research and Development
IERE	Institute of Electrical and Radio Engineers
	Institute of Electronic and Radio Engineers [Australia]
IES	Illuminating Engineering Society
	Institute for Earth Sciences [of ESSA]
	Institute of Environmental Sciences [US]
	Integral Error Squared
	Internal Environmental Simulator
	International Ecology Society
	Intrinsic Electric Strength
IESC	International Executive Service Corps
IESNA	Illuminating Engineering Society of North America
IET	Inelastic Electron Tunneling
IETC	Interagency Emergency Transportation Committee
IETG	International Energy Technology Group
IETS	Inelastic Electron Tunneling Spectroscopy
IEV	International Electrotechnical Vocabulary

I/EX	Instruction Execution
IF	Information Feedback
	Importance Factor
	Interface
	Interferon
	Intermediate Frequency
	Interstitial Free (Steel)
I/F	Interface
IFAC	Integrated Flexible Automation Center
	International Federation for Automatic Control
IFALPA	International Federation of Airline Pilot Associations
IFAP	International Federation of Agricultural Producers
IFATCA	International Federation of Air Traffic Controllers Associations
IFB	International Film Bureau
IFC	Image Flow Computer
	Instantaneous Frequency Correlation
	International Forging Congress
	International Formulation Committee [of ICPS]
IFCATI	International Federation of Cotton and Allied Textile Industries
IFCN	Interfacility Communication Network
IFCS	In-Flight Checkout System
	International Federation of Computer Sciences
IFD	Indentation Force Deflection
	Instantaneous Frequency Discriminator
	International Federation for Documentation
IFEMS	International Federation of Electron Microscope Societies
IFES	International Field Emission Symposium
IFF	Identification Friend or Foe
	If And Only If
	Interchange File Format
	Ionized Flow Field
	Isoelectric Focusing Facility
IFF/SIF	Identification Friend or Foe/Selective Identification Feature
IFH	Interferon, Human
IFHT	International Federation for (the) Heat Treatment (of Materials)
IFI	International Fastener Institute
IFIA	International Federation of Ironmongers and Iron Merchants Association
IFILE	Interface File
IFIP	International Federation for Information Processing
IFIPC	International Federation of Information Processing Congress
IFIPS	International Federation of Information Processing Societies
IFireE	Institution of Fire Engineers [UK]
IFL	Interfacility Link
	International Frequency List
IFLA	International Federation of Library Associations
IFME	International Federation of Medical Electronics
IFORS	International Federation of Operational Research Societies
IFLOT	Intermediate Focal Length Optical Tracer
IFO	Identified Flying Object
IFPI	International Federation of the Photographic Industry
IFPM	In-Flight Performance Monitor
IFPMM	International Federation of Purchasing and Materials Management
IFPVS	International Federation of Phonogram and Videogram Societies
IFR	Image-to-Frame Ratio
	Increasing Failure Rate
	In-Flight Refueling
	Instantaneous Frequency Receiver
	Instrument Flight Rules
	Internal Function Register
IFRB	International Frequency Registration Board [of ITU]
IFRF	International Flame Research Foundation
IFRU	Interference Frequency Rejection Unit
IFS	Interchange File Separator
	Intermediate Frequency Strip
	International Federation of Surveyors
IFSEM	International Federation of Societies for Electron Microscopy
IFSM	Inter-Fuel Substitution Model
IFT	Institute for Food Technologists [US]
	Interfacial Tension
	Intermediate Frequency Transformer
	International Foundation for Telemetering
IFTA	In-Flight Thrust Augmentation
IFTC	International Federation of Thermalism and Climatism
	International Film and Television Council
IFTE	Intermediate Forward Test Equipment
IFTS	In-Flight Test System
IFVME	Inspectorate of Fighting Vehicles and Mechanical Equipment [of MOD]
IFYGL	International Field Year for the Great Lakes
IG	Imperial Gallon
	Inertial Guidance
	Instructor's Guide
Ig	Immunoglobulin
IGA	Inner Gimbal Axis
	Intergranular Attack
IGAAS	Integrated Ground/Airborne Avionics System
IGC	Intergranular Corrosion
	International Geological Congress
	International Geophysical Committee
IGCAR	Indira Gandhi Center for Atomic Research [India]
IGCG	Inertial Guidance and Calibration Group [US Air Force]
IGCP	International Geological Correlation Program
IGE	Institution of Gas Engineers
IGES	Initial Graphics Exchange Specification
	International Graphics Exchange Standard
IGESUCO	International Ground Environment Subcommittee [of NATO]

IGF	Insulinlike Growth Factor
	Intergranular Fracture
IGFET	Insulated-Gate Field-Effect Transistor
IGI	Industrial Guest Investigator
IGIA	Interagency Group for International Aviation
IGM	Intergalactic Medium
IGN	Ignition
	International Geographic Institute
IGNITOR	Ignition Torus
IGOR	Intercept Ground Optical Recorder
IGPP	Institute of Geophysics and Planetary Physics [of SIO]
IGRC	International Gas Research Conference
IGS	Immunogold Staining
	Inert Gas-Shielded (Welding)
	Inertial Guidance System
	Information Generator System
	Integrated Graphics System
	Interchange Group Separator
IGSCC	Intergranular Stress Corrosion Cracking
IGSS	Immunogold Silver Staining
IGT	Institute of Gas Technology
	Insulated Gate Transistor
IGU	International Gas Union
	International Geographical Union
IGY	International Geophysical Year
IH	Industrialized Housing
	Interaction Handler
IHAS	Integrated Helicopter Avionics System
IHB	International Hydrographic Bureau
IHC	Interstate Highway Capability [US]
IHD	International Hydrological Decade
IHE	Institution of Highway Engineers
IHEA	Industrial Heating Equipment Association
IHET	Industrial Heat Exchanger Technology
IHF	Inhibit Halt Flip-Flop
	Intermediate High Frequency
IHFM	Institute of High-Fidelity Manufacturers
IHIS	Integrated Hospital Information System
IHP	Indicated Horsepower
	Inositol Hexaphosphate
IHPH	Indicated Horsepower-Hour
IHPhr	Indicated Horsepower-Hour
IHPVA	International Human Powered Vehicle Association
IHS	Information Handling Service
	Institute for Hydrogen Studies
	Integrated Hospital System
IHSI	Induction Heating Stress Improvement
IHTU	Interservice Hovercraft Trials Unit [of MOD]
IHW	Industrial and Hazardous Waste
IHX	Intermediate Heat Exchanger
IIA	Information Industry Association [US]
	International Institute of Agriculture
IIAES	Ion-Induced Auger Electron Spectroscopy
IIAILS	Interim Integrated Aircraft Instrumentation and Letdown System
IIAS	Interactive Instructional Answering System
	International Institute of Administrative Services

IIASA	International Institute for Applied Systems Analysis
IIC	International Institute of Communications
IID	Intermittent-Integrated Doppler
IIDQ	Isobutoxy Isobutoxycarbonyl Dihydroquinoline
IIE	Institute of Industrial Engineers
IIED	International Institute for Environment and Development
IIEE	Ion-Induced Electron Emission
IIHR	Iowa University Institute of Hydraulic Research [US]
I^2L	Integrated Injection Logic
IILS	International Institute for Labour Studies
IIM	Indian Institute of Metals
IIMA	Industrial Instruments Manufacturing Association
IIOC	Intelligent Input/Output Channel
IIOE	International Indian Ocean Expedition
IIP	Index of Industrial Production
	Instananeous Impact Point
IIPS	Interactive Instructional Presentation System
IIR	Infinite Impulse Response
	Institute of Intergovernmental Relations
	Integrated Instrumentation Radar
	International Institute of Refrigeration
	Isotactic Isoprene Rubber
I^2R	Resistance Heating
IIRC	Interrogation and Information Reception Circuit
IIRS	Institute for Industrial Research and Standards [US]
IIS	Institute of Industrial Supervisors
	Institute of Information Scientists
	Intrinsic Instruction Set
IISI	International Iron and Steel Institute
IISL	International Institute of Space Law
IISLS	Improved Interrogator Sidelobe Suppression
IISO	Institution of Industrial Safety Officers
IISRP	International Institute of Synthetic Rubber Producers
IISS	Integrated Information Support System
IIT	Illinois Institute of Technology [US]
IITA	International Institute of Tropical Agriculture [Nigeria]
IITRAN	Illinois Institute of Technology Translator
IITRI	Illinois Institute of Technology Research Institute [US]
IITV	Image Intensified Television
IIW	International Institute of Welding
IIX	Ion-Induced X-Rays
IJAJ	International Jitter Antijam
IJC	International Joint Commission
IJCAI	International Joint Conference on Artificial Intelligence
IJCRAB	International Joint Commission Research Advisory Board
IJJU	Intentional Jitter Jamming Unit
IJP	Ink Jet Printer
	Internal Job Processing

IK	Inverse Kinematics
IL	Illinois [US]
	Incorrect Length
	Indication Lamp
	Interface Loop
	Interior Length
	Interline
	Intermediate Language
	Interrupt Level
I&L	Installation and Logistics
ILA	Institute of Landscape Architects
	International Language for Aviation
ILAAS	Integrated Light Aircraft Attack System
	Integrated Light Attack Avionics System
ILAB	Bureau of International Labour Affairs [US]
ILAR	Institute of Laboratory Animal Resources
ILAS	Interrelated Logic Accumulating Scanner
	Instrument Landing Approach System
ILB	Inner Lead Bond
ILBM	Interleave Bit Map
ILBT	Interrupt Level Branch Table
ILC	Initial Launch Capability
	Instruction Length Code
	Instruction Length Counter
	Integrated Laminating Center
	Ion/Liquid Chromatography
ILCA	International Launching Class Association
ILCCS	Integrated Launch Control and Checkout System
Ile	Isoleucine
ILEED	Inelastic Low-Energy Electron Diffraction
ILF	Inductive Loss Factor
	Input Loading Factor
	Integrated Lift Fan
ILGB	International Laboratory of Genetics and Biophysics
ILI	Indiana Limestone Institute [US]
	Instant Lunar Ionosphere
ILIR	In-House Laboratories Independent Research
ILL	Illustration
	Institute Laue-Langevin [France]
Ill	Illinois [US]
ILLIAC	Illinois Integrator and Automatic Computer
ILLUM	Illumination
ILLUS	Illustration
ILLUST	Illustration
ILM	Information Logic Machine
ILMC	International Light Metal Congress
ILO	Injection-Locked Oscillator
	International Labour Office
	International Labour Organization [of UN]
ILP	Intermediate Language Processor
	International Lithosphere Program
ILPF	Ideal Low-Pass Filter
ILRT	Integrated Leakage Rate Test
ILS	Ideal Liquidus Structures
	Instrument Landing System
	Integrated Logistic Support
	Interlaminar Shear
	Ionization Loss Spectroscopy

ILSAP	Instrument Landing System Approach
ILSI	International Life Sciences Institute
ILSMT	Integrated Logistic Support Management Team
ILSO	Incremental Life Support Operation
ILSTAC	Instrument Landing System and TACAN
ILSW	Interrupt Level Status Word
ILTS	Institute of Low Temperature Science [Japan]
ILU	Illinois University [US]
ILW	Intermediate-Level Wastes
ILZ	International Lead and Zinc (Study Group)
ILZRO	International Lead and Zinc Research Organization
IM	Imaginary Part
	Imperial Measure
	Ingot Metallurgy
	Inner Marker
	Installation and Maintenance
	Institute of Meteorology
	Instrumentation and Measurements
	Integrated Modem
	Interceptor Missile
	Intermediate Missile
	Intermediate Modulus
	Intermodulation
	Interrupt Mask
	Item Mark
I&M	Inspection and Maintenance
I/M	Ingot Metallurgy
	Inspection and Maintenance
IMA	Ideal Mechanical Advantage
	Industrial Medical Association
	International Magnesium Association
	International Mineralogical Association
IMAG	Imaginary
IMAGE	Interactive Menu-Assisted Graphics Environment
IMAP	Integrated Mechanical Analysis Project [US]
IMarE	Institute of Marine Engineers
IMAS	Industrial Management Assistance Survey [US Air Force]
IMAW	International Molders and Allied Workers (Union)
IMC	Instructional Materials Center
	Integrated Maintenance Concept
	Interactive Medical Communications
	Instrument Meteorological Conditions
	Integrated Maintenance Concept
	International Maintenance Center
	International Maritime Committee
	International Micrographic Congress
IMCO	Intergovernmental Marine Consultative Organization
	International Maritime Countries Organization [now: IMO]
IMD	Institute for Marine Dynamics [Canada]
	Institute of Metals Division [of AIME]
	Intercept Monitoring Display
	Intermodulation Distortion
IMDC	Interceptor Missile Direction Center

IMDG	International Maritime Dangerous Goods
IMDO	Installation and Materiel District Office [of FAA]
IMDM	Iscove's Modified Dulbecco Medium
IME	Institute of Marine Engineers [UK]
	Institute of Mining Engineers
	Institution of Mechanical Engineers [UK]
	International Magnetosphere Explorer
IMECA	Independent Metallurgical Engineering Consultants of California [US]
IMechE	Institute of Mechanical Engineers
IMECO	International Measurement Congress
IMEP	Indicated Mean Effective Pressure
IMF	Institute of Metal Finishing
	International Monetary Fund
IMFI	Industrial Mineral Fiber Institute [US]
IMFP	Inelastic Mean Free Path
IMGCN	Integrated Missile Ground Control Network
IMH	Inlet Manhole
	Institute of Materials Handling
IMI	Ignition Manufacturers Institute
	Improved Manned Interceptor
IMIA	International Medical Informatics Association
IMIF	International Maritimes Industries Forum
IMinE	Institute of Mining Engineers
IMIP	Industrial Modernization Incentive Program
IMIR	Interceptor Missile Interrogation Radar
IMITAC	Image Input to Automatic Computer
IML	Inner Mold Line
	International Microgravity Laboratory [of NASA]
IMM	Institute of Mining and Metallurgy
	Integrated Maintenance Management
	Integrated Maintenance Manual
IMMA	Ion Microprobe Mass Analysis
IMME	Institute of Mining and Metallurgical Engineers
IMMP	Integrated Maintenance Management Plan
IMMS	International Material Management Society
IMMT	Integrated Maintenance Management Team
IMO	International Maritime Organization
IMP	Impact
	Imperial
	Imperfection
	Impulse
	Import
	Improvement
	Industrial Management Plan
	Information Management Program
	Integrated Microwave Products
	Intelligent Machine Prognosticator
	Interface Message Processor
	Interplanetary Monitoring Platform
	Interface Message Processor
	Interindustry Management Program
	Inosine Monophosphate
	Intrinsic Multiprocessing
IMPA	Information Management and Processing Association
	International Master Printers Association

IMPACT	Integrated Managerial Programming Analysis Control Technique
	Inventory Management Program and Control Technique
IMPATT	Impact Avalanche Transit Time Diode
IMPC	International Mineral Processing Congress
IMPCM	Improved Capability Missile
IMPE	International Meeting on Petroleum Engineering
IMPERF	Imperfect(ion)
IMPG	Impregnation
imp gal	imperial gallon [unit]
IMPI	International Microwave Power Institute
IMPL	Initial Microprogram Load
IMPR	Improvement
IMPRESS	Interdisciplinary Machine Processing for Research and Education in the Social Sciences
IMPRV	Improvement
IMPS	Integrated Master Programming and Scheduling System
IMPTS	Improved Programmer Test Station
IMR	Institute for Materials Research [of NBS]
	Institute of Mineral Research
	Internal Mold Release
	Interruption Mask Register
IMRA	Infrared Monochromatic Radiation
	International Marine Radio Association
IMRADS	Information Management Retrieval and Dissemination System
IMRI	Industrial Materials Research Institute
IMS	Industrial Management Service Group [ORF, Canada]
	Industrial Management System
	Industrial Mathematics Society [US]
	Information Management System
	Institute for Marine Science [US]
	Institute for Mathematical Statistics
	Instructional Management System
	Integrated Meteorological System [US Army]
	International Magnetospheric Studies
	International Metallographic Society
	Inventory Management and Simulator
	Ion Mobility Spectroscopy
IMSL	International Mathematical and Statistical Libraries
IMSR	Interplanetary Mission Support Requirements
IMSSCE	Interceptor Missile Squadron and Supervisory Control Equipment
IMSSS	Interceptor Missile Squadron Supervisory Station
IMS/VS	Information Management System/Virtual Storage
IMT	Industrial Materials Technology
IMTC	Instrumentation/Measurement Technology Conference [of IEEE]
IMTS	Improved Mobile Telephone Service
	International Machine Tool Show
IMU	Inertial Measurement Unit

	International Mathematical Union
IMW	International Map of the World
IN	Indiana [US]
	Indigo
	Inflammatory Response
	Inlet
	Input
	Interference-to-Noise (Ratio)
In	Indium
in	inch
INA	Institution of Naval Architects [UK]
	International Normal Atmosphere
INAA	Instrumental Neutron Activation Analysis
INAg	Ionospheric Network Advisory Group
INAS	Inertial Navigation and Attack System
INBD	Inboard
INC	Inclination
	Inclusion
	Incoming
	Incorporated
	Incorporation
	Increase
	Increment
	Input Control System
INCE	Institute of Noise Control Engineering
INCEND	Incendiary
INCH	Integrated Chopper
INCIN	Incinerator
INCL	Inclosure
	Inclusion
INCO	International Chamber of Commerce
INCOMEX	International Computer Exhibition
INCOR	Indian National Committee on Oceanic Research [India]
	Intergovernmental Conference on Oceanography
	Israel National Committee for Oceanographic Research
INCOSPAR	Indian National Committee for Space Research [India]
INCPT	Intercept
INCR	Increase
	Increment
	Interrupt Control Register
INCRA	International Copper Research Association
INCRE	Increment
INCUM	Indiana Computer Users Meeting [US]
IND	Index
	Indicator
	India(n)
	Indigo
	Indirect
	Induction
	Industry
	Independence
	Investigative New Drug
Ind	Indiana [US]
IndE	Industrial Engineer
INDEF	Indefinite
INDEP	Independent
INDEX	Inter-NASA Data Exchange

INDIAN	Interplanar Distances and Angles
INDOR	Internuclear Double Resonance
INDREG	Inductance Regulator
INDTR	Indicator-Transmitter
INDUSTR	Industrial
INDUSTL	Industrial
INEA	International Nuclear and Energy Association [also: IN&EA]
INEL	Idaho National Engineering Laboratory [US]
	Inelastic(ity)
	International Exhibition of Industrial Electronics
INEWS	Integrated Electronic Warfare System
INF	Information
	Intermediate-Range Nuclear Force
INFANT	Iroquois Night Fighter and Night Tracker
INFCE	International Nuclear Fuel Cycle Evaluation
INFL	Inflammability
INFO	Information
	Information Network and File Organization
INFOCEN	Information Center
INFOES	In-Flight Operational Evaluation of Space Systems
INFOL	Information-Oriented Language
INFRAL	Information Retrieval Automatic Language
ING	Intense Neutron Generator
INGO	International Nongovernmental Organization
InGPS	Indoleglycerolphosphate Synthetase
in Hg	inches of mercury [unit]
INIS	International Nuclear Information System [of IAEA]
INIT	Initial(ization)
INJ	Injection
INL	Internal Noise Level
in-lb	inch-pound
INLC	Initial Launch Capability
in lim	(in limine) - at the outset
in loc	(in locum) - in the place of
INM	International Nautical Mile
INMARSAT	International Maritime Satellite (Organization)
INMM	Institute of Nuclear Materials Management
INOSHAC	Indian Ocean and Southern Hemisphere Analysis Center
INP	Inert Nitrogen Protection
	Input
INPADOC	International Patent Documentation Center
INR	Impact Noise Reduction
	Interference-to-Noise Ratio
INREQ	Information on Request
INRO	International Natural Rubber Organization
INROWASP	In Rotating Water Spinning Process
INS	Inertial Navigation System
	Inspection
	Institute for Naval Studies
	Institute for Nuclear Studies [Japan]
	Insulation
	International Navigation System
	Interstation Noise Suppression
	Ion Neutralization Spectroscopy
	Iron Soldering

ins	inches
INSATRAC	Interception with Satellite Tracking
INSDOC	Indian National Scientific Documentation Center [India]
INSITE	Integrated Sensor Interpretation Techniques
in situ	(in situ) - in the natural or original position
INSJ	Institute for Nuclear Studies, Japan
INSOL	Insolubility
INSP	Inspection
	Inspector
INSPEC	Information Service in Physics, Electrotechnology and Control [of IEEE]
INSPEX	Engineering Inspection and Quality Control Conference and Exhibition
INSPN	Inspection
INSPTN	Inspection
INST	Instantaneous
	Instruction
	Instrument
	Institute
	Institution
INSTAR	Inertialess Scanning, Tracking and Ranging
INSTARS	Information Storage and Retrieval System
InstCE	Institute of Civil Engineers
INSTL	Installation
InstME	Institute of Mechanical Engineers
	Institute of Mining Engineers
InstMM	Institute of Mining and Metallurgy [UK]
INSTN	Institution
InstP	Institute of Physics
InstPet	Institute of Petroleum Engineers
INSTR	Instruction
	Instrument
InstWE	Institute of Water Engineers [UK]
INSULN	Insulation
INT	Integer
	Integration
	Interest
	Interior
	Internal
	International
	Interruption
	Intersection
	Interval
	Iodonitrotetrazolium
	Isaac Newton Telescope
INTCHG	Interchangeability
INTCO	International Code of Signals
INTELSAT	International Telecommunications Satellite (Organization)
INTEN	Intensity
INTER	Intermediate
	Interruption
INTERALIS	International Advanced Life Information System
INTERCOM	Intercommunication
INTERCON	International Convention [of IEEE]
INTERGALVA	International Galvanizing Conference
INTERMAG	International Magnetics Conference [of IEEE]
INTERP	Interpreter

INTERPLAS	International Plastics Exhibition and Conference
INTG	Integral
INTIP	Integrated Information Processing
INTIPS	Integrated Information Processing System
INTL	International
INTLK	Interlock
INTMT	Intermittent
INTOP	International Operations Simulation
INTPHTR	Interphase Transformer
INTR	Interior
	Interrupt
INTRAN	Input Translator
INTREX	Information Transfer Experiment
INTRO	Introduction
INTROD	Introduction
INTOR	International Tokamak Reactor
	International Torus
INU	Inertial Navigation Unit
INV	Inversion
	Invoice
INWATS	Inward Wide Area Telecommunication Service
in wg	inch water gauge [unit]
INX	Index Character
IO	In Order
	Input/Output
	Institute for Oceanography [of ESSA]
	Interpretative Operation
I-O	Input-Output
I/O	Input/Output
Io	Ionium
IOAP	Internally-Oxidized Alloy Powder
IOB	Input/Output Buffer
	Interorganization Board
IOBS	Input/Output Buffer System
IOC	Indirect Operating Costs
	Initial Operating Capability
	Initial Orbiting Capability [NASA]
	Input/Output Channel
	Input/Output Controller
	INTELSAT Operations Center
	Intergovernmental Oceanographic Commission [of UNESCO]
	International Organization Committee
IOCC	Input/Output Control Center
IOCE	Input/Output Control Element
IOCS	Input/Output Control System
IOCTF	IOC TDMA Facility
IOD	Input/Output Device
IOF	International Oceanographic Foundation
IOGA	Industry-Organized, Government-Approved
IOGEN	Input/Output Generation
IOL	Instantaneous Overload
	Intra-Ocular Lens
IOLS	Input/Output Label System
IOM	Input/Output Multiplexer
	Institute of Metals [UK]
	International Organization for Mycoplasmology
I/OM	Input/Output Multiplexer

ION	Institute of Navigation [US]
	Ionosphere and Aural Phenomena Advisory Committee [of ESRO]
IOOP	Input/Output Operation
IOP	Input/Output Processor
IOPAB	International Organization for Pure and Applied Biophysics
IOPB	International Organization of Plant Biosystematics
IOPS	Input/Output Programming System
IOR	Index of Refraction
	Indian Ocean Region
	Input/Output Register
IORB	Input/Output Record Block
IOREQ	Input/Output Request
IOS	Input/Output Selector
	Institute of Ocean Sciences
	International Organization for Standardization
IOT	In-Orbit Test Antenna
IOTA	Information Overload Testing Apparatus
IOU	Immediate Operation Use
	Input/Output Unit
	I owe you
IOUBC	Institute of Oceanography, University of British Columbia [Canada]
IOVST	International Organization for Vacuum Science and Technology
IOW	Institute of Welding
IOZ	Internal Oxidation Zone
IP	Ice Point
	Identification Point
	Imaginary Part
	Imipramine
	Impact Printer
	Incisoproximal
	Index of Performance
	Induced Polarization
	Industrial Park
	Industrial Production
	Information Processing
	Initial Phase
	Initial Point
	Inosine Phosphate
	Institute of Petroleum [UK]
	Instruction Processor
	Interface Processor
	Intermediate Pressure
	Internet Protocol
	Ion Pump
	Item Processing
	Powder Injection
I/P	Input
I&P	Indexed and Paged
IPA	Information Processing Association
	Information-Technology Promotion Agency [Japan]
	Integrated Photodetection Assembly
	Intermediate Power Amplifier
	International Phonetic Alphabet
	International Phonetic Association
	International Platform Association
	Isepentenyl Adenosine
	Isopropyl Alcohol
IPAC	Independent Petroleum Association of Canada
IPAD	Integrated Program for Aerospace Vehicle Design
	International Plastics Association Directors
IPADAE	Integrated Passive Action Detection Acquisition Equipment
IPAI	International Primary Aluminum Institute [UK]
IPC	Industrial Personal Computer
	Information Processing Center
	Information Processing Code
	Institute for Interconnecting and Packaging (Electronic) Circuits
	Institute for Printed Circuits
	Integrated Program for Commodities [of UN]
	Intermediate Processing Center
	Intermittent Position Control
	International Classification of Patents
	Interprocessor Communication
	Isopropylphenylcarbamate
IPCCS	Information Processing in Command and Control Systems
IPCEA	Insulated Power Cable Engineers Association [US]
IPCF	Interprogram Communication Facility
IPD	Insertion Phase Delay
	Isophoronediamine
IPDC	International Program for Development of Communications
IPDI	Isophorone Diisocyanate
IPDP	Industrial Programmed Data Processor
	Intervals of Pulsations of Diminishing Period
IPE	Industrial Plant Equipment
	Information Processing Equipment
	International Petroleum Exchange
	Interpret Parity Error
	Inverse Photoemission Experiment
IPETE	International Petroleum Equipment and Technology Exhibition
IPF	Indicative Planning Figures
	Inherent Power Factor
IPFM	Integral Pulse Frequency Modulation
IPGS	Industrial Postgraduate Scholarship [Canada]
IPI	Industrial Production Index
	Interchemical Printing Ink
IPL	Image Processing Laboratory [of NASA]
	Information Processing Language
	Initial Program Loader
	Initial Program Loading
	Integrated Payload
IPM	Illuminations per Minute
	Impulses per Minute
	Inches per Minute
	Inches Penetration per Month

	Incidental Phase Modulation
	Industrial Productivity Monitoring
	Input Position Map(per)
	Institute for Practical Mathematics [FRG]
	Interference Prediction Model
	Internal Polarization Modulation
	Interruptions per Minute
IPMA	In-Plant Printing Management Association [US]
IPMI	International Precious Metals Institute
IPN	Initial Priority Number
	Inspection Progress Notification
	Interpenetrating Network
	Interpenetrating Polymer Network
IPNL	Integrated Perceived Noise Level
IPNS	Intense Pulsed Neutron Source
IPO	Indolephenoloxidase
IPOEE	Institution of Post Office Electrical Engineers [UK]
IPP	Imaging Photopolarimeter
	Impact Prediction Point
	Institute of Plasma Physics [Japan]
IPPDSEU	International Plate Printers, Die Stampers and Engravers Union
IPPJ	Institute of Plasma Physics, Japan
IPPS	Institute of Physics and the Physical Society
IPQC	In-Process Quality Control
IPR	Inches per Revolution
	Inflow Performance Relationship
IPRC	International Personal Robot Congress (and Exposition)
IPRO	International Patent Research Office
IPS	Image Processing System
	Impact Polystyrene
	Impact Predictor System
	Inches per Second
	Installation Performance Specification
	Institute of Purchasing and Supply
	Instrumentation Power Supply
	Instrumentation Power System
	Intelligent Programming System
	International Pipe Standard
	Interruptions per Second
	International Planetarium Society
	Iron Pipe Size
IPSEP	International Project for Soft Energy Paths
IPSOC	Information Processing Society of Canada
IPSP	Inhibitory Postsynaptic Potential
IPSSB	Infomation Processing Systems Standards Board [of USASI]
IPST	International Practical Scale of Temperature
IPT	Indexed, Paged and Titled
	Internal Pipe Thread
IPTC	International Press Telecommunications Council
IPTG	Isopropylthiogalactopyranoside
IPTM	Interval Pulse Time Modulation
IPTO	Information Processing Technologies Office [of ARPA]
IPTS	International Practical Temperature Scale [of NBS]

IPU	Instruction Processing Unit
	Instruction Processor Unit
IPY	Inches Penetration per Year
IQ	Intelligence Quotient
iq	(idem quod) - the same as
IQC	International Quality Center [of EOQC]
IQEC	International Quantum Electronics Conference [IEEE]
iqed	(id quod erat demonstrandum) - that which was to be proved
IQI	Image Quality Indicator
IQS	Institute of Quantity Surveyors
IQSY	International Quiet Sun Year
IR	Industrial Relations
	Industrial Research
	Industrial Robot
	Informal Report
	Information Retrieval
	Infrared
	Infrared Radiation
	Inside Radius
	Insoluble Residue
	Instantaneous Relay
	Instruction Register
	Instrument Reading
	Insulation Resistance
	Intelligence Ratio
	Intermediate Register
	Internal Resistance
	Interrogator-Responder
	Interrupt Request
	Ireland
	Irish
	Isoprene Rubber
Ir	Iridium
IRAC	Interdepartmental Radio Advisory Committee
IRACQ	Infrared Acquisition Radar
	Instrumentation Radar and Acquisition Panel
IRAD	Independent Research and Development
IRAH	Infrared Alternate Head
IRAMS	Infrared Automatic Mass Screening
IRAN	Inspection and Repair as Necessary
IRAP	Industrial Research Assistance Program [Canada]
	Interagency Radiological Assistance Plan [USAEC]
IRAS	Infrared Astronomy Satellite
IRATE	Interim Remote Air Terminal Equipment
IRB	Infinitely Rigid Bear
	Infrared Brazing
IRBM	Intermediate Range Ballistic Missile
IRC	Industrial Relations Center
	Industrial Relations Council
	Instant Response Chromatography
	International Record Carrier
IRCM	Infrared Countermeasures
IRD	International Resource Development
IR&D	Independent Research and Development
IRDA	Industrial Research and Development Assistance

IRDB	Information Retrieval Database	**IRR**	Israel Research Reactor
IRDC	International Research Development Center	**IRRAD**	Infrared Range and Detection
IRDF	Indentation Residual Deflection Force	**IRRAS**	Infrared Reflection-Absorption Spectroscopy
IRDIA	Industrial Research and Development Incentives Act [Canada]	**IRRD**	International Road Research Documentation [of OECD]
IRDP	Industrial and Regional Development Program [Canada]	**IRREG**	Irregular
	Industrial Research Development Program	**IRRI**	International Rice Research Institute [Philippines]
IRE	Institute of Radio Engineers [of IEEE]	**IRRMP**	Infrared Radar Measurement Program
	Internal Reflection Element	**IRRS**	Infrared Reflection Spectroscopy
	Ireland	**IRS**	Independent Research Service
IRED	Infrared Emitting Diode		Infinitely Rigid System
IREX	International Research and Exchanges Board		Information Retrieval System
			Infrared Absorption Spectroscopy
IRF	International Road Federation		Infrared Soldering
IRFNA	Inhibited Red Fuming Nitric Acid		Interchange Record Separator
IRG	Interrange Instrumentation Group		Internal Revenue Service [US]
	Interrecord Gap		Interrecord Separator
IRGRD	International Research Group on Refuse Disposal		Investment Removal Salt
			Isotope Removal System
IRHD	International Rubber Hardness Degrees	**IRSS**	Inertial Reference Stabilization System
IRI	Industrial Research Institute [US]		Intelligent Remote Station Support
	Institution of the Rubber Industry [UK]	**IRSTD**	Infrared Search and Target Designation System
	International Robomation/Intelligence		
IRIA	Infrared Information and Analysis Center [US]	**IRSTS**	Infrared Search-Track System
		IRT	Index Return (Character)
IRICON	Infrared Vidicon Tube		Infrared Tracker
IRIG	Interrange Instrumentation Group [USDOD]		Institute of Reprographic Technology
IRIP	Industrial Research Institutes Program		Interrogator-Responder-Transponder
IRIS	Industrial Relations Information Service	**IRTE**	Institute of Road Transport Engineers [US]
	Infrared Information Symposia [US]	**IRTWG**	Interrange Telemetry Working Group
	Infrared Interferometer Spectrometer	**IRU**	Inertial Reference Unit
	Instant Response Information System		International Radium Unit
	Integrated Radar Imaging System		International Reference Unit
	Integrated Reconnaissance Intelligence System		International Road Transport Union
		IRW	Institute of Rural Water
	Interrogation and Location	**IS**	Incomplete Sequence (Relay)
IRL	Information Retrieval Language		Indian Standard
IRLS	Interrogation Recording and Location System [NASA]		Induction Soldering
			Information Sciences
IRM	Image Rejection Mixer		Information Separator
	Information Resources Management		Information Service
	Infrared Measurement		Information System
	Intermediate Range Monitor		Interference Suppressor
	Interim Research Memo(randum)		Interval Signal
	Isothermal Remanent Magnetization		Ion Source
IRMA	Information Revision and Manuscript Assembly		Island
			Isle
	Infrared Miss-Distance Approximeter		Isomer Shifts
IRMP	Infrared Measurement Program	**I&S**	Inspection and Survey
	Interservice Radiation Measurement Program	**ISA**	Information System Access
			Instrument Society of America
IROD	Instantaneous Readout Detector		Interrupt Storage Area
IROS	Increased Reliability Operational System	**ISAD**	Information Science and Automation Division
IRP	Initial Receiving Point		
IRPA	International Radiation Protection Association	**ISAM**	Indexed-Sequential Access Method
			Integrated Switching and Multiplexing
IRPAS	Infrared Photoacoustic Spectroscopy	**ISAP**	Information Sort and Predict
IRPBDS	Infrared Photothermal Beam Deflection Spectroscopy	**ISAR**	Information Storage and Retrieval
			International Society for Astrological Research
IRPM	Infrared Physical Measurement		

ISAS	Institute of Space and Aeronautical Sciences [Japan]
	Institute for Spectrochemistry and Applied Spectroscopy
	Institute of Space and Aeronautical Science [Japan]
	Institute of Space and Atmospheric Studies
ISB	Independent Sideband
	Infrared Security Barrier
	International Society of Biorheology
	International Society of Biometeorology
ISBB	International Society of Bioclimatology and Biometeorology
ISBN	International Standard Book Number
ISC	Information Society of Canada
	Initial Slope Circuit
	International Society for for Chronobiology
ISCAN	Inertialess Steerable Communication Antenna
ISCAS	International Symposium on Circuits and Systems [of IEEE]
ISCC	Intercrystalline Stress Corrosion Cracking
	International Semiconductor Conference
	International Service Coordination Center
	Intersociety Color Council [US]
ISCDS	International Stop Continental Drift Society
ISCES	International Symposium on Condensation and Evaporation of Solids
ISCLC	International Symposium of Column Liquid Chromatography
ISCLT	International Society of Clinical Laboratory Technologists
ISCOM	Indian Satellite for Communication Technology [India]
IScT	Institute of Science and Technology
ISD	Induction System Deposit
	Initial Selection Done
	International Society for Differentiation
	International Subscriber Dialing
	International Symbol Dictionary
IS&D	Integrate Sample and Dump
ISDF	Intercrystalline Structure Distribution Function
ISDN	Integrated Services Digital Network
ISDS	Inadvertent Separation and Destruct System
	Integrated Ship Design System [US Navy]
ISE	Induced Secondary Electron
	Institute of Space Engineering [Canada]
	Institution of Structural Engineers
	International Submarine Engineering
	Ion-Selective Electrode
I&SE	Installation and Service Engineering
ISEC	International Solvent Extraction Conference
ISEE	International Sun-Earth Explorer
ISEPP	International Sun-Earth Physics Program
ISES	International Solar Energy Society
ISF	Industrial Space Facility
	International Shipping Federation
ISFE	International Societies of Flying Engineers
ISFET	Ion-Sensitive Field-Effect Transistor
ISFMS	Indexed-Sequential File Management System

ISFR	International Society for Fluoride Research
ISG	Imperial Standard Gallon
	Instrumentation Selection Guide
ISGA	International Study Group for Aerograms
ISGE	International Society for Geothermal Engineering
ISHM	International Society for Hybrid Microelectronics
ISI	Indian Standards Institution [India]
	Indian Statistical Institute [India]
	Initial Spread Index
	Institute for Scientific Information [US]
	In-Service Inspection
	International Statistical Institute
	Ion Signal for Imaging
	Iron and Steel Institute
	Israel Standards Institute
ISIC	International Standard Industrial Classification
ISIP	Instantaneous Shut-In Pressure
ISIR	International Symposium on Industrial Robots
ISIS	Integrated Set of Information System [ILO, UN]
	International Satellites for Ionospheric Studies [US/Canada]
	International Science Information Service [US]
	International Shipping Information Service
ISISS	International Summer Institute in Surface Science
ISK	Insert Storage Key
ISL	Information Search Language
	Information System Language
	Instructional Systems Language
	Intelligent Systems Laboratory [of PARC, US]
	Interactive Simulation Language
	Intersatellite Link
ISLS	Improved Sidelobe Suppression
ISM	Industrial, Scientific and Medical
	Information Systems for Management
	Insulation System Module
ISMA	International Superphosphate Manufacturers Association
ISMC	International Switching Maintenance Center
ISMMS	Intrinsically Safe Mine Monitoring System
ISO	Individual System Operation
	International Science Organization
	International Standards Organization
	Isotropic
ISOC	Individual System/Organization Cost
ISODATA	Iterative Self-Organizing Data Analysis Technique
ISOM	Isometric
ISOTH	Isothermal
ISO WD	Isolation Ward
ISP	Independent Study Program
	Instruction Set Processor
	Interferometer Software Package
	International Society of Photogrammetry

	Italian Society of Physics
ISPA	International Society of Parametric Analysts
ISPC	International Sound Program Center
ISPE	International Society of Pharmaceutical Engineers
ISPEC	Insulation Specification
ISPEMA	Industrial Safety Personnel Equipment Manufacturers Association
ISPO	International Statistical Programs Office
ISPP	In-Service Professional Program
ISPRS	International Society of Photogrammetry and Remote Sensing
ISPS	Instruction Set Processor Specifications
ISR	Information Storage and Retrieval
	Interrupt Service Routine
	Intersecting Storage Ring
IS&R	Information Storage and Retrieval
ISRM	International Society for Rock Mechanics
ISRO	Indian Space Research Organization [India]
ISRS	International Safety Rating System
ISS	Ideal Solidus Structures
	Image Sharpness Scale
	Information Storage System
	Input Subsystem
	Institute of Systems Science [NUS, Singapore]
	Interlaminar Shear Strength
	International Society for Stereology
	International Student Service
	Interrupt Safety System
	Ionosphre Sounding Satellite
	Ion Scattering Spectroscopy
	Iron and Steel Society
	Issue
ISSC	Interdisciplinary Surface Science Conference
	International Ship Structures Congress
ISSCC	International Solid-State Circuits Conference
ISSMB	Information Systems Standards Management Board
ISSMFE	International Society for Soil Mechanics and Foundation Engineering
ISSN	International Standard Serial Number
ISSS	International Symposium on Surface Science
	International Society for the Study of Symbols
	International Society of Soil Science
ISSSE	International Society of Statistical Science in Economics
IST	Information Science and Technology
	Innovative Science and Technology
	Integrated System Test
	Integrated System Transformer
	International Skelton Tables
	International Standard Thread
ISTAR	Image Storage Translation and Reproduction
ISTC	Integrated System Test Complex
	International Steam Tables Conference [now: ICPS]
ISTF	Integrated Servicing and Test(ing) Facility
ISTFA	International Society for Testing and Failure Analysis

	International Symposium for Testing and Failure Analysis
ISTIM	Interchange of Scientific and Technical Information in Machine Language
ISTS	International Shock Tube Symposium
	International Symposium on Space Technology and Science
ISTVS	International Society for Terrain-Vehicle Systems
ISU	Inertial Sensor Unit
	Interface Switching Unit
	International Scientific Union
ISV	Independent Software Vendor
	Instantaneous Speed Variation
	International Scientific Vocabulary
ISVD	Information System for Vocational Decisions
ISVTNA	International Symposium on Vacuum Technology and Nuclear Applications
ISWG	(British) Imperial Standard Wire Gauge
IT	Immediate Transportation
	Indent Tab(ulation Character)
	Information Technology
	Information Theory
	Input Translator
	Institute of Technology
	Insulating Transformer
	Interrogator-Transponder
	Internal Thread
	Internal Translator
	International Tolerance
	In Transit
	Isometric Transition
	Isothermal Transformation
	Italian
	Italic
	Italy
	Item Transfer
ITA	Independent Television Authority [UK]
	Industrial Technical Adviser
	Industrial Transport Association
	Institute for Telecommunication and Aeronomy (of ESSA)
	International Tape Association
	International Trade Administration [US]
	International Tunneling Association
	International Typographic Association
ITAC	International Trade Affairs Committee
ITACS	Integrated Tactical Air Control System
ITAE	Integrated Time and Absolute Error
ITAL	Italian
	Italic
	Italy
ITARS	Integrated Terrain Access and Retrieval System
ITB	Intermediate Block
	Intermediate-Text-Block (Character)
ITC	Immense Technology Commitment
	Inland Transport Committee [of UN]
	Inter-American Telecommunications Network
	International Technology Council [of ABBE]

	International Teletraffic Congress
	International Television Center
	International Test Conference [US]
	International Tin Council
	International Trade Commission
	Interval Time Control
	Investment Tax Credit
	Ionic Thermoconductivity
IT&C	Industry, Trade and Commerce
ITCC	International Technical Communications Conference
ITCL	Item Class
ITCLC	Item Class Code
ITC/REE	(Ministry of) Industry, Trade and Commerce and Regional Economic Expansion [now: DRIE]
ITDSC	Item Description
ITDB	International Trade Data Bank
ITDE	Interchannel Time Displacement Error
ITDG	Intermediate Technology Development Group [UK]
ITE	Institute of Telecommunication Engineers
	Institute of Traffic Engineers [US]
	Institute of Transportation Engineers
	Integrated Test Equipment
	Intercity Transportation Efficiency
ITEA	International Test and Evaluation Association
ITEM	Interference Technology Engineer's Master
ITER	International Thermonuclear Experimental Reactor
ITEWS	Integrated Tactical Electronic Warfare System
ITF	Interactive Terminal Facility
	Interface File
ITFS	Instructional Television Fixed Service [US]
ITH	In-the-Hole (Drilling)
ITI	Industrial Technology Institute [US]
	Industrial Training Institute
ITIRC	IBM Technical Information Retrieval Center [US]
ITL	Industrial Test Laboratory [US Navy]
	Integrate-Transfer-Launch
	Inverse Time Limit
ITMC	International Transmission Maintenance Center
ITMJ	Incoming Trunk Message Junction
ITMT	Intermediate Thermomechanical Treatment
ITNS	Integrated Tactical Navigation System
ITNSA	Item Net Sales Amount
ITO	Indium-Tin Oxide
	International Trade Organization [of UN]
ITOS	Improved TIROS Operational Satellite
	Iterative Time Optimal System
ITP	Inosine Triphosphate
	Inspection Test Procedure
	Integral Thermal Process
	Integrated Test Program
ITPP	Institute of Technical Publicity and Publications
ITPR	Infrared Temperature Profile Radiometer

ITPS	Integrated Teleprocessing System
ITR	In-Core Thermionic Reactor
	INTELSAT Test Record
	Inverse Time Relay
	Ion Transfer Reaction
ITRC	International Tin Research Council
ITRI	International Tin Research Institute
ITRM	Inverse Thrromemanent Magnetization
ITS	Industrial Trade Show
	Information Transmission System
	Institute of Telecommunication Sciences [US]
	Insulation Test Specification
	Integrated Tracking System
	Integrated Trajectory System
	Intelligent Test System
	Interactive Training System
	Interface Test Set
	International Temperature Scale
	International Trade Show
	Invitation to Send
ITSA	Institute for Telecommunication Sciences and Aeronomy [of ESSA]
ITT	Impact Transition Temperature
	Institute for Textile Technology
	Intertoll Trunk
ITTFL	International Telephone and Telegraph Federal Laboratories [US]
ITU	International Telecommunication Union [of UN]
ITV	Improved TOW Vehicle
	Industrial Television
	Instructional Television
IU	Indiana University [US]
	Input Unit
	Instrumentation Unit
	Interference Unit
	International Unit
IUA	International Union of Architects
IUAPPA	International Union of Air Pollution Prevention Associations
IUB	International Union of Biochemistry
IUBS	International Union Biological Sciences
IUCAF	Inter-Union Commission on Allocation of Frequencies
IUCr	International Union on Crystallography
IUCSTP	Inter-Union Commission on Solar Terrestrial Physics
IUE	International Ultraviolet Explorer
IUEC	International Union of Elevator Constructors
IUFRO	International Union of Forest Research Organizations
IUG	ICES User Group
IUGG	International Union of Geophysics and Geodesy
IUGS	International Union of Geological Sciences
IULCS	International Union of Leather Chemists Societies
IUMP	International Upper Mantle Project
IUOE	International Union of Operating Engineers
IUP	Installed User Program

IUPAC	International Union of Pure and Applied Chemistry
IUPAP	International Union of Pure and Applied Physics
IUR	International Union of Railways
IUREP	International Uranium Resources Evaluation Project
IUS	Inertial Upper Stage
	Interchange Unit Separator
	Interim Upper Stage
IUSF	International Union for Surface Finishing
IUTAM	International Union of Theoretical and Applied Mechanics
IUTS	Inter-University Transit System
IUVSTA	International Union for Vacuum Science Techniques and Applications
IV	Current-Voltage
	Intermediate Voltage
	Iodine Value
IVAR	Insertion Velocity Adjust Routine [of NASA]
IVC	Intermediate Velocity Cloud
	International Vacuum Congress
IVD	Image Velocity Deceptor
	Innovative Vehicle Design
	Ion Vapor Deposition
IVDG	Innovative Vehicle Design Group
IVDS	Independent Variable Depth Sonar
IVDW	Integrated Voice/Data Workstation
IVHM	In-Vessel Handling Machine
IVHU	In-Vessel Handling Unit
IVI	Incremental Velocity Indicator
IVM	Initial Virtual Memory
IVMS	Integrated Voice Messaging System
IVMU	Inertial Velocity Measurement Unit
IVP	Initial Vapor Pressure
	Installation Verification Program
IVR	Integrated Voltage Regulator
IVV	Independent Verification and Validation
IVVS	In-Vessel Vehicle System
IW	Index Word
	Induction Welding
IWA	International Woodworkers of America
IWAHMA	Industrial Warm Air Heater Manufacturers Association
IWC	International Welding Conference
IWCS	Integrated Wideband Communications System
IWCS/SEA	Integrated Wideband Communications System/Southeast Asia
IWDA	Independent Wire Drawers Association
IWES	International Waste Energy System
IWFNA	Inhibited White Fuming Nitric Acid
IWG	Iron Wire Gauge
IWP	Interim Working Party
IWRA	International Water Resources Association
IWS	Institute of Work Studies
IWSA	International Water Supply Association
IWSc	Institute of Wood Science
IWSP	Institute of Work Study Practitioners
IWtC	International Wheat Council
IX	Index Register

IXC	Inter-Exchange Channel
IYSWIM	If you see what I mean

J

J	Jack
	January
	Joule
	Journal
	Jute
JA	Jump Address
JACC	Joint Automatic Control Conference
JACE	Joint Alternate Command Element
JAEC	Japan Atomic Energy Commission
JAERI	Japan Atomic Energy Research Institute
JAFNA	Joint Air Force - NASA [US]
JAGOS	Joint Air-Ground Operations System
JAIEG	Joint Atomic Information Exchange Group [USDOD]
JAIF	Japan Atomic Industrial Forum
JALPG	Joint Automatic Language Processing Group
JAMA	Japan Automobile Manufacturers Association
JAMTS	Japan Association of Motor Trade and Service
JAN	January
	Joint Army - Navy
JANAIR	Joint Army - Navy Aircraft Instrument Research [USDOD]
JANOT	Joint Army - Navy Ocean Terminal
JANS	Joint Army - Navy Specification [USDOD]
JAP	Japan
	Japanese
JAPEX	Japan Petroleum Exploration
JAPIA	Japan Auto Parts Industries Association
JAPIO	Japan Patent Information Organization
JARC	Joint Avionics Research Committee
JASC	Japan Sea Cable
JASDF	Japan Air Self-Defense Force
JASO	Japan Standards Organization
JATCRU	Joint Air Traffic Control Radar Unit
JATIS	Japan Technical Information Service
JATO	Jet-Assisted Takeoff
JB	Junction Box
JBMMA	Japan Business Machine Makers Association
JC	Joint Commission
JCA	Joint Commission on Accreditation
JCAE	Joint Committee on Atomic Energy [US]
JCAM	Joint Commission on Atomic Masses
JCAR	Joint Commission on Applied Radioactivity
JCASR	Joint Committee on Avionic Systems Research
JCB	Joint Communications Board
JCC	Joint Communications Center
	Joint Computer Conference
	Joint Control Centers
JCCCOMNET	Joint Coordination Center Communications Network

JCEC	Joint Communications Electronics Committee
JCEG	Joint Communications Electronics Group
JCEWG	Joint Communications Electronics Working Group
JCF	Joint Coordinating Forum
JCI	Joint Communications Instruction
JCII	Japan Camera Inspection Institute
JCL	Job Control Language
JCMB	Joint Committee on Medicine and Biology [of IEEE and ISA]
JCPDS	Joint Committee on Powder Diffraction Standards
JCS	Job Control Statement
	Joint Coordinate System
JCT	Journal Control Table
	Junction
JCTFI	Joint Committee for Training in the Foundry Industry
JCTN	Junction
JD	Joined
JDA	Japan Defense Agency
JDES	Joint Density of Electronic States
JDL	Job Description Language
	Job Descriptor Language
JDS	Job Data Sheet
JE	Jet Engine
	Job Entry
	Junction Exchange
	June
JEA	Joint Endeavor Agreement
JEBM	Jet Engine Base Maintenance
JEBM-RR	Jet Engine Base Maintenance Return Rate
JEC	Joint Economic Committee [US]
JECC	Japan Electronic Computer Center
JECS	Job Entry Central Service
JEDEC	Joint Electron Device Engineering Council [US]
JEIDA	Japan Electronic Industry Development Association
JEIPAC	JICST Electronic Information Processing Automatic Computer
JEL	Johnson Elastic Limit
JEM	Japan Experiment Module
	Joint Endeavor Manager
JEMC	Joint Engineering Management Conference
JEOCN	Joint European Operations Communications Network
JEOL	Japan Electro-Optics Laboratories
JEPIA	Japan Electronic Parts Industry Association
JEPOSS	Javelin Experimental Protection Oil Sands System
JEPP	Joint Emergency Planning Program
JEPS	Job Entry Peripheral Service
JEQ	Jump Equal
JERC	Joint Electronic Research Committee
JES	Job Entry Subsystem
	Job Entry System
JESAC	Joint Engineering Student Activity Committee
JESSI	Joint European Silicon Submicron Initiative

	Junior Engineers and Scientists Summer Institute [US]
JET	Joint Enroute Terminal
	Joint European Torus (Reactor)
JETDS	Joint Electronics Type Designation System
JETEC	Joint Electron Tube Engineering Council
JETRO	Japan External Trade Organization
JETS	Junior Engineering Technical Society [US]
JFAP	Joint Frequency Allocation Panel
JFET	Junction Field-Effect Transistor
JFKSC	John F. Kennedy Spaceflight Center [also: KSC]
JFL	Joint Frequency List
JFP	Joint Frequency Panel
JGN	Junction Gate Number
JHD	Joint Hypocenter Determination
JHU	Johns Hopkins University [US]
JIC	Japan Information Center
	Joint Industrial Council
	Joint Industry Conference
	Just-in-Case
JICST	Japan Information Center of Science and Technology
JIE	Junior Institute of Engineers
JIFDATS	Joint In-Flight Data Transmission System [USDOD]
JIFTS	Joint In-Flight Transmission System
JIII	Japan Institute of Invention and Innovation
JILA	Joint Institute for Laboratory Astrophysics [US]
JIM	Japan Institute of Metals
JIMA	Japan Industrial Management Association
JIMIS	Japan Institute of Metals International Symposium
JIMTOF	Japan International Machine Tool Fair
JIOA	Joint Intelligence Objectives Agency
JIP	Joint Input Processing
JIRA	Japan Industrial Robot Association
JIS	Japan Industrial Standards
JISC	Japan Industrial Standards Committee
JIT	Job Instruction Training
	Just-In-Time
JIT/TQC	Just-In-Time/Total Quality Control
JK	Jack
JLP	Jig Leg Plate
JMA	Japan Meteorological Agency
	Japan Microphotography Association
JMC	Joint Maritime Commission
	Joint Meteorological Committee
JMD	Jungle Message Decoder
JME	Jungle Message Encoder
JMED	Jungle Message Encoder-Decoder
JMI	Japan Metals Institute
JMP	Jump
JMSPO	Joint Meteorological Satellite Program Office
JN	Jam Nut
	Junction
JNACC	Joint Nuclear Accident Coordinating Center [US]
JNE	Jump Not Equal

JNL	Journal
JNR	Japan National Railways
JNR	Junior
JNSC	Japan Nuclear Safety Commission
	Joint Navigation Satellite Committee
JNSDA	Japan Nuclear Ship Development Agency
JO	Job Order
	Joint Organization
	Junction Office
JOBNO	Job Number
JOBLIB	Job Library
JOG	Joggle
JOI	Joint Oceanographic Institution
JOIDES	Joint Oceanographic Institution Deep Earth Samplings
JOVIAL	Jules' Own Version of International Algorithmic Language
JP	Jet Propulsion
	Jet Propellant
	Jet-Propelled
	Jig Pin
	Job Processing
	Job Processor
J&P	Joists and Planks
JPA	Job Pack Area
JPB	Joint Planning Board [US]
JPDR	Japan Power Demonstration Reactor
JPL	Jet Propulsion Laboratory [US]
JPNL	Judged Perceived Noise Level
JPPS	Joint Petroleum Products Subcommittee
JPRS	Joint Publications Research Service [US]
JPTF	Joint Parachute Test Facility [USDOC]
JR	Japan Rail [also: JNR]
	Junior
JRATA	Joint Research and Test Activitities
JRB	Jig Rest Button
JRC	Joint Research Center
JRC-ISPRA	Joint Research Center at Ispra [Italy]
JRD	Joint Research and Development
JRDOD	Joint Research and Development Objectives Document
JRIA	Japan Radioisotope Association
JRP	Jute-Reinforced Plastics
JRPS	Japan Reinforced Plastics Society
JS	Jam Strobe
	Jaswal-Sharma (Theory)
	Joint Services
J/S	Jamming to Signal
JSA	Job Safety Analysis
JSAM	JES Spool Access Method
JSC	Japan Science Council
	Johnson Space Center [NASA]
	Joint Security Control
JSCFA	Japan Steel Castings and Forgings Association
JSEM	Japan Society for Electron Microscopy
	Japan Society of Electrical Discharge Machining
JSEP	Joint Services Electronics Program [US]
JSFM	Japan Society for Strength and Fracture of Materials

JSHS	Junior Science and Humanities Symposium
JSIA	Joint Service Induction Area
JSL	Job Specification Language
JSME	Japan Society of Mechanical Engineers
JSMS	Japan Society of Materials Science
JSS	Joint Surveillance System
JSTARS	Joint Surveillance Target Attack Radar System
JT	Joint
JTAC	Joint Technical Advisory Committee
JTB	Joint Transportation Board [US]
JTC	Joint Telecommunications Committee
JTDS	Joint Track Data Storage
JTF	Joint Task Force
JTIDS	Joint Tactical Information Distribution System
JTIS	Japan Technical Information Service
JTRU	Joint Services Tropical Research Unit
JUDGE	Judged Utility Decision Generator
JUG	Joint Users Group
JUGFET	Junction-Gate Field-Effect Transistor
JUL	July
JUN	June
	Junior
JUNC	Junction
JUNE	Joint Utility Notification for Excavators
JUNR	Junior
JUPITER	Juvenescent Pioneering Technology for Robots
JUSE	Japanese Union of Scientists and Engineers
JW	Jacket Water
JWRI	Japan Welding Research Institute

K

K	Karat
	Keel
	Kelvin
	Key
	Kilo-Ohm
	Kip [= 1000 lbs]
	X-Ray
	2^{10} [= 1024 bytes of computer memory]
	(Kalium) - Potassium
	Lysine
k	kilo
KA	Kynurenic Acid
kA	kiloampere
KAEDS	Keystone Association for Educational Data Systems
KAFB	Kirkland Air Force Base [US]
KAIST	Korea Advanced Institute of Science and Technology
Kal	Kalamein
KALDAS	Kidsgrove ALGOL Digital Analog Simulation
KALDO	Kalling-Domnarvet (Steelmaking Process)
Kan	Kansas [US]

Kans	Kansas [US]	**K-M**	Kossel-Muellenstedt (Diffraction Patterns)
KB	Keyboard	**kM**	kilomega
kb	kilobit	**km**	kilometer
KBA	Ketobutyraldehydedimethyl Acetal	**kMc**	kilomegacycle
kbar	kilobar	**KMER**	Kodak Metal Etch Resist
KBD	Keyboard	**kmps**	kilometers per second
KBES	Knowledge-Base Expert Systems	**KMS**	Kloeckner-Metacon Steel Converter
kbps	kilobits per second	**KMT**	Kinetic-Molecular Theory
kc	kilocycle	**KN**	Knot
kcal	kilocalorie		Knurled Nut
KCL	Kirchhoff's Current Law	**kN**	kilonewton
kcps	kilocycles per second	**KNT**	Short-Cycle Gas Nitriding
kc/s	kilocycles per second	**KO**	Knockout
KD	Kiln-Dried	**kohm**	kilo-ohm
	Knocked Down	**KORF**	Korf Oxy-Refining Fuel
KDB	Korea Development Bank	**KORSTIC**	Korea Scientific and Technological Information Center
KDO	Ketodeoxyoctonate		
KDP	Potassium Dihydrogenphosphate	**KOTRA**	Korean Organization for Trade Advancement
KDR	Keyboard Data Recorder		
KDS	Keydata Station	**KP**	Kick Plate
	Kinetic-Dynamic Simulation		Key Pulsing
KE	Kinetic Ellipsometry		Key Punch
	Kinetic Energy	**KPA**	Klystron Power Amplifier
KEAS	Knots Equivalent Air Speed	**kPa**	kilopascal
Ken	Kentucky [US]	**KP&D**	Kick Plate and Drip
KET	Krypton Exposure Technique	**KPIC**	Key Phrase in Context
keV	kiloelectronvolt	**kpsi**	kilopounds per square inch
KEYBD	Keyboard	**KPSM**	Klystron Power Supply Modulator
KFRP	Kevlar Fiber-Reinforced Plastics	**Kr**	Krypton
KFT	Karl Fischer Titration	**KRM**	Kurzweil Reading Machine
kg	kilogram	**KRL**	Kurt Rossmann Laboratory [US]
kg-cal	kilogram-calorie	**KS**	Kansas [US]
kgf	kilogram-force		Ketosteroid
kgm	kilogram-meter		Kloeckner Steelmaking (Process)
kg per cu m	kilogram per cubic meter	**K/S**	Kurdjumov-Sachs (Relationship)
kgps	kilograms per second	**KSA**	Knob Shoe Assembly
kH	kilohenry [unit]	**KSC**	Kennedy Space Center [NASA]
KHN	Knoop Hardness Number	**KSDS**	Key-Sequenced Data Set
KHS	Knurled Head Screw	**KSE**	Keyboard Source Entry (Program)
kHz	kilohertz	**KSEA**	Korean Scientists and Engineering Association
KI	Keyboard Input		
KIAS	Knots Indicated Air Speed	**ksf**	kilopounds per square foot
KIFIS	Kollsman Integrated Flight Instrumentation System	**ksi**	kilopounds per square inch
		KSR	Keyboard Send and Receive
KIN	Kinescope	**KST**	Keyseat
kip	kilopounds	**kt**	karat [unit]
kip-ft	kilo foot-pounds	**KTFR**	Kodak Thin Film Resist
KIPO	Keyboard Input Printout	**KTMS**	Knapp Time Metaphor Scale
kips	kilo-instructions per second	**KTR**	Keyboard Typing Reperforator
	kilopounds per square inch	**KTS**	Key Telephone System
KIS	Keyboard Input Simulation	**KTSA**	Kahn Test of Symbol Arrangement
KISA	Korean International Steel Association	**KTSP**	Knots True Air Speed
KIST	Korean Institute for Science and Technology	**KTU**	Key Telephone Unit
		Ku	Kurchatovium
kJ	kilojoule	**KUR**	Kyoto University Reactor [Japan]
KK	One Million	**kV**	kilovolt
KL	Key Length	**kVA**	kilovolt-Ampere
kL	kiloliter	**kVAC**	Kilovolt Alternating Current
KLA	Klystron Amplifier	**kVAh**	Kilovolt-Ampere-hour
KLH	Keyhole Limpet Hemacyanine	**kVAhm**	Kilovolt-Amperehourmeter
KLO	Klystron Oscillator	**kvar**	kilovar
KM	Kirchoff Method		

kvarh	kilovarhour
kVDC	Kilovolt Direct Current
KVL	Kirchhoff's Voltage Law
KW	Keyword
kW	kilowatt
KWAC	Keyword and Context
kWe	Kilowatt of Electric Power
kWh	kilowatthour
kWhm	kilowatthourmeter
kwhr	kilowatthour
KWIC	Keyword in Context
KWIT	Keyword in Title
KWOC	Keyword Out of Context
KWR	Know-How Repeating
kWt	Kilowatt of Thermal Energy
KWY	Keyway
KX	Kilo-X-Unit
KY	Kentucky [US]
Ky	Kentucky [US]
KYS	Kloeckner-Youngstown Steelmaking Process

L

L	Inductance [symbol]
	Lambert [unit]
	Lamp
	Language
	Length
	Leucine
	Line
	Link
	Liquid
	Liter
	Load
	Lumen [unit]
l	left
	local
	long
LA	Lead Angle
	Lightning Arrester
	Line Adapter
	Linear Accelerator
	Longitudinal Acoustical
	Louisiana [US]
La	Lanthanum
	Louisiana [US]
LAA	Liters of Absolute Alcohol
LAAV	Light Airborne ASW (= Antisubmarine Warfare) Vehicle
LAAW	Light Assault Antitank Weapon
LAB	Laboratory
	Low-Altitude Bombing
	Low Angle Boundary
LABIL	Light Aircraft Binary Information Link
LABS	Low-Altitude Bombing System
LAC	Load Accumulator
	Local Arrangement Committee
	Lunar Aeronautical Chart

LACBWR	La Crosse Boiling Water Reactor
LACES	Los Angeles Council of Engineering Societies [US]
LACTAC	Los Angeles County Transportation Commission [US]
LACV	Logistics Air-Cushion Vehicle
LAD	Lactic Dehydrogenase
	Ladder
	Location Aid Device
	Logical Aptitude Device
	Lookout Assist Device
LADAR	Laser Detection and Ranging
LAE	Lead Angle Error
	Left Arithmetic Element
LAEDP	Large Area Electronic Display Panel
LAG	Load and Go (Technique)
LAGEOS	Laser Geodynamic Satellite [US]
LAH	Logical Analyzer of Hypothesis
LAHS	Low-Altitude, High-Speed
LAI	Leaf Area Index
LAINS	Low-Altitude Inertial Navigation System
LAK	Lymphokine-Activated Killer
LAL	Limulus Amebocyte Lysate
LALLS	Low-Angle, Laser-Light Scattering
LALO	Low-Altitude Observation
LALR	Lookahead Left-to-Right
LAM	Laminate(d)
	Load Accumulator with Magnetization
	Loop Adder and Modifier
LAMA	Local Automatic Message Accounting
	Locomotive and Allied Manufacturers Association
LAMCS	Latin American Communications System
LAMF	Los Alamos Meson Facility [US]
LAMMA	Laser Microprobe Microanalysis
	Laser Microprobe Mass Analysis
LAMP	Laser and Maser Patents
	Low-Altitude Manned Penetrator
LAMPF	Los Alamos Meson Physics Facility [US]
LAMS	Load Alleviation and Mode Stabilization
LAN	Local Area Network
LANDS	Language-Oriented Development System
LANG	Language
LANL	Los Alamos National Laboratories [US]
LANS	Load Alleviation and Stabilization
LANTIRN	Low-Altitude Navigation and Targeting Infrared for Night
LAP	Lesson Assembly Program
	Leucine Aminopeptidase
	List Assembly Program
LAPA	Leucocyte Alkaline Phosphatase Activity
	Lightweight Aggregate Producers Association
LAPDOG	Low-Altitude Pursuit Dive On Ground
LAPES	Low-Altitude Parachute Extraction System
LAPW	Linearized Augmented Plane Wave
LAQ	Lacquer
LAR	Low-Angle Reentry
LARA	Light-Armed Reconnaissance Aircraft
LARAM	Line-Addressable Random-Access Memory
LARC	Lighter, Amphibious Resupply Cargo Vessel

	Livermore Automatic Research Computer
LaRC	Langley Research Center [NASA]
LAREC	Los Alamos Reactor Econmics Code [US]
LARIAT	Laser Radar Intelligence Acquisition Technology
LARS	Laser Angular Rate Sensor
	Laser Articulated Robot System
LAS	Laboratory of Atmospheric Sciences [NSF, US]
	Landing and Approach System
	Large Astronomical Satellite [ESRO]
	Large Autoclave System
	Lithia-Alumina-Silicate
	Lithium Aluminosilicate
	Logic Analysis System
	Launch Auxiliary System
	Light-Activated Switch
LASA	Large Aperture Seismic Array
LASCO	Latin American Science Cooperation Office
LASCR	Light-Activated Silicon-Controlled Rectifier
LASCS	Light-Activated Silicon-Controlled Switch
LASE	Load at Specified Elongation
	Logic Analysis and Slave Emulation
LASER	Light Amplification by Stimulated Emission of Radiation [also: Laser]
LASH	Laser Antitank Semi-Active Homing
LASIL	Land and Sea Interaction Laboratory
LASL	Los Alamos Scientific Laboratory [US]
LASR	Laboratories for Astrophysics and Space Research [US]
	Litton Airborne Search Radar
LASRM	Low Altitude Short-Range Missile
	Low-Altitude Supersonic Research Missile
LASS	Local-Area Signaling Services
LASSO	Laser Search and Secure Observer
	Landing and Approach System, Spiral-Oriented
LASST	Laboratory for Surface Science and Technology [US]
LASV	Low-Altitude Supersonic Vehicle
LAT	Laser Acquisition and Tracking
	Lateral
	Latitude
	Local Area Transport
LATA	Local Access and Transport Area
LATAF	Logistics Activation Task Force
LATINCON	Latin American Convention [IEEE]
LATS	Litton Automated Test Set
	Long-Acting Thyroid Simulator
LAU	Line Adapter Unit
LAVA	Linear Amplifier for Various Applications
LAW	Light Antiarmor Weapon
	Light Antitank Weapon
	Logic Analysis Workstation
LB	Label
	Langmuir-Blodgett (Theory)
	Laser Brazing
	Light Bombardment
	Line Buffer
	Linoleum Base
	Local Battery

	Locating Button
	Long Barrel
	Lower Bound
lb	(libra) - pound
LBA	Lima Bean Agglutenin
	Linear Bounded Automaton
	Limestone-Building Algae
lb ap	pound, apothecaries'
lb av	pound, avoirdupois
LBBP	Laboratory of Blood and Blood Products [US]
LBC	Laser Beam Cutting
LBCM	Locator at the Back Course Marker
LBE	Lance Bubbling Equilibrium
lbf	pound (force)
lb-ft	pound-foot
lb-in	pound-inch
LBG	Load Balancing Group
LBL	Label
	Lawrence Berkeley Laboratory [US]
	Lima Bean Lectin
LBM	Laser Beam Machining
	Load Buffer Memory
lbm	pound (mass)
LBMA	Lumber and Building Materials Association
LBMAC	Lumber and Building Materials Association of Canada
LBMAO	Lumber and Building Materials Association of Ontario [Canada]
lbn	pounds, net
LBNP	Lower Body Negative Pressure (Boots)
LBO	Line Buildout (Network)
LBP	Laser Beam Printer
	Length between Perpendiculars
LBR	Laser Beam Recorder
	Lumber
lbs	pounds
LBT	Low Bit Test
lb t	pound, troy
LBTS	Land-Based Test Site
LBW	Laser Beam Welding
LBX	Local Bus Extension
LC	Inductance-Capacitance
	Labour Canada
	Landing Craft
	Lanthanum Chromite
	Launch Complex
	Lead-Covered
	Level Control
	Lethal Concentration
	Library of Congress [US]
	Life Cycle
	Line Carrying
	Line Connector
	Line Control
	Line of Communication
	Line of Contact
	Link Circuit
	Liquid Crystal
	Liquid Chromatography
	Location Counter

	Load Carrier		**LCMM**	Life Cycle Management Model
	Load Cell		**LCMS**	Logistics Command Management System
	Load Center		**LCN**	Load Classification Number
	Load Compensating (Relay)		**LCNR**	Liquid Core Nuclear Rocket
	Logistics Command		**LCNT**	Link Celestial Navigation Trainer
	Lomer-Cottrell (Dislocation)		**LCNTR**	Location Counter
	Low Carbon		**LCO**	Load Control and Optimization
	Lower Case		**LCP**	Language Conversion Program
	Lyman Continuum			Laser-Induced Chemical Processing
LCAC	Landing Craft Air Cushion			Link Control Procedure
LCAO	Linear Combination of Atomic Orbitals			Liquid Crystal Polymer
LCAO-MO	Linear Combination of Atomic Orbitals - Molecular Orbitals			Liquid Cyclone Process
				Logic-Controlled Protocolling
LCB	Launch Control Building		**LCR**	Inductance-Capacitance-Resistance
	Line Control Block			Lithium-Cooled Reactor
LCC	Landing Craft, Control		**LCRE**	Lithium-Cooled Reactor Experiment
	Launch Control Center		**LCRU**	Lunar Communications Relay Unit [NASA]
	Launch Control Console		**LCS**	Large-Capacity Storage
	Lead-Coated Copper			Large Core Storage
	Leadless Chip Carrier			Launch Control System
	Least-Cost, (Satisfactory Melting) Charge			Limiting Creep Stress
	Life Cycle Cost			Lincoln Calibration Sphere
	Liquid Crystal Cell			Loudness-Contour Selector
	Local Communications Complex			Low-Carbon Steel
	Local Communications Console		**LCSE**	Laser Communications Satellite Experiment [NASA]
	Logistics Coordination Center			
LCCC	Leadless Ceramic Chip Carrier		**LCSO**	Local Communications Service Order
LCCT	LaQue Center for Corrosion Technology		**LCT**	Linear Combination Technique
LCD	Least Common Denominator		**LCVD**	Laser Chemical Vapor Deposition
	Lowest Common Denominator			Least Voltage Coincidence Detection
	Liquid Crystal Display		**LCW**	Line Control Word
LCDTL	Load-Compensated Diode Transistor Logic		**LD**	Lactate Dehydrogenase
LCE	Land-Covered Earth			Lactic Dehydrogenase
	Launch Complex Equipment			Large Diameter
	Load Circuit Efficiency			Laser Diode
LCES	Least Cost Estimating and Scheduling			Lateral Direction
LCF	Launch Control Facility			Leak Detector
	Light Control Film			Lethal Dose
	Low-Cycle Fatigue			Letter Description
LCFS	Last Come, First Served			Level Discriminator
LCGO	Linear Combination of Gaussian Orbitals			Lift-Drag (Ratio)
LCH	Lens Culinaris			Light-Dark (Cycle)
LCHTF	Low-Cycle High-Temperature Fatigue			Limited
LCI	Load-Commutated Inverter			Line Drawing
LCJ	Low-Contaminant Jarosite (Process)			Line of Departure
LCL	Library Control Language			Linz/Donawitz (Steelmaking Process)
	Lifting Condensation Level			Lipid Droplet
	Linkage Control Language			List of Drawings
	Low Camera Length			Load
	Low Capacity Link			Logic Driver
	Lower Control Limit			Long Delay
	Lowest Car Load			Long Distance
LCL-CBED	Low Camera Length - Convergent Beam Electron Diffraction			Low Density
			LDA	Laser Doppler Anemometer
LCLU	Landing Control Logic Unit			Lead Development Association
LCLV	Liquid-Crystal Light Valve			Line Data Area
LCM	Large Core Memory			Line Driving Amplifier
	Lead-Coated Metal			Lithium Diisopropylamide
	Least Common Multiple			Localizer Directional Aid
	Liquid Curing Medium			Locate Drum Address
	Lowest Common Multiple			Logical Device Address

LD-AC	Linz-Donawitz/ARBED-CRM (Steelmaking Process)
LDB	Light Distribution Box
LDAM	Linked Direct Access Method
LDBA	Leakage Design Basis Accident
LDC	LASA (= Large Aperture Seismic Array) Data Center
	Latitude Data Computer
	Less Developed Country
	Light Direction Center
	Linear Detonating Cord
	Line Drop Compensator
	Liquid Dynamic Compaction
	Load Duration Curve
	Long Distance Communications
	Lower Dead Center
LDDS	Low-Density Data System
LDE	Laminar Defect Examination
	Language-Directed Editor
	Linear Differential Equation
LDEF	Long-Duration Exposure Facility
LDF	Light Distillate Feedstock
	Linear Discriminate Function
	Local Density Fluctuation
LDG	Landing
	Loading
LDH	Lactate Dehydrogenase
	Lactic Dehydrogenase
	Limiting Dome Height
LDH-A	Lactate Dehydrogenase-A
LDH-B	Lactate Dehydrogenase-B
LDH/PK	Lactic Dehydrogenase/Pyruvate Kinase
LDI	Lossless Discrete Integrator
LDL	Language Description Language
	Low-Density Lipoprotein
LDM	Limited-Distance Modem
	Linear-Delta Modulation
LDOS	Local Density of States
LDP	Language Data Processing
LDPE	Low-Density Polyethylene
LDR	Large Deployable Reflector
	Light-Dependent Resistor
	Limiting Draw(ing) Ratio
	Limiting Dome Ratio
	Loader
	Low Data Rate
LDRI	Low Data Rate Input
LDRS	LEM Data Reduction System
LDS	Large Disk Storage
	Library Delivery System
	Light Distillate Spirit
LDT	Linear Differential Transformer
	Logic Design Translator
	Long Distance Transmission
	Low-Density Telephony
LDTS	Low-Density Telephony Service
LDV	Laser Doppler Velocimetry
LDX	Long Distance Xerography
LE	Large End
	Launching Equipment
	Leading Edge

	Less or Equal
	Light Equipment
	Logic Element
	Logic Evaluator
	Low Efficiency
	Low Explosive
LEA	Long Endurance Aircraft
	Lycopersicon Esculentum Agglutinin
LEADS	Law Enforcement Automated Data System
LEAF	Leaflet
LEANS	Lehigh Analog Simulator
LEAP	Language for the Expression of Associative Procedures
	Lunar Escape Ambulance Pack [NASA]
LEAR	Logistics Evaluation and Review Technique
	Low-Energy Anti-Photon Ring
	Lower Echelon Automatic Switchboard
LEBCO	Low Energy Building Council
LEC	Liquid Encapsulated Czochralski
	Liquid Exclusion Chromatography
	Low-Energy Channeling
	Low-Energy Cure
LED	Light-Emitting Diode
LEDS	Low-Energy Dislocation Structure
LEDT	Limited-Entry Decision Table
LEED	Laser-Energized Explosive Device
	Low-Energy Electron Diffraction
LEELS	Low-Energy Electron Loss Spectrometry
LEF	Light-Emitting Film
LEFM	Linear Elastic Fracture Mechanics
LEG	Legal
LEG WT	Legal Weight
LEIB	Low-Energy Ion Beam
LEID	Low-Energy Ion Detector
LEIS	Low-Energy Ion Scattering
LEISS	Low-Energy Ion Scattering Spectroscopy
LEJ	Longitudinal Expansion Joint
LEK	Liquid Encapsulated Kyropoulos
LEL	Lower Explosion Limit
LELU	Launch Enable Logic Unit
LEM	Laboratory of Electromodeling
	Laser Exhaust Measurement
	Lunar Excursion Module
LEMA	Lifting Equipment Manufacturers Association
LEN	Length
LEO	Librating Equidistant Observer
	Low Earth Orbit
LEPS	Launch Escape Propulsion System
LEPT	Long Endurance Patrolling Torpedo
LeRC	Lewis Research Center [of NASA]
LERT	Linear-Extensional, Rotation and Twist (Robot)
LES	Launch Enabling System
	Launch Escape System
	Light-Emitting Switch
	Light Exposure Speed
	Lincoln Experimental Satellite [US]
	Local Excitatory State
	Loop Error Signal
	Louisiana Engineering Society [US]

LESA	Lake Erie Steam Association [US]		Length
	Lunar Exploration System for Apollo [NASA]		Linear Gate
			Line Graph
LESC	Light-Emitting Switch Control		Long
LESS	Lateral Epitaxy by Speeded Solidification	**LGA**	Light-Gun Amplifier
	Least Cost Estimating and Scheduling		Low Gloss/Automotive
LEST	Low-Energy Speed Transmission	**LGC**	Laboratory of Government Chemists
LET	Launch Escape Tower	**LGCP**	Lexical-Graphical Composer Printer
	Leading Edge Trigger	**LGE**	Large
	Letter		Lunar Geological Equipment [NASA]
	Linear Energy Transfer	**LGHM**	Low Gravity Heavy Media
	Lincoln Experimental Terminal [NASA]	**LGO**	Lamont Geological Observatory [US]
	Logical Equipment Table	**LGP**	Low Ground Pressure
LEU	Launch Enabling Unit	**LGR**	Ligroin
	Low-Enriched Uranium	**LGS**	Landing Guidance System
Leu	Leucine	**LGT**	Low Group Transmit
LEV	Level	**LGTH**	Length
LF	Ladle Furnace	**LGV**	Large Granular Vesicle
	Laminar Flow	**LH**	Left-Hand
	Launch Facility		Linear Hybrid
	Leapfrog Configuration		Liquid Hydrogen
	Leveling Foot		Litter Hook
	Limiting Fragmentation		Locating Head
	Linear Foot		Luteinizing Hormone
	Line Feed (Character)	**LHA**	Local Hour Angle
	Line Finder	**LHC**	Large Hadron Collider
	Linoleum Floor		Left-Hand Circular
	Load Factor	**LHChl**	Light-Harvesting Chlorophyll
	Logic Function	**LHCP**	Left-Hand Circular Polarization
	Low Frequency	**LHD**	Left-Hand Drive
LFC	Laminar Flow Control		Load-Haul-Dump (Mining)
	Local Forms Control	**LHNC**	Linearized Hypernetted Chain
LFCB	Load to Forms Control Buffer	**LHR**	Lower Hybrid Resonance
LFCM	Low-Frequency Cross-Modulation	**l-hr**	lumen-hour [unit]
LFD	Low-Frequency Disturbance	**LHS**	Left-Hand Side
LFE	Laboratory for Electronics	**LHSV**	Liquid Hourly Space Velocity
	Local Field Effect	**LHWA**	Linearized Hot-Wire Anemometer
LFFET	Low-Frequency Field-Effect Transistor	**LI**	Laser Interferometer
LFGS	Low Flow Gas Saver		Link
LFI	Laser-Fiber Illuminator		Liquid Injection
LFIM	Low-Frequency Instruments and Measurements [IEEE Committee]		Low Intensity
		Li	Lithium
LFM	Limited-Area Fine-Mesh Model	**LIA**	Laser Institute of America
	Low-Frequency Microphone		Laser Industries Association [US]
	Low-Powered Fan Marker		Lead Industries Association
LFN	Line Format Number		Linear Integrated Amplifier
LFO	Low-Frequency Oscillator		Lysine Iron Agar
LFPS	Low-Frequency Phase Shifter	**lib**	(liber) - book
LFQ	Light Foot Quantizer	**LIB**	Librarian
LFRD	Lot Fraction Reliability Definition		Library
LFS	Labour Force Survey		Line Interface Base
	Launch Facility Simulator	**LIC**	License
	Logical File Structure		Linear Integrated Circuit
LFSS	Landing Force Support Ship [US Navy]	**LICOR**	Lightning Correlation
LFSW	Landing Force Support Weapon [US Navy]	**LID**	Leadless Inverted Device
LFU	Least Frequently Used		Local Improvement District
LFV	Lunar Flying Vehicle [NASA]		Locked-In Device
LG	Landing Gear	**LIDAR**	Laser Intensity Direction and Ranging [also: Lidar]
	Landing Ground		
	Large		Light Detection and Ranging
	Large Grain	**LIED**	Linkage Editor

LIEF	Launch Information Exchange Facility [NASA]
LIF	Laser-Induced Fluorescence
	Low Insertion Force
LIFO	Last In, First Out
LIFT	Link Intellectual Functions Tester
	Logically Integrated FORTRAN Translator
LIL	Large Ionic Lithophile
	Lunar International Laboratory
LILO	Last In, Last Out
LIM	Limit(er)
	Linear Induction Motor
	Liquid Injection Molding
lim	limit value
LIMA	Laser-Induced Ion Mass Analyzer
LIMAC	Large Integrated Monolithic Array Computer
LIMB	Limestone Injection Multistage Burner
LIMIRIS	Laser Induced Modulation of Infrared in Silicon
LIMIT	Lot-Size Inventory Management Interpolation Technique
LIML	Limited-Information Maximum Likelihood
LIMP	Language-Independent Macroprocessor
	Lunar Interplanetary Monitoring Platform
LIMS	Laboratory Information Management System
	Laser Ionization Mass Spectrometer
LIM SW	Limit Switch
LIN	Library Information Network
	Linear
LINAC	Linear Accelerator
LINC	Laboratory Instrument Computer
LINET	Legal Information Network
lin ft	linear foot
LING	Linguistic(s)
LINLOG	Linear-Logarithmic
LINO	Linoleum
LINS	LORAN (= Long Range Navigation) Inertial System
LIOD	Laser In-Flight Obstacle Detector
LIOCS	Logical Input-Output Control System
LIPL	Linear Information Processing Language
LIPS	Logical Inferences per Second
LIQ	Liquid
LIR	Line Integral Refractometer
LIRA	Linen Industry Research Association
LIRF	Low-Intensity Reciprocity Failure
LIS	Laser-Induced Separation
	Library and Information Service
	Loop Input Signal
LISA	Library Systems Analysis
LISN	Line Impedance Stabilization Network
LISP	List Processing
	List Processor
LISSA	Life Insurance Software Systems of America
LIT	Liquid Injection Technique
	Logic Integrity Test
LITASTOR	Light Tapping Storage
LITE	Legal Information through Electronics
LITR	Low-Intensity Test Reactor

LITVC	Liquid Injection Thrust Vector Control
LITHO	Lithograph(y)
LISARDS	Library Information Search and Retrieval Data System
LJ	Leaf Jig
LK	Link
	Lock
LKR	Locker
LK WASH	Lock Washer
LL	Light Line
	Lines
	Live Load
	Local Loopback
	Loudness Level
	Low(er) Level
	Lower Limit
L/L	Latitude/Longitude
LLAD	Low-Level Air Defense
LLC	Liquid Live Culture
LLD	Low-Level Discriminator
LLDPE	Linear Low-Density Polyethylene
LLE	Laboratory for Laser Energetics
LLFM	Low-Level Flux Monitor
LLG	Logical Line Group
LLH	Low-Level Heating
LLL	Low-Level Language
	Low-Level Logic
LLLLLL	Laboratories Low-Level Linked List Language [also: L6]
LLLTV	Low Light Level Television
LLNL	Lawrence Livermore National Laboratory [US]
LLP	Line Link Pulse
LLR	Line of Least Resistance
	Load-Limiting Resistor
LLRES	Load-Limiting Resistor
LLRF	Lunar Landing Research Facility
LLRV	Lunar Landing Research Vehicle
LLS	Local Load Shearing
LLTT	Landline Teletype
LLTV	Low Light Level Television
	Lunar Landing Training Vehicle
LLW	Low-Level Waste
LLWS	Low-Level Wind-Shear
LLWSAS	Low-Level Wind Shear Alert System
LM	Life Member
	Light Microscopy
	Linear Monolithic
	List of Materials
	Logic Monitor
	Low-Melting
	Lunar Module [NASA]
L/M	Lines per Minute
	List of Material
lm	logarithmic mean area
	lumen [unit]
LMA	Laser Microspectral Analyzer
	Limited Motion Antenna
LMC	Large Magellanic Cloud
	Lime-Magnesium Carbonate
	Low-Pressure Molding Compound

LME	Liquid Metal Embrittlement		**LOB**	Launch Operations Branch [NASA]
	London Metal Exchange			Launch Operations Building [NASA]
LMEC	Liquid Metal Engineering Center [US]			Line of Balance
LMEIC	Life Member of the Engineering Institute of Canada		**LOBTA**	Lunar Orbiter Block Triangulation
			LOC	Launch Operations Center [NASA]
LMF	Linear Matched Filter			Liaison Officers Committee
	Low and Medium Frequency			Library of Congress [US]
LMFA	Light Metal Founders Association			Local
LMFBR	Liquid Metal Fast Breeder Reactor			Localizer
LMIE	Liquid Metal Induced Embrittlement			Locate
L/min	Liters per Minute			Location Counter
LMIS	Liquid Metal Ion Source			Loss of Coolant
LMLR	Load Memory Lockout Register			Oxygen Lance Cutting
LMM	Lactobacillus Maintenance Medium		**LOCA**	Loss of Coolant Accident
	Length Measuring Machine		**LOCATE**	Library of Congress Automation Techniques Exchange
	Light Meromyosin			
LMO	Life Members Organization		**LOCATS**	Lockheed Optical Communications and Tracking System
LMP	Larson-Miller Parameter			
	Large-Volume Microwave Plasma (Generator)		**LOCC**	Launch Operations Control Center
			loc cit	(loco citato) - in the place cited
	Lunar Module Pilot [NASA]		**LOCI**	Logarithmic Computing Instrument
LMPRT	Locally Most Powerful Rank Test		**LOCOS**	Local Oxidation of Silicon
LMR	Lowest Maximum Range		**LOCS**	Librascope Operations Control System
LMS	Laboratory Management Standards			Logic and Control Simulator
	Laser Mass Spectrometry		**LOD**	Launch Operations Directorate [NASA]
LMSA	Laser Microemission Spectroanalysis			Line of Direction
	Laser Microspectral Analysis			Locally One-Dimensional
LMSL	Lateral Monolayer Superlattice			Location Dependent
LMSS	Lunar Mapping and Survey System			Locked-On Device
LMT	Log Mean Temperature		**LODESMP**	Logistics Data Element Standardization and Management Process
LMTD	Logarithmic Mean Temperature Difference			
LMTO	Linear Muffin Tin Orbital		**LODESTAR**	Logically-Organized Data Entry, Storage, and Recording
LMTR	Limiter			
LMU	Logical Mining Unit		**LOERO**	Large Orbiting Earth Resources Observatory
LMW	Low Molecular Weight		**LOES**	Laser Optical Emission Spectroscopy
LN	Line		**LOF**	Local Oscillator Frequency
	Liquid Nitrogen			Loss of Flow
	Low-Noise			Loss of Fluid
ln	Naperian (natural) logarithm			Lowest Operating Frequency
LNA	Low-Noise Amplifier		**LOFAR**	Low-Frequency Acquisition and Ranging
LNF	Liposoluble Neutral Fraction		**LOFT**	Loss-of-Flow Test
LNFB	Linear Negative Feedback			Loss-of-Fluid Test
LNG	Liquefied Natural Gas		**log**	common logarithm [Base 10]
LNGC	Liquefied Natural Gas Carrier		**LOGACS**	Low-Gravity Accelerometer Calibration System
LNG	Long			
LNI	Local Network Interconnect(ion)		**LOGALGOL**	Logical Algorithmic Language
LNR	Low-Noise Receiver		**LOGAMP**	Logarithmic Amplifier
LNWT	Low and Nonwaste Technology		**LOGBALNET**	Logistics Ballistic Missile Network [US Air Force]
LO	Layout			
	Liaison Office(r)		**LOGEL**	Logic Generating Language
	Local Oscillator		**LOGFTC**	Logarithmic Fast Time Constant
	Lock-On		**LOGIPAC**	Logical Processor and Computer
	Longitudinal Optical		**LOGIT**	Logical Interference Tester
	Low Order		**LOGLAN**	Logical Language
	Lubricating Oil		**LOGRAM**	Logical Program
	Lunar Orbiter		**LOGTAB**	Logic Table
LOA	Lathyrus Odoratus Agglutinin		**LOH**	Light Observation Helicopter
	Length Overall		**LOHAP**	Light Observation Helicopter Avionics Package
LOAC	Low Accuracy			
LOAMP	Logarithmic Amplifier		**LOI**	Loss of Ignition
LOAS	Liftoff Acquisition System		**LOL**	Length of Lead

LOLA	Lunar Orbit Landing Approach	**LPA**	Limulus Polyphemus Agglutinin
LOLEX	Low-Level Extraction		Linear Pulse Amplifier
LOLITA	Language for the On-Line Investigation and Transformation of Abstractions		Link Pack Area
			Log Periodic Antenna
LOLO	Lift-On/Lift-Off (Vessel)	**LPAC**	Launching Program Advisory Committee [of ESRO]
LOM	Locator at the Outer Marker		
	Low-Frequency Outer Marker	**LPAS**	Low-Pressure Arc Spraying
LOMUSS	Lockheed Multiprocessor Simulation System	**LPATS**	Lightning Position and Tracking System
LONG	Longitude	**LPC**	Laboratory Pasteurized Count
LONGN	Longeron		Language Products Center [Canada]
LOOPS	LISP Object-Oriented Programming System		Linear Power Controller
LOP	Line of Position		Linear Predictive Coding
	Local Operational Plot		Loop Control
LOPAD	Logarithmic Outline Processing System for Analog Data	**LPCI**	Low-Pressure Coolant Injection System
		LPCVD	Low-Pressure Chemical Vapor Deposition
LOPAR	Low-Power Acquisition Radar	**LPD**	Landing Point Designator
LOPP	Lunar Orbiter Photographic Project		Language Processing and Debugging
LO-QG	Locked Oscillator-Quadrature Grid		Linear Phasing Device
LOR	Laboratory for Oil Recovery		Log Periodic Dipole
	Large Optical Reflector	**LPDA**	Log Periodic Dipole Array
	Lining-over-Refractory	**LPE**	Liquid Phase Epitaxy
	Loss on Reduction		Loop Preparation Equipment
	Lunar Orbital Rendezvous	**LPF**	Low-Pass Filter
LORADAC	Long-Range Active Detection and Communications System		Lymphocytosis-Promoting Factor
		LPG	Liquefied Petroleum Gas
LORAN	Long-Range Navigation [also: Loran]		List Program Generator
LORBI	Locked-On Radar Bearing Indicator	**LPGC**	Liquefied Petroleum Gas Carrier
LOREC	Long-Range Earth Communications	**LPGITC**	Liquefied Petroleum Gas Industry Technical Committee
LORL	Large Orbital Research Laboratory [NASA]		
LORO	Lobe on Receive-Only	**LPI**	Lines per Inch
LORPGAC	Long-Range Proving Ground Automatic Computer	**LPIA**	Liquid Propellant Information Agency [US]
		LPID	Logical Page Identifier
LORS	Lunar Optical Rendezvous System	**LPL**	Laser-Pumped Laser
LORTAN	Long-Range and Tactical Navigation [also: Lortan]		Light Proof Louver
			Linear Programming Language
LORV	Low-Observable Reentry Vehicle		Lipoprotein Lipid
LOS	Line-of-Sight		List Processing Language
	Loss of Signal		Long-Path Laser
LOSS	Landing Observer Signal System	**LP-LEC**	Low-Pressure Liquid Encapsulated Czochralski
	Large Object Salvage System		
LOT	Large Orbital Telescope	**LPM**	Laser Phase Macroscope
	Long Open Time		Lines per Minute
LOTIS	Logic, Timing, Sequencing		Lunar Portable Magnetometer [NASA]
LOWL	Low-Level Language	**LPN**	Logical Page Number
LOx	Liquid Oxygen [also: LOX]		Low-Pass Notch
LOZ	Liquid Ozone	**LPO**	Low-Power Output
LP	Laser Printer	**LPP**	Letter Processing Plant
	Library of Parliament [Canada]		Liquid Plasma Process
	License Program	**LPPD**	Low-Pressure Plasma Deposit
	Light Pen	**LPPS**	Low-Pressure Plasma Spraying
	Linear Programming	**LPRE**	Liquid Propellant Rocket Engine
	Line Printer	**LPS**	Launch Process System
	Liquefied Petroleum		Letter Processing Stream
	Long Play (Record)		Light Proof Shades
	Long Primer		Linear Pulse Sector
	Loop Control		Lines per Second
	Low-Pass (Filter)		Lipopolysaccharine
	Low Point	**LPSD**	Logically Passive Self-Dual
	Low Power	**LPSIRS**	Linear Potential Sweep Infrared Reflectance Spectroscopy
	Low Pressure		
	Low-Speed Printer	**LPT**	Laminated Plate Theory

	Line Printer
	Low-Pressure Turbine
LPTV	Large Payload Test Vehicle
	Low-Power Television
LPU	Life Preserver Unit
LPV	Light Proof Vent
lpW	lumens per watt
LQ	Letter-Quality
	Liquid
LQD	Liquid
LQDC	Lowest Quantitatively-Determined Concentration
LQDS	Liquids
LR	Landing Radar
	Left to Right
	Level Recorder
	Limited Recoverable
	Line Relay
	Local/Remote
	Load Ratio
	Load-Resistor (Relay)
	Logical Record
	Lunar Rover [NASA]
L/R	Locus of Radius
Lr	Lawrencium
LRA	Laser Research Association
LRB	Local Reference Beam
LRC	Langley Research Center [NASA]
	Large Rock Cavern
	Lewis Research Center [NASA]
	Light, Rapid, Comfortable (Train)
	Linguistics Research Center [US]
	Load Ratio Control
	Longitudinal Redundancy Check
LRCS	Long-Range Cruise Speed
LRE	Linear Resource Extension
LREP	Laser Rotating Electrode Process
LRI	Lighting Research Institute
	Long-Range Radar Input
LRIR	Low Resolution Infrared Radiometer
LRIS	Land-Related Information System
LRL	Lawrence Radiation Laboratory [US]
	Livermore Research Laboratory [US]
	Lunar Receiving Laboratory [NASA]
LRLTRAN	Lawrence Radiation Laboratory Translator
LRM	Limited Register Machine
	Lunar Reconnaissance Module
LRMTS	Laser Rangefinder and Marked Target Seeker
LRO	Long-Range Order(ing)
LRP	Land Reclamation Program
	Long-Range Path
LRPA	Long-Range Patrol Aircraft
LRPL	Liquid Rocket Propulsion Laboratory [US Army]
LRRP	Lowest Required Radiated Power
LRRS	Long-Range Radar Site
LRS	Laser Raman Scattering
	Linguistic Research System
	Log and Reporting System
	Long-Range Search

LRSM	Long-Range Seismographic Measurements
LRSS	Long-Range Survey System
LRT	Light Railway Transport
	Light Rapid Transit
LRTAP	Long-Range Transport of Air Pollutants
LRTAPP	Long-Range Transport of Air Pollutants Program
LRTF	Long-Range Technical Forecast
LRTL	Light Railway Transport League
LRTM	Long-Range Training Mission
LRTP	Long-Range Training Program
LRU	Least Recently Used (Rule)
	Line-Replaceable Unit
LRV	Light Rapid Vehicle
	Lunar Roving Vehicle [NASA]
LRWE	Long-Range Weapons Establishment [Australia]
LS	Land Surveyor
	Language Specification
	Large-Scale
	Laser System
	Least Significant
	Least Squares
	Level Switch
	Limit Switch
	Link Switch
	Load Sensing
	Lockscrew
	Loudspeaker
	Low-Speed
	Left Side
LSA	Large Surface Area
	Least Squares Approximation
	Lignosulfonic Acid
	Limited Space-Charge Accumulation
	Logistic Support Analysis
	Low Specific Activity
LSB	Least Significant Bit
	Least Significant Byte
	Lower Sideband
LSBR	Large Seed Blanket Reactor
LSC	Laser-Supported Combustion
	Least Significant Character
	Liquid Scintillation Counting
	Loop Station Connector
	Low Shear Continuous
LSc	Licentiate in Science
LSCC	Line-Sequential Color Composite
LSCE	Launch Sequence and Control Equipment
LSD	Language for Systems Development
	Large Screen Display
	Laser-Supported Detonation
	Last Significant Digit
	Launch Systems Data
	Least Significant Digit
	Lodolysergic Acid Diethylamide
	Logarithmic Series Distribution
	Logistics Support Division
	Low-Speed Data
	Lunar Surface Magnetometer
	Lysergic Acid Diethylamide

LSE	Language Sensitive Editor
	Launch Sequencer Equipment
	Line Signaling Equipment
	Longitudinal Section Electric
LSECS	Life Support and Environmental Control System
LSEP	Lunar Surface Experimental Package [NASA]
LSF	Lumped Selection Filter
LSFC	Lewis Spaceflight Center [of NASA]
LSFFAR	Low-Spin Folding Fin Aircraft Rocket
LSHI	Large-Scale Hybrid Integration
LSI	Large-Scale Integration
LSIC	Large-Scale Integrated Circuit
LSID	Local Session Identification
LSIG	Line Scan Image Generator
LSI-MOS	Large-Scale Integrated Metal-Oxide Semiconductor
LSL	Ladder Static Logic
LSLD	Local Store Loop Driver
LSM	Launcher Status Multiplexer
	Linearized Simulation Model
	Linear Synchronous Motor
	Line Switch Marker
	Local Synchronous Modem
	Logic-State Map
	Longitudinal Section Magnetic
	Long Span Mezzanine
	Lymphocyte Separation Media
LSMR	Landing Ship Medium Rocket
LSN	Linear Sequential Network
	Line Stabilization Network
	Low-Solids, Nondispersoid (Mud)
LSO	Line Signaling Oscillator
LSP	Large-Scale Production
	Laser Shock Processing
	Loop Splice Plate
	Low-Speed Printer
LSPET	Lunar Samples Preliminary Examination Team [NASA]
LSQA	Local System Queue Area
LSR	Laboratory for Space Research
	Least-Squares Regression
	Liquid Silicon Rubber
	Load Shifting Resistor
	Local-Shared Resources
	Low Spatial Resolution
LSR-AES	Low Spatial Resolution - Auger Electron Spectroscopy
LSS	Launch Status Summarizer
	Life Support System
	Loop Surge Suppressor
	Lunar Soil Simulator [NASA]
LSSD	Level-Sensitive Scan Design
LSSL	Lateral Surface Superlattice
LSSM	Lunar Scientific Survey Module [NASA]
LST	List
	Large Space Telescope
	Lyddane-Sachs-Teller (Relation)
LSTPC	List Price
LSTTL	Low-Power Schottky Transistor-Transistor Logic

LSU	Louisiana State University [US]
LSV	Lunar Surface Vehicle [NASA]
LSW	Lipshits-Slyozov-Wagner (Treatment)
LT	Laboratory Technology
	Language Translation
	Leak Testing
	Less than
	Letter Telegram
	Light
	Light Truck
	Line Telecommunications
	Line Telegraphy
	Link Trainer
	Logic Theorist
	Long Transverse
	Low Temperature
	Low Tension
	Low Torque
LTA	Lead Tetraacetate
	Lighter-than-Air
	Long-Term Arrangement
LTAS	Lighter-than-Air Ship
LTB	Lithium Tetraborate
LTBO	Linear Time Base Oscillator
LTC	Long Time Constant
LTCESFS	Low-Temperature Constant Energy Synchronous Fluorescence Spectroscopy
LTD	Limited
LTDE	Local Thermodynamic Equilibrium
LTDP	Long-Term Defense Program
LTDS	Laser Target Designation System
	Launch Trajectory Data System
	Low-Temperature Drawing Salt
LTE	Local Thermodynamic Equilibrium
LTFT	Low-Temperature Flow Test
LTG	Lightening
	Lightning
LTGH	Lightening Hole
LTH	Low-Temperature Herschel
	Luteotropic Hormone
LTHA	Low-Temperature High-Activity
LTHR	Leather
LTI	Light Transmission Index
	Low-Technology Intensive
LTIC	Laminated Timber Institute of Canada
LTL	Less than Truckload
LTM	Low-Temperature Mixing
LTMR	Laser Target Marker Ranger
LTOM	Less than One Minute
LTOS	Less than One Second
LTP	Library Technology Project
	Long-Tailed Pair
LTPD	Lot Tolerance Percent Defective
LTPWHT	Low-Temperature Postweld Heat Treatment
LTR	Long-Term Revitalization
LTR	Letter
LTRS	Letter Shift
LTS	Launch Telemetry Station
	Launch Tracking System
	Long-Term Stability
	Low-Temperature Sensitization

LTTAT	Long Tank Thrust Augmented Thor
LTTL	Low-Power Transistor-Transistor Logic
LTTMT	Low-Temperature Thermomechanical Treatment
LTU	Line Termination Unit
LTWA	Log Tape Write Ahead
LTV	Light Vessel
	Long Tube Vertical
LU	Logical Unit
Lu	Lutetium
LUB	Logical Unit Block
LUB	Lubrication
LUBR	lubrication
LUCID	Language for Utility Checkout and Instrumentation Development
LUE	Linear Unbiased Estimator
LUF	Lowest Usable Frequency
LUG	Light Utility Glider
LUHF	Lowest Usable High Frequency
LU-LU	Logical Unit - Logical Unit (Session)
LUM	Lumber
LUME	Light Utilization More Efficient
LUNR	Land Use and Natural Resources
LUSS	Linear Ultrasonic Scanning System
LUT	Launcher Umbilical Tower
	Look-up Table
LUVO	Lunar Ultraviolet Observatory [NASA]
LUW	Logical-Unit-of-Work
LV	Large-Volume
	Laser Vision
	Launch Vehicle
	Linear Velocity
	Low Velocity
	Low Voltage
	Low Volume
LVA	Launch Vehicle Availability
LVCD	Least Voltage Coincidence Detector
LVD	Low-Velocity Detonation
LVDA	Launch Vehicle Data Adapter
LVDC	Launch Vehicle Digital Computer
LVDT	Linear Variable Differential Transformer
	Linear Variable Displacement Transducer
	Linear Variable Displacement Transformer
	Linear Velocity Displacement Transformer
	Linear Voltage Differential Transformer
LVE	Linear Viscoelasticity
LVHV	Low-Volume High-Velocity
LVL	Low-Velocity Layer
LVM	Long Vertical Mark
LVOD	Launch Vehicles Operations Division [NASA]
LVP	Low-Voltage Protection
LVPS	Low-Voltage Power Supply
LVR	Ladle Vacuum Refinement
	Low-Voltage Release
LVRE	Low-Voltage Release Effect
LVRJ	Low-Volume Ramjet
LVS	Low-Velocity Scanning
LVSEM	Low-Voltage Scanning Electron Microscopy
LVT	Landing Vehicle Tracked
LVTR	Low VHF (= Very High Frequency) Transmitter-Receiver

LVZ	Low-Viscosity Zone
Lw	Lawrencium
LWB	Lower Bound
LWBR	Light-Water Breeder Reactor
LWC	Lightweight Concrete
	Loop Wiring Concentrator
LWD	Larger Word
LWIC	Lightweight Insulating Concrete
LWF	Lightweight Fighter (Plane)
LWL	Limited War Laboratory [US Army]
	Load Water Line
LWM	Low Water Mark
LWOST	Low Water of Spring Tide
LWR	Light Water Reactor
LWRHU	Lightweight Radioisotope Heater Unit
LWS	Loire-Wendel-Sprunck (Process)
lx	lux [unit]
LYRIC	Language for Your Remote Instruction by Computer
Lys	Lysine
LZT	Lead Zirconate Titanate
	Local Zone Time

M

M	Machine
	Magnification
	Mantissa
	March
	Master
	Magnetization
	Magnetron
	May
	Mechanical
	Memory
	Meridian
	(Meridies) - Noon
	Metal
	Meter [instrument]
	Methionine
	Metric
	Million
	Mischmetal
	Mixture
	Mode
	Modem
	Moment
	Monitor
	Monday
	Motor
	Multiplier
	(Mille) - Thousand
m	mass
	meter [unit]
	manual
	minute
	month
MA	Main Amplifier

	Maleic Anhydride
	Maritime Administration
	Martensite/Retained Austenite
	Massachusetts [US]
	Mass Analyzer
	Master of Arts
	Matched Angles
	Mathematical Association [UK]
	Mechanical Advantage
	Mechanical Alloying
	Megampere [unit]
	Memory Address
	Message Assembler
	Meter Angle
	Methyl Acrylate
	Microalloy
	Mill Anneal (Process)
	Milliammeter
	Modify Address
	Moving Average
	Multichannel Analyzer
Ma	Masurium
mA	milliampere [unit]
MAA	Macroaggregate
	Manitoba Architectural Association [Canada]
	Master of Arts in Architecture
	Mathematical Association of America
	Modeling Association of America
	Mixed Aluminum Alloy
MA-A	Medium Abrasive-Abrasive
MAAC	Model Aeronautics Association of Canada
MAACS	Multi-Address Asynchronous Communication System
MAAD	Material Availability Date
MAALOX	Magnesium-Aluminum Hydroxide
MAArch	Master of Arts in Architecture
MAARC	Magnetic Annular Arc
MAAW	Medium Antitank Assault Weapon
MAB	Materials Advisory Board [NRC, US]
	Monomethylaminoazobenzene
MAb	Monoclonal Antibody
MABA	Meta-Aminobenzoic Acid
MABS	Maritime Application Bridge System
MAC	Machine-Aided Cognition
	Maintenance Allocation Chart
	Man and Computer
	Man against Computer
	Manufacturing Advisory Committee
	Material Acquisition and Control
	Maximum Admissible Concentration
	Maximum Allowable Concentration
	Mean Aerodynamic Chord
	Metal-Arc Cutting
	Microgravity Advisory Committee
	Military Airlift Command [US Air Force]
	Mineralogical Society of Canada
	Mining Association of Canada
	Multiple Access Computer
	Multiple Access Computing
MACC	Modular Alter and Compose Console

MACCS	Manufacturing Cost Collection System
	Manufacturing and Cost Control System
MACDAC	Machine Communication with Digital Automatic Computer
MACE	Management Applications in Computer Environments
MACH	Machine
	Machinery
	Machinist
MACMIS	Maintenance and Construction Management Information System
MACRF	Macro Instruction Form
MACROL	Macrolanguage
MACS	Media Account Control System
	Medium Altitude Communications Satellite
	Member of the American Chemical Society
	Military Aeronautical Communication System
	Multiple-Technique Analytical Computer System
	Multiproject Automated Control System
	Multipurpose Accelerated Cooling System
	Multipurpose Acquisition Control System
MACSMB	Measurement and Automatic Control Standards Management Board
MACU	Material Cost per Unit
	Michigan Association of Colleges and Universities [US]
MACV	Multipurpose Airmobile Combat-Support Vehicle
MAD	Machine ANSI Data
	Magnetic Airborne Detector
	Magnetic Anomaly Detection
	Magnetic Azimuth Detector
	Maintenance, Assembly and Disassembly
	Many Acronymed Device [the HEMT]
	Mass Analyzer Detector
	Michigan Algorithmic Decoder
	Multi-Aperture Device
	Multiple Access Device
	Multiple Audio Distribution
	Multiply and Add
	Mutually Assured Destruction
MADA	Multiple-Access Discrete Address
MADAEC	Military Application Division of the Atomic Energy Commission
MADAM	Moderately Advanced Data Management
	Multipurpose Automatic Data Analysis Machine
MADAR	Malfunction Analysis Detection and Recording
	Malfunction Data Recorder
MADDAM	Macromodule and Digital Differential Analyzer Machine
MADDIDA	Magnetic Drum Digital Differential Analyzer
MADE	Magnetic Device Evaluator
	Microalloy Diffused Electrode
	Minimum Airborne Digital Equipment
	Multichannel Analog-to-Digital Data Encoder
MADGE	Microwave Aircraft Digital Guided Equipment

MADIS	Millivolt Analog-Digital Instrumentation System
MADM	Manchester Automatic Digital Machine
MADP	Main Air Display Plot
MADRE	Magnetic Drum Receiving Equipment
	Martin Automatic Data-Reduction Equipment
MADREC	Malfunction Detection and Recording
MADS	Machine-Aided Drafting System
	Missile Attitude Determination System
MADT	Microalloy Diffused-Base Transistor
MAE	Magnetic After-Effect
	Master of Aeronautical Engineering
MAECON	Mid-America Electronics Conference
MAESTRO	Machine-Assisted Educational System for Teaching by Remote Operation
MAF	Major Academic Field
	Mixed Amine Fuel
	Multi-Apertured Fabrics
MAFD	Minimum Acquisition Flux Density
MAG	Macrogenerator
	Magazine
	Magnesium
	Magnet
	Magnetism
	Magnetometer
	Magnitude
	Maritime Air Group [Canadian Forces]
	Metal Active-Gas (Welding)
MAGAMP	Magnetic Amplifier
MAGI	Multi-Array Gamma Indicator
MAGIC	Machine-Aided Graphic Input to Computer
	Machine for Automatic Graphics Interface to a Computer
	Matrix Algebra General Interpretative Coding
	Michigan Automatic General Integrated Computation
	MIDAC (= Michigan Digital Automatic Computer) Automated General Integrated Computation
MAGICS	Manufacturing, Accounting and General Information Control System
MAGIS	Marine Air-Ground Intelligence System [US]
MAGLEV	Magnetic Levitation
MAGLOC	Magnetic Logic Computer
MAGMOD	Magnetic Modulator
MAGN	Magnet(ism)
	Magnetron
MAGPIE	Machine Automatically Generating Production Inventory Evaluation
MAgricSc	Master of Agricultural Sciences
MAgrSc	Master of Agricultural Sciences
MAGS	Measurement and Graphics System
MAI	Mobile Arctic Island
MAIBC	Member of the Architectural Institute of British Columbia [Canada]
	Multiple Aircraft Identification Display
MAIDS	Multipurpose Automatic Inspection and Diagnostic System
MAIEE	Member of the American Institute of Electrical Engineers

MAIG	Matsushita Atomic Industrial Group [Japan]
MAIME	Member of the American Institute of Mining and Metallurgical Engineers
MAINT	Maintenance
MAIntDes	Master of Arts in Interior Design
MAIP	Matrix Algebra Interpretative Program
MAIR	Molecular Airborne Intercept Radar
MAJ	Majority
MAJAC	Monitor, Anti-Jam and Control
MAK	Methylated Albumin Kieselguhr
MAL	Macro Assembly Language
MALE	Multi-Aperture Logic Element
MALL	Malleability
MALLAR	Manned Lunar Landing and Return
MALS	Maximum-Intensity Approach Light System
MALT	Mnemonic Assembly Language Translator
MAM	Methylazoxymethanol
	Multiple Access to Memory
MAMA	Methylazoxylmethanol Acetate [also: MAMAc]
MAMAc	Methylazoxylmethanol Acetate [also: MAMA]
MAMBA	Martin Armoured Main Battle Aircraft
MAMBO	Mediterranean Association for Marine Biology and Oceanography [Malta]
MAMF	Masters' Association of Metals Finishers
MAMI	Machine-Aided Manufacturing Information
MAMIE	Magnetic Amplification of Microwave Integrated Emissions
MAMMAX	Machine-Made and Machine Aided Index
MAMOS	Marine Automatic Meteorological Observing Station
MAN	Manual
	Metropolitan Area Network
	Microwave Aerospace Navigation
	Molecular Anatomy
Man	Manitoba [Canada]
MANDRO	Mechanically Alterable NDRO
MANF	Manifold
MANFEP	University of Manitoba Finite Element Program
MAN HR	Man Hour
MANIP	Manual Input
MANIAC	Mathematical Analyzer Numerical Integrator and Computer
	Mechanical and Numerical Integrator and Computer
MANOVA	Multivariate Analysis of Variance
MANSCETT	Manitoba Society of Certified Engineering Technicians and Technologists [Canada]
MANTRAC	Manual Angle Tracking Capability
MANUV	Maneuvering
MAO	Monoamine Oxidase
MAP	Machinist Apprentice Program
	Macro Arithmetic Processor
	Macro Assembly Program
	Magneto-Abrasive Powder
	Maintenance Analysis Procedure
	Manifold Absolute Pressure
	Manufacturing Automation Protocol
	Mathematical Analysis without Programming

	Message Acceptance Pulse
	Military Assistance Program
	Minimum Audible Pressure
	Missed Approach Point
	Monoaluminumphosphate
	Multiple Allocation Procedure
MAPAO	Mines Accident Prevention Association of Ontario [Canada]
MAPC	Metropolitan Area Planning Committee
MAPCHE	Mobile Automatic Programmed Checkout Equipment
MAPED	Machine-Aided Program for Preparation of Electrical Designs
MAP-EPA	Manufacturing Automation Protocol - Enhanced Performance Architecture
MAPI	Machinery and Allied Products Institute [US]
MAPICS	Manufacturing Accounting and Production Information Control System
MAPID	Machine-Aided Program for Preparation of Instruction Data
MAPORD	Methodology Approach to Planning and Programming Operational Requirements Research and Development [US Air Force]
MAPP	Mathematical Analysis of a Perception and Preference
	Methylacetylene Propadiene
MAPPS	Management Association of Private Photogrammetric Surveyors
MApplS	Master of Applied Science
MAppS	Master of Applied Science
MAPS	Middle Atlantic Planetarium Society
	Multicolor Automatic Projection System
	Multivariate Analysis and Prediction of Schedules
MAQL	Mine Air Quality Laboratory
MAR	Machine-Readable Cataloguing
	Malfunction Array Radar
	March
	Marine
	Maritime
	Memory Address Register
	Micro-Analytical Reagent
	Minimal Angle of Resolution
	Multifunction Array Radar
MARAD	Marine Administration [US]
MARAIRMED	Maritime Air Forces Mediterranean [NATO]
MARC	Machine-Readable Cataloguing
	Material Accountability Recoverability Code
MARCAS	Maneuvering Reentry Control and Ablation Studies
MARCEP	Maintainability and Reliability Cost-Effectiveness Program
MArch	Master of Architecture
MARCIA	Mathematical Analysis of Requirements for Career Information Appraisal
MARCO	Machine Referenced and Coordinated Outline
MARDAN	Marine Differential Analyzer
MAREA	Member of the American Railway Engineering Association

MARGEN	Management Report Generator
MARKAR	Mapping and Reconnaissance Ku-Band Airborne Radar
MARLIS	Multi-Aspect Relevance Linkage Information System
MARNAF	Marquardt Navair Fuel
MARPAC	Maritime Forces Pacific [DND]
MARS	Machine Retrieval System
	Magnetic Airborne Recording System
	Management Analysis Reporting Service
	Manned Aerodynamic Reusable Spaceship
	Manned Astronomical Research Station
	Marconi Automatic Relay System
	Market Analysis Research System
	Martin Automatic Reporting System
	Memory-Address Register Storage
	Military Affiliated Radio System
	Military Amateur Radio System
	Mirror Advanced Reactor Study
	Multi-Access Retrieval System
	Multi-Aperture Reluctance Switch
	Multiple Artillery Rocket System [US]
MARSAT	Maritime Satellite
MART	Maintenance Analysis Review Technique
	Mean Active Repair Time
	Mississippi Aerial River Transit System
	Mobile Automatic Radiation Tester
MARTEC	Martin Thin-Film Electronic Circuit
MARTEL	Missile Antiradar and Television
MARV	Maneuverable Antiradar Vehicle
	Multi-Element Articulated Research Vehicle
MAS	Magnesia-Alumina-Silicate
	Management Advisory Services
	Microbeam Analysis Society [US]
	Microfilm Advisory Service
	Military Agency for Standardization [of NATO]
	Military Alert System
	Multi-Aspect Signaling
mAs	Milliampere Second
MASA	Master of Advanced Studies in Architecture
	Military Automotive Supply Agency [US]
MASc	Master of Agricultural Sciences
	Master of Applied Science
MASCOT	Motorola Automatic Sequential Computer Operator Test
MASDC	Military Aircraft Storage and Deposition Center
MASE	Materials, Applications and Services Exposition [US]
MASER	Materials Science Experiment Rocket
	Microwave Amplification of Stimulated Emission of Radiation [also: Maser]
MASG	Missile Auxiliary Signal Generator
MASH	Manned Antisubmarine Helicopter
MASIS	Management and Scientific Information System
MASK	Maneuvering and Seakeeping
MASM	Member of the American Society for Metals
MASME	Member of the American Society of Mechanical Engineers

MASRT	Marine Air Support Radar Teams	**MAUS**	Mobile Automated Ultrasonic Scanner
MASRU	Marine Air Support Radar Unit [USDOD]	**MAUTEL**	Microminiaturized Autonetics Telemetry
MASS	Maintenance Activities Subsea Surface	**MAVAR**	Modulating Amplifier using Variable Reactance
	Michigan Automatic Scanning System		
	Mobile Army Sensor System	**MAVES**	Manned Mars and Venus Exploration Studies
	Monitor and Assembly System		
	Multi-Axial Span System	**MAVIN**	Machine-Assisted Vendor Information Network
	Multiple Access Sequential Selection		
Mass	Massachusetts [US]	**MAW**	Marine Aircraft Wing
MASSDAR	Modular Analysis, Speedup, Sampling, and Data Reduction		Medium Anti-Armor Weapon
			Medium Antitank Assault Weapon
MASSTER	Mobile Army Sensor System Test Evaluation and Review		Mission-Adaptable Wing
		MAWP	Maximum Allowable Working Pressure
MAST	Magnetic Annular Shock Tube	**MAWS**	Mobile Aircraft Weighing System
	Material Status	**MAX**	Maximum
MASTER	Matching Available Student Time to Educational Resources	**MAX WT**	Maximum Weight
		MB	Magnetron Branch
	Miniaturized Sink-Rate Telemetering Radar		Main Battery
	Multiple Access Shared Time Executive Routine		Manitoba [Canada]
			Megabyte
MASU	Metal Alloy Separation Unit		Memory Buffer
MAT	Master of Arts in Teaching		Missile Bomber
	Material		Mixing Box
	Matrix		Mobile
	Mechanical Aptitude Test		Motor Boat
	Mechanical Assembly Technique	**Mb**	Megabit
	Mechanically-Agitated Tank	**MBA**	Marine Biological Association [US]
	Microalloy Transistor		Master of Business Administration
MATA	Michigan Aviation Trades Association [US]		Modulated Bayard-Alpert (Gauge)
MATCON	Microwave Aerospace Terminal Control	**MBAS**	Methylene Blue Active Substance
MATD	Mine and Torpedo Detector	**MBBA**	Methoxybenzylidenebutylaniline
MATE	Modular Automatic Test Equipment	**MBC**	Main Beam Clutter
	Multiple-Access Time-Division Experiment		Master of Building Construction
	Multisystem Automatic Test Equipment	**MBD**	Magnetic Bubble Device
MatE	Materials Engineer	**M bd ft**	Board Foot Measure [unit]
MATH	Mathematical	**MBE**	Missile Borne Equipment
	Mathematician		Molecular Beam Epitaxy
	Mathematics	**MBF**	Materials Business File
MATIC	Multiple Area Technical Information Center		Molecular Beam Facility
MATICO	Machine Applications to Technical Information Center Operations		Thousand Board Feet [unit]
		MBFR	Mutual Balanced Force Reduction
MATILDA	Metering and Totalizing Instrument for Load Demand Assessment	**MBH**	Thousand BTU per Hour
		MBI	Multibeam Image
MATL	Material	**MBIAC**	Missouri Basin Inter-Agency Committee [US]
MATLAN	Matrix Language		
MA-TPM	Maritime Administration - Transport Planning Mobilization [US]	**MBiChem**	Master of Biological Chemistry
		MBiEng	Master of Biological Engineering
MATPS	Machine-Aided Technical Processing System	**MBiPhy**	Master of Biological Physics
MATRIC	Matriculation	**MBiS**	Master of Biological Science
MATRS	Miniature Airborne Telemetry Receiving Station	**MBITS**	Monitored Burn-In Test System
		MBK	Multiple Beam Klystron
MATS	Military Air Transport Service	**MBL**	Marine Biological Laboratory
MAT SCI	Materials Sciences	**MBM**	Magnetic Bubble Memory
MATT	Missile ASW (= Antisubmarine Warfare) Torpedo Target		Multipurpose Boring Machine
			Thousand Board Feet [unit]
MATZ	Military Airfield Traffic Zone	**MBP**	Mechanical Balance Package
MAU	Marine Amphibious Unit		Mid-Boiling Point
	Multi-Attribute Utility		Myelin Basic Protein
MAUD	Master of Arts in Urban Design	**MBPS**	Milk and Beef Production System
MAUDE	Morse Automatic Decoder	**Mbps**	Megabits per second [unit]
MAuE	Master of Automotive Engineering	**MBQ**	Modified Binary Code

MBR	Marker Beacon Receiver
	Member
	Memory Buffer Register
MBRE	Memory Buffer Register, Even
MBRO	Memory Buffer Register, Odd
MBRV	Maneuverable Ballistic Reentry Vehicle
MBS	Magnetron Beam Switching
	Master of Business Studies
	Methacrylate-Butadiene-Styrene
	Multiple Batch Station
MBSA	Methylated Bovine Serum Albumin
MBSc	Master of Business Sciences
MBT	Main Battle Tank
	Mercaptobenzothiazole
	Metal-Base Transistor
MBTH	Methylbenzothiazolinone Hydrazone
MBTU	Million British Thermal Units
MBU	Memory Buffer Unit
MBV	Modified Bauer-Vogel (Process)
MBWO	Microwave Backward-Wave Oscillator
MC	Magnetic Card
	Magnetic Core
	Main Channel
	Manifold Control
	Manual Control
	Maritime Commission [US]
	Master Control
	Medium Curing
	Member of Council
	Memory Control
	Metal Carbide
	Metal Clad (Cable)
	Meter-Candle
	Methylcellulose
	Military Committee
	Military Computer
	Missile Control
	Moisture Content
	Molded Compound
	Momentary Contact
	Monitor Call
	Monte Carlo (Method)
	Motorcycle
	Multichip
	Multiple Contact
Mc	Megacycle
M&C	Monitor and Control
MCA	Manufacturing Chemists Association [US]
	Material Coordinating Agency
	Mechanical Contractors Association
	Methylcholanthrene
	Model Cities Administration [US]
	Modified Conventional Alloy
	Multichannel Analyzer
	Multiple Crevice Assembly
MCAA	Mechanical Contactors Association of America
MCAE	Mechanical Computer-Aided Engineering
MCAR	Machine Check Analysis and Recording
MCAS	Micro-Controlled Airflow System
MCASI	Member of the Canadian Aeronautics and Space Institute

MCBF	Mean Cycles between Failures
MCC	Main Communications Center
	Management Control Center
	Manned Control Car [USAEC]
	Metric Commission of Canada
	Microcrystalline Cellulose
	Miscellaneous Common Carrier
	Mission Control Center [of NASA]
	Mixing Cross-Bar Connectors
	Multicomponent Circuit
	Multi-Crossover Cryotron
	Multiple Computer Complex
MCC-H	Mission Control Center - Houston [of NASA]
MCCIS	Military Command, Control and Information System
MCCR	Ministry of Consumer and Commercial Relations
MCCS	Monte Carlo Computer Simulation
MCCU	Multiple Communications Control Unit
MCD	Magnetic Circular Dichroism
	Measurement Control and Display
	Metals and Ceramics Division [of US Air Force]
	Months for Cyclical Dominance
MCDP	Microprogrammed Communication Data Processor
MCDS	Management Control Data System
MCE	Master of Civil Engineering
MCEB	Military Communications Electronics Board [US]
MCEng	Master of Civil Engineering
MCerE	Master of Ceramic Engineering
MCES	Member of the Civil Engineering Society
MCEWG	Military Communications Electronics Working Group
MCF	Metal-Coated Fiber
	Micro-Complement Fixation
	Monolithic Crystal Filter
	Mutual Coherence Function
	Thousand Cubic Feet
MCFD	Million Cubic Feet per Day
MCFH	Million Cubic Feet per Hour
MCG	Man-Computer Graphics
	Materials Coordinating Group
	Microwave Command Guidance
	Mobile Command Guidance
McGU	McGill University [Canada]
MCH	Machine
	Machine Check Handler
	Methylcyclohexane
MChA	Master of Applied Chemistry
MCHC	Mean Cellular Hemoglobin Concentration
MChE	Master of Chemical Engineering
MCI	Malleable Cast Iron
MCIA	Microcomputer Investors Association
MCIC	Metals and Ceramics Information Center [US]
MCID	Multipurpose Concealed Intrusion Detector
MCIDAS	Man-Computer Interactive Data Access System

MCIM	Member of the Canadian Institute of Mining and Metallurgy
MCIMM	Member of the Canadian Institute of Mining and Metallurgy
MCIS	Maintenance Control Information System
MCL	Manufacturing Control Language
	Mathematics Computation Laboratory [US]
	Mid-Canada Line
	Maximum Contamination Level
MCLK	Master Clock
MCLOS	Manual Command Line-of-Sight
MCM	Machine for Coordinating Multiprocessing
	Magnetic Core Memory
	Million Cubic Meters
	Mine Countermeasures
	Monte Carlo Method
	Moving Coil Motor
	Thousand Circular Mils
	Thousand Cubic Meters
MCMES	Member of the Civil and Mechanical Engineering Society
MCMG	Man-Carrying Motion Generator
MCO	Molding and Cost Optimization
MCOM	Mathematics of Computation
MCompSc	Master of Computer Science
MCP	Master Control Program
	Memory-Centered Processor
	Message Control Program
	Microchannel Plate
	Military Construction Program
	Multichannel Communications Program
MCPA	Methylchlorotolyloxyacetic Acid
MCPBA	M-Chloroperoxybenzoic Acid
MCPHA	Multichannel Pulse-Height Analyzer
MCR	Magnetic Character Reader
	Magnetic Character Recognition
	Master Change Record
	Master Control Routine
	Military Compact Reactor [USAEC]
	Minimum Creep Rate
MCROA	Marine Corps Reserve Officers Association
MCRR	Machine Check Recording and Recovery
MCRT	Maximum Cruise Thrust
	Multichannel Rotary Transformer
MCS	Marine Corps School [US]
	Manufacturing Consulting Service
	Master Control System
	Medical Computer Services
	Message Control System
	Method of Constant Stimuli
	Microwave Carrier Supply
	Military Communications Station
	Missile Control System
	Mobile Communications System
	Modular Computer Systems
	Multinational Character Set
	Multiple Cloning Site
	Multiple Console Support
	Multiprogrammed Computer System
	Multipurpose Communications and Signaling

Mc/s	Megacycles per second
MCSA	Microcomputer Software Association [of ADAPSO]
MCSC	Member/Customer Service Center [of ASM]
MCSS	Military Communications Satellite System
MCT	Mass Culturing Technique
	Maximum Continuous Thrust
	Mechanical Comprehension Test
	Mercury Cadmium Telluride
	Mobile Communications Terminal
MCTC	Maritime Cargo Transportation Conference [of MTRB]
MCTI	Metal Cutting Tool Institute
MCTR	Message Center
MCU	Magnetic Card Unit
	Memory Control Unit
	Message Construction Unit
	Microcomputer Unit
	Microprogram Control Unit
	Ministry of Colleges and Universities [Canada]
	Multiplexer Control Unit
	Multiprogrammed Control Unit
MCUG	Military Computers Users Group
MCVD	Modified Chemical Vapor Deposition
MCW	Modulated Continuous Wave
MCY	Motorcycle
MD	Doctor of Medicine
	Magnetic Disk
	Magnetic Drum
	Main Drum
	Manual Data
	Manual Damper
	Maryland [US]
	Mechanical Drafting
	Message Data
	Messages per Day
	Meteorology Department
	Mine Disposal
	Molecular Dynamics
	Monitor Display
	Motor Drive
	Movement Directive
	Multinominal Distribution
	Municipal District
M-D	Modulation-Demodulation
Md	Mendelevium
	Maryland [US]
MDA	Maintainability Design Approach
	Manufacturing Defects Analyzer
	Methylene Dianiline
	Minimum Descent Altitude
	Monochrome Display Adapter
	Multidimensional Access
	Multidimensional Analysis
	Multidimensional Array
	Multidocking Adapter
MDAA	Mutual Defense Assistance Act
MDAC	Multiplying Digital-to-Analog Converter
MDAE	Multidisciplinary Accident Engineering
MDAL	McDonell Douglas Aerophysics Laboratory [US]

MDAP	Machine and Display Application Program
	Mutual Defense Assistance Program
MDB	Maritime Development Board
MDC	Machinability Data Center [US]
	Main Display Console
	Main Distribution Center
	Maintenance Data Collection
	Maintenance Dependency Chart
	Materials Dissemination Center
	Materials Distribution Center
	Minimum Detectable Concentration
	Missile Development Center [US Air Force]
	Multiduty Collector
MDCC	Master Data Control Console
MDCE	Monitoring and Duplicate Control Equipment
MDCS	Maintenance Data Collection System
	Master Digital Command System
MDCK	Madin-Darby Canine Kidney
MDCS	Maintenance Data Collection System
MDD	Machine Dependence Data
MDDR	Mean Depth of Deformation Rate
MDE	Magnetic Decision Element
	Missile Display Equipment
	Modular Design of Electronics
MDEA	Methyldiethanolamine
MDes	Master of Design
MDF	Macrodefect-Free
	Main Distributing Frame
	Medium-Frequency Direction Finding
	Mild Detonating Fuse
	Misorientation Distribution Function
MDFNA	Maximum Density Fuming Nitric Acid
MDFY	Modify
MDH	Malic Dehydrogenase
	Maximum Diameter Heat
MDI	Diphenylmethane Diisocyanate
	Magnetic Direction Indicator
	Manual Data Input
	Methylene Diisocyanate
	Methylene Diphenyl Diisocyanate
	Microdielectrometry
	Miss Distance Indicator
MDiEng	Master of Diesel Engineering
MDIF	Manual Data Input Function
MDI-PA	Diphenylmethane Diisocyanate - Polyamide
MDIU	Manual Data Input Unit
MDL	Macro Description Language
	Maritime Dynamics Laboratory
	Master Data Library
	Mine Defense Laboratory [USDOD]
	Minimum Detectable Limit
	Module
MDM	Maximum Design Meter
	Metal-Dielectric-Metal (Filter)
	Minimum Detectable Mass
MDMS	Methylene Dimethanesulfonate
MD/NC	Mechanical Drafting/Numerical Control
MDO	Medium Density Overlaid
MDP	Methylenedioxyphenyl
	Missile Data Processor

MDPE	Medium-Density Polyethylene
MDPR	Mean Depth-of-Penetration
MDR	Manual Data Room
	Memory Data Register
	Miscellaneous Data Recorder
	Mission Data Reduction
	Multichannel Data Recorder
MDRP	Mean Depth Rate of Penetration
MDS	Magnetic Detector System
	Main Device Scheduler
	Maintenance Data System
	Malfunction Detection System
	Management Decision System
	Master Drum Sender
	Memory Disk System
	Micro Disk Storage
	Microprocessor Development System
	Minimum Detectable Signal
	Modern Data Systems
	Mohawk Data System
	Multipoint Distribution Service
MDSE	Merchandise
MDSS	Meteorological Data Sounding System
MDT	Maximum Diameter of the Thorax
	Mean Downtime
	Mobile Data Terminal
	Modified Data Tag
	Mountain Daylight Time
MDTA	Manpower Development and Training Act
MDU	Message Decoder Unit
	Mine Disposal Unit
MDW	Measured Daywork
ME	Magnetic Estimation
	Maine [US]
	Male Elbow (Pipe Section)
	Manufacturing Engineer
	Master of Engineering
	Maximum Effort
	Mechanical Efficiency
	Mechanical Engineer(ing)
	Melt Extraction
	Mercaptoethanol
	Metabolizable Energy
	Methyl [also: Me]
	Military Engineer(ing)
	Mining Engineer(ing)
	Moessbauer Effect
	Muzzle Energy
Me	Metal
	Methyl [also: ME]
	Maine [US]
MEA	Maintenance Engineering Analysis
	Malt Extract Agar
	Master of Engineering Administration
	Materials Experiment Assembly [NASA]
	Minimum Enroute Altitude
	Monoethanolamine
MEAB	Maintenance Engineering Analysis Board
MEAC	Manufacturing Engineering Application Center
MEAL	Master Equipment Allowance List

	Master Equipment Authorization List
MEAR	Maintenance Engineering Analysis Record
	Maintenance Engineering Analysis Report
MEARA	Manitoba Environmental Assessment and Review Agency [Canada]
MEAS	Measure
MEC	Market Economy Country
	Materials Engineering Center
	Meteorology Engineering Center [NOSC, US]
MeC	Metal Carbide
MECA	Molecular Emission Cavity Analysis
MECC	Minnesota Educational Computing Consortium [US]
MECCA	Mechanized Catalogue
MECH	Mechanic
	Mechanical
	Mechanics
	Mechanism
MechE	Mechanical Engineer
MECI	Manufacturing Engineering Certification Institute [US]
MECL	Motorola Emitter-Coupled Logic
MECOMSAG	Mobility Equipment Command Scientific Advisory Group
MECR	Maintenance Engineering Change Request
MECU	Master Engine Control Unit
MED	Median
	Medium
	Microelectronic Device
	Mobile Energy Depot
MEDAB	Methyl Dimethylaminoazobenzene
MEDAC	Medical Electronic Data Acquisition and Control
	Medical Equipment Display and Conference
MEDAL	Micromechanized Engineering Data for Automated Logistics
MEDDA	Mechanized Defense Decision Anticipation
MEDes	Master of Environmental Design
MEDIA	Magnavox Electronic Data Image Apparatus
MEDLARS	Medical Literature Analysis and Retrieval System
MEDLINE	MEDLARS On-Line System
MEDS	Marine Environmental Data Service
	Medical Evaluation Data System
MEDSMB	Medical Standards Management Board
MEE	Master of Electrical Engineering
MEECN	Minimum Essential Emergency Communications Network [US]
MEED	Medium Energy Electron Diffraction
MEEng	Master of Electrical Engineering
MEF	Multiple Effect Flash (Evaporator)
MEG	Magneto-Encephalography
	Megohm
	Message Expediting Group
meg	megabyte
MEGW	Megawatt
MEHQ	Methyl Ether of Hydroquinone
MEI	Manual of Engineering Instructions
	Metals Engineering Institute [of ASM]
	Ministry of the Electronics Industry [PR China]

MEIC	Member of the Engineering Institute of Canada
MEIS	Medium Energy Ion Scattering
	Metal-Epitaxial Insulator-Semiconductor
	Multispectral Electrooptical Imaging Scanner
MEISFET	Metal-Epitaxial Insulator-Semiconductor Field-Effect Transistor
MEIU	Mobile Explosives Investigation Unit
MEK	Methyl Ethyl Ketone
MEKP	Methyl Ethyl Ketone Peroxide
MEL	Many-Element Laser
	Marine Engineering Laboratory [US Navy]
	Marine Ecology Laboratory [Canada]
	Materials Evaluation Laboratory [of IMR]
	Maximum Excess Loss
MELBA	Multipurpose Extended Life Blanket Assembly
MELEC	Microelectronics
MELEM	Microelement
MELVA	Military Electronic Light Valve
MEM	Mars Excursion Module
	Maximum Entropy Method
	Memory
	Memorandum
	Methoxyethoxymethyl
	Minimum Essential Medium
	Modified Eagle's Medium
	Modified Effective Modulus
MEMA	Microelectronic Modular Assembly
	Machinery and Equipment Manufacturers' Association
MEMAC	Machinery and Equipment Manufacturers Association of Canada
MEMB	Membrane
MEMO	Memorandum
MEMP	Maximization of Expected Maximum Profits
MEMS	Mechanism Modeling System
MEN	Multiple Event Network
MENEX	Maintenance Engineering Exchange
MEng	Master of Engineering
	Mining Engineer(ing)
MEngSc	Master of Engineering and Science
MEO	Major Engine Overhaul
	Management Engineering Office
MeO	Metal Oxide
MEP	Management Engineering Program
	Master of Engineering Physics
	Mean Effective Pressure
	Mean Effective Potential
	Mission Effects Projector [NASA]
	Motor-Evoked Potential
MEPCB	Mechanical Earth Pressure Counterbalance
MEPF	Multiple Experiment Processing Furnace
MEPP	Mineral Economics and Policy Program
MER	Manned Earth Reconnaissance
	Minimum Energy Requirement
	Multiple Ejector Rack
MERA	Molecular Electronics for Radar Applications
	Molecular Electronics for Radar Arrays
MERDC	Mobility Equipment Research and Development Center [US]

MERDL	Medical Equipment Research and Development Laboratory [US Army]
MEREA	Member of the Electrical Railway Engineering Association [US]
MERGE	Mechanized Retrieval for Greater Efficiency
MERGV	Martian Exploratory Rocket Glide Vehicle
MERL	Municipal Environmental Research Laboratory [US]
MERM	Material Evaluation Rocket Motor
MERMUT	Mobile Electronics Robot Manipulator and Underwater Television
MERS	Mobiltity Environmental Research Study
MES	Manual Entry Subsystem
	Master of Environmental Sciences
	Master of Environmental Studies
	McMaster Engineering Society [Canada]
	Mechanical Engineering Society
	Metal-Gate Schottky
	Moessbauer Effect Spectroscopy
	Michigan Engineering Society [US]
	Morpholineethanesulfonic Acid
MESA	Manned Environmental System Assessment [NASA]
	Marine Ecosystems Analysis
	Miniaturized Electrostatic Accelerometer
	Mining Enforcement and Safety Administration [US]
	Modular Experiment Platform for Science and Applications
MESc	Master of Engineering Sciences
MESFET	Metal-Gate Schottky Field-Effect Transistor
	Metal-Semiconductor Field-Effect Transistor
MESG	Maximum Experimental Safe Gap
	Minimum Electrical Spark Gap
MESG	Message
MESNA	Mercaptoethanesulfonic Acid
MESS	Monitor Event Simulation System
MESTIND	Measurement Standards Instrumentation Division [of ISA]
MESUCORA	Measurement, Control Regulation and Automation
MET	Management Engineering Team
	Mechanical Engineering Technology
	Metal
	Metaphor
	Metaphysics
	Meteorology
	Metronome
	Metropole
	Microelectronics Technology
	Modfified Expansion Tube
	Multi-Emitter Transistor
Met	Methionine
META	Maintenance Engineering Training Agency [US Army]
	Method for Extracting Text Automatically
METAL	Metallurgical
	Metallurgist
	Metallurgy
METALL	Metallurgical
	Metallurgist

	Metallurgy
METAPLAN	Methods of Extracting Text Automatically Programming Language
METCOM	Metropolitan Consortium for Minorities in Engineering
METCUT	Metal Cutting Exhibition
MetE	Metallurgical Engineer
METEOR	Meteorological
	Meteorologist
	Meteorology
METEOSAT	Meteorological Satellite
METH	Methyl
METH AL	Methyl Alcohol
METL	Multi-Element, Two-Layer
METLO	Meteorological Electronic Technical Liaison Office [US Navy]
METRIC	Multi-Echelon Technique for Recoverable Item Control
METS	Modular Engine Test System
	Modular Environmental Test System
METTP	Marine Engineering Technician Training Plan
METWK	Metalwork
MeV	Megaelectronvolt
MEW	Microwave Early Warning
MEWS	Missile Electronic Warfare System
MEX	Mexican
	Mexico
MEXE	Military Engineering Experimental Establishment [MOD]
MEZZ	Mezzanine
MF	Magnetic Foot [unit]
	Magnetomotive Force
	Maintenance Factor
	Mass Fragmentography
	Master File
	Measuring Force
	Medium Frequency
	Melamine Formaldehyde
	Microfiche
	Microfilm
	Microfiltration
	Molecular Filter
	Motor Field
	Multifrequency
	Multifunction
M&F	Male and Female [fasteners]
MFA	Multifont Adapter
	Multifurnace Assembly
MFB	Mill Fixture Base
	Mixed Functional Block
MFBS	Multifrequency Binary Sequence
MFC	Magnetic-Tape Field Scan
	Manual Frequency Control
	Microfunctional Circuit
	Monochromator Focussing Circle
	Multifunction Controller
MFCM	Multifunction Card Machine
MFCU	Multifunction Card Unit
MFD	Magnetic Frequency Detector
	Magnetofluid Dynamics

	Multifunction Detector
	Multifunction Display
mfd	microfarad [unit]
MFED	Maximum Flat Envelope Delay
MFEng	Master of Forestry Engineering
MFG	Manufacturing
MFH	Maximum Fork Height
MFI	Mobile Fuel Irradiator
	Melt Flow Indexer
MFK	Mill Fixture Key
MFKP	Multifrequency Key Pulsing
MFL	Mantitoba Federation of Labour [Canada]
	Mobile Foundry Laboratory
MFLD	Message Field
MFLOPS	Million Floating-Point Operations per Second
MFM	Magnetic-Field Meter
	Modified Frequency Modulation
	Multistage Frequency Multiplier
MFN	Most Favoured Nation [tariffs]
MFO	Mixed Function Oxidase
MFOD	Manned Flight Operations Division
MFP	Mean Free Path
MFPB	Mineral Fiber Products Bureau
MFR	Manipulator Foot Restraint
	Manufacture
	Manufacturer
	Multifrequency Receiver
MFRS	Manufacturers
MFS	Manned Flying System
	Marine Finish Slate
	Multifunction Sensor
MFSA	Metal Finishing Suppliers Association
MFSK	Multiple Frequency Shift Keying
MFT	Metallic Facility Terminal
	Miniature Fluorescent Tube
	Minimum Film Formation Temperature
	Monolayer Formation Time
	Multiposition Frequency Telegraphy
	Multiprogramming with a Fixed Number of Tasks
MFTF	Mirror Fusion Test Facility
MFTG	Manufacturing Technology Group [of IEEE]
MFTRS	Magnetic Flight Test Recording System
MFTR	Multifrequency Transmitter
MFV	Military Flight Vehicle
MG	Machine-Glazed (Paper)
	Machine Gun
	Master Generator
	Melt Growth
	Message Generator
	Methylene Glycol
	Mining
	Mixed Grain
	Motor Generator
	Multigage
	Mycoplasma Gallisepticum
Mg	Magnesium
	Megagram [unit]
mg	milligram [unit]
MGA	Mycoplasma Gallisepticum Agglutinin
	Methylene Glycol Anion

MGC	Malachite Green Carbinol
	Manual Gain Control
	Missile Guidance and Control
MGCC	Missile Guidance and Control Computer
MGCR	Maritime Gas-Cooled Reactor
MGD	Magnetogasdynamics
	Million Gallons per Day
MGE	Master of Geological Engineering
MGG	Memory Gate Generator
MGGB	Modular Guided Glide Bomb
MGL	Malachite Green Leucocyanite
	Matrix Generator Language
MGM	Mechanics of Granular Materials
MGO	Malachite Green Oxalate
	Megagauss Oersted [unit]
MGP	Magnetic and Graphic Products
	Methyl Green-Pyronin (Solution)
MGR	Manager
MGS	Market Grade Stainless
MH	Magnetic Head
	Main Hatch
	Maleic Hydrazide
	Manhole
	Medium Hardness
	Message Handler
	Modified Huffman
mH	millihenry [unit]
MHA	Modified Handling Authorized
	Mueller Hinta Agar
MHC	Martin Hard Coat
MHCP	Mean Horizontal Candlepower
MHD	Magnetohydrodynamics
MHDF	Medium and High Frequency Direction Finding
MHE	Materials Handling Equipment
MHEA	Materials Handling Engineers Association
MHEDA	Materials Handling Equipment Distributors Association [US]
MHF	Massive Hydraulic Fracturing
	Medium High Frequency
	Mixed Hydrazine Fuel
MHHW	Mean Higher High Water
MHiE	Master of Highway Engineering
MHKW	Midvale-Heppenstall-Kloeckner-Werke (Process)
MHL	Messo Heated Ladle
MHLW	Mean Higher Low Water
MHR	Member of the House of Representatives [US]
MHSCP	Mean Hemispherical Candlepower
MHT	Mean High Tide
	Mild Heat Treatment
	Museum of History and Technology [US]
MHV	Manned Hypersonic Vehicle
MHVDF	Medium, High and Very High Frequency Direction Finding
MHW	Mean High Water
MHWS	Mean High Water Spring
MHz	Megahertz
MI	Machine-Independent
	Malleable Iron
	Manual Input

	Metabolic Index
	Metals Information [of ASM]
	Michigan [US]
	Micro-Instruction
	Mile [also: mi]
	Mill
	Miller Indices
	Miller Integrator
	Military Intelligence
	Mineral Insulated
	Multiple Interaction
mi	mile [also: MI]
MIA	Metal Interface Amplifier
MIAC	Minimum Automatic Computer
MIACF	Meander Inverted Autocorrelated Function
MIACS	Manufacturing Information and Control System
MIAS	Member of the Institute of Aeronautical Sciences
	Microprobe Image Analysis System
	Municipal Industrial Abatement Strategy [Canada]
MIB	Manual Input Buffer
MIBC	Methyl Isobutylcarbinol
MIBK	Methyl Isobutylketone
MIBL	Masked Ion Beam Lithography
MIC	Management Information Center
	Macro Instruction Compiler
	Master Interrupt Control
	Message Identification Code
	Methyl Isocyanate
	Michigan Instructional Computer
	Micrometer [instrument]
	Microphone
	Microscope
	Microwave Integrated Circuit
	Minimum Ignition Current
	Minimal Inhibitory Concentration
	Minimum Ignition Current
	Minimum Inhibiting Concentration
	Missing Interruption Checker
	Monitoring, Identification and Correlation
	Monolithic Integrated Circuit
	Multilayer Interconnection Computer
	Mutual Interference Chart
MICA	Macro Instruction Compiler Assembler
MICAM	Microammeter
MICE	Member of the Institute of Civil Engineers
MICELEM	Microphone Element
MIChemE	Member of the Institute of Chemical Engineers
MICG	Mercury Iodide Crystal Growth
Mich	Michigan [US]
MICOM	Missile Command [US Army]
MICR	Magnetic Ink Character Recognition
MICRO	International Symposium on Microscopy
	Microcomputer
	Multiple Indexing and Console Retrieval Options
microamp	microampere [unit]
micro-in	micro-inch [unit]

MICROMIN	Microminiature
MICROPAC	Micromodule Data Processor and Computer
MICS	Management Information and Control System
	Mineral-Insulated Copper-Sheathed
MICV	Mechanized Infantry Combat Vehicle [US Army]
MID	Master of Industrial Design
	Message Input Descriptor
MIDA	Maritime Industrial Development Area
MIDAC	Management Information for Decision Making and Control
	Michigan Digital Automatic Computer
MIDAR	Microwave Detection and Ranging [also: Midar]
MIDAS	Map Information Display and Analysis System
	Measurement Information Data Analytic System
	Micro-Imaged Data Addition System
	Missile Intercept Data Acquisition System
	Modified Integration Digital Analog Simulator
	Modulator Isolation Diagnostic Analysis System
MIDES	Missile Detection System
MIDIP	Military Industry Data Interchange Procedure
MIDOP	Missile Doppler
MIDOR	Miss Distance Optical Recorder
MIDOT	Multiple Interferometer Determination of Trajectories
MIDS	Movement Information Distribution Station
MIE	Master of Industrial Engineering
	Metal-Induced Embrittlement
MIEE	Member of the Institute of Electrical Engineers
MIEEE	Member of the Instutute of Electrical and Electronics Engineers
MIES	McMaster (University) Institute for Energy Studies [Canada]
MIFAS	Mechanized Integrated Financial Accounting System
MIFI	Missile Flight Indicator
MIFL	Master International Frequency List
MIFR	Master International Frequency Register
MIFS	Multiplex Interferometric Fourier Spectroscopy
MIG	Magnetic Injection Gun
	Metal Inert-Gas (Welding)
	Mikoyan and Gurevish [Russian fighter plane]
MIGS	Metal-Induced Gap State
MIH	Miles per Hour
MIK	Methyl Isobutyl Ketone
MIKE	Measurement of Instantaneous Kinetic Energy
	Microphone
MIKER	Microbalance Inverted Knudsen Effusion Recoil
MIKES	Mass-Analyzed Ion Kinetic Energy Spectroscopy

MIL	Mileage
	Military
	Milling
MILA	Merritt Island Launch Area [NASA]
MILADGRU	Military Advisory Group [US]
MILDIP	Military Industry Logistics Data Interchange Procedure
MILECON	Military Electronics Conference
MILES	Multiple Integrated Laser Engagement System
MILL	Million
MIL-E-CON	Military Electronics Conference
MIL-HDBK	Military Handbook
MIL-I	Military Specification on Interfaces
MILREP	Military Representative
MILS	Missile Location System
MIL-SPEC	Military Specification
MILSTAAD	Military Standard Activity Address
MIL-STD	Military Standard
MILSTICCS	Military Standard Item Characteristics Coding Structure
MILSTRIP	Military Standard Requisitioning and Issue Procedure [US Army]
MIM	Metal Injection Molding
	Metal-Insulator-Metal
	Modified Index Method
MIMarE	Member of the Institute of Marine Engineers
MIMD	Multiple Instruction, Multiple Data
MIME	Member of the Institution of Mining Engineers [UK]
MIMechE	Member of the Institute of Mechanical Engineers
MIMinE	Member of the Institute of Mining Engineers
MIMM	Member of the Institute of Mining and Metallurgy
MIMO	Man In, Machine Out
MIMR	Magnetic Ink Mark Recognition
MIMS	Multi-Item Multi-Source
MIN	Minimum
	Mineral
	Mining
min	minute
MINA	Member of the Institute of Naval Architects
MINAC	Miniature Linear Accelerator
	Miniature Navigation Airborne Computer
MINDAC	Marine Inertial Navigation Data Assimilation Computer
MINDD	Minimum Due Date
MINEAC	Miniature Electronic Autocollimator
MINGEL	Martin Integrated Neutral Graphics and Engineering Language
MINI	Minicomputer
MINIAPS	Minimum Accessory Power Supply
MINIMARS	Mini-Mirror Advanced Reactor Study
MINIMAX	Minimizing the Maximal Error
MINIRAR	Minumum Radiation Requirements
MINI-SUBLAB	Miniature Submarine Laboratory
Minn	Minnesota [US]
MINNEMAST	Minnesota Mathematics and Science Teaching Project [US]
MINPRT	Miniature Processing Time
MINPROC	Mineral Processing Database [of CANMET]
MINS	Miniature Inertial Navigation System
MINSD	Minimum Planned Start Date
MINSOP	Minimum Slack Time per Operation
MInstCE	Member of the Institute of Civil Engineers
MInstME	Member of the Institute of Mining Engineers
MInstMM	Member of the Institute of Mining and Metallurgy
MInstWE	Member of the Institute of Water Engineers
MINT	Materials Identification and New Item Control Technique
MINTECH	Ministry of Technology [UK]
MINTEC	Mining Technology Database [of CANMET]
MIN WT	Minimum Weight
MIO	Map Information Office [of USGS]
MIOP	Multiplexing Input/Output Processor
MIP	Manual Input Processing
	Manual Input Program
	Matrix Inversion Program
	Mean Indicated Pressure
	Minimum Impulse Pulse
	Missile Impact Predictor
	Most Important Person
MIPE	Magnetic Induction Plasma Engine
	Modular Information Processing Equipment
MIPIR	Missile Precision Instrumentation Radar
MIPPT	McMaster Institute for Polymer Production Technology [Canada]
MIPS	Millions of Instructions per Second
	Microcomputer Image Processing System
	Minimum Inventory Production System
	Missile Impact Predictor Set
MIR	Memory Information Register
	Memory Input Register
	Multiple Internal Reflection
	Music Information Retrieval
MIRA	Mechanical Industrial Relations Association
	Motor Industry Research Association [UK]
MIRACL	Management Information Report Access without Computer Languages
MIRACODE	Microfilm Retrieval Access Code
MIRAGE	Microelectronic Indicator for Radar Ground Equipment
MIRAN	Missile Ranging
MIRD	Medical Internal Radiation Dose
MIRE	Member of the Institute of Radio Engineers
MIRF	Multiple Instantaneous Response File
MIRFAC	Mathematics in Recognizable Form Automatically Compiled
MIRG	Metabolic Imaging Research Group
MIRO	Mining Industry Research Organization
MIROC	Mining Industry Research Organization of Canada
MIROS	Modulation Inducing Retrodirective Optical System
MIRPS	Multiple Information Retrieval by Parallel Section
MIRR	Material Inspection and Receiving Report
MIRS	Manpower Information Retrieval System

MIRT	Molecular Infrared Track
MIRU	Move-In, Rig-Up
MIRV	Multiple Independent-Oriented Reentry Vehicle
	Multiple Individually-Targeted Reentry Vehicle
MIS	Management Information Science
	Management Information Service
	Management Information System
	Metal-Insulator-Semiconductor
	Metering Information System
	Modified Initial System
MISA	Municipal Industrial Strategies for Abatement
MISC	Miscellaneous
MISD	Multiple-Instruction, Single-Data (Stream)
MISDAS	Mechanical Impact System Design for Advanced Spacecraft
MISFET	Metal-Insulator-Semiconductor Field-Effect Transistor
MISHAP	Missiles High-Speed Assembly Program
MISP	Medical Information Systems Program
	Mineral Investment Stimulation Program [Canada]
MISS	Mechanical Interruption Statistical Summary
	Missile Intercept Simulation System
	Mobile Integrated Support System
	Multi-Item, Single-Source
Miss	Mississippi [US]
MISS-D	Minuteman Integrated Schedules Status and Data System
MISSIL	Management Information System Symbolic Interpretative Language
MISTRAM	Missile Trajectory Measurement System
MIT	Maupa Institute of Technology [Phillipines]
	Massachusetts Institute of Technology [US]
	Master Instruction Tape
MITE	Microelectronic Integrated Test Equipment
	Miniaturized Integrated Telephone Equipment
	Missile Integration Terminal Equipment
	Multiple Input Terminal Equipment
MITEA	Modular Integrated Towed Electronics Array
MITI	Minstry of International Trade and Industry [Japan]
MITM	Military Industry Technical Manual
MITO	Minimum Interval Takeoff
MITOC	Missile Instrumentation Technical Operations Communications
MITOL	Machine-Independent Telemetry-Oriented Language
MITR	Massachusetts Institute of Technology Reactor [US]
MITT	Ministry of Industry, Trade and Technology [Canada]
MIU	Malfunction Insertion Unit
	Message Injection Unit
MIUS	Modular Integrated Utility System
MIX	Magnetic Ionization Experiment
	Master Index

	Methyl Isobutylxanthine
	Mixture
MJ	Megajoule [unit]
MK	Mask
	Mark
MKG	Marking
	Meter-Kilogram System
MKS	Meter-Kilogram-Second System
MKSA	Meter-Kilogram-Second-Ampere System
MKT	Market
MKTG	Marketing
MkWh	Thousand Kilowatthours
ML	Machine Language
	Magnetic Latching
	Major Lobe
	Main Lobe
	Material List
	Materials Laboratory
	Maximum Likelihood
	Mean Level
	Memory Location
	Methods of Limit
	Mold Line
	Monolayer
	Multilayer
M-L	Metallic-Longitudinal
mL	millilambert [unit]
	milliliter [unit]
MLC	Maneuver Load Control
	Multilayer Ceramic
	Multilayer Circuit
MLCA	Multiline Communications Adapter
MLCAEC	Military Liaison Committee to the Atomic Energy Commission [US]
MLCB	Multilayer Circuit Board
MLD	Masking Level Difference
	Minimum Lethal Dose
	Mixed Layer Depth
	Multilayer Dielectric
MLDG	Molding
MLE	Maximum Likelihood Estimate
	Maximum Likelihood Estimator
	Molecular-Layer Epitaxy
MLEV	Manned Lifting Entry Vehicle
MLF	Medium Longitudinal Fascicule
	Multilateral Force
MLG	Main Landing Gear
	Methyl-L-Glutamate
MLGW	Maximum Landing Gross Weight
MLH	Maximum Likelihood
MLHCP	Mean Lower Hemispherical Candlepower
MLHW	Mean Lower High Water
MLI	Magnetic Level Indicator
	Marker Light Indicator
MLL	Manned Lunar Landing
MLLP	Manned Lunar Landing Program [of NASA]
MLLW	Mean Lower Low Water
MLO	Manned Lunar Orbiter [of NASA]
MLP	Machine Language Program
MLPCB	Machine Language Printed Circuit Board
MLPWB	Multilayer Printed Wiring Board
MLR	Main Line of Resistance

	Memory Lockout Register
	Monodisperse Latex Reactor
	Mixed Leucocyte Reaction
	Multiple Linear Regression
MLRG	Marine Life Research Group [of SIO]
MLRS	Monodisperse Latex Reactor System
MLS	Machine Literature Searching
	Manned Lunar Surface [NASA]
	Master of Library Science
	Metal-Langmuir Semiconductor
	Microwave Landing System
	Missile Location System
	Mudline Suspension System
	Multilanguage System
MLSFET	Metal-Langmuir Semiconductor Field-Effect Transistor
MLSNPG	Microwave Landing System National Planning Group [US]
MLSS	Mixed Liquid Suspended Solids
MLT	Mass-Length-Time
	Mean Length per Turn
	Mean Logistical Time
	Mean Low Tide
MLU	Memory Loading Unit
MLV	Medium Logistics Vehicle
MLW	Mean Low Water
MLWS	Mean Low Water Spring
MLWK	Millwork
MM	Main Memory
	Maintenance Manual
	Mass Memory
	Master Monitor
	Materials Measurement
	Methyl Methacrylate
	Micromechanics
	Middle Marker
	Mischmetall
	Monostable Multivibrator
	MOS (= Metal-Oxide Semiconductor) Monolithic
mm	millimeter
MMA	Manual Metal-Arc (Welding)
	Methyl Methacrylate
	Multiple Module Access
MMath	Master of Mathematics
MMAU	Master Multiattribute Utility
MMB	Modulated Molecular Beam
MMC	Magnesium Methyl Carbonate
	Maximum Material Condition
	Maximum Metal Condition
	Metal-Matrix Composite
	Metal-Reinforced Matrix Composite
MMCIAC	Metal-Matrix Composites Information Analysis Center
MMD	Mass Median Diameter
	Moving Map Display
MME	Master of Mechanical Engineering
MMechEng	Master of Mechanical Engineering
MMES	Member of the Mechanical Engineering Society
MMet	Master of Metallurgy

MMetEng	Master of Metallurgical Engineering
MMF	Magnetomotive Force
	Man-Made Fiber
	Minimum Mass Fraction
	Moving Magnetic Features
MMFITB	Man-Made Fibers Industry Training Board
MMFS	Manufacturing Message Format Standard
MMH	Methylmercury Hydroxide
	Monomethylhydrazine
mm Hg	millimeters of mercury [unit]
mm Hg L	millimeters of mercury liters [unit]
MMI	Man-Machine Interface
MMIC	Monolithic Microwave Integrated Circuit
	Motorcycle and Moped Industry Council
MMIEng	Master of Mining Engineering
MMIJ	Mining and Metallurgical Institute of Japan
m/min	meters per minute
MMIS	Materials Management Information System
MML	Material Mechanics Laboratory
MMLS	Microgravity Materials Science Laboratory
MMM	Maintenance and Material Management
	Manned Mars Mission [NASA]
	Mars Mission Module
	Multimission Module
MMMI	Meat Machinery Manufacturers Institute [US]
MMMIS	Maintenance and Material Management Information System
MMMSocAm	Member of the Mining and Metallurgical Society of America
MMOD	Micromodule
MMP	Minimum Miscibility Pressure
	Modified Modular Plug
	Monitor Metering Panel
	Multiplex Message Processor
MMPF	Microgravity and Materials Processing Facility
MMPR	Missile Manufacturers Planning Reports
MMPT	Man-Machine Partnership Translation
MMR	Modified Modified Read
	Multiple Match Resolver
MMRBM	Mobile Medium-Range Ballistic Missile
MMRIM	Mat-Molding Reaction Injection Molding
MMRRI	Mining and Mineral Resources Research Institute [US]
MMS	Magnetic Median Surface
	Man-Machine System
	Mass Memory Store
	Materials Management System
	Metallurgy and Materials Science
	Methyl Methanesulfonate
	Missile Monitoring System
	Module Management System
MMSA	Mining and Metallurgical Society of America
MMSE	Minimum-Mean-Squared Error
MMSI	Medium Metal-Support Interaction
MMSL	Microgravity Materials Science Laboratory
MMSocAm	Mining and Metallurgical Society of America
MMT	Methyl-M-Tyrosine

	Multiple Mirror Telescope	**MOC**	Magnetic-Optic Converter
MMTC	Mechanical Maintenance Training Center		Master Operational Controller
MMTS	Methyl Methylthiomethyl Sulfoxide		Master Operations Console
MMTT	Multiple Mechanicothermal Treatment		Master Operations Control
MMU	Memory Management Unit		Memory Operating Characteristic
	Modular Maneuvering Unit		Minimum Operational Characteristics
MMW	Main Magnetization Winding		Mission Operation Computer
	Millimeter Wave	**MOCA**	Methylenebis-O-Chloroaniline
m mu	millimicron [unit]		Minimum Obstruction Clearance Altitude
MN	Main	**MOCVD**	Metallo-Organic Chemical Vapor Deposition
	Meganewton [unit]	**MOCR**	Mission Operations Control Room
	Minnesota [US]	**MOD**	Message Output Descriptor
	Mnemonic Symbol		Metallo-Organic Deposition
	Motor Number		Ministry of Defense [UK]
Mn	Manganese		Model
mN	millinewton [unit]		Moderation
MNA	Methoxyneuraminic Acid		Modern
	Methyl Nadic Anhydride		Modification
	Multi-Network Area		Modulation
MNBLE	Modified Nearly Best Linear Estimator		Modulation-Doped
MNDX	Mobile Nondirector Exchange		Modulator
MNE	Master of Nuclear Engineering		Module
MNL	Manual		Modulus
	Minnesota National Laboratory [US]	**MODA**	Motion Detector and Alarm
MNNG	Methylnitronitroguanidine	**MODAC**	Mountain System Digital Automatic
MNOS	Metal-Nitride-Oxide Semiconductor		Computer
	Metal-Nitride Oxide Silicon	**MODEM**	Modulator-Demodulator [also: Modem]
MNOSFET	Metal-Nitride Oxide Semiconductor Field-	**MODEST**	Missile Optical Destruction Technique
	Effect Transistor	**MODFET**	Modulation-Doped Field-Effect Transistor
MNP	Methyl-N-Pyrrolidine	**MODI**	Modified Distribution Method
MNPT	Medium National Pipe Thread	**MODICON**	Modular-Dispersed Control
MNR	Minimum Noise Routes	**MODILS**	Modular Instrument Landing System
	Ministry of Natural Resources [Canada]	**MODS**	Major Operation Data System
MNRL	Mineral	**MOE**	Measure of Effectivenes
MNS	Metal-Nitride Semiconductor		Ministry of Energy [Canada]
MNSFET	Metal-Nitride Semiconductor Field-Effect	**MOERO**	Medium Orbiting Earth Resources
	Transistor		Observatory
MNT	Mononitrotoluene	**MOF**	Maximum Operating Frequency
MNU	Methyl Nitrosourea	**MOGA**	Microwave and Optical Generation and
MO	Manual Output		Amplification
	Manual Operation	**MOI**	Moment of Inertia
	Masonry Opening	**MOL**	Machine-Oriented Language
	Mass Observation		Manned Orbiting Laboratory
	Master Oscillator		Ministry of Labour [Canada]
	Metal(lo)organic		Molecule
	Metal Oxide		Mol Nuclear Research Center [located in
	Mining Office		Mol, Belgium]
	Missouri [US]	**mol**	mole [unit]
	Mixed Oxide	**mol%**	mole percent
	Mobile Object	**MOLAB**	Mobile Lunar Laboratory [of NASA]
	Molecular Orbital		Moving Lunar Laboratory
	Month	**MOLDS**	Multiple On-Line Debugging System
Mo	Molybdenum	**MOLE**	Molecular Optical Laser Examiner
	Missouri [US]	**MOLECTRONICS** Molecular Electronics
MOA	Matrix Output Amplifier	**MOLSINK**	Molecular Sink of Outer Space
MOBAT	Modified Battalion Antitank	**MOL WT**	Molecular Weight
MOBIDAC	Mobile Data Acquisition System	**MOM**	Metal-Oxide-Metal
MOBIDIC	Mobile Digital Computer	**MOMS**	Metalorganic Magnetron Sputtering
MOBL	Macro-Oriented Business Language	**MON**	Monday
MOBS	Multiple Orbit Bombardment System		Monitor
MOBULA	Model Building Language		Monument

	Motor Octane Number
MONO	Monostable
MONOCL	Monoclinic
Mont	Montana [US]
MoOPH	Oxodiperoxymolybdenumhexamethyl-phosphoramide
MOOSE	Manned Orbital Operations Safety Equipment
	Man Out of Space Easiest
MOP	Melt Optimization Program
	Memory Organization Packet
	Multiple On-Line Programming
MOPA	Master Oscillator Power Amplifier
MOPB	Manually-Operated Plotting Board
MOPEG	Methoxyphenylglycol
MOPR	Manner of Performance Rating
MOPS	Military Operation Phone System
	Missile Operations
	Morpholinepropanesulfonic Acid
MOPSO	Morpolinohydroxypropanesulfonic Acid
MOPTS	Mobile Photographic Tracking Station
MOR	Magnetic Optical Rotation,
	Malate Oxidoreductase
	Mars Orbital Rendezvous
	Modulus of Rupture
	Molten Alkali Resistance
MORE	Maintenance, Repair, Overhaul Excellence
MORL	Manned Orbital Research Laboratory
MORP	Meteorite Observation and Recovery Project
MORS	Military Operations Research Symposium
MOR T	Morse Taper
MOS	Management Operating System
	Manufacturing Operating System
	Material Ordering System
	Metal-Oxide Semiconductor
	Metal-Oxide-Silicon
	Ministry of Supply [US]
	Mission Operating System [of NASA]
	Months
MOSAIC	Macro Operation Symbolic Assembler and Information Compiler
	Metal-Oxide Semiconductor Advanced Integrated Circuit
	Ministry of Supply Automatic Integrator and Computer
	Mobile System for Accurate ICBM Control
MOSAR	Modulation Scan Array Receiver
MOSERD	Ministry of State for Economic and Regional Development
MOSFET	Metal-Oxide Semiconductor Field-Effect Transistor
MOS/LSI	Metal-Oxide Semiconductor for Large-Scale Integration
MOSM	Metal-Oxide Semimetal
MOSRAM	Metal-Oxide Semiconductor Random Access Memory
MOSST	Ministry of State for Science and Technology [Canada]
MOST	Management Operation System Technique
	Metal-Oxide Semiconductor Transistor
	Military Operational and Support Truck

	Ministry of Science and Technology [Korea]
	Multipurpose Observation Sizing Technique
MOT	Manned Orbital Telescope [NASA]
	Ministry of Transport
	Ministry of Overseas Trade [UK]
	Motor
MOTARDES	Moving Target Detection System
MOTMX	Methoxy Trimethylxanthine
MOTNE	Meteorological Operational Telecommunications Network of Europe
MOTOR	Mobile-Oriented Triangulation of Reentry
MOTS	Module Test Set
MOTU	Mobile Optical Tracking Unit
MOU	Memorandum of Understanding
MOV	Metal-Oxide Varistor
MOVECAP	Movement Capability
MOVPE	Metalorganic Vapor Phase Epitaxy
MOWASP	Mechanization of Warehousing and Shipment Processing
MP	Machine Pistol
	Magnifying Power
	Mathematical Programming
	Main Phase
	Measurement Pragmatic
	Mechanical Part
	Mechanical Properties
	Medium-Pressure
	Melting Point
	Metallized Paper
	Methylpurine
	Microperoxidase
	Microprobe
	Microprocessor
	Minimum Phase
	Miscellaneous Paper
	Miscellaneous Publication
	Missing Pulse
	Multiphase
	Multiplier Phototube
	Multipole
	Multiprocessing
	Multiprocessor
	Multipulse
	Multipurpose
	Myeloperoxidase
mP	millipoise [unit]
MPA	Maclura Pomifera Agglutinin
	Master of Public Administration
	Metal Powder Association
	Methylphosphoric Acid
	Multiphoton Absorption
	Multiple-Period Average
MPa	Megapascal [unit]
mPa	millipascal [unit]
MPAA	Motion Picture Association of America
MPACS	Management Planning and Control System
MPAS	Metal Powder Association Standard
MPB	Material Performance Branch [of US Air Force]
MPBE	Molten Plutonium Burn-Up Experiment
MPC	Metals Properties Council

	Multipurpose Communications
MPCC	Multiprotocol Communications Controller
MPD	Magnetoplasmadynamics
	Map Pictorial Display
	Materials Physics Division [of US Air Force]
	Maximum Permissible Dose
	Multiphoton Decomposition
MPDA	Metaphenylene Diamine
MPDC	Mechanical Properties Data Center [US]
MPDI	Marine Products Development Irradiator
MPDSM	Micro Powder Diffraction Search/Match
MPE	Mathematical and Physical Sciences and Engineering
	Mannesmann-Pfannen-Entschwefelung [= Mannesmann ladle degassing]
	Maximum Permissable Exposure
	Microwave Plasma Etching
MPeEng	Master of Petroleum Engineering
MPEMA	Ethyl-P-Tolylmalonamide
MPEP	Manual of Patent Examining Procedures
MP/EP	Missing Pulse/Extra Pulse
MPFP	Melt Processible Fluoropolymer
MPG	Microwave Pulse Generator
	Miles per Gallon
	Miniature Precision Gyrocompass
MPH	Miles per Hour
MPHPS	Miles per Hour per Second
MPhy	Master of Physics
MPI	Magnetic Particle Inspection
	Mannose Phosphate Isomerase
	Max-Planck Institute
	Mean Point of Impact
MPIC	Message Processing Interrupt Count
MPIF	Metal Powder Industries Federation
MPIO	Mission and Payload Integration Office [of NASA]
MPIP	Miniature Precision Inertial Platform
MPIS	Multilateral Project Information System
MPL	Maximum Permissible Level
	Maximum Print Line
	Multipurpose Processing Language
	Multischedule Private Link
MPLC	Medium-Pressure Liquid Chromatography
MPLEA	Multipurpose Long Endurance Aircraft
MP-LPC	Multipulse Linear Predictive Coding
MPLR	Maximum Permissible Leakage Rate
	Mininum Pressure Live Roller (Conveyor)
MPLX	Multiplex(er)
MPLXR	Multiplexer
MPM	Maximum Pionization Method
	Microprocessor Module
	Monocycle Position Modulation
	Multistand Pipe Mill
mpm	meters per minute [also: m/min]
MPMA	Montford Point Marine Association [US]
MPN	Most Probable Number
MPO	Maximum Power Output
	Mono Power Amplifier
	Myeloperoxidase
MPP	Message Processing Program
	Mission Planning Program

	Most Probable Position
MPPH	Methylphenylphenylhydantoin
MPPL	Multipurpose Processing Language
	Multipurpose Programming Language
MPR	Materials and Processing Report
	Microseismic Processor Recorder
MPRE	Medium Power Reactor Experiment
	Multipurpose Research Reactor
MPS	Mathematical Programming System
	Materials Processing in Space
	Member of the Pharmaceutical Society
	Methylacetylene Propadiene Stabilized
	Mission Preparation Sheet
	Mucopolysaccaride
	Multiple Partition Support
	Multiprocessing System
	Multiprogramming System
MPSI	Thousand Pounds per Square Inch
MPSX	Mathematical Programming System, Extended
MPT	Male Pipe Thread
	Metallurgical Plant and Technology
	Mouse-Protection Test
	Multiple Pure Tones
MPTA	Mechanical Power Transmission Association
	Metal Powder Technology Association
MPTR	Mobile Position Tracking Radar
MPTS	Mobile Photographic Tracking Station
MPU	Microprocessing Unit
	Microprocessor Unit
MPW	Male Pipe Weld
	Modified Plane Wave
MPX	Multiplex(er)
MPY	Multiplier
MQ	Multiplier-Quotient
MQC	Manufacturing Quality Control
MQR	Multiplier-Quotient Register
MQW	Multiquantum Well
MR	Magazine Records
	Machine Rifle
	Map Reference
	Marble
	Mask Register
	Matching Record
	Measurement Range
	Memorandum Report
	Memory Register
	Mercury-Redstone [NASA]
	Message Repeat
	Methylresorcinol
	Mineral Rubber
	Mine-Run
	Miscellaneous Report
	Moderating Ratio
	Modified Read
	Moisture Resistance
	Molasses Residuum
	Monitor Recorder
mR	milliroentgen [unit]
MRA	Microgravity Research Association
	Minimum Reception Altitude

MRAD	Mass Random Access Disk
MRad	Master of Radiology
MRADS	Mass Random Access Data Storage
MRaEng	Master of Radio Engineering
MRAIC	Member of the Royal Architectural Institute of Canada
MR-ATOMIC	Multiple Rapid Automatic Test of Monolithic Integrated Circuit
MRAD	Mass Random Access Disk
MRB	Magnetic Recording Borescope
MRBM	Medium-Range Ballistic Missile
MRC	Maintenance Requirement Card
	Marine Resources Council [US]
	Mathematics Research Center
	Meteorological Research Committee [UK]
MRCA	Multirole Combat Aircraft
MRDE	Mining Research and Development Establishment
MR&DF	Malleable Research and Development Foundation [US]
MRE	Microbiological Research Establishment [MOD]
	Mining Research Establishment
	Multiple-Response Enable
	Microbiological Research Establishment [UK]
MREM	Minerals Resources Engineering and Management
MRF	Multipath Reduction Factor
MRG	Materials Research Grant
	Materials Research Group
MRH	Mechanical Recording Head
	Mobile Remote Handler
MRI	Machine Records Installation
	Magnetic Resonance Imaging
	Magnetic Rubber Inspection
	Mean Recurrence Interval
	Member of the Royal Institute
	Midwest Research Institute [US]
	Miscellaneous Radar Input
	Monopulse Resolution Improvement
MRIR	Medium-Resolution Infrared Radiometer
MRL	Manufacturing Reference Line
	Materials Research Laboratories [NSF, US]
	Multiple Rocket Launcher
MRM	Master of Resource Management
	Metabolic Rate Monitor
mR/min	milliroentgen per minute [unit]
MRMU	Mobile Remote Manipulating Unit [US Air Force]
MRN	Meteorological Rocket Network [of NASA]
	Minimum Reject Number
mRNA	Messenger Ribonucleic Acid [also: M-RNA]
MRO	Maintenance, Repair and Operating Supplies
	Memory Resident Overlay
	Midrange Objectives
MRP	Malfunction Reporting Program
	Manufacturer's Resource Planning
	Manufacturing Requirements Planning
	Manufacturing Resource Planning

	Materials Requirement Planning
	Materials Requirement Program
	Materials Resource Planning
	Metal Refining Process
MRPC	Mercury Rankine Power Conversion [USAEC]
MRPPS	Maryland Refutation Proof Procedure System
MRR	Mechanical Reliability Report
	Multiple Restrictive Requirement Quality
MRRV	Maneuvering Reentry Research Vehicle
MRS	Manned Repeater Station
	Materials Research Society
	Mobile Roof Support
MRSA	Mandatory Radar Service Area
MRT	Mean Repair Time
	Mobile Radar Target
	Modified Rhyme Test
	Multiple Requesting Terminal Program
MRU	Machine Records Unit
	Material Recovery Unit
	Message Retransmission Unit
	Mobile Radio Unit
MRV	Maneuverable Reentry Vehicle
	Maximum Relative Variation
MRWC	Multiple Read-Write Compute
MS	Machine Selection
	Machine Steel
	Macromodular System
	Magnetic Storage
	Mail Steamer
	Main Switch
	Manuscript
	Mass Spectrometry
	Mass Spectroscopy
	Mass Storage
	Master of Science
	Master Switch
	Material Specification
	Materials Science
	Maximum Stress
	Mean Square
	Medium Setting
	Megasample
	Melt Spinning
	Memory System
	Metallurgical Society [US]
	Meteoritical Society
	Methanesulfonate Salt
	Metric System
	Mild Steel
	Military Standard
	Minesweeper
	Mississippi [US]
	Mobile System
	Moessbauer Spectroscopy
	Molar Substitution
	Monitoring System
	Motor Ship
	Multispectral
M&S	Maintenance and Supply

ms	millisecond [unit]		**MS&E**	Materials Science and Engineering
MSA	Mechanical Signature Analysis		**msec**	millisecond [unit]
	Microgravity Science and Applications		**MSEE**	Master of Science in Electrical Engineering
	Mineralogical Society of America			Mean Square Error Efficiency
	Minimum Sector Altitude		**MSEI**	Mean Square Error Inefficiency
	Modem Signal Analyzer		**MSEM**	Master of Science and Engineering of Mines
	Multiplication Stimulating Activity		**MSEMPR**	Missile Support Equipment Manufacturers
MSAM	Multiple Sequential Access Method			Planning Reports
MSAP	Minislotted Alternating Priorities		**MSERD**	Ministry of State for Economic and Regional
MSAR	Mine Safety Appliance Research			Development
MSAT	Mobile Satellite		**MSF**	Mass Storage Facility
	Multipurpose Satellite			Master Source File
MSATA	Motorcycle, Scooter and Allied Trades			Multistage Flash
	Association [US]		**MSFC**	Marshall Spaceflight Center [NASA]
MSB	Methylstyrylbenzene		**MSFD**	Multistage Flash Distillation
	Most Significant Bit		**MSFH**	Manned Spaceflight Headquarters [NASA]
MSBY	Most Significant Byte		**MSFN**	Manned Spaceflight Network [NASA]
MSC	Macro Selection Compiler		**MSG**	Manufacturers Standard Gauge
	Manned Spacecraft Center [of NASA]			Mapper Sweep Generator
	Mass Storage Control			Message
	Mesitylenesulfonyl Chloride			Miscellaneous Simulation Generator
	Microscopical Society of Canada			Missing
	Mile of Standard Cable			Monosodium Glutamate
	Military Sealift Command		**MSGE**	Master of Science in Geological Engineering
	Minesweeper Coastal		**MSH**	Melanocyte Stimulating Hormone
	Monolithic Crystal Filter			Mesitylenesulfonyl Hydrazide
	Most Significant Character		**MSHA**	Mine Safety and Health Administration
	Multisystem Coupling		**MSI**	Medium-Scale Integration
MSc	Master of Science			Metal-Support Interaction
MScA	Master of Applied Science		**MSIE**	Master of Science in Industrial Engineering
MSCC	Manned Spaceflight Control Center [US Air		**MSIO**	Mass Storage Input/Output
	Force]		**MSIS**	Manned Satellite Inspection System
MScCE	Master of Science in Civil Engineering			Manufacturing Systems Integration Service
MScCS	Master of Science in Computer Sciences		**MSK**	Minimum Shift Keying
MSCE	Main Storage Control Element		**MSL**	Map Specification Library
	Master of Science in Civil Engineering			Materials Science Laboratory
MScE	Master of Science in Engineering			Mean Sea Level
MSCF	Thousand Standard Cubic Feet			Mineral Sciences Laboratory [of CANMET]
MSCP	Mass Storage Control Protocol		**MSLD**	Mass Spectrometer Leak Detector
	Mean Spherical Candlepower		**MSM**	Metal-Semiconductor-Metal
MSCR	Machine Screw			Thousand Feet Surface Measure
MSCS	Mass Storage Control System		**MSMB**	Mass-Spectrometric Molecular Beam
MSCT	Mass Storage Control Table			Mechanical Standards Management Board
MSCTC	Mass Storage Control Table Create		**MSME**	Master of Science in Mechanical
MSD	Main Storage Database			Engineering
	Mass Selective Detector		**MSMLCS**	Mass Service Main Line Cable System
	Materials Science Division		**MSM-PD**	Metal-Semiconductor-Metal Photodiode
	Mean Solar Day		**MSMV**	Monostable Multivibrator
	Missile Systems Division [US Air Force]		**MSNT**	Mesitylenesulfonylnitrotriazole
	Most Significant Digit		**MSO**	Methionine Sulfoximine
MSDB	Main Storage Database		**MSOC**	Methylsulfonylethyloxycarbonyl
MSDR	Materials Science Double Rack		**MSOS**	Mass Storage Operating System
MSDS	Marconi Space and Defense System		**MSP**	Microelectronics Support Program
	Material Safety Data Sheet			Modular System Program
MSDT	Maintenance Strategy Diagramming		**MSPetEng**	Master of Science in Petroleum Engineering
	Technique		**MSPLT**	Master Source Program Library Tape
MSE	Master of Sanitary Engineering		**MSPR**	Master Spares Positioning Resolver
	Master of Science in Engineering			Minimum Space Platform Rig
	Materials Science and Engineering		**MSR**	Machine Status Register
	Mean Square Error			Machine Stress Rating
	Modern Shipping Equipment			Magnetic Slot Reader

	Mass Storage Resident		Mode Transducer
	Missile Site Radar		Montana [US]
	Molten Salt Reactor		Mount
MSRad	Master of Science in Radiology		Mountain
MSRE	Molten Salt Reactor Experiment		Multiple Transfer
MSREF	Metal Scrap Research and Education Foundation	**MTA**	Manitoba Trucking Association [Canada]
			Mobility Test Article
MSRL	Marine Sciences Research Laboratory		Multiple Terminal Access
MSRS	Multiple Stylus Recording System		Multiterminal Adapter
MSRT	Missile System Readiness Test	**MTAC**	Mathematics Tables and other Aids to Computation
MSS	Management Science Systems		
	Manufacturers Standardization Society	**MTB**	Methyl-tert-Butylether
	Mass Storage System		Methylthymol Blue
	Military Supply Standard [USDOD]		Motor Topedo Boat
	Mobile Servicing Station	**MTBCF**	Mean Time between Confirmed Failures
	Multispectral Scanner	**MTBE**	Methyl tert-Butyl Ether
MS/s	Megasamples per second	**MTBF**	Mean Time between Failures
MSSC	Mass Storage System Communicator	**MTBM**	Mean Time between Maintenance
MSSE	Master of Science in Sanitary Engineering	**MTBMA**	Mean Time between Maintenance Actions
MSSEng	Master of Science in Sanitary Engineering	**MTBSTFA**	Methyl-tert-Butyldimethylsilyltrifluoro-acetamide
MSSG	Message		
MSSR	Mars Soil Sample Return	**MTC**	Maintenance Time Constraint
MSST	Ministry of State for Science and Technology [Canada]		Master of Textile Chemistry
			Master Tape Control
MSStEng	Master of Science in Structural Engineering		Measurement Technology Conference
MST	Master of Science and Technology		Memory Test Computer
	Master of Science in Teaching		Ministry of Transportation and Communications [Canada]
	Master Station		
	Mean Solar Time		Missile Test Center
	Minimum Spanning Tree		Mission and Traffic Control
	Mobile Service Tower		Modulation Transfer Curve
	Monolithic Systems Technology	**MTCA**	Multiple Terminal Communications Adapter
	Mountain Standard Time	**MTCC**	Master Timing and Control Circuit
	Multisubscriber Time Sharing	**MTCF**	Mean Time to Catastrophic Failure
MS&T	Methodical Structures and Textures	**MTCU**	Magnetic Tape Control Unit
MSTFA	Methyltrimethylsilyltrifluoroacetamide	**MT-CVD**	Medium Temperature - Chemical Vapor Deposition
MSTR	Master		
MSTS	Multisubscriber Time Sharing System	**MTCXO**	Mathematically Temperature-Compensated Crystal Oscillator
MSU	Message Switching Unit		
	Michigan State University [US]	**MTD**	Manufacturing Technology Division [of US Air Force]
MSUDC	Michigan State University Discrete Computer		
			Master of Transport Design
MSVC	Mass Storage Volume Control		Mean Temperature Difference
MSVI	Mass Storage Volume Inventory		Minimal Toxic Dose
MSW	Master Switch		Mounted
	Municipal Solid Waste		Moving Target Detector
MT	Machine Translation		Multiple Target Detection
	Magnetic Particle Testing	**MTDR**	Machine Tool Design and Research
	Magnetic Tape	**MTDS**	Manufacturing Test Data System
	Magnetic Tube		Marine Tactical Data System [US Navy]
	Magnetotellurics	**MTE**	Maximum Tracking Error
	Male Tee (Pipe Section)		Michigan Test of English
	Master Timer	**MTech**	Master of Technology
	Maximum Torque	**MTF**	Mean Time to Failure
	Mean Tide		Mechanical Time Fuse
	Mean Time		Medium Time to Failure
	Mechanical Transport		Mississippi Test Facility [of NASA]
	Medium Temperature		Modulation Transfer Function
	Message Table	**MTFA**	Modulation Transfer Function Analyzer
	Methyltyrosine	**MTG**	Meeting
	Metric Ton		Methane-to-Gasoline

	Methylthiogalactoside	MTRS	Magnetic Tape Recorder Start
	Mounting	MTS	Machine-Tractor Station
	Multiple-Trigger Generator		Magnetic Tape Station
MTH	Magnetic Tape Handler		Magnetic Tape System
	Methylthiohydantoin		Maintenance Test Station
	Month		Marine Technology Society [US]
MTHPA	Methyltetrahydrophtalic Anhydride		Material Testing System
MTI	Machine Tool Industry		Mechanical Testing System
	Materials Technology Institute		Member of Technical Staff
	Medium-Technology Intensive		Message Telecommunication Service
	Metal Treating Institute [US]		Message Transport Service
	Moving Target Indicator		Michigan Terminal System
MTIC	Moving Target Indicator Coherent		Missile Tracking System
MTIL	Maximum Tolerable Insecurity Level		Mobile Telephone Service
MTIRA	Machine Tool Industry Research Association [UK]		Mobile Tracking Station [NASA]
			Modem Test System
MTL	Manufacturing and Technology Laboratory		Motor-Operated Transfer Switch
	Master Tape Loading		Mountains
	Material		Multichannel Television Sound
	Materials Technology Laboratory [US Army]	MTSC	Magnetic Tape Selective Composer [also: MT/SC]
	Merged-Transistor Logic		
MTLP	Master Tape Loading Program	MTSD	Military Transmission Systems Department [of NORAD]
MTM	Methods-Time Measurement		
MTMA	Methods-Time Measurement Association	MTSE	Magnetic Trap Stability Experiment
MTMASR	MTMA for Standards and Research	MT/SC	Magnetic Tape Selective Composer [also: MTSC]
MTMC	Methylthio-M-Cresol		
MT-MF	Magnetic Tape to Microfilm	MTSMB	Material and Testing Standards Management Board
MTMTS	Military Traffic Management and Terminal Service [USDOD]		
		MTSS	Military Test Space Station
MTN	Mountain	MTST	Magnetic Tape Selective Typewriter
MTNS	Metal Thick Oxide Semiconductor	MTT	Magnetic Tape Terminal
MTNS	Mountains		Magnetic Tape Transport
MTO	Master Terminal Operator		Mechanicothermal Treatment
	Metal Turnover		Methylthiozoltetrazolium
MTOE	Million Tonnes of Oil Equivalent		Microwave Theory and Techniques [IEEE Society]
MTOGW	Maximum Takeoff Gross Weight		
MTOP	Molecular Total Overlap Population	MTTA	Machine Tool Trades Association
MTOS	Metal Thick Oxide Silicon	MTTD	Mean Time to Diagnosis
MTP	Mechanical Thermal Pulse	MTTF	Mean Time to Failure
	Methylthiophenol	MTTFF	Mean Time to First Failure
	Minimum Time Path	MTTR	Maximum Time to Repair
	Multiply Twinned Particle		Mean Time to Repair
MTPA	Methoxytrifluoromethylphenylacetic Acid		Mean Time to Restore
MTPD	Metric Tonnes per Day	MTU	Magnetic Tape Unit
MTPS	Magnetic Tape Programming System		Master Terminal Unit
MTPT	Minimal Total Processing Time		Methylthiouracil
MTPY	Million (Short) Tons per Year		Michigan Technological University [US]
MTR	Magnetic Tape Recorder		Multiplexer and Terminal Unit
	Mass Transit Railway	MTV	Marginal Terrain Vehicle
	Materials Testing Reactor	MTVAL	Master Tape Validation
	Materials Testing Report	MTW	Male Tube Weld
	Migration Traffic Rate	MTX	Methotrexate
	Missile Tracking Radar		Methylpteroylglutamic Acid
	Monitor	MU	Machine Unit
	Motor		Memory Unit
	Moving Target Reactor		Methylene Unit
	Multiple Track Radar		Missouri University [US
	Multiple Tracking Range		Multiple Unit
MTRB	Maritime Transportation Research Board [US]	mu	micron [unit]
		muA	microampere [unit]
MTRE	Magnetic Tape Recorder End	MUBIS	Multiple Beam Interval Scanner

MUCHA	Multiple Channel Analysis
MUDPAC	Melbourne University Dual-Package Analog Computer
MUF	Materials Utilization Factor
	Maximum Usable Frequency
muF	microfarad [unit]
MUGB	Methylumbelliferyl Guanidinobenzoate
muH	microhenry [unit]
MUI	Module Interface Unit
mu-in	micro-inch [unit]
mu-in-rms	micro-inch root mean square
MULPIC	Multipurpose Interrupted Cooling Process
MULT	Multiple
	Multiplication
	Multiplier
MULTEWS	Multiple Electronic Warfare Surveillance
MULTICS	Multiplexed Information and Computer Service
MUM	Methodology for Unmanned Manufacturing
MUMS	Mobile Utility Module System
MUMMS	Marine Corps Unified Materiel Management System [US]
mu mu	micromicron [unit]
MUPS	Multiple Utility Peripheral System
MURA	Midwestern University Research Association [US]
MURB	Multiple Unit Residential Buildings
	Multiple User Residential Building
MURC	Missouri University Research Center [US]
MUSA	Multiple-Unit Steerable Antenna
MUSAP	Multisatellite Augmentation Program
MUSC	Microgravity User Support Center
MUSE	Multiple Stream Evaluator
MUT	Mean Uptime
MUTMAC	Methylumbelliferyltrimethylammonium-cinnamate Chloride
muW	microwatt [unit]
MUX	Multiplex(er)
MV	Mean Value
	Mean Variation
	Measured Value
	Measured Vector
	Measured Voltage
	Medium Vacuum
	Medium Voltage
	Megavolt [unit]
	Methyl Violet
	Million Volts
	Motor Vehicle
	Motor Vessel
	Multivibrator
	Muzzle Velocity
Mv	Mendelevium
mV	millivolt [unit]
MVA	Machine Vision Association [of SME]
	Megavoltampere
	Mevalonic Acid
MVA/SME	Machine Vision Association of the Society of Manufacturing Engineers [US]
MVB	Multivesicular Body
	Multivibrator

MVBR	Multivibrator
MVC	Manual Volume Control
	Multiple Variate Counter
MVD	Map and Visual Display
MVDA	Memory Volume Discount Addendum
MVDF	Medium and Very High Frequency Direction Finding
MVE	Multivariate Exponential Distribution
MVF	Martensite Volume Fraction
MVI	Motor Vehicle Inspection
MVLUE	Minimum Variance Linear Unbiased Estimate
	Minimum Variance Linear Unbiased Estimator
MVM	Mariner versus Mercury
	Minimum Virtual Memory
	Modified Volume Module
MVMA	Motor Vehicle Manufacturers Association
MVMC	Motor Vehicle Maintenance Conference
MVMT	Movement
MVP	Matrox Vision Processor
MVPCB	Motor Vehicle Pollution Control Board [US]
MVS	Magnetic Voltage Stabilizer
	Minimum Visual Signal
	Motor Vehicle Specification
	Multiple Virtual Storage
	Multiprogramming with Virtual Storage
MVT	Multiprogramming with a Variable Number of Tasks
MVTA	Motor Vehicle Transport Act
MVU	Minimum Variance Unbiased
MVUE	Minimum Variance Unbiased Estimate
	Minimum Variance Unbiased Estimator
MVULE	Minimum Variance Unbiased Linear Estimate
	Minimum Variance Unbiased Linear Estimator
MW	Megawatt
	Medium Wave
	Microwave
	Molecular Weight
	Music Wire
mW	milliwatt
MWB	Metropolitan Water Board
	Multiprogram Wire Broadcasting
MWCO	Molecular Weight Cutoff
MWD	Measurement While Drilling
	Megawatt Day
	Molecular Weight Distribution
MWDDEA	Mutual Weapons Development Data Exchange Program
MWDP	Mutual Weapons Development Program
MWD/t	Megawatt Days per Ton [unit]
MWE	Megawatt of Electric Power
MWG	Music Wire Gauge
MWH	Megawatt of Heat
MWh	Megawatt hour
MWHGL	Multiple Wheel Heavy Gear Load
MWI	Message Waiting Indicator
MWL	Milliwatt Logic
MWO	Medicine White Oil

MWP	Maneuvering Work Platform [NASA]
	Maximum Working Pressure
MWR	Magnetic (Tape) Write Memory
	Mean Width Ratio
MWS	Microwave Station
MWT	Master of Wood Technology
MWt	Megawatt of Thermal Energy [also: MWT]
MWV	Maximum Working Voltage
MX	Multiplex
MXC	Multiplexer Channel
MXD	Mixed
MXR	Mask Index Register

N

N	Asparagine
	Navigation
	Navy
	Newton
	Nitrogen
	Node
	North
	November
	Number
	Nylon
n	net
	normal
	numeric
NA	(American) National Acme Thread
	Naturally Aspirated
	Naval Architect
	Nicotinic Acid
	Nitric Acid
	Nonabrasive
	Nonacosadiynoic Acid
	Noradrenaline
	North America
	Not Assigned
	Not Available
	Numerical Aperture
N/A	Not Applicable
Na	(Natrium) - Sodium
nA	nanoampere [unit]
NAA	Naphthaleneacetic Acid
	Naphthylacetic Acid
	North Atlantic Assembly [of NATO]
	Neutron Activation Analysis
NAAD	Nicotine Acid Adenine Dinucleotide
NAAG	NATO Army Armament Group
NAAFI	Navy, Army, and Air Force Institute [US]
NAAL	North America Aerodynamic Laboratory [US]
NAAQS	National Ambient Air Quality Standards
NAAS	National Association of Academies of Science [US]
NAATS	National Association of Air Traffic Specialists [US]
NAB	National Aircraft Beacon

	National Alliance of Businessmen [US]
	National Association of Broadcasters [US]
NABDC	National Association of Blueprint and Diazotype Coaters [US]
NABE	National Association for Business Education [US]
NABER	National Association of Business and Educational Radio [US]
NABSC	National Association of Building Service Contractors
NABTS	North American Basic Teletext Specification
NAC	Nacelle
	North Atlantic Council [of NATO]
NACA	National Advisory Committee for Aeronautics [of NASA]
	National Agricultural Chemicals Association [US]
NACAA	National Association of Computer-Assisted Analysis [US]
NACATTS	North American Clear Air Turbulence Tracking System
NACC	National Automatic Controls Conference [US]
NACD	National Association of Container Distributors [US]
NACE	National Association of Corrosion Engineers [US]
NACEIC	National Advisory Council on Education for Industry and Commerce
NACIS	North American Cartographic Information Society [US]
NACK	Negative Acknowledgement
NACME	National Action Council for Minorities in Engineering
NACOA	National Advisory Committee on Oceans and Atmosphere [US]
NACOLADS	National Council on Libraries, Archives and Documentation Services [Jamaica]
NACS	Northern Area Communications System
NAD	Nicotinamide Adenine Dinucleotide
	No-Acid Descaling
NADA	National Automobile Dealers' Association [US]
NADC	NATO Air Defense Committee
	Naval Air Development Center [US Navy]
NADDRG	North American Deep Drawing Research Group
NADEEC	NATO Air Defense Electronic Environment Committee
NADEFCOL	NATO Defense College
NADGE	NATO Air Defense Ground Environment
NADGECO	NATO Air Defense Ground Environment Consortium
NADH	Nicotinamide Adenine Dinucleotide, Reduced Form
NADIDE	Nicotinamide Adenine Dinucleotide
NADMC	Naval Air Development and Material Center [US]
NADP	Nicotinamide Adenine Dinucleotide Phosphate
NADPH	Nicotinamide Adenine Dinucleotide Phosphate, Reduced Form

NADWARN	Natural Disaster Warning System [US]
NAE	National Academy of Engineering [US]
	National Aeronautical Establishment [Canada]
NAEC	National Aeronautical Establishment Canada
	National Aerospace Education Council [US]
	Naval Air Engineering Center [US Navy]
NAECON	National Aerospace Electronics Conference [of IEEE]
NAEDS	National Association of Educational Data Systems
NAEP	National Association of Environmental Professionals
NAES	Naval Air Experimental Station [US Navy]
NAESC	National Association of Energy Service Companies
NAET	National Association of Educational Technicians
NAEW	NATO Early Warning
NAF	Naval Aircraft Factory
	Naval Avionics Facility [US Navy]
NAFAG	NATO Air Force Armaments Group
NAFEC	National Administrative Facilities Experimental Center [US]
	National Aviation Facilities Experimental Center [US]
NAFI	Naval Avionics Facility, Indianapolis [US Navy]
NAG	Numerical Algorithms Group
NAGE	National Association of Government Engineers [US]
NAI	New Alchemy Institute
NAIC	National Astronomy and Ionosphere Center [US]
NAIG	Nippon Atomic Industry Group [Japan]
NAIOP	Navigational Aid Inoperative for Parts
NAIPRC	Netherlands Automatic Information Processing Research Center
NAIT	Northern Alberta Institute of Technology [Canada]
NAK	Negative Acknowledgement (Character)
NAL	National Accelerator Laboratory [USAEC]
	National Aeronautical Laboratory [India]
	Naval Aeronautical Laboratory [US Navy]
NALM	National Association of Liftmakers
NAM	National Association of Manufacturers [US]
	N-Acridinylmaleimide
	Network Access Machine
	Nautical Air Miles
N AM	North America
NAMA	National Automatic Merchandising Association
NAMC	Naval Air Material Center [US Navy]
NAME	National Association of Marine Engineers [Canada]
NAMES	NAVDAC Assembly, Monitor, Executive System
NAMF	National Association of Metal Finishers
NAMFI	NATO Missile Firing Installation
NAMG	Narrow-Angle Mars Gate [NASA]
NAMMO	NATO Multi-Role (Combat Aircraft Development and Production) Management Organization
NAMRAD	Non-Atomic Military Research and Development
NAMRI	North American Manufacturing Research Institute [of SME]
NAMS	National Association of Marine Surveyors
NAMSA	NATO Maintenance and Supply Agency
NAMSO	NATO Maintenance and Supply Organization
NAMTC	Naval Air Missile Test Center [US Navy]
NAN	Neuraminidase
	Neutron Activation Analysis
NANCO	National Association of Noise Control Officials
NAND	Not And
NANEP	Navy Air Navigation Electronic Project [US Navy]
NANWEP	Navy Numerical Weather Prediction [US]
NAO	North Atlantic Oscillation
NAOS	North Atlantic Ocean Station
NAP	Network Access Pricing
NA-P	Nonabrasive-Polishing
NAPA	National Asphalt Pavement Association [US]
	National Association of Purchasing Agents [US]
NAPCA	National Air Pollution Control Administration
	National Association of Pipe Coating Applicators [US]
NAPD	National Association of Plastics Distributors
NAPE	National Association of Power Engineers
NAPHCC	National Association of Plumbing, Heating, and Cooling Contractors [US]
NAPL	National Association of Printers and Lithographers [US]
NAPLPS	North American Presentation Level Protocol Syntax
	North American Presentation Level Protocol System
NAPM	National Association of Purchasing Management
NAPP	National Aerospace Productivity Program
NAPS	Night Aerial Photographic System
	Nimbus Automatic Programming System
NAPSS	Numerical Analysis Problem Solving System
NAPU	Nuclear Auxiliary Power Unit
NAPWPT	National Association of Professional Word Processing Technicians
NAR	Net Assimilation Rate
NARATE	Navy Radar Automatic Test Equipment
NARBA	North American Regional Broadcasting Agreement
NARDIS	Navy Automated Research and Development Information System [US Navy]
NAREC	Naval Research Electronic Computer
NARF	Naval Air Rework Facility
	Nuclear Aerospace Research Facility [US Air Force]

NARI	National Association of Recycling Industries [US]
	Nuclear Aerospace Research Institute [US Air Force]
NARL	National Aerospace Research Laboratory
NARM	National Association of Relay Manufacturers [US]
NARMC	National Association of Regional Media Centers
NARS	National Archives and Records Service [US]
	North Atlantic Radio System
NARTE	National Association of Radio and Telecommunication Engineers
NARTS	National Association for Radio Telephone Systems [US]
	Naval Aeronautics Test Station [US]
	Naval Air Rocket Test Station [US]
NARUC	National Association of Regulatory Utility Commissioners [US]
NAS	National Academy of Sciences [US]
	National Aircraft Standards [US]
	National Airspace System [US]
	National Astrological Society
	National Avionics Society [US]
NASA	National Aeronautics and Space Administration [US]
NASA/JPL	NASA Jet Propulsion Laboratory [US]
NASARR	North American Search and Range Radar
NASCAS	NAS/NRC Committee on Atmospheric Sciences [US]
NASCO	National Academy of Sciences' Committee on Oceanography [US]
NASCOM	NASA Communications Network
NASDA	National Space Development Agency [Japan]
NASDAQ	National Association of Security Dealers' Automated Quotations
NASIS	National Association for State Information Systems [US]
	National Sourcing Information System
NASL	Naval Applied Sciences Laboratory [US]
NAS/NRC	National Academy of Sciences/National Research Council [US]
NASP	National Airport System Plan
NA-SP	Nonabrasive-Slightly Polishing
NASTRAN	NASA Structural Analysis (Program)
NASW	National Association of Science Writers [US]
NASWF	Naval Air Special Weapons Facility [US]
NAT	Nation
	National
	Natural
	Nature
	Node Attached Table
	Normal Allowed Time
NATA	National Association of Testing Authorities [Australia]
	North American Telecommunications Association
	Northern Air Transport Association
NATAS	North American Thermal Analysis Society
NATC	Naval Air Test Center [US]

NATCS	National Air Traffic Control Service [of BOT]
NATEC	Naval Air Technical Evaluation Center [UK]
NATEL	Nortronics Automatic Test Equipment Language
NATES	National Analysis of Trends in Emergency Systems
NATF	Naval Air Test Facility [US]
NATL	Naval Aeronautical Turbine Laboratory [US]
NATL	National
nat log	natural logarithm
NATM	New Austrian Tunneling Method
NATO	North Atlantic Treaty Organization
NAT ORD	Natural Order
NATS	National Activity to Test Software
	National Association of Temporary Services [US]
NATSOPA	National Society of Operative Printers and Assistants
NATSPG	North Atlantic Systems Planning Group
NAT/TFG	North Atlantic Traffic Forecasting Group
NATTS	National Association of Trade and Technical Schools [US]
	Naval Air Test Turbine Station [US Navy]
NATU	Naval Aircraft Torpedo Unit [US Navy]
NAU	Network Addressable Unit
NAUS	National Airspace Utilization Study
NAUT	Nautics
NAV	Navigation
	Naval
NAVA	National Audio Visual Association [US]
NAVAIDS	Navigational Aids
NAVAIR	Naval Air Systems Command [US Navy]
NAVAPI	North American Voltage and Phase Indicator
NAVAR	Radar Air Navigation and Control System [also: Navar]
NAVARHO	Navigation Aid, Rho Radio Navigation System
NAVASCOPE	NAVAR Airborne Radarscope
NAVASCREEN	NAVAR Ground Screen
NAVC	National Audiovisual Center
NAVCM	Navigation Countermeasures
NAVCOM	Naval Communication
NAVCOMMSTA	Naval Communications Station [US Navy]
NAVDAC	Navigation Data Assimilation Center
	Navigation Data Assimilation Computer
NAVDOCKS	Navy Yards and Docks Bureau [US]
NAVELECSYSCOM	Naval Electronics System Command [US Navy]
NAVELEX	Naval Electronics System Command [US Navy]
NAVEX	Navigation Experiment Package [NASA]
NAVFEC	Naval Facilities Engineering Command [US Navy]
NAVMAT	Office of Naval Materiel [US Navy]
NAVMINDEFLAB	Navy Mine Defense Laboratory [US Navy]
NAVOCEANO	Naval Oceanographic Office [US Navy]
NAVORD	Naval Ordnance

NAVPERS	Naval Personnel
NAVPHOTOCEN	Naval Photographic Center [US Navy]
NAVS	Navigation System
NAVSAT	Navigation Satellite [US Navy]
NAVSEA	Naval Sea Systems Command [US Navy]
NAVSEC	Naval Ship Engineering Center [US Navy]
NAVSHIPS	Naval Ship Systems Command [US Navy]
NAVSPASUR	Naval Space Surveillance System [US Navy]
NAVSUP	Naval Supply Systems Command [US Navy]
NAVTRADEVCEN	Naval Training Device Center [US Navy]
NAVWEPS	Bureau of Naval Weapons [US Navy]
NAW	Non Acid Washed
NAWAS	National Warning System
NAWLT	Nitric Acid Weight Loss Test
NB	Narrowband
	Nebraska [US]
	New Brunswick [Canada]
	No Bias
	Nopol Benzyl
	(Nota Bene) - Note Well; Take Notice
Nb	Niobium
NBA	Narrowband Allocation
	National Brassfounders Association
	National Building Agency
	N-Bromoacetamide
NBAA	National Business Aircraft Association [US]
NBAC	National Biotechnology Advisory Committee
NBACSTT	New Brunswick Association of Certified Survey Technicians and Technologists [Canada]
NBB	N-Butylboronate
NBC	Narrowband Conducted
	National Building Code
	Norwegian Bulk Carrier
NBCC	National Building Code of Canada
NBCV	Narrowband Coherent Video
NBD	Negative Binomial Distribution
	Nitrobenzoxadiazole
NBD-F	Fluoronitrobenzofurazane
NBDI	Nitrobenzyldiisopropylisourea
NBDL	Narrowband Data Line
NBE	N-Bromoacetylethylenediamine
NBEPC	New Brunswick Electric Power Commission [Canada]
NBFL	New Brunswick Federation of Labour [Canada]
NBFM	Narrowband Frequency Modulation
NBFU	National Board of Fire Underwriters
NBG	Near Band Gap
NBHA	Nitrobenzylhydroxylamine
NBIS	New Brunswick Information Service [Canada]
NBL	Naval Biological Laboratory [US Navy]
	New Brunswick Laboratory [AEC, Canada]
NBLE	Nearly Best Linear Estimator
NBLS	New Brunswick Land Surveyors [Canada]
NBMPR	Nitrobenzylthioribofuranosylpurine
NBO	Network Buildout
NBOMB	Neutron Bomb [also: N-BOMB]

NBP	Normal Boiling Point
NBR	Narrowband Radiated
	Nitrile-Butadiene Rubber
	Number
NBRF	National Biochemical Research Foundation
NBRI	National Building Research Institute
NBRPC	New Brunswick Research and Productivity Council [Canada]
NBS	National Bureau of Standards [US]
	Natural Black Slate
	N-Bromosuccinimide
	New British Standard [UK]
	Numeric Backspace (Character)
NBSCETT	New Brunswick Society of Certified Engineering Technicians and Technologists [Canada]
NBSD	Night Bombardment Short Distance
NBSFS	National Bureau of Standards Frequency Standard
NBSR	National Bureau of Standards Reactor
NBT	Narrow-Beam Transducer
	Nitro Blue Tetrazolium
	Null-Balance Transmissometer
NBTDR	Narrowband Time-Domain Reflectometry
NBTG	Nitrobenzylthioguanosine
NBTL	Naval Boiler and Turbine Laboratory [US Navy]
NC	(American) National Coarse Thread
	Network Control(ler)
	No Charge
	No Coil
	No Connection
	Noise Criteria
	Non-Crystalline
	Normally Closed
	North Carolina [US]
	Nuclear Capability
	Numbering Counter
	Numerical Control
N/C	Numerical Control
nC	nanocoulomb [unit]
NCA	National Coal Association
	National Computer Association
	Naval Communications Annex
	Northwest Computing Association [US]
NCAPC	National Center for Air Pollution Control [PHS, US]
NCAR	National Center for Atmospheric Research [NSF, US]
	National Conference on the Advancement of Research
NCASI	National Council for Air and Stream Improvement [US]
NCAT	National Center for Appropriate Technology [USDOE]
NCB	National Coal Board [UK]
	Naval Communications Board
N-CBZ	N-Carbobenzoxy
NCC	National Computer Conference
	National Computing Center
	Network Computer Center

NCCA	National Coil Coaters Association
NCCAT	National Committee for Clean Air Turbulence [US]
NCCCC	Naval Command, Control and Communications Center [US Navy]
NCCDPC	NATO Command, Control and Data Processing Committee
NCCIS	NATO Command Control and Information System
NC/CNC	Numerical Control/Computer Numerical Control
NCCU	National Conference of Canadian Universities
NCD	Nicotinamide Cytosine Dinucleotide
NCDC	Nitrocarboxyphenyldiphenylcarbamate
NCDEAS	National Committee of Deans of Engineering and Applied Sciences
NCDRH	National Center for Devices and Radiological Health
NCEA	North Central Electric Association
NCEB	National Council for Environmental Balance
	NATO Communications Electronic Board
NCEE	National Council of Engineering Examiners
NCEFT	National Commission on Electronic Funds Transfer
NCEL	Naval Civil Engineering Laboratory [US Navy]
NCET	National Council for Educational Technology [US]
NCF	Nominal Characteristics File
NCFP	National Conference on Fluid Power [US]
NCF	National Clayware Federation
NCFM	National Committee on Fluid Mechanics
NCFMF	National Committee for Fluid Mechanic Films [US]
NCG	Network Control Group
	Nickel-Coated Graphite Fiber
	Non-Condensable Gas
NCGA	National Computer Graphics Association
NCGG	National Committee for Geophysics and Geodesy
NCGR	National Council for Geocosmic Research
	National Council on Gene Resources
NCHEML	National Chemical Laboratory [MINTECH]
NCHRP	National Cooperative Highway Research Program [of AASHO]
NCHS	National Center for Health Statistics
NCHVRFE	National College for Heating, Ventilating, Refrigeration and Fan Engineering
NCI	National Computer Institute [US]
	National Computing Industries
	Naval Counterintelligence
	Northeast Computer Institute [US]
NCISC	Naval Counterintelligence Support Center [US]
NCL	National Central Library
	National Chemical Laboratory
NCLIS	National Commission on Libraries and Information Science
NCLT	Night Carrier Landing Trainer
NCMA	National Ceramic Manufacturers Association [US]
	National Concrete Masonry Association [US]
NCMRD	National Center for Management Research and Development
NCMRED	National Council on Marine Resources and Engineering Development [US]
NCO	North Canadian Oils
NCOR	National Committee for Oceanographic Research
NCP	Network Control Program
NCPC	Northern Canada Power Commission
NCP-L	Network Control Program - Local
NCP-LR	Network Control Program - Local/Remote
NCP-R	Network Control Program - Remote
NCPS	National Commission on Product Safety [US]
NCPUA	National Committee on Pesticide Use in America [US]
NCPWI	National Council on Public Works Improvement [US]
NCQR	National Council for Quality and Reliability
NCR	National Cash Register
	No Carbon Required
NCRE	Naval Construction Research Establishment
NCRH	National Center for Radiological Health [PHS, US]
NCRL	National Chemical Research Laboratory
NCRP	National Committee on Radiation Protection [US]
NCRPM	National Council on Radiation Protection and Measurement
NCRUCE	National Conference of Regulatory Utility Commission Engineers
NCRV	National Committee for Radiation Victims
NCS	National Communications System
	National Computer Systems
	National Corrosion Service
	Naval Communication Station
	Naval Control of Shipping
	N-Chlorosuccinimide
	Network Control Station
	Network Computing System
	Netherlands Computer Society
	Noncrystalline Solid
	Numerical Category Scaling
	Numerical Control Society
NCSC	North Carolina State College [US]
NCSE	North Carolina Society of Engineers [US]
NCSL	National Conference on Standards Laboratories [US]
NCSPA	National Corrugated Steel Pipe Association [US]
NCSU	North Carolina State University [US]
NCTA	National Cable Television Association [US]
NCTM	National Council of Teachers of Mathematics
NCTS	Northeast Corridor Transportation System [US]
NCTSI	National Council of Technical Service Industries [US]
NCU	Navigation Computer Unit
NCUC	Nuclear Chemistry Users Committee

NCUG	Nevada COBOL Users Group [US]
NCUR	National Committee for Utilities Radio
NCW	Non-Communist World
NCWM	National Conference on Weights and Measures
ND	North Dakota [US]
	Nucleotidase
	No Data
	Normal Direction
	Not Detectable
	Not Determined
	Not Done
Nd	Neodymium
NDA	Naphthaline Dicarboxylic Acid
NDAB	Numerical Data Advisory Board
NDAC	Nuclear Defense Affairs Committee [of NATO]
NdAD	Nicotinamide Deoxyadenosine Dinucleotide
N Dak	North Dakota [US]
NDB	Nondirectional Beacon
NDC	National Data Communication
	National Design Council
	Negative Differential Conductivity
	NORAD Direction Center
	Normalized Device Coordinates
NDE	Nondestructive Evaluation
	Nonlinear Differential Equation
NDF	Neutral Detergent Fiber
	No Defect Found
	Nonlinear Distortion Factor
NDGA	Nordihydroguaiaretic Acid
NDI	Nickel Development Institute [Canada]
	Nondestructive Inspection
	Non-Development Item
	Numerical Designation Index
NDIR	Nondispersive Infrared
NDL	Network Definition Language
	Nuclear Defense Laboratory [US Army]
NDM	Normal Disconnected Mode
nDNA	Native Deoxyribonucleic Acid
N-DNP	N-Dinitrophenyl
NDP	Normal Diametral Pitch
	Nucleoside Diphosphatase
NDPK	Nucleoside Diphosphate Kinase
NDPS	National Data Processing Service
NDR	Network Data Reduction
	Nondestructive Read
	Nondestructive Readout
NDRO	Nondestructive Readout
NDS	Nonparametric Detection Scheme
NDT	Network Description Template
	Newfoundland Daylight Time [Canada]
	Nil Ductility Transition
	Nondestructive Testing
NDTA	National Defense Transportation Association [US]
NDTC	Nondestructive Testing Center
NDTE	Nondestructive Testing Equipment
NDTT	Nil Ductility Transition Temperature
NDUV	Nondispersive Ultraviolet
Nd-YAG	Neodymium-Doped Yttrium Aluminum Garnet

NE	Nebraska [US]
	New England [US]
	No Effect
	Norepinephrine
	Northeast
	Not Equal
	Nuclear Explosive
Ne	Neon
NEA	National Education Association [US]
	National Electronic Association [US]
	Nuclear Energy Agency [OECD]
NEAA	Non-Essential Amino Acid
NEAC	Nippon Electric Automatic Computer
NEACP	National Emergency Airborne Command Post [USDOD]
NEADAI	National Education Association Department of Audiovisual Instruction
NEAP	National Energy Audit Program
NEAT	National Cash Register Electronic Autocoding Technique
NEB	National Energy Board [Canada]
	Noise Equivalent Bandwidth
Neb	Nebraska [US]
Nebr	Nebraska [US]
NEBS	New Exporters to Border States [Canada]
NEBSS	National Examinations Board in Supervisory Studies
NEC	National Electric Code [US]
	National Electronics Conference [US]
	National Electronics Council [UK]
	National Engineering Consortium
	Not Elsewhere Classified
NECIES	Northeast Coast Institution of Engineers and Shipbuilders [UK]
NECPA	National Emergency Command Post Afloat [USDOD]
NECPUC	New England Conference of Public Utility Commissioners [US]
NECS	National Electric Code Standards [US]
	Nationwide Educational Computer Service [US]
NECTA	National Electrical Contractors Trade Association [US]
NEDN	Naval Environmental Data Network [US Navy]
NEDSA	Non-Erasing Deterministic Stack Automation
NEDU	Navy Experimental Diving Unit
NEEDS	New England Educational Data System [US]
NEELS	National Emergency Equipment Locator System [of EPS]
NEEP	Nuclear Electronic Effects Program [US]
NEF	National Energy Foundation
	(American) National Extra Fine Thread
	Noise Exposure Forecast
NEFA	Non-Esterified Fatty Acid
NEFD	Noise Equivalent Flux Density
NEFO	National Electronics Facilities Organization [US]
NEG	Negative
NEHA	National Environmental Health Association [US]

NEI	Noise Exposure Index
	Not Elsewhere Indicated
NEIC	National Earthquake Information Center [US]
	National Energy Information Center [US]
NEIS	National Engineering Information System
NEL	National Engineering Laboratory [UK]
	Navy Electronics Laboratory [US Navy]
NELAT	Navy Electronics Laboratory Assembly Tester
NELC	Navy Electronics Laboratory Center
NELCON	National Electronics Conference [of IEEE]
NELCON NZ	National Electronics Conference, New Zealand [of IEEE]
NELEX	Naval Electronics Systems Command Headquarters [US Navy]
NELIAC	Navy Electronics Laboratory International Algorithmic Compiler [US Navy]
NELMA	Northeastern Lumber Manufacturers Association [US]
NELPA	Northwest Electric Light and Power Association
NEM	N-Ethylmaleimide
	Nitrogen Ethylmorpholine
	Not Elsewhere Mentioned
NEMA	National Electrical Manufacturers Association [US]
NEMAG	Negative Effective Mass Amplifiers and Generators
NEMI	National Elevator Manufacturing Industry [US]
NEMP	Nuclear Electromagnetic Pulse
NEMRIP	New England Marine Resources Information Program [US]
NEMS	Nimbus E Microwave Spectrometer
NEOF	No Evidence of Failure
NEP	National Energy Plan
	National Emphasis Program
	National Energy Program
	Never-Ending Program
	Noise Equivalent Power
NEPA	National Environmental Policy Act [US]
NEPCON	National Electronic Packaging and Production Conference [US]
NEPD	Noise Equivalent Power Density
NEPIS	N-Ethylphenylisoxazolium Sulfonate
NEPTUNE	Northeastern Electronic Peak Tracing Unit and Numerical Evaluator
NERC	National Electric Reliability Council [US]
	National Electronics Research Council [US]
	Natural Environment Research Council
	North American Electric Reliability Council
NEREM	Northeast Electronics Research and Engineering Meeting [US]
NERO	National Energy Resources Organization
	Na (= Sodium) Experimental Reactor of 0 (= Zero) (Energy)
	Near Earth Rescue Operation [NASA]
NERV	Nuclear Emulsion Recovery Vehicle
NERVA	Nuclear Engine for Rocket Vehicle Applications

NES	National Employment Service
	Noise Equivalent Signal
	Not Elsewhere Specified
NESC	National Electrical Safety Code [US]
	National Environmental Satellite Center [ESSA]
	Naval Electronics Systems Command [US Navy]
NESP	National Environment Studies Project
NESS	National Environmental Satellite Service [US]
NESSEC	Naval Electronic Systems Security Engineering Center
NEST	Naval Experimental Satellite Terminal
NESTOR	Neutron Source Thermal Reactor
NESTS	Non-Electric Stimulus Transfer System
NET	National Educational Television
	Net Equivalent Temperature
	Network
	Next European Torus
	Noise Equivalent Temperature
NETA	National Environmental Training Association
NETC	National Emergency Transportation Center [US]
NETE	Naval Engineering Test Establishment
NETFS	National Educational Television Film Service [US]
NETH	Netherlands
NETSET	Network Synthesis and Evaluation Technique
NEUT	Neutral
Nev	Nevada [US]
NEWAC	NATO Electronic Warfare Advisory Committee
NEWRADS	Nuclear Explosion Warning and Radiological Data System
NEWS	Naval Electronic Warfare Simulator
NEWWA	New England Water Works Association
NF	(American) National Fine Thread
	National Formulary [US]
	Near Face
	Newfoundland [Canada]
	Noise Figure
	Noise Frequency
	Nonferrous
	Nose Fuse
N/F	No Funds
nF	nanofarad [unit]
NFAC	Naval Facilities Engineering Command Headquarters [US Navy]
NFAIS	National Federation of Abstracting and Information Services [US]
NFB	National Film Board [Canada]
NFC	National Fire Code
	National Freight Consortium [UK]
	Not Favorably Considered
NFCB	National Federation of Community Broadcasters
NFCIT	National Federation of Cold Storage and Ice Trades

NFDC	National Flight Data Center
NFDM	Non-Fat Dry Milk
NFE	Nearly Free Electron
NFEA	National Federated Electrical Association
NFEC	National Food and Energy Council
	Naval Facilities Engineering Command [US Navy]
NFETM	National Federation of Engineers' Tools Manufacturers
NFF	No Fault Found
NFFS	Nonferrous Founders' Society
NFL	Newfoundland Federation of Labour [Canada]
Nfld	Newfoundland [Canada]
NFM	Narrow Frequency Modulation
	Non-Fat Milk
	Nonferromagnetic
NFMES	National Fund for Minority Engineering Students
N-FMOC	N-Fluoroenylmethoxycarbonyl
NFMS	Non-Fat Milk Solids
NFMSAEG	Naval Fleet Missile System Analysis and Evaluation Group [US Navy]
NFO	National Freight Organization
NFPA	National Fire Protection Association [US]
	National Fluid Power Association [US]
	National Forest Products Association [US]
NFPEDA	National Farm and Power Equipment Dealers Association [US]
NFR	No Further Requirement
NFS	Network File System
NFSAIS	National Federation of Science Abstracting and Indexing Services
NFSR	National Foundation for Scientific Research
NFSWMM	National Federation of Scale and Weighing Machine Manufacturers
NFT	Nutrient Film Technique
NG	Narrow Gauge
	Natural Gas
	No Good
	Nitroglycerine
ng	nanogram [unit]
NGAA	Natural Gas Association of America
NGC	New Galactic Catalogue
NGCC	National Guard Computer Center [US]
NGD	Nicotinamide Guanine Dinucleotide
NG-EGDN	Nitroglycerine Ethylene Glycol Dinitrate
NGGLT	Natural Gas and Gas Liquids Tax
NGL	Natural Gas Liquid
NGM	Neutron-Gamma Monte Carlo
NGMA	National Gas Measurement Association [US]
NGO	Nongovernmental Organization
NGPA	Natural Gas Policy Act
NGR	Nongrain-Raising (Stain)
	Nuclear Gamma Ray Resonance
NGRS	Narrow Gauge Railway Society
NGS	National Geographic Society
	Nominal Guidance Scheme
NGSDS	National Geophysical and Solar Terrestrial Data Center [US]
NGT	Nonsymmetrical Growth Theory

NGTE	National Gas Turbine Establishment [UK]
NGV	Natural Gas Vehicle
NH	New Hampshire [US]
	Nonhygroscopic
nH	nanohenry [unit]
NHA	National Hydropower Association
NHAM	National Hose Assemblies Manufacturers (Association)
NHB	Hydroxynitrobenzoic Acid
NHD	Nevada Highway Department [US]
NHE	Normal Hydrogen Electrode
NHI	National Health Institute
NHLA	National Hardwood Lumber Association [US]
NHM	No Hot Metal
NHMO	NATO Hawk Management Office
NHNP-E	N-Hydroxynaphthaloxypropylethylenediamine
NHP	Nominal Horsepower
	Numeric Handprinting
NHPLO	NATO Hawk Production and Logistics Organization
NHPMA	Northern Hardwood and Pinewood Manufacturers Association [US]
NHQ	Naphthohydroquinone
NHRDP	National Health Research and Development Program
NHRI	National Hydrology Research Institute
NHRL	National Hurricane Research Laboratory [US]
NHTSA	National Highway Traffic Safety Administration [US]
NI	Negative Input
	Noise Index
	Northern Ireland
	Numerical Index
Ni	Nickel
NIAC	National Industry Advisory Committee [FCC, US]
	National Information and Analysis Center [US]
NIAE	National Institute of Agricultural Engineering
NIAG	NATO Industrial Advisory Group
NIAL	Nested Interactive Array Language
NIAM	Netherlands Institute for Audiovisual Media
NIAE	National Institute of Agricultural Engineering [US]
NIB	Hydroxyiodonitrobenzoic Acid
	Node Initialization Block
	Non-Interference Basis
NIBS	National Institute of Building Sciences
NIC	National Indicational Center
	National Industrial Council
	National Inspection Council
	Nearly Instantaneous Companding
	Negative Impedance Converter
	Network Information Center
	Newly Industrialized Countries
	Nineteen-Hundred Indexing and Cataloguing
	Not in Contact

NI-CAD	Nickel-Cadmium
NICAP	National Investigations Committee on Aerial Phenomena [US]
NICB	National Industrial Conference Board [US]
NICE	National Institute for Computers in Engineering
	National Institute of Ceramic Engineers [of ACS]
	Normal Input-Output Control Executive
NICEIC	National Inspection Council for Electrical Installation Contracting
NICET	National Institute for Certification in Engineering Technologies
NICOA	National Independent Coal Operators Association [US]
NICOL	Nineteen-Hundred Commercial Language
NICS	NATO Integrated Communication System
NICSMA	NICS Management Agency
NICSO	NICS Organization
NID	Nuclear Instruments and Detectors [IEEE Committee]
NIDA	Northeastern Industrial Development Association [US]
	Numerically Integrated Differential Analyzer
NiDI	Nickel Development Institute [Canada]
NIER	National Industrial Equipment Reserve
NIF	Noise Improvement Factor
NIFES	National Industrial Fuel Efficiency Service
NIFTE	Neon Indicating Functional Test Equipment
NIGP	National Institute of Governmental Purchasing [US]
NIH	National Institute of Health [US]
	Not Invented Here
NIM	Network Interface Machine
	Nuclear Instrument Module
NIMMS	Nineteen-Hundred Integrated Modular Management System
NIMP	New and Improved Materials and Processes
NIMPHE	Nuclear Isotope Monopropellant Hydrazine Engine
NINA	National Institute Northern Accelerator
NINIA	Nephelometric Inhibition Immunoassay
NIO	National Institute of Oceanography [UK]
NIOBE	Numerical Integration of the Boltzmann Transport Equation
NIOSH	National Institute for Occupational Safety and Health
NIP	Hydroxyiodonitrophenylacetic Acid
	Nipple
	Nonimpact Printer
	Nucleus Initialization Procedure
	Nucleus Initialization Program
NIPCC	National Industrial Pollution Control Council
NIPHLE	National Institute of Packaging, Handling and Logistic Engineers
NIPO	Negative Input, Positive Output
NIPS	Naval Intelligence Processing System [US Navy]
	NMCS (= National Military Command System) Information Processing System
NIPTS	Noise-Induced Permanent Threshold Shift
NIR	Near Infrared
	Nickel-Iron Refinery
NIRA	Near Infrared Analysis
NIRIM	National Institute for Research in Inorganic Materials [Japan]
NIRNS	National Institute for Research in Nuclear Science
NIRO	Nike-Iroquois Rocket
NIRTS	New Integrated Range Timing System
NIS	Not in Stock
NISARC	National Information Storage and Retrieval Center
NISC	National Industrial Space Committee [US]
NISEE	National Information Service for Earthquake Engineering
NI-SIL	Nickel-Silver
NIT	Net Ingot Ton
	New Information Technology
	Nonlinear Inertialess Three-Pole
NITEP	National Incinerator Testing and Evaluation Program [Canada]
NITP	National Industrial Training Program
NITROS	Nitrostarch
NIU	Northern Illinois University [US]
NIW	Naval Inshore Warfare
NJ	Network Junction
	New Jersey [US]
NJAC	National Joint Advisory Council
NJC	National Joint Committee
	National Joint Council
NJCC	National Joint Computer Committee [now: AFIPS]
NJCEC	NATO Joint Communications and Electronic Committee
NJIT	New Jersey Institute of Technology [US]
NJP	Network Job Processing
NJPMB	Navy Jet-Propelled Missile Board [US Navy]
NK	Natural Killer
NKCF	Natural Killer Cytotoxic Factor
NL	National Library
	New Line (Character)
nl	(non licet) - it is not permitted
	(non liquet) - it is not clear
NLAE	National Laboratory for the Advancement of Education
N LAT	North Latitude
NLC	National Library of Canada
NLEFM	Nonlinear Elastic Fracture Mechanics
NLG	Noise Landing Gear
NLGA	National Lumber Grades Authority
NLGI	National Lubricating Grease Institute [US]
NLI	Nonlinear Interpolating
NLL	National Lending Library [US]
NLM	Noise Level Monitor
NLMA	National Lumber Manufacturers Association [now: NFPA]
NLO	Nonlinear Optics
NLOGF	National Lubricating Oil and Grease Federation

NLQ	Near-Letter Quality
NLP	Nonlinear Programming
NLR	Noise Load Ratio
NLRB	National Labour Relations Board [US]
NLS	No Load Speed
NLT	Not Less Than
	Net Long Ton
NLTE	Nonlocal Thermodynamic Equilibrium
NM	Nautical Miles
	New Mexico [US]
	Network Management
	Noise Margin
	Noise Meter
	No Measurement
	Nonmetal(lic)
	Not Measured
	Nuclear Magnetron
nm	nanometer [unit]
NMA	Nadic Methyl Anhydride
	National Management Association
	National Microfilm Association [US]
	Network Management Agent
NMAA	National Machine Accountants Association [now: DPMA]
NMAB	National Materials Advisory Board
NMAP	National Metric Advisory Panel [US]
NMC	National Meteorological Center [US]
	Naval Materials Command [US Navy]
	Naval Missile Center [US Navy]
	Network Measurement Center
	Nonmetallic Cable
NMCC	National Military Command Center [US]
NMCL	Naval Missile Center Laboratory [US Navy]
NMCS	National Military Command System
	Nuclear Materials Control System
NMCSSC	National Military Command System Support Center
NMDSC	Naval Medical Data Service Center [US Navy]
NME	Noise-Measuring Equipment
NMEA	National Marine Electronics Association
NMEL	Navy Marine Engineering Laboratory [US]
	Nuclear Mechano-Electronic Laboratory
NMERI	National Mechanical Engineering Research Institute
N Mex	New Mexico [US]
NMF	National Microelectronics Facility
NMFC	National Motor Freight Classification
NMH	No-Mar Hammer
	Nautical Miles per Hour
NMI	NASA Management Instruction
	Nautical Miles [also: NM and nmi]
NMM	Nautical Miles per Minute
NMMD	Nuclear Materials Management Department
NMN	Nicotinamide Mononucleotide
NMNH	Nicotinamide Mononucleotide, Reduced Form
NMOS	Negative Metal-Oxide Semiconductor
	N-Channel Metal-Oxide Semiconductor
NMP	N-Methylpenazine
NMPDN	National Materials Property Data Network

NMPH	Nautical Miles per Hour
NMPM	Nautical Miles per Minute
NMPS	Nautical Miles per Second
NMR	National Military Representative [NATO]
	National Missile Range [US]
	Normal Mode Rejection
	Nuclear Magnetic Resonance
NMRA	National Mine Rescue Association [US]
NMRR	Normal-Mode Rejection Ratio
NMS	Nautical Miles per Second
	Naval Meteorological Service [US Navy]
NMSC	Nonmartensitic Structural Component
NMSD	Next Most Significant Digit
NMSI	National Maximum Speed Limit
N-MSOC	N-Methylsulfonylethyloxycarbonyl
NMSS	National Multipurpose Space Station
NMT	Not More Than
NMTBA	National Machine Tool Builders Association [US]
NNC	National Nomination Committee
NNE	North-Northeast
NNI	Noise Number Index
NNP	Net National Product
	Net Naturalization Potential
NNS	Near Net Shape
NNSS	Navy Navigational Satellite System [US Navy]
NNW	North-Northwest
NO	Negative Output
	Nitric Oxide
	Normally Open
	North
	(Numero) - Number
No	Nobelium
NOA	National Oceanographic Association [US]
NOAA	National Oceanographic and Atmospheric Administration [US]
NOBO	Nonobjecting Beneficial Owner
NOC	National On-Line Circuit
NOCO	Noise Correlation
NODAC	Naval Ordnance Data Automation Center [US]
NODC	National Oceanographic Data Center [US]
	Non-Oil Developing Countries
NOE	Not Otherwise Enumerated
NOEB	NATO Oil Executive Board
NOEB-E	NATO Oil Executive Board - East
NOEB-W	NATO Oil Executive Board - West
NOESS	National Operational Environmental Satellite System
NOEU	Naval Ordnance Experimental Unit [US Navy]
NOF	National Optical Font
	NCR Optical Font
	Network Operations and Facilities
NOFI	National Oil Fuel Institute [US]
NOHC	National Open Hearth Committee [US]
NOHP	Not Otherwise Herein Provided
NOI	Not Otherwise Indexed
NOIA	National Ocean Industries Association
	Newfoundland Ocean Industries Association [Canada]

NOIBN	Not Otherwise Indicated by Name
	Not Otherwise Indicated by Number
NOISE	National Organization to Insure a Sound-Controlled Environment
NOL	National Ordnance Laboratory [US]
	Net Operating Loss
NOM	Nomenclature
	Nominal
NOMA	National Office Management Association
	National Organization of Minority Architects
NOMAD	Naval Oceanographic Meteorological Automatic Device
	Nominal Michigan Algorithmic Decoder
NOMEN	Nomenclature
NOML	Nominal
NOMSS	National Operational Meteorological Satellite System
NOM STD	Nominal Standard
NONADD	Nonadditivity
NONCOHO	Noncoherent Oscillator
NON COND	Noncondensing
N-on-P	Negative on Positive
non seq	(non sequitur) - it does not follow
NOO	Naval Oceanographic Office [US Navy]
NO-OP	No Operation
NOP	No Operation
	Not Otherwise Provided
NOPI	Negative Output, Positive Input
NOPOL	No Pollution
NOR	Nitrogen Oxide Reduction
	Normal
	North
	Norway
	Norwegian
	Not Or
NORAD	North American Aerospace Defense Command [new definition]
	North American Air Defense Command [old definition]
NORC	Naval Ordnance Research Computer
NORM	Normalization
	Not Operationally Ready Maintenance
NORP	New Oil Reference Pricing
NORS	Not Operationally Ready Supply
NORSAT	Norwegian Domestic Satellite
NORW	Norway
	Norwegian
NOS	National Ocean Service [US]
	National Ocean Survey [US]
	Naval Ordnance Station [US Navy]
	Nosing
	Not Otherwise Specified
	Numbers
NOSA	National Occupational Safety Association
NOSC	Naval Ocean Systems Center
NOSMO	Norden Optics Setting, Mechanized Operation
NOSS	National Orbiting Space Station
	Nimbus Operational Satellite System
NOTU	Naval Operational Training Unit

NOV	November
NOx	Oxides of Nitrogen
NP	(American) National Pipe (Thread)
	Nameplate [also: N/P]
	Naval Publication
	Neutrino Patents
	Nondeterministic Polynomial
	No Pin
	Not Provably
	Nucleoside Phosphorylase
Np	Neper
	Neptunium
NPA	National Petroleum Association [US]
	National Pilots' Association
	National Planning Association [US]
	National Productivity Award [Canada]
	Normal Pressure Angle
	Numbering Plan Area
	Numerical Production Analysis
nPa	nanopascal [unit]
NPAC	National Pipeline Agency of Canada
NPC	NASA Publication Control
	Naval Photographic Center [US Navy]
NPCF	National Pollution Control Federation [US]
NPD	National Power Demonstration
NPDS	Nuclear Particle Detection System
NPE	Natural Parity Exchange
NPEC	Nuclear Power Engineering Committee [of IEEE]
NPEGE	Nonyl Phenyl Eicosa-Ethylene Glycol Ether
NPES	National Printing Equipment Show [US]
NPFO	National Power Field Office [US Army]
NPG	Neopentyl Glycol
	Nuclear Planning Group [NATO]
NPIN	Negative-Positive-Intrinsic (Transistor)
NPIRI	National Printing Ink Research Institute [US]
NPL	National Physical Laboratory [UK]
	National Physics Laboratory [US]
	New Programming Language
NPLG	Navy Program Language Group [US Navy]
NPLO	NATO Production and Logistics Organization
NPN	Negative-Positive-Negative (Semiconductor)
	Nonprotein Nitrogen
	Normal Propyl Nitrate
NPNP	Negative-Positive-Negative-Positive (Semiconductor)
NPO	Naphthylphenyloxazole
NPPA	Northwest Pulp and Paper Association [US]
NPR	National Public Radio
	Net Profit Royalty
	Noise Pollution Ratio
	Noise Power Ratio
	Nozzle Pressure Ratio
	Numeric Position Readout
NPRA	National Personal Robot Association
	National Petroleum Refiners Association
NPRCG	Nuclear Public Relations Contact Group
NPRD	Nonelectronic Parts Reliability Data
NPRM	Notice of Proposed Rule Making
NPRZ	Nonpolarized Return-to-Zero Recording

NPS	Nitrided Pressureless Sintering
	Nitrophenylsulfonylamino Acid
	Nominal Pipe Size
	Nuclear and Plasma Sciences [IEEE Society]
NPSC	(American) National Straight Pipe Thread in Pipe Couplings
NPSD	Neutron Power Spectral Density
NPSF	(American) National Straight Pipe Thread for Dryseal Joints
NPSH	(American) National Straight Pipe Thread for Hose Couplings
	Net Positive Suction Head
NPSHA	Net Positive Suction Head Available
NPSHR	Net Positive Suction Head Required
NPSL	(American) National Straight Pipe Thread for Locknuts
NPSM	(American) National Straight Pipe Thread for Mechanical Joints
NPT	(American) National Taper Pipe Thread
	Network Planning Technique
	Nonproliferation Treaty
	Normal Pressure and Temperature
NPTF	(American) National Taper Pipe Thread for Dryseal Joints
NPTR	(American) National Taper Pipe Thread for Railing Fixtures
NPU	Naval Parachute Unit
NPV	Net Present Value
	Nitrogen Pressure Valve
NQ	Nitroquinoline
NQO	Nitroquinoline Oxide
NQR	Nuclear Quadruple Resonance
NR	Natural Rubber
	Navigational Radar
	Negative Resistance
	Noise Rating (Curve)
	Noise Ratio
	Nonreactive
	Nonrecoverable
	Non-Return (Valve)
	Not Recommended
	Nuclear Reactor
NRA	National Railway Association
	Naval Radio Activity
	Network Resolution Area
	Nuclear Reaction Analysis
NRAO	National Radio Astronomy Observatory [US]
NRB	Nuclear Reactors Branch [of AEC]
	Nuclear Resonance Broadening
NRC	National Replacement Character
	National Research Council
	Noise Rating Curve
	Noise Reduction Coefficient
	Notch Root Contraction
	Nuclear Regulatory Commission [US]
NRCC	National Research Council of Canada
NRCD	National Reprographic Center for Documentation
NRCST	National Referral Center for Science and Technology [US]
NRD	Negative Resistance Diode

	Network Resource Dictionary
NRDC	National Research Development Council
	Natural Resources Defense Council
NRDL	Naval Radiological Defense Laboratory [US Navy]
NRDS	Nuclear Rocket Detection System
	Nuclear Rocket Development Station [US]
NRE	Naval Research Establishment [now: DREA]
NRECA	National Rural Electric Cooperative Association [US]
NRFSA	Navy Radio Frequency Spectrum Activity
NRIM	National Research Institute for Metals
NRIP	Number of Rejected Initial Pickups
NRJ	Non-Reciprocal Junction
NRL	National Reference Library
	Naval Research Laboratory [US Navy]
NRLM	National Research Laboratory of Metrology [Japan]
NRLSI	National Reference Library for Science and Invention [UK]
NRM	Natural Remanent Magnetization
	Normal Response Mode
NRMA	Nuclear Records Management Association
NRMCA	National Ready Mixed Concrete Association
NRML	Normal
NRMS	Network Reference and Monitor Station
NRN	Negative Run Number
NRP	Normal Rated Power
NRPRA	Natural Rubber Producers Research Association
NRRA	National Resource Recovery Association
NRRFSS	National Research and Resource Facility for Submicron Structures [US]
NRRS	Naval Radio Research Station
NRS	Naval Radio Station
	Nonrising Stem
NRT	Net Register Ton
	Nonradiating Target
NRTS	National Reactor Testing Station [US]
NRTSC	Naval Reconnaissance and Technical Support Center [US Navy]
NRU	National Research Universal
	Nuclear Reactor Universal
NRX	National Research Experimental
	NERVA Reactor Experimental [AEC]
	Nuclear Reactor Experimental
NRZ	Nonreturn-to-Zero Recording
NRZC	Nonreturn-to-Zero Change Recording
NRZI	Nonreturn-to-Zero Inverted Recording
	Nonreturn-to-Zero I (= One)
NRZL	Nonreturn-to-Zero Level
NRZM	Nonreturn-to-Zero Mark Recording [also: NRZ (M)]
NRZ1	Nonreturn-to-Zero Change on One [also: NRZ-1]
NS	National Special Thread
	National Standard
	Navigation System
	Near Side
	New Style
	Nickel Steel

	Nonsequenced
	Notch Strength
	Not Specified
	Not Sufficient
	Nova Scotia [Canada]
	Nuclear Systems
	Sodium (= Na) Silicate
Ns	Nimbo-Stratus
ns	nanosecond [unit]
NSA	National Security Agency
	National Standards Association [US]
	National Shipping Authority [US]
	Netherlands Society for Automation
	Nonenyl Succinic Anhydride
	Nonsequenced Acknowledgement
	Nuclear Suppliers Association
NSAA	Nova Scotia Association of Architecture [Canada]
NSAI	Nonsteroidal Anti-Inflammatory (Drug)
NSB	National Science Board [US]
	Naval Standardization Board [US Navy]
NSBE	National Society of Black Engineers [US]
NSC	National Safety Council
	National Security Council
	NATO Supply Center
	Numerical Sequence Code
	Numeric Space Character
NSD	Network Status Display
NSDA	National Space Development Agency [Japan]
NSEC	Naval Ships Engineering Center [US Navy]
nsec	nanosecond [unit]
NSEF	Navy Security Engineering Facility [US Navy]
NSEIP	Norwegian Society for Electronic Information Processing
NSERC	Natural Sciences and Engineering Research Council [Canada]
NSF	National Sanitation Foundation
	National Science Foundation [US]
	Naval Supersonic Facility [US Navy]
	Not Sufficient Funds
NSFA	National Science Foundation Act [US]
NSFD	Note of Structural and Functional Deficiency
NSFL	National Science Film Library
	Nova Scotia Federation of Labour [Canada]
NSFS	Net Suction Fracture Strength
NSG	NASA Grant [US]
	Naval Security Group [US Navy]
NSGC	Naval Security Group Command
NSI	National Space Institute
	Nonsatellite Identification
	Nonsequenced Information
	Nonstandard Item
NSIA	National Security Industrial Association [US]
NSIF	Near Space Instrumentation Facility [US]
NSL	National Science Library [now: CISTI]
	National Science Laboratories [US]
	Northrop Space Laboratories [US]
	Naval Supersonic Laboratory [US Navy]

NSLS	National Synchrotron Light Source
NSM	Network Status Monitor
NSMB	Nuclear Standards Management Board
NSMRSE	National Study of Mathematics Requirements for Scientists and Engineers
NSOEA	National Stationery and Office Equipment Association [US]
NSP	NASA Support Plan
	Network Support Plan
NSPE	National Society of Professional Engineers [US]
NSPI	National Society for Programmed Instruction [US]
NSPIE	National Society for the Promotion of Industrial Education [US]
NSPP	Nuclear Safety Pilot Plant [of ORNL]
NSPS	National Society of Professional Surveyors
	New Source Performance Standards
NSPV	Number of Scans per Vehicle
NSQCRE	National Symposium on Quality Control and Reliability in Electronics [US]
NSR	Notch Strength Ratio
	Notch Stress R
NSRB	National Security Resources Board [US]
NSRC	Natural Science Research Council
NSRDB	Nova Scotia Resources Development Board [Canada]
NSRDC	Naval Ship Research and Development Center [US Navy]
NSRDS	National Standard Reference Data System [of OSRD]
NSRF	Nova Scotia Research Foundation [Canada]
NSRG	Northern Science Research Group
NSRS	Naval Supply Radio Station
NSRT	Near Surface Reference Temperature
NSS	National Standards System
	Navy Secondary Standards
NSSA	National Science Supervisors Association [US]
NSSC	Naval Sea System Command [US Navy]
	Naval Supply Systems Command [US Navy]
NSSCC	National Space Surveillance Control Center [US]
NSSDC	National Space Science Data Center [US]
NSSL	National Severe Storms Laboratory [US]
NSSS	National Space Surveillance System [US]
	Nuclear Steam Supply System
NST	Network Support Team
	Newfoundland Standard Time [Canada]
	Non-Slip Tread
NSTA	National Science Teachers Association [US]
NSTB	National Safety Transportation Board
NSTIC	Naval Scientific and Technical Information Center [US Navy]
NSTL	National Space Technology Laboratory [of NASA]
NSTS	National Space Transportation System [US]
NSV	Net Sales Value
	Nuclear Service Vessel
NSWC	Naval Surface Weapons Center [US Navy]
NT	Non-Tight

	Northwest Territories [Canada]
	Nortriptyline
	No Transmission
	Numbering Transmitter
NTA	National Technical Association
	Nitrilotriacetate
	Nitrilotriacetic Acid
N-t-B	Nitroso-tert-Butane
N-t-BOC	N-tert-Butoxycarbonyl
NTC	National Telemetering Conference
	National Training Center [US Army]
	Naval Telecommunications Command [US Navy]
	Negative Temperature Coefficient
	Network Transmission Committee [of VITEAC]
NTCA	National Telephone Cooperative Association
NTD	Neutron Transmutation Doping
NTD&P	National Tool, Die and Precision Machining Association
NTDS	Naval Tactical Data System [US Navy]
NTE	Navy Teletypewriter Exchange [US Navy]
NTEP	New Technology Employment Program
NTF	No Trouble Found
NTHRN	Northern
NTI	Noise Transmission Impairment
NTIA	National Telecommunications and Information Administration [US]
NTIAC	Nondestructive Testing Information Analysis Center [US]
NTIS	National Technical Information Service [US]
NTMA	National Tooling and Machining Association
NTN	National Telecommunications Network
NTOTC	National Training and Operational Technology Center
NTP	National Toxicology Program [US]
	Normal Temperature and Pressure
NTPC	National Technical Processing Center [US]
NTS	National Topographic System
	Navigation Technology Satellite
	Negative Torque Signal
	Nevada Test Site [of NASA]
	Notch Tensile Strength
	Not to Scale
NTSA	National Technical Services Association [US]
NTSB	National Transportation Safety Board [DOT, US]
NTSC	National Television System Committee [US]
	National Thermal Spray Conference [US]
NTTCIW	National Technical Task Committee on Industrial Wastes [US]
NTU	Nephelometric Turbidity Unit
	Number of Transfer Units
NTV	Network Television
	Nonlinear Thickness Variation
NTW	Nonpressure Thermit Welding
NTWK	Network
NTWS	Nontrack while Scan
NT WT	Net Weight
NU	No Umbra (Technique)

	Number Unobtainable
	Numeral
NUCA	National Utility Contractors' Association [US]
NUCLEX	International Nuclear Industrial Fair and Technical Meeting
NUDAC	Nuclear Data Center
NUDET	Nuclear Detection
NUDETS	Nuclear Detection System
NUL	Null (Character)
NULACE	Nuclear Liquid Air Cycle Engine
NUM	Number
	Numeric(s)
NUME	Numerical Methods in Engineering
NUMETA	Numerical Methods in Engineering - Theory and Applications
NUOS	Naval Underwater Ordnance Station [US Navy]
NUPAD	Nuclear Powered Active Detection System
NUR	National Union of Railwaymen
	Net Unduplicated Research
NUS	National University of Singapore
	Nuclear Utility Service
NUSC	Naval Underwater Systems Center [US Navy]
NUSL	Naval Underwater Sound Laboratory [US Navy]
NUT	National Union of Teachers
	Nutrient
NUWC	Naval Undersea Warfare Center [US Navy]
NUWES	Naval Undersea Warfare Engineering Station [US Navy]
NV	Nevada [US]
	New Version
	Nonvolatile
nV	nanovolt [unit]
NVBO	Natural Valence-Band Offset
NVM	Nonvolatile Matter
NVR	No Voltage Release
NVRAM	Nonvolatile Random-Access Memory
NVSAVS	National Vacuum Symposium of the American Vacuum Society
NVSD	Night Vision System Development
NW	Net Weight
	Northwest
	Nuclear Weapon
nW	nanowatt [unit]
N/W	Nishiyama-Wassermann (Relationship)
NWAC	National Weather Analysis Center
NWAG	Naval Warfare Analysis Group [US Navy]
NWC	National Water Commission [US]
	Naval Weapons Center [US Navy]
NWDS	Number of Words
NWDSEN	Number of Words per Entry
NWEF	Naval Weapons Evaluation Facilities [US Navy]
NWEP	Nuclear Weapons Effects Panel
NWG	National Wire Gauge
NWIS	National Network of Minority Women in Science
NWH	Normal Working Hours

NWK	Network
NWL	Naval Weapons Laboratory [US Navy]
NWMA	National Woodwork Manufacturing Association [US]
NWP	Numerical Weather Prediction
NWPCA	National Wooden Pallet and Container Association [US]
NWPF	National Water Purification Foundation [US]
NWO	Nonwoven Oriented
NWRA	National Water Resources Association
NWRC	National Weather Records Center [of ESSA]
NWRF	Naval Weather Research Facility [US Navy]
NWRI	National Water Research Institute
NWS	National Weather Service
	North Warning System [US/Canada]
NWSC	National Weather Satellite Center [US]
NWSD	Naval Weather Service Division [US Navy]
NWSSG	Nuclear Weapons System Satellite Group [US]
NWT	Net Weight
	Non-Waste Technology
	Northwest Territories [Canada]
NWTFL	Northwest Territories Federation of Labour [Canada]
NXDO	Nike X Development Office
NY	New York (State) [US]
NYADS	New York Air Defense Sector [US]
NYAP	New York Assembly Program
NYARTCC	New York Air Route Traffic Control Center [US]
NYCBAN	New York Center Beacon Alphanumerics [of FAA]
NYCX	New York Commodity Exchange [US]
NYD	Navy Yard
	Not Yet Detected
	Not Yet Determined
NYIT	New York Institute of Technology [US]
NYNS	New York Naval Shipyard [US Navy]
NYP	Not Yet Published
NYSU	New York State University [US]
NYU	New York University [US]
NZ	New Zealand
NZDSIR	New Zealand Department of Scientific and Industrial Research
NZNCOR	New Zealand National Committee on Oceanic Research
NZSI	New Zealand Standards Institute

O

O	Ocean
	October
	Octopole
	Ohio [US]
	Operand
	Operation
	Operator
	Order

	Output
	Oxide
	Oxygen
OA	Office Automation
	Office of Administration [US]
	Oil-to-Air
	Omniantenna
	Operations Analysis
	Output Axis
	Outside Air
	Overall
	Overflow Area
O/A	Outer Anchorage
OAA	Ontario Association of Architects [Canada]
OAC	Operations Analysis Center
	Optimally Adaptive Control
OACETT	Ontario Association of Certified Engineering Technicians and Technologists
OACUL	Ontario Association of College and University Libraries [now: OCULA]
OAD	Operations Analysis Division
OAEM	Ontario Approved Educational Microcomputers
OAF	Origin Address Field
OAFU	Observers' Advanced Flying Unit
OAI	Outside Air Intake
OAIDE	Operational Assistance and Instructive Data Equipment
OALS	Observer Air Lock System
OAME	Orbital Altitude Maneuvering Electronics
OAMP	Optical Analog Matrix Processing
OAMS	Orbital Altitude Maneuvering System
OAO	Orbiting Astronomical Observatory
	Orthogonalized Atomic Orbital
OAPEC	Organization of Arab Petroleum Exporting Countries
OAQPS	Office of Air Quality Planning and Standards [of EPA]
OAR	Office of Aerospace Research [US Air Force]
	Operator Authorization Record
OARAC	Office of Aerospace Research Automatic Computer
OARC	Ordinary Administrative Radio Conference
OART	Office of Advanced Research and Technology [of NASA]
OAS	Orbit Adjust Subsystem
	Organization of American States [US]
OASF	Orbiting Astronomical Support Facility
OASIS	Ocean All Source Information System
	Office Automation Services and Information Systems
OASPL	Overall Sound Pressure Level
OASV	Orbital Assembly Support Vehicle
OAT	Office for Advanced Technology [US Air Force]
	Operational Air Traffic
OATRU	Organic and Associated Terrain Research Unit [Canada]
OAW	Oxyacetylene Welding
OB	Oil Bearing
	Output Buffer

O-B	Octal-to-Binary
OBA	Oxygen Breathing Apparatus
OBAWS	On-Board Aircraft Weighing System
OBC	Optical Bar Code
OBD	Open Blade Damper
OBES	Orthonormal Basis of an Error Space
OBF	One-Bar Function
OBGS	Orbital Bombardment Guidance System
OBI	Office of Basic Instrumentation
	Omnibearing Indicator
	Open-Back Inclinable (Press)
OBIC	Optical-Beam-Induced Current
OBIFCO	On-Board In-Flight Checkout
OBJ	Object
OBM	Ontario Basic Mapping [Canada]
	Oxygen-Bottom Blown Maxhuette (Steelmaking)
OBN	Out-of-Band Noise
OBOS	Oil Bulk-Ore Ship
OBR	Outboard Recorder
OBS	Observation
	Observatory
	Observer
	Obsolescence
	Omnibearing Selector
OBSPL	Octave-Band Sound Pressure Level
OBSS	Ocean Bottom Scanning Sonar
OBU	Offshore Banking Unit
OC	Official Classification
	On Center
	Open Circuit
	Open Coil (Annealing)
	Operating Characteristic
	Operational Characteristic
	Operational Computer
	Outside Circumference
	Over Center
	Overcurrent
	Oxygen Chemisorption
	Oxygen Cutting
OCA	Operational Control Authority
OCAL	On-Line Cryptanalytic Aid Language
OCAM	Ontario Center for Advanced Manufacturing [Canada]
OCAPT	Ontario Center for Automotive Parts Technology [Canada]
OCAS	On-Line Cryptanalytic Aid System
	Organization of Central American States
OCB	Oil Circuit Breaker
OCC	Operational Computer Complex
	Operator Control Command
	Other Common Carrier
	Output Control Character
OCCA	Oil and Color Chemists Association
OCDU	Optics Coupling Display Unit
OCE	Ocean Covered Earth
OCEAN	Oceanographic Coordination Evaluation and Analysis Network
OCF	Orientation Correlation Function
	Owens-Corning Fiberglass
OCGS	Ontario Council on Graduate Studies [Canada]

OCI	Operator Control Interface
	Optically-Coupled Insulator
	Optically-Coupled Isolator
OCIA	Organic Crop Improvement Association [Canada]
OCIMF	Oil Companies International Marine Forum
OCIS	OSHA Computerized Information System
OCL	Operation Control
	Operation Control Language
	Operational Control Level
OCLC	On-Line Computer Library Center [US]
OCM	Ontario Center for Microelectronics [Canada]
OCMA	Oil Companies Material Association
OCO	Open-Close-Open
	Operation Capability Objectives
OCP	Obstacle Clearance Panel
	Office of Commercial Programs
	Operating Control Procedure
	Operational Checkout Procedure
OCR	Oil Circuit Recloser
	Optical Character Reader
	Optical Character Reading
	Optical Character Recognition
	Organic-Cooled Reactor
	Overconsolidated Ratio
	Overcurrent Relay
	Overhaul Component Requirement
OCRA	Office Communications Research Association
OCR-A	Optical Character Recognition - Font A
OCR-B	Optical Character Recognition - Font B
OCRD	Office of the Chief of Research and Development [US Army]
OCRI	Ottawa-Carleton Research Institute [Canada]
OCRS	Ontario Center for Remote Sensing [Canada]
OCRUA	Optical Character Recognition Users Association [US]
OCS	Office Communications System
	Office of Commodity Standards [of NBS]
	Office of Communication Systems [US Air Force]
	Onboard Checkout System
	Optical Character Scanner
	Optical Contact Sensor
	Output Control System
	Overseas Communications Service [India]
OCSA	Ontario Council of Safety Associations [Canada]
OCT	Octagon
	Octal
	Octane
	October
	Office of Critical Tables [NAS/NRC, US]
	Operational Cycle Time
	Ornithine Carbamyltransferase
OCTAHDR	Octahedron
OCTG	Oil Country Tubular Goods
OCTV	Open Circuit Television
OCU	Ontario Council of Universities [Canada]

	Operational Control Unit
OCUA	Ontario Council on University Affairs [Canada]
OCUB	Osmium Collidine Uranylenbloc
OCUFA	Ontario Confederation of University Faculty Association [Canada]
OCUL	Ontario Council on University Libraries [Canada]
OCULA	Ontario College and University Library Association [Canada]
OCV	Open Circuit Voltage
OD	Olive Drab
	On Demand
	Operations Directive
	Optical Density
	Output Data
	Outside Diameter
	Overdose
1D	One-Dimensional [also: 1-D]
ODA	Official Development Assistance
	Operational Data Analysis
	Operational Design and Analysis
	Oxydianiline
ODAL	Octadecenal
ODD	Operator Distance Dialing
ODDRE	Office of the Director of Defense Research and Engineering [USDOD]
ODE	Ordinary Differential Equation
	Oxygen Defect Electron
ODESSA	Ocean Data Environmental Science Services Acquisition
ODF	Orientation Distribution Function
ODIN	Optical Design Integration
ODLRO	Off-Diagonal Long-Range Order
ODM	Orbital Determination Module
ODMR	Optical Detection of Magnetic Resonance
	Optical Double Magnetic Resonance
ODN	Own Doppler Nullifier
ODOP	Offset Doppler
ODP	Ocean Drilling Program
	Optical Data Processor
	Original Document Processing
ODPCS	Oceanographic Data Processing and Control System
ODPEX	Offshore Drilling and Production Exhibition
ODR	Optical Data Recognition
ODRN	Orbiting Data Relay Network
ODS	Octadecyltrimethyloxysilane
	Oxide Dispersion Strengthened
	Oxygen Dispersion Strengthened
ODT	Octal Debugging Technique
	Outside Diameter Tube
ODU	Output Display Unit
ODVAR	Orbit Determination and Vehicle Attitude Reference
OE	Office of Education [US]
	Open End
Oe	Oersted [unit]
OEA	Organization of European Aluminum Smelters
OEB	Ontario Energy Board [Canada]

OEC	Ontario Economic Council [Canada]
OECD	Organization for Economic Cooperation and Development
OECO	Outboard Engine Cutoff
OED	Orbiting Energy Depot
OEDC	Ontario Engineering Design Competition [Canada]
OEEC	Organization for European Economic Cooperation [now: OECD]
OEF	Origin Element Field
OEG	Open-End (Wave-)Guide
	Operations Evaluation Group
OEIC	Optoelectronic Integrated Circuit
OEM	Original Equipment Manufacturer
	Other Equipment Manufacturer
OEMI	Office Equipment Manufacturers Institute
OEMS	Open-Site EMI (= Electromagnetic Interference) Measurement System
OEP	Odd-Even Predominance
OEQ	Order of Engineers of Quebec [Canada]
OER	Office of Exploratory Research [US]
OERD	Office of Energy Research and Development
OES	Optical Emission Spectroscopy
	Orbital Escape System [NASA]
	Order Entry System
OEW	Operating Empty Weight
OF	Outside Face
	Overflow
	Oxygen-Free
O/F	Orbital Flight
OFA	Oil-Immersed Forced Air-Cooled (Transformer)
OFB	Operational Facilities Branch [NASA]
OFC	Operational Flight Control
	Oxyfuel Gas Cutting
	Oxygen-Free Copper
OFCA	Ontario Federation of Construction Associations [Canada]
OFC-A	Oxyacetylene Cutting
OFC-H	Oxyhydrogen Cutting
OFC-N	Oxynatural Gas Cutting
OFC-P	Oxypropane Cutting
OFF	Office
	Officer
OFG	Optical Frequency Generator
OFHC	Oxygen-Free High Conductivity (Copper)
OFHIC	Oxygen-Free High Conductivity (Copper)
OFI	Optical Fiber Identifier
OFL	Ontario Federation of Labour [Canada]
	Optical Fault Locator
	Overflow
OFLP	Oxygen-Free Low-Phosphorus (Copper)
OFO	Office of Flight Operations [NASA]
OFP	Operating Force Plan
	Oscilloscope Face Plane
OFPU	Optical Fiber Production Unit
OFR	On-Frequency Repeater
	Overfrequency Relay
OFS	Oxygen-Free with Silver (Copper)
OFSD	Operating Flight Strength Diagram
OFT	Optical Fiber Thermometry

OFTDA	Office of Flight Tracking and Data Acquisition [NASA]	**OIDC**	Ontario Industrial Development Council [Canada]
OFV	Overflow Valve	**OIDO**	Ocean Industries Development Office
OFW	Oxyfuel Gas Welding	**OIDPS**	Oversea Intelligence Data Processing System
OFXLP	Oxygen-Free Extra Low Phosphorus (Copper)	**OIFC**	Oil-Insulated and Fan-Cooled
OFZ	Obstacle-Free Zone	**OIG**	Optically Isolated Gate
OG	Outer Gimbal	**OII**	Office of Invention and Innovation [of NBS]
	Oxygen (Converter) Gas (Process)	**OIL TURP**	Oil of Turpentine
OGA	Outer Gimbal Axis	**OIP**	Oil-in-Place
OGE	Operational Ground Equipment	**OIRCA**	Ontario Industrial Roofing Contractors Association
OGI	Oculogyral Illusion	**OIRM**	Office of Information Resources Management [US]
OGM	Outgoing Message		
OGMC	Ordnance Guided Missile Center	**OIS**	Office of Information Services [of FAA]
OGO	Orbiting Geophysical Observatory		Office Information System
OGR	Outgoing Repeater		Operational Intercommunication System
	Oxygen-Gas Recovery System	**OISA**	Office of International Science Activities
OGRC	Office of Grants and Research Contracts		Office of International Scientific Affairs [US]
OGS	Ontario Geological Survey [Canada]	**OISC**	Oil-Insulated and Self-Cooled
	Ontario Graduate Scholarship [Canada]	**OIWC**	Oil-Immersed and Water-Cooled
OGT	Outgoing Trunk	**OJT**	On-the-Job Training
OGU	Outgoing Unit	**OK**	Oklahoma [US]
OH	Ohio [US]		All Right
	Oil Hardening	**OKA**	Otherwise Known As
	Open Hearth (Furnace)	**Okla**	Oklahoma [US]
	Operational Hardware	**OL**	On-Line
O-H	Octal-Hexidecimal		Open Loop
OHA	Outside Helix Angle		Operating Location
	Oxygen Hemoglobin Affinity		Overflow Level
OHC	Occupational Health Center		Overhead Line
	Overhead Camshaft		Overlap
OHD	Over-the-Horizon Detector		Overload
OHF	Occupational Health Facility	**O/L**	Operations/Logistics
OHM	Ohmmeter		Overload
OHP	Overhead Projector	**OLAC**	On-Line Accelerated Cooling
	Oxygen at High Pressure	**OLABS**	Offshore Labrador Biological Studies
OHRD	Ontario Hydro Research Division [Canada]	**OLB**	Outer Lead Bond
OHS	Occupational Health and Safety	**OLBM**	Orbital-Launched Ballistic Missile
OHSGT	Office of High Speed Ground Transportation	**OLC**	On-Line Computer
OHT	Oxygen at High Temperature		Outgoing Line Circuit
OH&T	Oil Hardened and Tempered	**OLD**	Open Loop Damping
OHTB	Ontario Highway and Transport Board [Canada]	**OLE**	On-Line Enquiry
		OLERT	On-Line Executive for Real Time
OHTDC	Ontario Hydro Tritium Dispersion Code	**OLF**	One Hundred Linear Feet
OHTE	Ohmically-Heated Toroidal Experiment		Orbital Launch Facility
OHV	Overhead Valve	**OLM**	On-Line Monitor
OHW	Oxyhydrogen Welding		Optical Light Microscopy
OI	Oil Insulation	**OLMA**	Ontario Lumber Manufacturers Association [Canada]
	Operations Instruction		
OIA	Ocean Industries Association [US]	**OLO**	Orbital Launch Operations
OIB	Operating Impedance Bridge	**OLP**	Oxygen Lance Powder (Converter)
	Operations Integration Branch [NASA]	**OLPARS**	On-Line Pattern Analysis and Recognition System
OIC	Operations Instrumentation Coordinator		
	Optical Integrated Circuit	**OLPS**	On-Line Programming System
OICAP	Ocean Industries Capital Assistance Program	**OLRT**	On-Line Real Time
		OLS	Ontario Land Surveyor [Canada]
OICC	Ontario Institute of Chartered Cartographers [Canada]		Ordinary Least Squares
		OLSC	On-Line Scientific Computer
OICD	Office of International Cooperation and Development	**OLSE**	Ordinary Least-Squares Estimator
		OLSS	On-Line Software System
OICR	Ontario Institute for Computer Research [Canada]	**OLT**	On-Line Test

OLTEP	On-Line Test Executive Program
OLTP	On-Line Transaction Processing
OLTS	On-Line Test System
OLTT	On-Line Terminal Test
OLVP	Office of Launch Vehicle Programs
	Office of Launch Vehicles and Propulsion
OM	Optical Master
	Optical Microscopy
	Organic Matter
	Organometallic
	Outer Marker
O&M	Operation and Maintenance
	Organization and Method
OMA	Ontario Mining Act [Canada]
	Ontario Mining Association [Canada]
O&M, A	Operation and Maintenance, Army
OMAC	Occupational Medical Association of Canada
OMAT	Office of Manpower Automation and Training
OMB	Office of Management and Budget [US]
	Outer Marker Beacon
OMBI	Observation, Measurement, Balancing and Installation
OMCVD	Organometallic Chemical Vapor Deposition
OME	Office of Management Engineers
	Office of Minerals Exploration [US]
	Ontario Ministry of Energy [Canada]
	Optimum Mineral Extraction
OMEF	Office Machines and Equipment Federation
OMEP	Ontario Mineral Exploration Program [Canada]
OMETA	Ordnance Management Engineering Training Agency
OMIBAC	Ordinal Memory Inspecting Binary Automatic Computer
OMIS	Optical Microscope Inspection System
OML	Ordnance Missile Laboratories [US]
OMNITENNA	Omnirange Antenna
OMP	Office of Metric Programs
	Orotidine Monophosphate
OMPA	Octamethylpyrophosphoramide
OMPR	Optical Mark Page Reader
OMPRA	One-Man Propulsion Research Apparatus [NASA]
OMPT	Observed Man Point Trajectory
OMR	Optical Mark Reader
	Optical Mark Reading
	Optical Mark Recognition
OMRV	Operational Maneuvering Reentry Vehicle
OMS	Operational Meteorological Satellite [NASA]
	Optoelectronic Measuring System
	Orbital Maneuvering System
	Output per Man Shift
	Output Multiplex Synchronizer
OMSF	Office of Manned Spaceflight [of NASA]
OMT	Orthomode Transducer
OMTS	Organizational Maintenance Test Station
OMU	Optical Measuring Unit
OMV	Orbital Maneuvering Vehicle
OMVPE	Organometallic Vapor-Phase Epitaxy
ON	Octane Number

	Office of Naval Research [US Navy]
	Ontario [Canada]
ONI	Operator Number Identification
ONR	Office of Naval Research [US Navy]
Ont	Ontario [Canada]
ONWARD	Organization of Northwest Authorities for Rationalized Design [US]
OO	Object-Oriented
	Oceanographic Office
OOD	Opposite Oriented Diffusion
OOG	Out of Gauge
OOIP	Original Oil-in-Place
OOK	On-Off Keying
OOL	Operator-Oriented Language
OOPS	Off-Line Operating Simulator
OP	Observation Plane
	Old Process (Patenting)
	Operand
	Operating Procedure
	Operation
	Operational Priority
	Operation Part
	Operator
	Operator Panel
	Oppenheimer-Phillips (Process) [also: O-P]
	Orthogonal Polynomial
	Out of Print
	Output
O/P	Output
1P	One Pole
OPA	Optoelectronic Pulse Amplifier
	Orthophthalaldehyde
OPADEC	Optical Particle Decoy
OPAIT	Ontario Program for the Advancement of Industrial Technology [Canada]
OPAMP	Operational Amplifier
op cit	(opere citato) - in the work cited
OPBA	Ontario Public Buyers Association
OPC	Optical Phase Conjugation
OPCOM	Operations Communications
OPCODE	Operation Code
OPCON	Optimizing Control
OPCR	One Pass Cold Rolled
OPD	Operand
	Optical Particle Detector
OPDAC	Optical Data Converter
OPDAR	Optical Detection and Ranging
OPE	Operations Project Engineer
	Oxygen Plasma Etched
OPEC	Organization of Petroleum Exporting Countries
OPEP	Orbital Plane Experimental Package [NASA]
OPER	Operation
OPEX	Operational and Executive
OPF	Osmium Potassium Ferrocyanide
OPFET	Optical Field-Effect Transistor [also: OpFET]
OPG	Oxalate, Peroxide and Gluconic Acid
1PH	Single Phase
OPI	Office of Public Information

	Ontario Petroleum Institute [Canada]	**O&R**	Ocean and Rail
OP&I	Office of Patents and Inventions [US]	**ORA**	Office of Research Analysis [US Air Force]
OPIM	Order Processing and Inventory Monitoring	**ORACLE**	Oak Ridge Automatic Computer and Logical
OPK	Optokinetics		Engine [US]
OPLE	Omega Position Location Experiment	**ORAE**	Operational Research and Analysis
	[NASA]		Establishment
OPLIN	Ontario Public Libraries Information	**ORAN**	Orbital Analysis
	Network	**ORATE**	Ordered Random Access Talking Equipment
OPM	Office of Personnel Management	**ORAU**	Oak Ridge Associated Universities [US]
	Operations per Minute	**ORB**	Omnidirectional Research Beacon
	Operator Programming Method		Operational Research Branch [Canada]
	Orbits per Minute	**ORBID**	On-Line Retrieval of Bibliographic Data
	Output Position Map	**ORBIS**	Orbiting Radio Beacon Ionospheric Satellite
OPNG	Opening		[NASA]
OPO	Optical Parametric Oscillator	**ORBIT**	On-Line Real-Time Branch Information
OPP	Octal Print Punch		Technique
	Open Pore		Oracle Binary Internal Translator
	Oriented Polypropylene	**ORC**	Operational Research Committee [Canada]
OPP CE	Opposite Commutator End		Operations Research Center [US]
OPPOSIT	Optimization of a Production Process by an		Ordnance Rocket Center
	Ordered Simulation and Iteration		Oxidation Reduction Cycle
	Technique	**ORCA**	Ontario Royal Commission on Asbestos
OPP PE	Opposite Pulley End		[Canada]
OPR	Operation	**ORCAL**	Orange County Manufacturing and
	Optical Page Reading		Metalworking Conference and Exposition
OPS	On-Line Process Synthesizer		[US]
	Operational Paging System	**ORCO**	Organization Committee [of ISO]
	Oriented Polystyrene	**ORD**	Office of Research and Development
	Overhead Positioning System		Once-Run Distillate
OPSCAN	Optical Scanning Users Group		Operational Readiness Data
OPSCON	Operations Control		Operational Research Division
OPSF	Orbital Propellant Storage Facility		Optical Rotary Dispersion
OPSK	Optimum Phase-Shift Keying		Order
OPSKS	Optimum Phase-Shift Keyed Signal		Ordnance
OPT	Ophthaldialdehyde	**ORDEF**	Ontario Research Development Foundation
	Optical		[Canada]
	Optics	**ORDENG**	Ordnance Engineering
	Option	**ORDVAC**	Ordnance Variable Automatic Computer
OPTA	Optimal Performance Theoretically	**ORE**	Ocean Resources Engineering
	Attainable		Operational Research Establishment [now:
OPTAR	Optical Automatic Ranging		DORE]
OPTIM	Order Point Technique for Inventory	**Ore**	Oregon [US]
	Management	**Oreg**	Oregon [US]
OPTUL	Optical Pulse Transmission Using Laser	**OREO**	Orbiting Radio Emission Observatory
OPU	Operations Priority Unit		[NASA]
OPUS	Octal Program Updating System	**ORE RES**	Ore Reserve
OPW	Operating Weight	**ORF**	Ontario Research Foundation [Canada]
	Orthogonalized Plane Wave		Orifice
OQ	Oil Quenching	**ORG**	Operations Research Group
OR	Operational Readiness		Organization
	Operations Requirement		Origin
	Operations Research	**ORGDP**	Oak Ridge Gaseous Diffusion Plant
	Operating Room		[USAEC]
	Orange	**ORGN**	Organization
	Oregon [US]	**ORI**	Ocean Research Institute [Japan]
	Output Register		Operational Readiness Inspection
	Outside Radius	**OR&IE**	Operations Research and Industrial
	Overhaul and Repair		Engineering
	Overload Relay	**ORIG**	Origin
	Overrun		Original
	Oxide Removal	**ORIT**	Operational Readiness Inspection Team

ORKID	Orientation Determination from Kikuchi Diagrams
ORL	Orbital Research Laboratory [NASA]
	Ordnance Research Laboratory
ORLY	Overload Relay
ORN	Orange
	Ornament
ORNL	Oak Ridge National Laboratory [of USAEC]
	Oak Ridge Nuclear Laboratory [of USAEC]
ORO	Operations Research Office [US]
OROM	Optical Read-Only Memory
ORP	Optional Response Poll
	Oxidation Reduction Potential
	Oxidation Reduction Probe
ORR	Oak Ridge Research Reactor [US]
ORRAS	Optical Research Radiometrical Analysis System
ORS	Object Recognition Systems
	Octahedral Research Satellite [NASA]
	On-Site Reclamation System
	O-Ring Seal
ORSA	Operations Research Society of America
ORS-BR	O-Ring Seal, Braze Type
ORS-BT	O-Ring Seal, Bite Type
ORSORT	Oak Ridge School of Reactor Technology [USAEC]
ORT	Operational Readiness Test
	Overland Radar Technology
ORTAG	Operations Research Technical Assistance Group [US Army]
ORTHO	Orthogonal
ORV	Ocean Range Vessel [US Air Force]
	Orbital Rescue Vehicle
OS	Oblique Sounding
	Oil Shale
	Oil Switch
	Old Style
	ON Switch
	Operating System
	Operational Sequence
	Optical Scanning
	Organosilicon Device
Os	Osmium
OSA	Olefin-Modified Styrene-Acrylonitrile
	On-Stream Analysis
	Open Systems Architecture
	Optical Society of America
OSAF	Origin Subarea Field
OSAM	Overflow Sequential Access Method
OSC	Ontario Science Center [Canada]
	Optically Sensitive Controller
	Organosilicon
	Oscillation
	Oscillator
OSCAR	Optimum Systems Covariance Analysis Results
	Orbiting Satellite Carrying Amateur Radio
OSD	Office of the Secretary of Defense [US]
	Operational Sequence Diagram
	Operational Systems Development
OS&D	Over, Short and Damage Report

OSDP	On-Site Data Processor
OSE	Operational Support Equipment
OSEE	Optically-Simulated Electron Emission
OSERP	Oil Sands Environmental Research Program
OSF	Office of Spaceflight [NASA]
	One Hundred Square Feet [also: osf]
OSFM	Office of Spacecraft and Flight Missions [NASA]
OSFP	Office of Spaceflight Programs [NASA]
OSH	Oil Sands and Heavy Oil
OSHA	Occupational Safety and Health Act
	Occupational Safety and Health Administration [US]
OSI	Office of Scientific Intelligence
	Open Systems Interconnect(ion)
	Open Systems Interface
OSIC	Optimization of Subcarrier Information Capacity
OSIL	Operating System Implementation Language
OSIS	Office of Science Information Service [of NSF, US]
OSL	Outstanding Leg
OSM	Office of Surface Mining
	Oil-Sands Mining
	Option Select Mode
OS/MFT	Operating System/Multiprogramming a Fixed Number of Tasks
OSMV	One-Shot Multivibrator
OS/MVS	Operating System/Multiprogramming with Virtual Storage
OS/MVT	Operating System/Multiprogramming a Variable Number of Tasks
OSO	Orbital Solar Observatory [US]
	Orbiting Satellite Observer
	Orbiting Scientific Observatory
	Ore-Slurry-Oil Ship
OSP	Office of Scientific Personnel [NRC-NAS, US]
	Optical Signal Processing
OSPk	Offset Keyed Quadrature Phase Shift Keying
OSR	Office of Scientific Research
	Oil/Steam Ratio
	Operational Scanning Recognition
	Optical Solar Reflector
	Output Shift Register
	Over-the-Shoulder Rating
OSRD	Office of Scientific Research and Development
	Office of Standard Reference Data [of NBS]
OSRMD	Office of Scientific Research, Mechanics Division
OSS	Ocean Science and Surveys
	Ocean Surveillance Satellite
	Office of Space Sciences
	Office of Statistical Standards
	Orbital Space Station
OSSA	Office for Space Science and Applications [NASA]
OSSS	Orbital Space Station System [NASA]
OST	Office of Science and Technology [of PSAC]

	Operating System Toolbox	**OTP**	Office of Telecommunications Policy [US]
	Operational System Test		Operational Test Procedure
OSTAC	Ocean Science and Technology Advisory		Overtemperature Protection
	Committee [of NSIA]	**OTR**	Optical Tracking
OSTI	Office of Scientific and Technical	**OTRAC**	Oscillogram Trace Reader
	Information [of DES]	**OTS**	Office of Technical Services [US]
	Organization for Social and Technological		Office of Technological Services
	Innovation [US]		Orbital Technology Satellite
OSTP	Office of Science and Technology Policy		Orbital Test Satellite
	[US]		Orbital Transport System
OSTS	Office of State Technical Service [US]	**OTSR**	Optimum Track Ship Routing
OSU	Ohio State University [US]	**OTT**	One-Time Tape
OSURF	Ohio State University Research Foundation	**OTU**	Office of Technology Utilization [NASA]
	[US]		Operational Taxonometric Unit
OSV	Orbital Support Vehicle		Operational Test Unit
OSW	Office of Saline Water [US]		Operational Training Unit
	Office of Solid Wastes [US]	**OTV**	Operational Television
OS&Y	Outside Screw and Yoke		Orbital Transfer Vehicle
OT	Oil-Tight	**OUB**	Occasional-Use Bands
	On Truck	**OUF**	Oxygen Utilization Factor
	Open Top	**OULCS**	Ontario University Libraries Cooperation
	Operating Time		System [Canada]
	Overall Test	**OUST**	Office of Underground Storage Tanks [EPA,
OTA	Office of Technology Assessment		US]
	Ontario Trucking Association [Canada]	**OUT**	Outgoing
OTANS	Offshore Trade Association of Nova Scotia		Outlet
	[Canada]		Output
OTBD	Outboard	**OUTA**	Ontario Urban Transit Association [Canada]
OTC	Office of Telecommunications	**OUTBD**	Outboard
	Offshore Technology Conference	**OUTLIM**	Output Limiting Facility
	Operational Test Center	**OUTRAN**	Output Translator
	Overhead Travelling Crane	**OV**	Oil of Vitriol
	Overseas Telecommunications Commission		Orbiting Vehicle
	Over-the-Counter		Overvoltage
	Oxytetracycline	**OVA**	Ovalbumin
OTCA	Overseas Technical Cooperation Agency	**OVBD**	Overboard
	[Japan]	**OVD**	Outside Vapor Deposition
OTCCC	Open-Type Control Circuit Contact	**OVERS**	Orbital Vehicle Reentry Simulator
OTCR	Office of Technical Cooperation and	**OVFL**	Overflow
	Research [US]	**OVH**	Oval Head
OTD	Original Transmission Density	**OVHD**	Overhead
OTDA	Office of Tracking and Data Acquisition	**OVHL**	Overhaul
OTDR	Optical Time Domain Reflectometer	**OVLD**	Overload
OTE	Operational Test Equipment	**OVP**	Overvoltage Protection
OTEC	Ocean Thermal Energy Conversion	**OVRHD**	Overhead
OTES	Optical Technology Experiment System	**OVV**	Overvoltage
OTF	Optical Transfer Function	**OW**	Oil-Immersed Water-Cooled (Transformer)
OTH	Over-the-Horizon		Operating Weight
OTH/B	Over-the-Horizon Backscatter		Orderwire
OTHR	Over-the-Horizon Radar	**O/W**	Oil/Water (Emulsion)
OTIU	Overseas Technical Information Unit	**OWC**	Oil/Water Contact
OTJ	On-the-Job	**OWE**	Operating Weight - Empty
OTL	Order Trunk Line	**OWF**	Optimal Work Function
OTLP	Zero Transmission Level Point		Optimum Working Frequency
OTLT	Outlet	**OWFEA**	Ontario Wholesale Farm Equipment
OTM	Office of Telecommunications Management		Association [Canada]
	[US]	**OWG**	Oil, Water, Gas
OTMA	Office Technology Management Association	**OWM**	Office of Weights and Measures [of NBS]
OTMJ	Outgoing Trunk Message Junction	**OWP**	Oil Well Pumper
OTN	Operational Teletype Network	**OWPR**	Ocean Wave Profile Recorder
O to O	Out to Out	**OWRC**	Ontario Water Resources Commission
			[Canada]

OWRR	Office of Water Resources Research [US]
OWRT	Office of Water Research and Technology
OWS	Ocean Weather Station
	Operational Weather Support
	Operators' Workstation
	Orbital Workshop Station [NASA]
OXD	Oxidized
OXS	Oxygen Sensor
OXY	Oxygen
oz	ounce [unit]
oz ap	ounce, apothecaries
oz av	ounce, avoirdupois
OZD	Observed Zenith Distance
oz fl	ounce, fluid
oz-ft	ounce-foot
oz-in	ounce-inch
oz t	ounce, troy

P

P	Page
	Paper
	Parallel
	Parity
	Part
	Peak
	Penetration
	Pentode
	Perimeter
	Permanence
	Permeability
	Phosphorescence
	Phosphorus
	Pitch
	Plate
	Plywood
	Poise
	Pointer
	Polarization
	Pole
	Polishing
	Port
	Positive
	Power
	President
	Pressure
	Printer
	Probability
	Procedure
	Process
	Processor
	Program
	Proline
	Proton
	Punch
p	page
	poor
	pint [unit]

PA	Pack Area
	Pad Abort
	Particular Average
	Pending Availability
	Pennsylvania [US]
	Phtalic Anhydride
	Physics Abstracts
	Pilotless Aircraft
	Plasma-Arc
	Point of Aim
	Polyallomer
	Polyamide
	Power Amplifier
	Preamplifier
	Preplaced Aggregate
	Press Association
	Pressure Angle
	Probability of Acceptance
	Product Analysis
	Product Assurance
	Program Access (Key)
	Program Action (Key)
	Program Address
	Program Analysis
	Program Attention (Key)
	Program Authorization
	Principle of Adding
	Production Adjustment
	Public Address
	Pulse Amplifier
	Puromycin Aminonucleoside
Pa	Pascal
	Protactinium
	Pennsylvania [US]
pA	picoampere
pa	(per annum) - each year
PAA	Phase Antenna Array
	Polyacrylamide
	Polyacrylic Acid
PAABA	P-Acetamidobenzoic Acid
PAABS	Pan-American Association of Biochemical Societies
PAAC	Program Analysis Adaptable Control
PAA-PS	Polyacrylic Acid - Polysulfone
PAB	Aminopentyl Benzimidazole
	P-Aminobenzyl
	Primary Application Block
	Pulse Adsorption Bed
PABA	P-Aminobenzoic Acid
PABLA	Problem Analysis by Logical Approach
PABST	Primary Adhesively Bonded Structures Technology
PABX	Private Area Exchange
	Private Automatic Branch Exchange
PAC	Packaged Assembly Circuit
	Pacific
	Packaging Association of Canada
	Pedagogic Automatic Computer
	Personal Analog Computer
	Perturbed Angular Correlation
	Plasma-Arc Cutting

	Polycyclic Aromatic Hydrocarbon
	Portable Air Compressor
	President's Advisory Council [US]
	Process Automation Computer
	Program Authorized Credentials
	Programmable Automatic Comparator
	Public Archives Canada
PACA	Polyamide Carboxylic Acid
PACAF	Pacific Air Force [US]
PACC	Products Administration and Contract Control
PACCT	PERT (= Production Evaluation and Review Technique) and Cost Correlation Technique
PACE	Packaged Cram Executive
	Passive Attitude Control Experiment
	Physics and Chemistry Experiment
	Precision Analog Computing Equipment
	Preflight Acceptance Checkout Equipment
	Prelaunch Automatic Checkout Equipment
	Process Automation and Control Executive
	Programmed Automatic Communications Equipment
	Programming Analysis Consulting Education
	Projects to Advance Creativity in Education
PACED	Program for Advanced Concepts in Electronic Design
PACE-LV	Preflight Acceptance Checkout Equipment for Launch Vehicle
PACER	Portfolio Analysis, Control, Evaluation and Reporting
	Process Assembly Case Evaluator Routine
	Program-Assisted Console Evaluation and Review
	Program of Active Cooling Effects and Requirements
PACE-SC	Preflight Acceptance Checkout Equipment for Spacecraft
PACF	Periodic Autocorrelation Function
PACFORNET	Pacific Coast Forest Research (Information) Network
PACIA	Particle Counting Immunoassay
PACM	Pulse Amplitude Code Modulation
PACOR	Passive Correlation and Ranging
PACS	Pacific Area Communications System
	Physics and Chemistry Classification Scheme
	Project Analysis and Control System
PACT	Pay Actual Computer Time
	Production Analysis Control Technique
	Program for Automatic Coding Techniques
	Programmed Analysis Computer Transfer
	Programmed Automatic Circuit Tester
PACV	Patrol Air Cushion Vehicle
PACVD	Plasma-Assisted Chemical Vapor Deposition
PAD	Packet Assembly/Disassembly
	Perturbed Angular Distribution
	Pixel Access Definition
	Plastics Analysis Division [of SPE]
	Polyaperture Device
	Positioning Arm Disk

	Post-Activation Diffusion
	Power Amplifier Device
	Propellant-Actuated Device
PADA	Pyridineazodimethylaniline
PADAR	Passive Detection and Ranging
PADD	Petroleum Administration for Defense Districts
PADE	Pad Automatic Data Equipment
PADIS	Pan-African Documentation and Information System [Ethiopia]
PADL	Pilotless Aircraft Development Laboratory [US Navy]
PADLOC	Passive-Active Detection and Location
PADRE	Patient Automatic Data Recording Equipment
	Portable Automatic Data Recording Equipment
PADS	Passive-Active Data Simulation
	Performance Analysis and Design Synthesis
	Precision Aerial Display System
PAE	Polyarylene Ether
PAEM	Program Analysis and Evaluation Model
PAF	Picric Acid Formaldehyde
	Platelet Activating Factor
	Pre-Atomized Fuel
PAFC	Phase-Locked Automatic Frequency Control
PAG	Polyalkylene Glycol
	Prealbumin Globulin
	Protein A-Colloidal Gold
	Protein Advisory Group [US]
PAGE	PERT Automated Graphical Extension
	Polyacrylamide Gel Electrophoresis
PAGEOS	Passive Geodetic Earth Orbiting Satellite [NASA]
PAGICEP	Petroleum and Gas Industry Communications Emergency Plan [FCC, US]
PAGS	Prior Austenite Grain Size
PAH	Polycyclic Aromatic Hydrocarbon
PAI	Polyamideimide
	Process Analytical Instrumentation
	Programmer Appraisal Instrument
PAID	Personnel and Accounting Integrated Data (System)
PAIR	Performance and Integration Retrofit
PAIT	Program for Advancement of Industrial Technology [Canada]
PAKEX	Packaging Exhibition
PAL	Pedagogic Algorithmic Language
	Permanent Artificial Lighting
	Permissive Action Link
	Phase Alternation Line
	Phase Attenuation by Line
	Process Assembly Language
	Production and Application of Light
	Programmable Array Logic
	Programmed Application Library
	Psychoacoustic Laboratory
PAL-D	Phase Alternation Line Delay
PALM	Phillips Automated Laboratory Management
	Precision Altitude and Landing Monitor

PAM	Pamphlet
	Payload Assist Module
	Peripheral Adapter Module
	Photoacoustic Microscopy
	Plan-Applier Mechanism
	Pole Amplitude Modulation
	Portable Activity Monitor
	Pozzolan Aggregate Mixture
	Process Automatic Monitor
	Pulse Amplitude Modulation
	Pyridine Aldoxime Methiodide
PAMA	Pulse-Address Multiple Access
PAMAC	Parts and Materials Accountability Control
PAMD	Periodic Acid Mixed Diamine
PAMELA	Plan-Applier Mechanism for English Language Analysis
PAM-FM	Pulse Amplitude Modulation - Frequency Modulation
PAMI	Prairie Agricultural Machinery Institute [Canada]
PAML	Program Authorized Materials List
PAMPER	Practical Application of Midpoints for Exponential Regression
PAMS	Pad Abort Measuring System
PAN	Polyacrylonitrile
	Positional Alcohol Nystagmus
	Pyridylazonaphthol
PANA	Pan-African News Agency
PANAFTEL	Pan-African Telecommunication Network
PANS	Procedures for Air Navigation Services
PANSDOC	Pakistan National Scientific and Technical Documentation Center
PANSMET	Procedures for Air Navigation Services - Meteorology
PANT	Pantograph
PAO	Phenylarsine Oxide
PAP	P-Aminophenylalanine
	Peroxidase-Anti-Peroxidase
	Phosphoadenosinephosphate
PAPA	Programmer and Probability Analyzer
PAPD	Periodate Dimethylphenylenediamine
PAPI	Polymethylene Polyphenyl Isocyanate
PAPM	Pulse Amplitude Phase Modulation
PAPRICAN	Paper Research Institute of Canada
PAPS	Periodic Acid Phenylhydrazine Schiff
	Periodic Armaments Planning System
	Phosphoadenosine Phosphosulfate
	Physics Auxiliary Publication Service
PAR	Paragraph
	Parallel
	Parenthesis
	Parity
	Peak-to-Average Ratio
	Performance Analysis and Review
	Perimeter Acquisition Radar
	Precision Approach Radar
	Processor Address Register
	Production Automated Riveting
	Professional Abstracts Registries
	Program Appraisal and Review
	Progressive Aircraft Rework

	Pyridylazoresorcinol
PARA	Paragraph
PARADE	Passive-Active Ranging and Determination
PARAL	Parallel
PARAMI	Parsons Active Ring Around Miss-Indicator
PARAMP	Parametric Amplifier
PARASEV	Paraglider Search Vehicle
PARASYN	Parametric Synthesis
PARC	Palo Alto Research Center [US]
	Progressive Aircraft Reconditioning Cycle
PARD	Parts Application Reliability Data
	Periodic and Random Deviation
	Precision Annotated Retrieval Display System
PARDOP	Passive Ranging Doppler
PARIS	Pulse Analysis-Recording Information System
	Portable Automated Remote Inspection System
PARL	Prince Albert Radar Laboratory [of DRTE]
PARL	Parliament
PARM	Program Analysis for Resource Management
PARM	Parameter
PAROS	Passive Ranging of Submarines
PARR	Procurement Authorization and Receiving Report
PARS	Paragraphs
PARSAC	Particle Size Analog Computer
PARSEV	Paraglider Search Vehicle
PART	Parts Allocation Requirement Technique
PARTAC	Precision Askania Range Target Acquisition
PARTEI	Purchasing Agents of the Radio, TV and Electronics Industry
PARTN	Partition
PARTNER	Proof of Analog Results through a Numerical Equivalent Routine
PARV	Paravane
PAS	P-Aminosalicylic Acid
	Periodic Acid-Schiff
	Phase Address System
	Phenylaminosalicylate
	Photoacoustic Spectroscopy
	Polyacrylsulfone
	Positron Annihilation Spectroscopy
	Pressure-Assisted Sintering
	Primary Alert System
	Process Analysis Services
	Program Address Storage
	Public Address System
	Pyrotechnic Arming Switch
PASCAL	PASCAL (Programming Language) [named after Blaise Pascal]
	Philips Automatic Sequence Calculator
PASE	Power-Assisted Storage Equipment
PASEM	Program of Assistance to Solar Equipment Manufacturers
PA-SM	Periodic Acid-Silver Methenamine
PASS	Passage
	Passenger
	Passive
	Private Automatic Switching System

	Production Automated Scheduling System		Planning Board
	Program Aid Software System		Playback
	Program Alternative Simulation System		Plotboard
	Purchasing Activities Support System		Plugboard
PASSIM	Presidential Advisory Staff on Scientific Information Management [US]		Polybutylene
			Precipitation Body
PAT	Parametric Artificial Talker		Publications Board
	Patent		Purplish Blue
	Personalized Array Translator		Pull Box
	Picric Acid Turbidity		Push Button
	Prediction Analysis Technique	**P/B**	Peak-to-Background (Ratio)
	Production Acceptance Test	**Pb**	(Plumbum) - Lead
	Proficiency Analytical Testing	**PBB**	Polybrominated Biphenyl
	Program Aptitude Test	**PBBI**	Polybutadiene Bisimide
	Programmable Automatic Tester	**PBBO**	Biphenylylphenylbenzoxazole
	Prophylaminohydroxy Tetrahydronaphthalene	**PBCS**	Post-Boost Control System
		PBD	Parallel Blade Damper
PATA	Pneumatic All-Terrain Amphibian		Phenylbiphenylyloxadiazole
PATC	Professional, Administrative, Technical, and Clerical	**PBDG**	Push-Button Data Generator
		PBDS	Photothermal Beam Deflection Spectroscopy
PATCA	Phase-Lock Automatic Tuned Circuit Adjustment	**PBE**	Proton Balance Equation
			Pulsed Bridge Element
PATCO	Professional Air Traffic Controllers Organization	**PBEC**	Pacific Basin Economic Committee
		PBEIST	Planning Board for European Inland Surface Transport
PATD	Patented		
PATE	Programmed Automatic Test Equipment	**PBF**	Potential Benefit Factor
PATH	Performance Analysis and Test Histories		Power Burst Facility [AEC]
PATHFINDER	Pathological Element Finder	**PBFA**	Particle Beam Fusion Accelerator
PATI	Passive Airborne Time-Difference Intercept	**PBH**	Planar-Buried Heterostructure
PAT OFF	Patent Office	**PBHP**	Pounds per Brake Horsepower [also: pbhp]
PATRA	Packaging and Allied Trades Research Association [now: PIRA]	**PBI**	Plant Biological Institute [Canada]
			Plant Biotechnology Institute
PATRIC	Pattern Recognition Interpretation and Correlation		Polybenzimidazole
			Process Branch Indicator
PATS	Precision Altimeter Techniques Study		Protein Bound Iodine
PATSY	Programmer's Automatic Testing System	**PBIT**	Parity Bit
PATT	Pattern	**PBM**	Permanent Benchmark
	Project for the Analysis of Technology Transfer	**PBN**	Butylphenylnitrone
		PBNA	Phenyl-Beta-Naphthylamine
PATTERN	Planning Assistance through Technical Evaluation of Relevance Numbers	**PBOS**	Planning Board for Ocean Shipping
		PBP	Pushbutton Panel
PATU	Pan-African Telecommunications Union	**PBPS**	Post-Boost Propulsion System
PATWAS	Pilots Automatic Telephone Weather Answering Service	**PBR**	Polished-Bore Receptacle
		PBRF	Plum Brook Reactor Facility
PAU	Pan-American Union	**PBS**	Pacific Biological Station [Canada]
	Pilotless Aircraft Unit		Phosphate Buffered Saline
	Precision Approach - UNICOM		Public Broadcasting Service
PAV	Phase Angle Voltmeter	**PBSTA**	Push Button Station
	Position and Velocity	**PBT**	Polybutylene Terephthalate
PAVE	Position and Velocity Extraction		Polyphenylene Benzobisthiazole
PAVT	Position and Velocity Tracking	**PBTF**	Polybromotrifluoroethylene
PAW	Plasma-Arc Welding	**PBTP**	Polybutylene Terephthalate
PAWOS	Portable Automatic Weather Observable Station	**PBV**	Post-Boost Vehicle
		PBW	Parts by Weight
PAWS	Programmed Automatic Welding System	**PBX**	Private Branch (Telephone) Exchange
PAX	Private Area Exchange	**PBZT**	Polyphenylene Benzalthiazole
	Private Automatic Exchange	**PC**	(Single) Paper, (Single) Cotton (Wire)
PAYT	Payment		Parametric Cubic
PB	Peripheral Buffer		Parts Catalogue
	Phenobarbital		Path Control
	Piperonyl Butoxide		Percent

	Personal Computer
	Phase Coherent
	Phase Conjugation
	Photocell
	Photochromatic
	Photoconductance
	Photoconductive Cell
	Photoconductor
	Phthalocyanine
	Piece
	Pitch Circle
	Pitch Control
	Plastocyanine
	Pocket Calculator
	Point of Curve
	Polycarbonate
	Portable Computer
	Price
	Printed Circuit
	Pre-Emphasis Circuit
	Processor Controller
	Professional Communication
	Program Coordination
	Program Counter
	Programmable Control(ler)
	Pulsating Current
	Pulse Compression
	Pulse Controller
	Punched Card
	Pyrurate Carboxylase
pC	picocoulomb
	picocurie [unit]
P-C	Plastic-Carbon (Replica)
PCA	Performance and Coverage Analyzer
	Personal Computers Association
	Phaseolus Coccineus Agglutinin
	Polar Cap Absorption
	Polycrystalline Alumina
	Portland Cement Association
	Precision Cleaning Agent
	Private Communications Association [US]
	Pyrotechnic Control Assembly
	Pyrrolidonecarboxylic Acid
PCAC	Partially Conserved Axial-Vector Current
PCAM	Probe Card Assembly Machine
	Punched Card Accounting Machine
PCAO	Pollution Control Association of Ontario [Canada]
PCB	Page Control Block
	Polychlorinated Biphenyl
	Power Circuit Breaker
	Printed Circuit Board
	Process Control Block
	Program Communication Block
PCBC	Partially Conserved Baryon Current
PCBS	Positive Control Bombardment System
PCC	Panama Canal Commission
	Parametric Channel Controller
	Partial Crystal Control
	Peripheral Control Computer
	Point of Common Coupling
	Point of Compound Curve
	Portland Cement Concrete
	Power Control Center
	Pyridinium Chlorochromate
PCCC	Phoenix Conference on Computers and Communications [US]
PCCD	Peristaltic Charge-Coupled Device
PCCS	Photographic Camera Control System
	Positive Control Communications System
PCD	Polycrystalline Diamond
	Power Control and Distribution
	Process Control Division
	Production Common Digitizer
	Procurement and Contracts Division [of NASA]
PCDC	Punched Card Data Processing
PCE	Power Conditioning Electronics
	Punched Card Equipment
	Pyrometric Cone Equivalent
PCEA	Pacific Coast Electric Association
PCEM	Process Chain Evaluation Model
PCF	Potential Controlled Flotation
	Pounds per Cubic Foot
	Power Cathode Follower
	Pulse-to-Cycle Fraction
PCG	Phonocardiogram
	Planning and Control Guide
PCGN	Permanent Committee on Geographical Names
PCGS	Protein Crystal Growth System
PCH	Proton Channeling
	Punch
PCI	Panel Call Indicator
	Peripheral Command Indicator
	Pilot Controller Integration
	Plant Control Interface
	Portable Cesium Irradiator [AEC]
	Power Conversion International (Congress)
	Product Configuration Identification
	Program-Controlled Interruption
PCIC	Petroleum and Chemical Industry Conference
PCIV	Prestressed Cast Iron Pressure Vessel
PCL	Printed Circuit Lamp
	Process Control Language
	Programmasyst Control Language
PCLO	Printed Circuit Board Layout
PCM	Passive Countermeasures
	Phase-Change Material
	Pitch Control Motor
	Plug Compatible Mainframe
	Pulse Code Modulation
	Pulse Count Modulation
	Punched Card Machine
PCMA	Personal Computer Management Association
PCMB	P-Chloromercuribenzoate
PCMC	Preparatory Commission for Metric Conversion
PCMD	Pulse Code Modulation Digital
PCME	Pulse Code Modulation Event
PCM-FM	Pulse Code Modulation - Frequency Modulation

PCMH	Professional Certified in Materials Handling	
PCMI	Photochromic Micro-Image	
PCMIA	Pittsburgh Coal Mining Institute of America	
PC MK	Piece Mark	
PCMM	Professional Certified in Materials Management	
PCM/PAM	Pulse Code Modulation/Pulse Amplitude Modulation	
PCMS	Punched Card Machine System	
PCNB	Pentachloronitrobenzene	
PCOS	Primary Communications Oriented System	
	Process Control Operating System	
PCP	Parallel Cascade Processor	
	Parallel Circular Plate	
	Photon-Coupled Pair	
	Polychloroprene	
	Primary Control Program	
	Process Control Processor	
	Processor Control Program	
	Program Change Proposal	
	Programmable Communication Processor	
	Project Control Plan	
	Pulse Comparator	
	Punched Card Punch	
pCp	Cytidinebiphosphate	
PCPA	P-Chlorophenylalanine	
PC/PET	Polycarbonate/Polyethylene Terephthalate	
PCR	Photoconductive Relay	
	Pickled and Cold Rolled	
	Post-Column Reaction	
	Precision Control Relay	
	Procedure Change Request	
	Program Change Request	
	Program Control Register	
	Program-Controlled Request	
	Punched Card Reader	
PCRCA	Pickled, Cold Rolled and Close Annealed	
PCS	Pieces	
	Planning Control Sheet	
	Plastic-Clad Silica	
	Pointing Control System	
	Portable Computer System	
	Power Conditioning System	
	Preferred Character Set	
	Primary Coolant System	
	Print Contrast Scale	
	Print Contrast Signal	
	Print Contrast System	
	Process Control System	
	Punched Card System	
PCSA	Personal Computing Systems Architecture	
PCSC	Power Conditioning, Switching and Control	
PCSD	Printer Control Sequence Description	
PCSIR	Pakistan Council of Scientific and Industrial Research	
PCSP	Polar Continental Shelf Project	
PCT	Percent	
	Photochemical Transfer	
	Photon-Coupled Transistor	
	Planning and Control Techniques	
	Portable Camera-Transmitter	

	Program Control Table
PCTA	Polymer of Cyclohexanedimethonol Terephthalic Acid
PCTFE	Polychlorotrifluoroethylene
PCTM	Pulse Count Modulation
PCTS	Photocapacitance Transient Spectroscopy
PCU	Passenger Car Unit
	Physical Control Unit
	Pound Centigrade Unit
	Power Control Unit
	Power Conversion Unit
	Program Control Unit
	Progress Control Unit
PCV	Packed Cell Volume
	Pollution Control Valve
	Positive Crankcase Ventilation
PCW	Pulsed Continuous Wave
PD	Paid
	Pallet Decoupler
	Peripheral Device
	Permissable Dosage
	Photodiode
	Physical Distribution
	Pitch Diameter
	Point Detonating
	Polar Distance
	Positive Displacement
	Postal District
	Potential Difference
	Power Density
	Power Distribution
	Power Doubler
	Preliminary Design
	Priority Directive
	Prime Driver
	Principal Distance
	Prism Diopter
	Probability of Detection
	Problem Definition
	Problem Determination
	Procedure Division
	Projected Display
	Propellant Dispersion
	Proportional Derivative
	Proportional Differential
	Pulse Doppler
	Pulse Driver
Pd	Palladium
PDA	Photodiode Array
	Post-Deflection Acceleration
	Precision Drive Axis
	Probability Discrete Automaton
	Probability Distribution Analyzer
	Prospectors and Developers Association
	Pulse Discrimination Analysis
	Pump Drive Assembly
PDAC	Prospectors and Developers Association of Canada
PDAID	Problem Determination Aid
PDAP	Polydiallylphthalate
PDAPS	Pollution Detection and Prevention System

PDB	Paradichlorobenzene
PDC	(Single) Paper Double Cotton (Wire)
	Polycrystalline Diamond Compact
	Power Distribution Control
	Premission Documentation Change
	Printing Density Controller
	Probability of Detection and Conversion
	Pyridinium Dichromate
PDCE	Paramagnetic Design and Cost Effectiveness
PDCPD	Polydicyclopentadiene
PDD	Partial Discharge Detector
	Physical Damage Division
	Program Design Data
	Projected Data Display
	Pulse Delay Device
PDDI	Product Data Definition Interface
PDDP	Product Design and Development Program
PDE	Partial Differential Equation
PDES	Product Data Exchange Specification
	Product Definition Exchange Specification
PDF	Pair Distribution Function
	Point Detonating Fuse
	Postdoctoral Fellow
	Powder Diffraction File
	Probability Density Function
	Probability Distribution Function
	Protected Difference Fat
PDG	Precision Drop Glider
	Pregnanediol Glucoronide
PDGDL	Plasma Dynamics and Gaseous Discharge Laboratory
PDGS	Precision Delivery Glides System
	Product Design Graphics System
PDH	Philibert-Duncumb-Heinrich (Equation)
	Phosphate Dehydrogenase
PDI	Perfect Digital Invariant
	Picture Description Instruction
	Plastic Drum Institute
PDIO	Photodiode
PDIR	Peripheral Data Set Information Record
PDL	Procedure Definition Language
pdl	poundal [unit]
PDM	Physical Distribution Management
	Polynomial Discriminant Method
	Predictive Maintenance
	Project Development Methodology
	Protected Difference Milk
	Pulse Delta Modulation
	Pulse Duration Modulation
PDME	Pendant Drop Melt Extraction
PDM-FM	Pulse Duration Modulation - Frequency Modulation
PDMP	Product Development Management Program
PDN	Public Data Network
PDO	Program Directive Operations
PDP	Positive Displacement Pump
	Product Development Program
	Program Definition Phase
	Programmed Data Processor
	Project Definition Phase
PDPC	Position Display Parallax-Corrected

Pd-PEI	Palladium-Polyethylenimine
PDPS	Parts Data Processing System
PDQ	Product Demand Quotation
	Programmed Data Quantizer
PDR	Performance Data Rate
	Periscope Depth Range
	Pilot's Display Recorder
	Pounder
	Power Directional Relay
	Precision Depth Recorder
	Preliminary Data Report
	Preliminary Design Review
	Priority Data Reduction
	Processed Data Recorder
	Processing, Distributing and Retailing
	Production, Distribution and Retailing
	Program Discrepancy Report
	Program Drum Recording
PDS	Partitioned Data Set
	Photothermal Deflection Spectroscopy
	Power Density Spectrum
	Power Distribution System
	Procedures Development Simulator
	Program Data Set
	Program Data Source
	Programmable Data Station
	Programmable Device Support
	Programmable Distribution System
	Propellant Dispersion System
PD/S	Problem Definition/Solution
PDSMS	Point-Defense Surface Missile System
PDT	Pacific Daylight Time
	Picture Description Test
	Programmable Data Terminal
	Pyridyldiphenyltriazine
PDU	Pilot's Display Unit
	Power Distribution Unit
	Pressure Distribution Unit
	Protocol Data Unit
PDW	Partially Delactosed Whey
PDWP	Partially Delactosed Whey Powder
PE	Page-End (Character)
	Peripheral Equipment
	Permanent Echo
	Petroleum Engineer(ing)
	Phase Encoding
	Photoelectric
	Pinion End
	Plasma Etching
	Polyester
	Polyether
	Polyethylene
	Potential Energy
	Primary Electron
	Prince Edward Island [Canada]
	Probable Error
	Processor Element
	Professional Engineer
	Programming Environment
	Pseudoelastic
	Pulley End

P/E	Price/Earnings (Ratio)
PEA	Pennsylvania Electric Association [US]
	Polyethylacrylate
PEAC	Program Evaluation and Audit Committee
PEACU	Plastic Energy Absorption in Compression Unit
PEAR	Plasma Extended Arc Reactor [Canada]
PEARL	Package for Efficient Access to Representations in LISP
	Performance Evaluation of Amplifiers from a Remote Location
PEC	Photoelectric Cell
	Plant Equipment Code
	Platform Electronic Cards
	Polyester Carbonate
	Polyphenylene Ether Copolymer
	Previous Element Coding
	Program Exception Code
PECAN	Pulse Envelope Correlation Air Navigation
PECBI	Professional Engineers Conference Board for Industry
PECS	Portable Environmental Control System
PECVD	Plasma-Enhanced Chemical Vapor Deposition
PED	Pedestal
	Personnel Equipment Data
	Pulse Edge Discrimination
PEDN	Planned Event Discrepancy Notification
PEDRO	Pneumatic Energy Detector with Remote Optics
PEDS	Peltier Effect Diffusion Separation
P&EE	Proving and Experimental Establishment [UK]
PEEK	Polyetherether Ketone
PEEM	Photoelectron Emission Microscopy
	Photoelectron Emission Electron Microscopy
	Photoemission Electron Microscopy
PEEP	Pilot's Electronic Eye-Level Presentation
PEER	Professional Engineering Employment Registry [US]
PEF	Physical Electronics Facility
PEFR	Peak Expiratory Flow Rate
PEG	Polyethylene Glycol
PEH	Polyphenylene Ether Homopolymer
PEI	Polyetherimide
	Polyethylenimine
	Porcelain Enamel Institute
	Preliminary Engineering Inspection
	Prince Edward Island [Canada]
PEIC	Periodic Error Integrating Controller
PEIFL	Prince Edward Island Federation of Labour [Canada]
PEILS	Prince Edward Island Land Surveyors [Canada]
PEL	Permissible Exposure Limits
PEL	Picture Element
PELSS	Precision Emitter Location Strike System
PEM	Photoelastic Modulator
	Photoelectromagnetic
	Photoelectron Microscopy
	Plant Engineering and Maintenance

	Production Engineering Measure
	Program Execution Monitor
	Proto-Environmental Model
PEMA	Ethylphenylmalonamide Monohydrate
	Process Equipment Manufacturers Association [US]
	Procurement Equipment and Missiles - Army [US]
PEMAC	Professional Engineers Manpower Assessment Committee
PEMD	Program for Export Market Development
PEN	Penicillin
PENA	Primary Emission Neutron Activation
PEN-B	Penicillin, Benzalthine Salt
PENCIL	Pictorial Encoding Language
PEng	Professional Engineer
PEN-K	Penicillin, K (= Potassium) Salt
Penn	Pennsylvania [US]
PEN-NA	Penicillin, Na (= Sodium) Salt
Penna	Pennsylvania [US]
PENNSTAC	Penn(sylvania) State University Automatic Computer
PEN-P	Penicillin, Procain Salt
PENRAD	Penetration Radar
PENT	Pentode
PEO	Polyethylene Oxide
PEOS	Propulsion and Electrical Operating System
PEP	Partitioned Emulation Programming
	Peak Envelope Power
	People for Energy Process
	Phosphoenolpyruvate
	Phosphoenolpyruvic Acid
	Photoelectric Potential
	Planar Epitaxial Passivated (Transistor)
	Planetary Ephemeris Program
	Political and Economic Planning
	Porsche Experimental Prototype
	Program Evaluation Procedure
	Propulsion and Energetics Panel
PEPAG	Physical Electronics and Physical Acoustics Group
PEPCK	Phosphoenolpyruvate Carboxykinase
PEP-L	Partitioned Emulation Programming - Local
PEP-LR	Partitioned Emulation Programming - Local/Remote
PEPP	Planetary Entry Parachute Program
	Professional Engineers in Private Practice
PEPR	Precision Encoding and Pattern Recognition (Device)
PER	Period
	Person
	Program Event Recording
	Preliminary Engineering Report
	Protein Efficiency Ratio
	Pseudoequilibrium Ratio
PERA	Planning and Engineering for Repairs and Alterations
	Production Engineering Research Association
PERC	Percussion
PERC DR	Percussion Drilling

PERCOS	Performance Coding System
PERD	Panel on Energy Research and Development [Canada]
PEREF	Propellant Engine Research Environmental Facility
PERF	Perfection
	Perforation
	Performance
PERFS	Perforations
PERGO	Project Evaluation and Review with Graphic Output
PERI	Photoengravers Research Institute [US]
	Protein Engineering Research Institute [Japan]
PERIF	Peripheral
PERL	Perkin Elmer Robot Language
PERM	Permanent
	Permeability
	Program Evaluation for Repetitive Manufacture
PERMINVAR	Permeability Invariant
PERP	Perpendicular
	Process Evaluation/Research Planning
PERS	Personnel
PERT	Performance Evaluation and Review Technique
	Production and Evaluation Review Technique
	Program Evaluation Research Task
	Program Evaluation Review Technique
PERTCO	Program Evaluation Review Technique with Cost
PERU	Production Equipment Records Unit
PES	Photoelectric Scanner
	Photoelectron Spectroscopy
	Photoemission Spectroscopy
	Polyether Sulfone
	Power Engineering Society [of IEEE]
	Programmer Electronic Switch
PESC	Passivated Emitter Solar Cell
PESDA	Printing Equipment Supply Dealers Association
PEST	Parameter Estimation by Sequential Testing
PESTF	Proton Event Start Forecast
PET	Patterned Epitaxial Technology
	Pentaerythrital
	Peripheral Equipment Tester
	Personal Electronic Transactor
	Petroleum
	Physical Equipment Table
	Plasma Edge Technique
	Polyethylene Terephthalate
	Portable Executive Telephone
	Position-Event-Time
	Positron Emission Tomography
	Precision End Trimmed
	Production Environmental Testing
PETA	Pentaerythritol Triacrylate
PETC	Pittsburgh Energy Technology Center [US]
PETE	Pneumatic End-to-End
PetE	Petroleum Engineer

PETG	Glycol-Modified Polyethylene Terephthalate
PETN	Pentaerythritol Tetranitrate
PETP	Polyethylene Terephthalate
PETRO	Petroleum
PETT	Portable Ethernet Transceiver Tester
PEW	Percussion Welding
PEX	Projectable Excitation
p ex	(par exemple) - for example
PF	Packing Factor
	Penetration Fracture (Test)
	Perchlorylfluoride
	Phenol-Formaldehyde
	Plug-Flow Transport
	Point of Frog
	Pole Figure
	Powder Forging
	Power Factor
	Profile
	Program Function
	Protection Factor
	Pulse Frequency
	Punch Off
pF	picofarad
PFA	Perfluoroalkoxy
	Polyfluoroalkoxy
	Pulverized Fuel Ash
PFBA	Polyperfluorobutylacrylate
PFBC	Pressurized Fluidized Bed Combustion
PFBT	Performance Functional Board Tester
PFC	Pack Feed and Converter
	Phase Frequency Characteristics
	Planar Flow Casting
	Private First Class
PFCS	Primary Flight Control System
PFD	Personal Flotation Device
	Power Flux Density
	Primary Flash Distillate
PFIR	Part Fill-In and Ram
PFIX	Power Failure Interrupt
PFK	Phosphofructokinase
	Plastic Fluted Knob
	Program Function Key
PFL	Propulsion Field Laboratory [US]
PFM	Power Factor Meter
	Pulse Frequency Modulation
PFN	Permanent File Name
	Pulse Forming Network
PFP	Performic Acid Phosphotungsite
	P-Fluorophenylalanine
	Plutonium Finishing Plant [USDOE]
	Post Flight Processor
PFPA	Pitch Fiber Pipe Association
PFPE	Perfluoropolyether
PFR	Parts Failure Rate
	Plug Flow Reactor
	Programmed Film Reader (System)
	Prototype Fast (Breeder) Reactor
PFRS	Portable Field Recording System
PFS	Propellant Field System
PFT	Paper, Flat Tape
PFZ	Precipitate-Free Zone

PG	Page	**PHC**	Petroleum Hydrocarbon
	Phase Gradient		Physical Hydrogen Cracking
	Polyethylene Glycol	**PhC**	Pharmaceutical Chemist
	Portugal	**PHCVD**	Photochemical Chemical Vapor Deposition
	Power Gain		Photosensitized Chemical Vapor Deposition
	Power Gate	**PHD**	Plastohydrodynamic(s)
	Power Generation		Pulse Height Discriminator
	Pressure Gauge	**PhD**	Doctor of Philosophy
	Pulse Generator	**PHE**	Plate Heat Exchange
	Pyrolytic Graphic	**Phe**	Phenylalanine
pg	picogram	**PHEMA**	Polyhydroxyethyl Methacrylate
PGA	Pin Grit Array	**PHEN**	Phenolic
	Professional Graphics Adapter	**PHENO**	Precise Hybrid Elements for Nonlinear
	Pressure Garment Assembly		Operation
	Propane Gas Association	**P&HEP**	Plasma and High-Energy Physics [IEEE
	Pteroylglutamic Acid		Group]
PGAA	Prompt Gamma-Ray Activation Analysis	**PHH**	Phillips Head (Screw)
PGAL	Proof Gallons	**PHI**	Phosphohexose Isomerase
PGC	Professional Graphics Controller		Position and Homing Indicator
	Pulsed Gas Chromatography	**PHIG**	Programmer's Hierarchical Interactive
	Pyrolysis Gas Chromatography		Graphics
PGE	Platinum-Group Element	**PHIGS**	Programmer's Hierarchical Interactive
PGEC	Professional Group on Electronic Computers		Graphics System
PGEN	Program Generator	**PHIN**	Position and Homing Inertial Navigator
PGeol	Professional Geologist	**PHLODOT**	Phase Lock Doppler Tracking
PGEWS	Professional Group on Engineering Writing	**PHM**	Patrol Hydrofoil Missile
	and Speech [of IEEE]		Phase Meter
PGH	Patrol Gunboat, Hydrofoil	**PHODEC**	Photometric Determination of Equilibrium
PGI	Phosphoglucoisomerase		Constants
PGK	Phosphoglycerate Kinase	**PHOENIX**	Plasma Heating Obtained by Energetic
PGLIN	Page and Line		Neutral Injection Experiment
PGM	Phosphoglycerate Mutase	**PHON**	Phonetics
	Platinum Group Metal	**PHONET**	Phonetics
	Precision-Guided Munitions	**PHONO**	Phonograph(ic)
	Program		Phonography
PGNCS	Primary Guidance and Navigation Control	**PHOT**	Photograph(ic)
	System		Photography
PGOS	Petroleum, Gas and Oil Shale	**PHOTO**	Photograph
PGR	Precision Graphic Recorder	**PHP**	Parts, Hybrids, and Packaging [IEEE]
PGRF	Pulse Group Repetition Frequency		Pounds per Horsepower
PGRT	Petroleum and Gas Revenue Tax		Pump Horsepower
PGS	Power Generation System	**PHR**	Physical Record
	Power Generator Section		Pound-Force per Hour
PGT	Page Table	**PHRAN**	Phrasal Analyzer
	Planetary Gear Train	**PHRED**	Phrasal English Diction
PGU	Pressure Gas Umbilical	**PHS**	Pulse Height Selection
PGW	Pressure Gas Welding	**PHSE**	Phase
PH	Phase	**PHT**	Phototube
	Phasemeter		Pyrrolidone Hydrotribromide
	Power House	**PHTC**	Pulse Height to Time Converter
	Precipitation Hardening	**PHW**	Pressurized Heavy Water
Ph	Phenyl	**PHWR**	Pressurized Heavy Water Reactor
pH	(Potential of Hydrogen) - Negative	**PHYS**	Physical
	Logarithm of Hydrogen-Ion Activity		Physicist
PHA	Phaseolus Vulgaris Agglutinin		Physics
	Phytohemagglutinin	**PHYSBE**	Physiological Simulation Benchmark
	Pulse Height Analyzer		Experiment
PHARM	Pharmaceutical	**PHYSCHEM**	Physical Chemistry
	Pharmacology	**PHYSMET**	Physical Metallurgy
	Pharmacy	**PhysS**	Physical Society
PH BRZ	Phosphor Bronze	**PI**	Paper Insulation

	Parallel Input
	Particular Integral
	Performance Index
	Pilotless Interceptor
	Plastics Institute
	Point Initiating
	Point Insulating
	Point of Intersection
	Polyimide
	Position Indicator
	Positive Input
	Principal Investigator
	Priority Interrupt
	Productivity Index
	Program Indicator
	Program Interface
	Program Interrupt(ion)
	Programmed Instruction
	Proportional Integral
	Propyl Isome
Pi	Inorganic Phosphate
PIA	Peripheral Interface Adapter
	Petroleum Incentives Administration
	Plastics Institute of America
	Pre-Installation Acceptance
	Printing Industries of America
PIAC	Petroleum Industry Advisory Council
PIAD	Plastics in Automotive Design [of SME]
PIAM	Petroleum Industry Application of Microcomputers
PIANC	Permanent International Association of Navigation Congresses
PIARC	Permanent International Association of Road Congresses
PIB	Petroleum Information Board
	Polar Ionosphere Beacon
	Polyisobutylene
	Polytechnic Institute of Brooklyn [US]
	Pyrotechnic Installation Building
PIBAC	Permanent International Bureau of Analytical Chemistry
PIBAL	Polytechnic Institute of Brooklyn, Aerodynamics Laboratory [US]
PIBMRI	Polytechnical Institute of Brooklyn, Microwave Research Institute [US]
PIBUC	Pilot Backup Control
PIBS	Polar Ionosphere Beacon Satellite [NASA]
PIC	Particle-in-Cell
	Photographic Interpretation Center
	Plastic-Insulated Cable
	Plastic-Insulated Conductor
	Polyethylene-Insulated Conductor
	Polymer Impregnated Concrete
	Power Information Center
	Process Interface Control
	Program Information Code
	Program Interrupt Control
PICA	Power Industry Computer Applications
	Public Interest Computer Association
PICAC	Power Industry Computer Applications Conference

PICAO	Provisional International Civil Aviation Organization
PICLS	Purdue International and Computational Learning System [US]
PICS	Personnel Information Communication System
	Production Information and Control System
PID	Photo Ionization Detector
	Program Information Department [of IBM]
	Proportional Integral Derivation
	Proportional-Integral-Derivative
	Proportional Integral Differential
	Pseudo Interrupt Device
PIDA	Phenylindan Dicarboxylic Acid
PIE	Plug-in Electronics
	Pulse Interference Emitting
PIF	Payload Integration Facility
PIFEX	Programmable Image Feature Extractor
PIGA	Pendulous Integrating Gyro Accelerometer
PIGME	Particle Induced Gamma Ray Emission
	Programmed Inert-Gas Multi-Electrode (Welding)
PIL	Percentage Increase in Loss
	Pilot
	Pitt Interpretive Language
PILC	Paper-Insulated Lead-Covered
PILE	Product Inventory Level Estimator
PILOT	Permutation Indexed Literature of Technology
	Piloted Low-Speed Test
PILP	Program for Industrial Laboratory Projects [Canada]
PIM	Peak Integration Method
	Precision Indicator of the Meridian
	Program Integration Manual
	Pulse Interval Modulation
PIMA	Prairie Implement Manufacturers Associations
PIMISS	Pennsylvania Interagency Management Information Support System
PIN	Personal Identification Number
	Plant Information Network
	Position Indicator
	Positive-Intrinsic-Negative (Transistor)
	Power Input
	Product Identification Number
PIND	Particle Impact Noise Detector
PINFET	Positive-Intrinsic-Negative Field-Effect Transistor
PINI	Plug-in Neutral Injector
PINO	Positive Input - Negative Output
PINS	Portable Inertial Navigation System
	Precise Integrated Navigation System
PINSAC	PINS Alignment Console
PIO	Parallel Input/Output
	Pilot Induced Oscillation
	Position Iterative Operation
PIOCS	Parallel Input/Output Control System
PIOSA	Pan-Indian Ocean Science Association
PIOU	Parallel Input/Output Unit
PIP	Payload Integration Plan

	Peripheral Interchange Program
	Personal Identification Project
	Petroleum Incentive Program
	Plant-in-Place
	Plug-in Programmer
	Pollution Information Program
	Predicted Impact Point
	Probabilistic Information Processing
	Problem Input Preparation
	Product Improvement Plan
	Programmable Integrated Processor
	Project on Information Processing
	Pulsed Integrating Pendulum
PIPA	Pulsed Integrating Pendulum Accelerometer
PIPE	Points, Income, Personnel Expense
PIPES	Piperazinebisethanesulfonic (Acid)
PIPO	Parallel-in, Parallel-out
PIPR	Plant-in-Place Records
PIRA	Paper Industries Research Association
PIRD	Program Instrumentation Requirement Document
PIRN	Preliminary Interface Revision Notice
PIROGAS	Plasma Injection of Reducing Overheated Gas System
PIRT	Precision Infrared Tracking
	Precision Infrared Triangulation
	Public Information Retrieval Terminal
PIRV	Programmed Interrupt Request Vector
PISH	Program Instrumentation Summary Handbook
PISO	Parallel-in, Serial-out
	Polyimidesulfone
PISW	Process Interrupt Status Word
PIT	Peripheral Input Tape
	Processing of Index Terms
	Program Instruction Tape
PITA	Parison Inflation Thinning Analysis
PITAC	Pakistan Industrial Technical Assistance Center
PITB	Petroleum Industry Training Board
PITC	Phenylisothiocyanate
PITS	Passive Intercept Tracking System
PIU	Path Information Unit
	Plug-in Unit
	Programmable Interface Unit
PIUMP	Plug-in Unit Mounting Panel
PIV	Peak Inverse Voltage
	Product Inspection Verification
PIX	Picture
PIXE	Particle-Induced X-Ray Emission
	Particle-Induced X-Ray Excitation
	Proton-Induced X-Ray Emission
PIXEL	Picture Element [also: pixel]
PIXI	Professional Industrial X-Ray Imaging
PK	Pack
	Peak
	Pyruvate Kinase
pk	peck [unit]
PKD	Partially Knocked Down
PKG	Package
	Packing

PK/LDH	Pyruvate Kinase/Lactic Dehydrogenase
PKU	Phenylketonuria
PK/VAL	Peak/Valley (Ratio)
PL	Parting Line
	Parts List
	Payload
	Perception of Light
	Phase Line
	Phospholipid
	Photoluminescence
	Pile
	Place
	Plant
	Plate
	Plug
	Production Language
	Program Library
	Programming Language
	Proportional Limit
PLA	Physiological Learning Aptitude
	Polylactic Acid
	Polylalamine
	Programmable Logic Array
	Programmed Logic Array
	Proton Linear Accelerator
PLAAR	Packaged Liquid Air-Augmented Rocket
PLACE	Programming Language for Automatic Checkout Equipment
PLAD	Parachute Low-Altitude Delivery
PLADS	Parachute Low-Altitude Delivery System
PLAN	Program Language Analyzer
	Programming Language Nineteen Hundred
PLANET	Planned Logistics Analysis and Evaluation Technique
PLANIT	Programming Language for Interactive Teaching
PLANN	Plant Location Assistance Nationwide Network [US]
PLANS	Program Logistics and Network Scheduling System
PLAP	Pulsed-Laser Atom Probe
PLAQ	Planned Quantity
PLAS	Plaster
PLASTEC	Plastics Technical Evaluation Center [US Army]
PlasTIPS	Plastics Training and Information to Promote Safety
PLAT	Pilot Landing Aid Television
PLATF	Platform
PLATO	Programmed Logic for Automatic Teaching Operation
PLBG	Plumbing
PLC	Pattern Length Coding
	Power Line Carrier
	Power Line Conditioner
	Power Line Cycle
	Prime Level Code
	Programmable Logic Controller
PLCA	Pipeline Contractors Association
PLCAC	Pipeline Contractors Association of Canada
PLCB	Pseudo-Line Control Block

PLCC	Plastic Leaded Chip Carrier
PLD	Phase-Lock Demodulator
	Programmable Logic Device
	Pulse-Length Discriminator
PLDTS	Propellant Loading Data Transmission System
PLEA	Pacific Lumber Exporters Association [US]
PLENCH	Pliers and Wrench (Combination)
PLF	Parachute Landing Fall
	Plant Load Factor
PL GL	Plate Glass
PLH	Pituitaries Luteinizing Hormone
PL/I	Programming Language/I (I = One)
PLIANT	Procedural Language Implementing Analog Techniques
PLIB	Program Library
PLIM	Post-Launch Information Message
PLL	Peripheral Light Loss
	Phase-Locked Loop
PLM	Pulse Length Modulation
PL/M	Programming Language - Microprocessor
PLMBG	Plumbing
PLMBR	Plumber
PLNG	Planning
PLO	Phase-Locked-Oscillator
	Program Line Organization
	Project Line Organization
PLOD	Planetary Orbit Determination
PLOO	Pacific Launch Operations Office [NASA]
PLOP	Pressure Line of Position
PLOTG	Plotting
PLP	Pattern Learning Parser
	Periodate-Lysine-Paraformalin
PLPS	Propellant Loading and Pressurization System
PLRACTA	Position Location Reporting and Control Tactical Aircraft
PLRS	Position Location Reporting System
PLRTY	Polarity
PLS	Positive Lubrication System
	Private-Line Service
	Propellant Loading System
	Pure Live Seed
PLSS	Portable Life-Support System
	Prelaunch Status Simulator
PLST	Plastering
PLSTG	Plastering
PLSTC	Plastic
PLT	Pilot
	Plant
	Program Library Tape
PLTG	Plating
PLU	Primary Logical Unit
PLUOT	Parts Listing and Used On Technique
PLUS	Program Library Update System
PLUTO	Pipeline under the Ocean
	Plutonium Loop Testing Reactor
PLV	Pitch Line Velocity
PLYWD	Plywood
PLZT	Pb (= Lead)-Lanthanum-Zirconate-Titanate (Ceramics)

PM	Paramagnet
	Performance Management
	Performance Monitoring
	Permanent Magnet
	Permanent Mold
	Phase Modulation
	Physical Metallurgy
	Photomultiplier
	Plane Matching (Interface)
	Plasma Membrane
	Pounds per Minute
	Powder Metallurgy
	Preventive Maintenance
	Procedures Manual
	Pulsating Mixing
	Pulse Modulation
	Pyrometallurgy
P/M	Powder Metallurgy
Pm	Promethium
pm	picometer
	(post meridiem) - afternoon
	(post mortem) - after death
PMA	Permanent Magnet Association
	Permanent Management Arrangements
	Pharmaceutical Manufacturers Association
	Polymetylacrylate
	Precision Measurement Association [US]
	Production Monitoring Analysis
PM-A	Polishing Medium Abrasive
PMAC	Pharmaceutical Manufacturers Association of Canada
	Purchasing Management Association of Canada
PMAD	Polymer Modifiers and Additives Division
PMAF	Pharmaceutical Manufacturers Association Foundation
PMB	Physical Metallurgy Branch
PMBC	Pilot Make Busy Call
	Plywood Manufacturers of British Columbia [Canada]
PMBX	Private Manual Branch Exchange
PMC	Polymer-Reinforced Matrix Composites
	Program Management Control
	Program Marginal Checking
	Pseudo Machine Code
PMCG	Pyrrolidylmethyl Cyclopentylphenylglycolate
PMCS	Pulse-Modulated Communications System
PMD	Parts Manufacturing Division
	Post Mortem Dump [computer]
	Program Module Dictionary
PMDA	Pyromellitic Dianhydride
PMDI	Polymethylene Diisocyanate
PMDR	Phosphorescence Microwave Double Resonance
PMDS	Plant Management and Maintenance Design Engineering Show
PME	Photomagnetoelectric
	Photomagnetoelectric Effect
	Precision Measuring Equipment
	Protective Multiple Earthing
PMEE	Prime Mission Electronic Equipment

PMEL	Precision Measuring Equipment Laboratory		Precision Measurement Unit
PMG	Prediction Marker Generator	PMVI	Periodic Motor Vehicle Inspection
PMGS	Predictable Model Guidance Scheme	PMW	Project Manager Workbench
PMH	Production per Man-Hour	PMX	Physical Modeling Extension
PMHS	Polymethylhydrosiloxane	PN	Paranodal
PMI	Phosphomannose Isomerase		Paranode
	Postmaintenance Inspection		Partition Number
	Preventive Maintenance Inspection		Part Number
	Private Memory Interconnect		Performance Number
	Private Memory Interface		Positive-Negative (Semiconductor)
	Pseudo Matrix Isolation		Pseudonoise
PMIA	Precious Metals Industry Association	PNA	Pacific/North American
P/MIA	Powder Metallurgy Industries Association [also: PMIA]	PNBH	P-Nitrobenzaldehyde Hydrazone
		PNC	Parity Nonconservation
PMIP	Postmaintenance Inspection Pilot		Programmed Numerical Control
PMIS	Personnel Management Information System	PNCH	Punch
	Plant Management Information System	PND	Pictorial Navigation Display
PML	Polymer Microdevice Laboratory	PNdB	Perceived Noise Level in Decibels
PMM	Polymethylmethacrylate	PNDC	Parallel Network Digital Computer
PMMA	Paper Machinery Makers Association	PNE	Pacific National Exhibition
	Polymethylmethacrylate	PNEC	Proceedings of the National Electronics Conference
PMMC	Permanent Magnetic Movable Coil		
PMMI	Packaging Machinery Manufacturing Institute	PNEU	Pneumatic
			Pneumatics
PMN	Premanufacture Notification	PNF	Phosphonitrilic Fluoroelastomer
PMOS	P-Channel Metal-Oxide Semiconductor	PNGCS	Primary Navigation, Guidance, and Control System
	Positive Metal-Oxide Semiconductor		
PMP	Parts-Material-Packaging	PNH	Pan Head
	Phased Manufacturing Program	PNIP	Positive-Negative-Intrinsic-Positive (Transistor)
	Polymethylpentene		
	Premodulation Processor	PNL	Pacific Naval Laboratory [now: DREP]
	Pressure Measurement Package		Panel
	Preventive Maintenance Plan		Perceived Noise Level
	Program Management Plan	PNM	Pulse Number Modulation
	Project Master Plan	PNMT	Phenylethanolamine N-Methyltransferase
PMPA	Powder Metallurgy Parts Association [also: P/MPA]	PNP	Pick-and-Place (Robot)
			Positive-Negative-Positive (Semiconductor)
PMPO	Pulse-Modulated Power Oscillator		Precision Navigation Processor
PMR	Pacific Missile Range [US Air Force]	PNPG	P-Nitrophenylglycerine
	Proton Magnetic Resonance	PNPN	Positive-Negative-Positive-Negative (Semiconductor)
PMRL	Physical Metallurgy Research Laboratories [Canada]		
	Pulp Manufacturers' Research League	PNT	Paint
PMS	Paramethylstyrene		Plasma Nitriding
	Phenazinemethosulfate	PNTD	Personnel Neutron Threshold Detector
	Physical and Mathematical Sciences	PO	Parallel Output
	Processor-Memory-Switch		Parking Orbit
	Program Management System		Patent Office
	Project Management System		Planetary Orbit
	Project Map System		Polymerizable Oligomers
	Public Message Service		Polyolefin
PMSF	Phenylmethanesulfonylfluoride		Positive Output
PMSRP	Physical and Mathematical Sciences Research Paper		Postal Order
			Post Office
PMT	Phenylmercaptotetrazole		Power Oscillator
	Photomultiplier Tube		Printout
	Program Master Tape		Production Order
PMTHP	Project Mercury Technical History Program [NASA]		Program Objective
			Punch On
PMU	Parametric Measurement Unit		Purchase Order
	Portable Memory Unit		Pushout
		P&O	Paints and Oils

	Pickled and Oiled
Po	Polonium
POB	Post Office Box
POBATO	Propellant on Board at Takeoff
POBN	Pyridyloxide Butylnitrone
POC	Metal Powder Cutting
	Process Operator Console
	Procurement Outlook Conference
	Product of Combustion
	Purgeable Organic Carbon
POCP	Program Objectives Change Proposal
POCS	Patent Office Classification System [US]
POD	Point of Origin Device
	Post Office Department [US]
	Preflight Operations Division
PODA	Priority-Oriented Demand Assignment
PODEM	Path-Oriented Decision Making
PODS	Post-Operative Destruct System
POE	Port of Embarkation
POF	Point of Failure
	Pulsed Optical Feedback
POGO	Polar Orbiting Geophysical Observatory
	Programmer-Oriented Graphic Operation
POI	Parking Orbit Injection
	Program of Instruction
POISE	Panel on In-Flight Scientific Experiments
	Photosynthetic Oxygenation Illuminated by Solar Energy
POL	Pacific Oceanographic Laboratory
	Petroleum, Oil and Lubricants
	Poland
	Pole
	Polarization
	Polish
	Problem-Oriented Language
	Procedure-Oriented Language
	Process-Oriented Language
POLANG	Polarization Angle
POLYTRAN	Polytranslation Analysis and Programming
POM	Pivaloyloxymethyl Chloride
	Polyoxymethylene
POMM	Preliminary Operating and Maintenance Manual
POMS	Panel on Operational Meteorological Satellites
POMSEE	Performance, Operational and Maintenance Standards for Electrical Equipment
PONA	Paraffins, Olefins, Naphthenes and Aromatics
PONI	Positive Output, Negative Input
POO	Post Office Order
POP	Percentage of Precipitation
	Perpendicular Ocean Platform
	Population
	Power On/Off Protection
	Printing Out Paper
	Programmed Operators and Primitives
	Program Operating Plan
POPO	Push-On, Pull-Off
POPOP	Phenyloxazolyl-Phenyloxazolyl-Phenyl
POPS	Pantograph Optical Projection System

POPSI	Precipitation and Off-Path Scattered Interference
POPSO	Piperazine Hydroxypropanesulfonic Acid
POR	Pacific Ocean Region
	Payable on Receipt
	Problem-Oriented Routine
PORC	Porcelain
PORM	Plus or Minus
PORS	Portable Oil Reclamation System
PORT	Photo-Optical Recorder Tracker
	Portable
	Portugal
	Portuguese
POS	Point of Sale
	Position
	Positive
	Primary Operating System
	Product-of-Sums
POSI	Parallel-out, Serial-in
POSS	Passive Optical Satellite Surveillance
	Photo-Optical Surveillance Subsystem
	Polar Orbiting Satellite System [US]
	Possible
POSSUM	Polar Orbiting Satellite System - University of Michigan [US]
POST	Power-On Self-Test
POSTP	Postprocessor
POT	Plain Old Telephone
	Potential
	Potentiometer
	Pottery
POTC	PERT (= Program Evaluation Review Technique) Orientation and Training Center [USDOD]
POTF	Polychromatic Optical Thickness Fringes
POTW	Publicly-Owned Treatment Works
POT W	Potable Water
POUT	Power Output
POWD	Powder
POWS	Pyrotechnic Outside Warning System
POV	Peak Operating Voltage
POWTECH	International Power and Bulk Solids Technology Exhibition and Conference
PP	Pages
	Panel Point
	Parallel Processor
	Peripheral Processor
	Physical Properties
	Plane Parallel
	Polypropylene
	Postprocessor
	Pounds Pressure
	Power Pack
	Preprocessor
	Present Position
	Pressureproof
	Print/Punch
	Professional Purchaser
	Program Preparation
	Propylene Plastic
	Push-Pull

	Pyrophosphate
P/P	Point-to-Point
P-P	Peak-to-Peak
pp	(punctum proximum) - near point
PPA	Phenylpyruvic Acid
	Photo-Peak Analysis
	Polyphenylacetylene
	Professional Programmers Association
pPa	picopascal [unit]
PPAAR	Princeton University/Pennsylvania University Army Avionics Research [US]
PPB	Powder Particle Boundary
	Professional Public Buyer
	Pyrethrins Piperonyl Butoxide
ppb	parts per billion
PPBS	Planning, Programming and Budgeting System
PPC	Paperboard Packaging Council [US]
	Pollable Protocol Converter
	Pulsed Power Circuit
PPD	Diphenyloxadiazole
	Pulse-Type Phase Detector
PPDC	Programming Panels and Decoding Circuits
PPDD	Plan Position Data Display
PPDR	Pilot Performance Description Record
PPE	Polyphenylether
	Premodulation Processing Equipment
	Problem Program Efficiency
PPEM	Parallel Plate Electron Multiplier
PPF	Payload Processing Facility [US Air Force]
PPFRT	Prototype Preliminary Flight Rating Test
PPG	Pacific Proving Grounds [USAEC]
	Preprinted Gothics
	Program Pulse Generator
	Propulsion and Power Generator
PPH	Pulses per Hour
pph	parts per hundred
PPHMDSO	Plasma-Polymerized Hexamethyldisiloxane
PPI	Pictorial Position Indicator
	Planar Plug-In
	Plan-Position Indicator
	Plastic Pipe Institute [US]
	Plastic Pronged Knob
	Purpose Parallel Interface
P&PI	Pulp and Paper Industry
PPi	Inorganic Pyrophosphatase
PPL	Polymorphic Programming Language
	Preferred Parts List
PPM	Periodic Permanent Magnet
	Peak Program Meter
	Planned Preventive Maintenance
	Press Piercing Mill
	Pulse-Phase Modulation
	Pulse-Position Modulation
	Pulses per Minute
ppm	parts per million
PPM/MPM	Press Piercing Mill/Multistand Pipe Mill
PPMS	Program Performance Measurement System
PPO	Diphenyloxazole
	Polyphenoloxidase
	Polyphenylene Oxide

	Precedence Partition and Outdegree
PPP	Peak Pulse Power
	Poly-P-Phenylene
3P	Pollution Prevention Pays (Program) [Canada]
PPPL	Princeton Plasma Physics Laboratory [US]
PPQA	Pageable Partition Queue Area
PPR	Paper
	Photo-Plastic Recording
PPRA	Precedence Partition and Random Assignment
PPRBD	Paperboard
PPRI	Pulp and Paper Research Institute [US]
PPRIC	Pulp and Paper Research Institute of Canada
PPS	Parallel Processing System
	Parcel Processing System
	Phosphorous Propellant System
	Plant Protection System
	Polyphenylene Sulfide
	(Post Postscriptum) - Second Postscript
	Primary Propulsion System
	Pulses per Second
PPT	Pattern Processing Technology
	Phenylalanine-Pyruvate Transaminase
	Polymer Production Technology
	Processing Program Table
	Punched Paper Tape
ppt	parts per trillion
PPTP	Poly-P-Phenyleneterephtalamide
PPTS	Pyridinium-P-Toluenesulfonate
PPU	Peripheral Processing Unit
	Probe Processing Unit
PPV	Polarized Platen Viewer
PPW	Plasma Powder Welding
PPWC	Pulp, Paper and Woodworkers of Canada
PQ	Permeability Quotient
	Physical Quality
	Plastoquinone
	Province of Quebec [Canada]
PQAA	Province of Quebec Association of Architects [Canada]
PQGS	Propellant Quantity Gauging System
PQM	Proto Qualification Model
PQR	Procedure Qualification Record
PR	Pair
	Paper Tape Reader
	Pattern Recognition
	Penicillium Roqueforti
	Periodic Reverse
	Physical Reader
	Pitch Ratio
	Prefix
	Preliminary Record
	Price
	Printer
	Private Renter
	Program Register
	Program Requirement
	Progress Record
	Project Report

	Proportional Representation
	Pseudorandom
	Public Relations
	Puerto Rico [US]
	Pulse Rate
	Pulse Ratio
	(Punctum Remotum) - far point
	Purplish Red
Pr	Prandtl Number
	Praseodymium
PRA	Photochemical Research Associates
	Port Rear Access
	Precision Axis
	Production Reader Assembly
	Program Reader Assembly
	Prompt Radiation Analysis
	Public Roads Administration
PRADOR	PRF Ranging Doppler Radar
PRAI	Phosphoribosyl Anthranilate Isomerase
	Project Research Applicable in Industry
PRB	Panel Review Board
	Primary Resonator Block
PRBAL	Previous Balance
PRBS	Pseudorandom Binary Sequence
PRC	People's Republic of China
	Pierce
	Pittsburgh Research Center [US]
	Point of Reverse Curve
PRCS	Precured Resin-Coated Sand
PRCST	Precast
PRD	Printer Description
	Printer Dump
	Program Requirements Data
	Program Requirements Document
PRDP	Power Reactor Development Program
PRDPEC	Power Reactor Development Program Evaluation Committee
PRE	Prefix
	Protein Relaxation Enhancement
PrE	Printer Emulator
PREAMP	Preamplifier
PREC	Preceding
PRECIP	Precipitate
PREDICT	Prediction of Radiation Effects by Digital Computer Techniques
PREF	Preface
	Prefix
	Propulsion Research Environmental Facility
PREFAB	Prefabrication
PRELIM	Preliminary
PRELORT	Precision Long-Range Tracking (Radar)
PREMID	Programmable Remote Identification
PREP	Plasma Rotating Electrode Process
	Preparation
	Programmed Educational Package
PRES	President
	Program Reporting and Evaluation System
PRESS	Pacific Range Electromagnetic Signature Studies
	Pressure
PRESTO	Program for Rapid Earth-to-Space Trajectory Optimization

	Program Reporting and Evaluation System for Total Operations
PRF	Proof
	Pulse Repetition Frequency
PRFD	Pulse Recurrence Frequency Discrimination
PRFCS	Pattern Recognition Feedback Control System
PRFL	Pressure-Fed Liquid
PRFS	Pulse Recurrence Frequency Stagger
PRI	Photo Radar Intelligence
	Plastics and Rubber Institute
	Priority
	Project Readout Indicator
	Pulse Rate Indicator
PRIDE	Programmed Reliability in Design Engineering
prim	primary [chemicals]
PRIME	Precision Integrator for Meteorological Echoes
	Precision Recovery including Maneuverable Entry
	Program Independence, Modularity and Economy
	Programmed Instruction Form Management Education
PRIN	Principle
PRINCE	Parts Reliability Information Center [of NASA]
	Programmed Reinforced Instruction Necessary to Continuing Education
PRINS	Process Information System
PRIO	Priority
PRIS	Propeller Revolution Indicator System
PRISE	Program for Integrated Shipboard Electronics
PRISM	Programmable Interactive System
	Programmed Integrated System Maintenance
PRL	Parallel
	Photoreactivating Light
	Prairie Regional Laboratory [NRC, Canada]
	Print Lister
PRM	Power Range Monitor
	Pressure Remanent Magnetization
	Pulse Rate Modulation
PRMLD	Premold
PRN	Pseudorandom Noise
	Printer
	Pulse Ranging Navigation
PRNTR	Printer
PRO	Professional
Pro	Proline
PROB	Problem
PROC	Process(ing)
	Processor
	Proceedings
	Procedure
	Programming Computer
PROCLIB	Procedure Library
PROCOMP	Process Computer
	Program Compiler

PROD	Product(ion)
PRODAC	Production Advisers Consortium [UK]
	Programmed Digital Automatic Control
PROF	Professional
Prof	Professor
PROFAC	Propulsive Fluid Accumulator
PROFILE	Programmed Functional Indices for Laboratory Evaluation
PROFIT	Program for Financed Insurance Techniques
	Programmed Reviewing, Ordering, and Forecasting Inventory Technique
PROFS	Professional Office System
PROG	Program(mer)
	Progression
PROGDEV	Program Device
PROGR	Programmer
PROJ	Project
	Projectile
PROJACS	Project Analysis and Control System
PROLAN	Processed Language
PROLOG	Programming in Logic
PROM	Pockels Readout Optical Memory
	Programmable Read-Only Memory
PROMIS	Process Management Information System
PROMPT	Production Reviewing, Organizing and Monitoring of Performance Techniques
	Program Monitoring and Planning Technique
	Program Reporting, Organization and Management Planning Technique
PROMT	Precision Optimized Measurement Time
PRONTO	Program for Numerical Tools Operation
PROP	Performance Review for Operational Programs
	Planetary Rocket Ocean Platform
	Profit Rating of Projects
	Propeller
	Property
	Proposition
PROPHOS	Bisdiphenylphosphinopropane
PROS	Professional Reactor Operator Society
PROSPER	Profit Simulation Planning and Evaluation of Risk
PROSPRO	Process Systems Program
PROSUS	Program on Submicrometer Structures [US]
PROT	Protection
PROTECT	Probabilities Recall Optimizing the Employment of Calibration Time
pro tem	(pro tempore) - temporarily
PROV	Provision
	Province
PROVER	Procurement for Minimum Total Cost through Value Engineering and Reliability
	Procurement, Value, Economy, Reliability
PROWORD	Procedure Word
PROXI	Projection by Reflection Optics of Xerographic Images
PROXYL	Tetramethylpyrrolidine-N-Oxyl
PRP	Platelet-Rich Plasma
	Preparation
	Pseudorandom Pulse

	Pulse Repetition Period
PRPM	Primary Power Monitor
PRPN	Propane
PRPP	Phosphorylribose Pyrophosphate
PRPQ	Programming Request Price Quotation
PRR	Pulse Repetition Rate
PRS	Pattern Recognition Society
	Pattern Recognition System
	Portable Rework System
	Press
PRT	Pattern Recognition Technique
	Personal Rapid Transit
	Phosphoribosyl Transferase
	Platinum Resistance Thermometer
	Portable Remote Terminal
	Primary Ranging Test
	Printer
	Program Reference Table
	Pulse Recurrence Time
	Pulse Repetition Time
PRTC	Ports Canada
PRTG	Printing
PRTLND CEM	Portland Cement
PRTM	Prague Ring Tunneling Method
PRTOT	Prototype Real-Time Optical Tracker
PRTR	Plutonium Recycling Test Reactor
	Printer
PRTS	Pseudorandom Ternary Sequence
PRTY	Priority
PRU	Programs Research Unit
PRUF	Program of Research by Universities in Forestry
	Program Request under Format
PRV	Peak Reverse Voltage
	Pressure Reducing Valve
PRVT	Production Reliability Verification Testing
PRW	Percent Rated Wattage
PS	Parity Switch
	Passenger Steamer
	Payload Specialist
	Permanent Stress
	Phase Separation
	Phase Shift
	Phasing System
	Physical Sciences
	Planetary Society
	Plasma Screen
	Plasma Spray(ing Technique)
	Point of Switch
	Polystyrene
	(Post Scriptum) - Postscript
	Potentiometer Synchro
	Power Source
	Power Supply
	Presentation Services
	Pressure Switch
	Problem Solution
	Problem Specification
	Program Store
	Programming System
	Proportional Space

	Proton Synchrotron	**PSGC**	Puget Sound Governmental Conference [US]
P/S	Processed and Sintered	**PSI**	Packetnet Systems Interface
P-S	Pressure-Sensitive		Pakistan Standards Institution
P&S	Port and Starboard		Phase-Shifting Interferometry
ps	picosecond		Pounds per Square Inch
PSA	Parametric Semiconductor Amplifier		Preprogrammed Self-Instruction
	Petroleum Services Association		Present Serviceability Index
	Pisum Sativum Agglutinin	**PSIA**	Pounds per Square Inch, Absolute
	Power Servo Amplifier	**PSID**	Pounds per Square Inch, Differential
	Prefix Storage Area	**PSIG**	Pounds per Square Inch, Gauge
	Pressure-Sensitive Adhesive	**PSIM**	Power System Instrumentation and
	Pressure-Sensitive Adsorption		Measurement [IEEE Committee]
	Problem Specification Analyzer	**PSK**	Phase-Shift Key(ing)
	Pushdown Stack Automaton	**PSL**	Photographic Science Laboratory
PSAC	Petroleum Services Association of Canada		Physical Sciences Laboratory
	President's Science Advisory Committee		Problem Specification Language
	[US]		Problem Statement Language
PSAD	Prediction, Simulation, Adaption, Decision	**P SL**	Pipe Sleeve
PSALI	Permanent Supplementary Artificial	**PSL/PSA**	Problem Statement Language/Problem
	Lighting Installation		Specification Analyzer
	Permanent Supplementary Artificial	**PSM**	Parallel Slit Map
	Lighting of Interiors	**PSMD**	Photoselective Metal Deposition
PSAR	Preliminary Safety Analysis Report	**PSMDE**	Pseudostationary Mercury Drop Electrode
	Programmable Synchronous/Asynchronous	**PSMR**	Parts Specification Management for
	Receiver		Reliability
PSAT	Programmable Synchronous/Asynchronous	**PSMS**	Permanent Section of Microbiological
	Transmitter		Standardization [of IAMS]
PSB	Parallel System Bus	**PSN**	Print Sequence Number
	Persistent Slip Band		Private Satellite Network
	Program Specification Block		Programmable Sampling Network
PSBLS	Permanent Space Based Logistics System		Public Switched Network
PSC	Power Supply Circuit	**PSNS**	Programmable Sampling Network Switch
	Power System Communication		Puget Sound Naval Shipyard [US]
	Program Sequence Control	**PSO**	Pilot Systems Operator
	Public Service Commission		Polysulfone
	Pulse Signaling Circuit	**P SOL**	Partly Soluble
PSCF	Primary System Control Facility	**PSP**	Paper Tape Space
PSCC	Power System Computation Conference		Planet Scan Platform
PSD	Phase-Sensitive Demodulator		Planned Schedule Performance
	Photon-Stimulated Ion Desorption		Planned Standard Programming
	Position-Sensitive Detector		Plasma Spraying
	Positive Sensitive Detector		Power System Planning
	Power Spectral Density		Programmable Signal Processor
	Prevention of Significant Deterioration	**PSPC**	Position Sensitive Proportional Counter
	Pulse Shape Discrimination	**PSPO**	Public Safety Project Office [NRC, Canada]
PSDC	Purge Sample, Detect and Calibrate	**PSR**	Power System Relaying [IEEE Committee]
PSDF	Propulsion System Development Facility		Pre-Soak Rail
PSDN	Packed-Switched Data Network	**PSRB**	Post-Sintered Reaction-Bonded
PSE	Packet Switching Exchange	**PSRBSN**	Post-Sintered Reaction-Bonded Silicon
	Power System Engineering		Nitride
PSECT	Phototype Section	**PSRP**	Physical Sciences Research Paper
PSEE	Photostimulated Exoelectron Emission	**PSS**	Packet Switching Service
psec	picosecond		Personal Signaling System
PSEP	Passive Seismic Experiment Package		Power Supply System
PSEUD	Pseudonym		Production Support System
PSF	Passive Solar Foundation		Proprietary Software Systems
	Permanent Signal Finder		Propulsion Support System
	Polysulfone	**PSSC**	Physical Sciences Study Committee
	Pounds per Square Foot	**PSSD**	Position-Sensitive Scintillation Detector
PSG	Phosphorus-Doped SiO_2 Glass	**PSSG**	Physical Sciences Study Group
	Phosphosilicate Glass	**PSSN**	Pressure Sintered Silicon Nitride

PST	Pacific Standard Time
	Paired Selected Ternary
	Partition Specification Table
	Photostress Technology
	Point of Spiral Tangent
	Polished Surface Technique
	Polycrystal Scattering Topography
	Post-Stimulus Time
	Pressure Sensitive Tape
	Provincial Sales Tax [Canada]
PSTA	Packaging Sciences and Technology Abstract
PSTC	Pressure Sensitive Tape Council [US]
PSU	Pennsylvania State University [US]
	Port Sharing Unit
	Power Supply Unit
	Primary Sampling Unit
PSV	Program Status Vector
	Pulverization Sous Vede
PSVT	Passivate
PSW	Pipe Socket Weld
	Program Status Word
PSZ	Partially-Stabilized Zirconia
PT	(Liquid) Penetrant Testing
	Paper Tape
	Part
	Patrol Torpedo
	Payment
	Performance Test
	Pint
	Pipe Tap
	Point
	Point of Tangency
	Point of Tangent
	Port
	Positional Tolerancing
	Potential Transformer
	Pressure - Temperature (Diagram)
	Propellant Transfer
	P-Terphenyl
	Pulse Time(r)
	Punched Tape
P&T	Posts and Timbers
Pt	Platinum
pt	pint [unit]
	point [unit]
PTA	Phenyltrimethylammonium
	Phosphotungstic Acid
	Planar Turbulence Amplifier
	Plasma Transferred Arc
	Programmed Time of Arrival
	Power Transfer Assembly
	Pulse Torquing Assembly
PTAB	Photographic Technical Advisory Board [US]
PTACV	Prototype Tracked Air Cushion Vehicle
PTAH	Phosphotungstic Acid Hematoxylin
PTB	Patrol Torpedo Boat
PTBR	Punched Tape Block Reader
PTC	Pacific Telecommunications Council
	Passive Thermal Control
	Personnel Transfer Capsule

	Phenylthiocarbamide
	Plasma Thromboplastin Component
	Positive Temperature Coefficient
	Postal Telegraph Cable
	Power Toggle Clamp
	Programmed Transmission Control
	Pulse Time Code
PTCR	Pad Terminal Connection Room
	Positive Temperature Coefficient of Resistivity
PTCS	Propellant Tanking Computer System
PTD	Painted
PTDA	Power Transmission Distributors Association [US]
PTEC	Plastics Technical Evaluation Center [US Army]
PTERM	Physical Termination
PTF	Phase Transfer Function
	Program Temporary Fix
PTFE	Polytetrafluoroethylene
PTFS	Precision Temperature Forcing System
PTG	Painting
	Printing
PTG STD	Petrograd Standard
PTH	Parathyroid Hormone
	Phenylthiohydantoin
	Plated Through-Hole
PTI	Presentation of Technical Information
PTIC	Patent and Trademark Institute of Canada
PTIO	Pesticides Technical Information Office
PTK	Plastic Tapered Knob
PTL	Power Transmission Line
	Process and Test Language
PTLBD	Particleboard
PTLU	Pretransmission Line-Up
PTM	Performance Test Model
	Proof Test Model
	Pulse Time Modulation
	Pulse Time Multiplex
PTMEG	Polytetramethane Glycol
PTML	PNPN (= Positive-Negative-Positive-Negative) Transistor Magnetic Logic
PTMO	Polytetramethane Oxide
PTMT	Polytetramethylene Terephthalate
	Preliminary Thermomechanical Treatment
PTN	Partition
PTO	Patent and Trademark Office
	Please Turn Over
	Power Takeoff
PTP	Paper Tape Punch
	Point-to-Point (Programming)
	Preferred Target Point
pTpT	Phosphoryl-Thymidylyl-Thymidine
PTR	Paper Tape Reader
	Part Throttle Reheat
	Pool Test Reactor
	Position Track Radar
	Power Transformer
	Pressure Tube Reactor
	Printer

	Processor Tape Read
PTRM	Partial Thermoremanent Magnetization
PTS	Permanent Threshold Shift
	Pneumatic Test Set
	Power Transient Suppressor
	Program Test System
	Propellant Transfer System
	Pure Time Sharing
PTT	Post, Telephone and Telegraph
	Precipitation-Time-Temperature (Curve)
	Program Test Tape
	Prothrombin Time
	Push to Talk
PTTRN	Pattern
PTV	Passenger Transfer Vehicle
	Predetermined Time Value
PTVA	Propulsion Test Vehicle Assembly
PTW	Pressure Thermit Welding
PTX	Pressure - Temperature - Concentration (Diagram)
PTY	Party
PU	Physical Unit
	Pick Up
	Polyurethane
	Power Unit
	Processing Unit
	Propellant Utilization
	Propulsion Unit
	Purdue University [US]
Pu	Plutonium
PUB	Publication
PUBN	Publication
PUC	Public Utilities Commission
PUCP	Physical Unit Control Point
PUCK	Propellant Utilization Checkout Kit
PUCOT	Piezoelectric Ultrasonic Composite Oscillator Technique
PUCS	Propellant Utilization Control System
PUFA	Polyunsaturated Fatty Acid
PUFFT	Purdue University Fast FORTRAN Translator [US]
PUG	PASCAL Users Group
PUGS	Propellant Utilization Gauging System
PULL B SW	Pull Button Switch
PULV	Pulverizer
PUN	Phenolic Urethane No-Bake (Binder)
	Punch
PUNC	Practical, Unpretentious, Nomographic Computer
	Punctuation
PUP	Peripheral Unit Processor
	Plutonium Utilization Program
PUR	Polyurethane
	Purchase
	Purdue University Research [US]
PURCH	Purchase
PUREX	Plutonium-Uranium Reduction and Extraction
PURP	Purple
PURPA	Public Utilities Regulatory Policy Act [US]
PURSUIT	Purchaser/Supplier Information Transfer

PURV	Powered Underwater Research Vehicle
PUSAS	Proposed USA (= United States of America) Standard
PUSS	Pilot's Universal Sighting System
PUT	Programmable Unijunction Transistor
PV	Photovoltaic
	Pilot Vessel
	Pipe Ventilated
	Plan View
	Pore Volume
	Positive Volume
	Present Value
	Pressure Vessel
	Pressure - Volume (Diagram)
	Pressure - Velocity (Diagram)
P/V	Peak-to-Valley (Ratio)
PVA	Pivalic Acid
	Plan View Area
	Polyvinyl Alcohol
PVAC	Polyvinyl Acetate
PVAL	Polyvinyl Alcohol
PVB	Polyvinyl Butyral
PVC	Pigment Volume Concentration
	Polyvinyl Chloride
	Position and Velocity Computer
	Pulse Voltage Converter
PVD	Paravisual Director
	Physical Vapor Deposition
	Plan Video Display
	Plan View Display
PVDC	Polyvinylidene Chloride
PVDF	Polyvinylidene Fluoride
PVF	Polyvinyl Fluoride
PVFM	Polyvinylformal
PVHD	Peripheral Vision Horizon Device
PVK	Polyvinyl Carbazole
PVME	Polyvinyl Methyl Ether
PVOR	Precision VHF (= Very High Frequency) Omnirange
PVP	Present Value Profit
	Polyvinylpyrrolidone
PVPDC	Polyvinylpyridinium Dichromate
PVP-I	Polyvinylpyrrolidone Iodine
PVR	Precision Voltage Reference
PVRC	Pressure Vessel Research Committee
PVS	Performance Verification System
	Program Validation Services
PVT	Physical Vapor Transport
	Polyvinyl Toluene
	Pressure Viscosity Test
	Pressure - Volume - Temperature (Diagram)
PVTX	Pressure - Volume - Temperature - Composition (Diagram)
PVTOS	Physical Vapor Transport of Organic Solutions
PVW	Process Validation Wafer
PW	Packed Weight
	Password
	Plain Washer
	Prime Western
	Printed Wiring

	Pulsewidth
pW	picowatt
PWB	Printed Wire Board
	Printed Wiring Board
PWC	Pulse-Width Coded
PWD	Powder
	Power Distributor
	Public Works Department
	Pulsewidth Discriminator
PWE	Pulse-Width Encoder
PWF	Present Worth Factor
PWHT	Postweld Heat Treatment
PWI	Pilot Warning Indicator
	Pilot Warning Instrument
	Proximity Warning Indicator
PWL	Piecewise Linear
	Power Level
PWM	Pokeweed Mitogen
	Pulse-Width Modulation
	Pulse-Width Multiplier
PWM-FM	Pulse-Width Modulation - Frequency Modulation
PWP	Personal Word Processor
	Programmable Weld Positioner
pWp	Picowatt, psophometrically weighted
pWp0/pWpP	Picowatt, psophometrically weighted measured at a point of 0 (= zero) reference level
PWR	Power
	Pressurized Water Reactor
PWR-FLECHT	Pressurized Water Reactor - Full Length Emergency Core Heat Transfer
PWT	Propulsion Wind Tunnel
pwt	pennyweight [unit]
PX	Post Exchange
	Pressure - Concentration (Diagram)
PXSTR	Phototransistor
PYPH	Polyphase
PYR	Pyrometer
Pyr	Pyridine
PYRO	Pyrotechnic(s)
PYROMET	Pyrometallurgy
PZ	Pick-Up Zone
PZC	Potential of Zero Charge
PZT	Lead-Zirconate-Titanate
	Piezoelectric Transducer
	Piezoelectric Translator

Q

Q	Glutamine
	Quality
	Quantity
	Quebec [Canada]
	Query
	Queue
	Question
	Quire [sheets of paper]

	Quotient
q	quasi
	quart
	quarterly
	quarto
QA	Quality Assurance
	Query Analyzer
QADS	Quality Assurance Data System
QAE	Quarternary Aminoethyl
QAK	Quick-Acting Knob
QAM	Quadrature Amplitude Modulation
	Queued Access Method
QAM-PAM	Quadrature Amplitude Modulation - Pulse Amplitude Modulation
QAMA	Quebec Asbestos Mining Association [Canada]
QAO	Quality Assurance Office [of UN]
QA/QC	Quality Assurance/Quality Control
QAT	Quality Action Team
QAVC	Quiet Automatic Volume Control
QB	Quick-Break
QBE	Query by Example
Q-BOP	Quiet-Quick-Quality Basic Oxygen (Steelmaking) Process
QBS	Quebec Bureau of Statistics [Canada]
QC	Quality Control
	Quasi-Cleavage
	Quiesce-Completed
QCB	Queue Control Block
QCE	Quality Control Engineer(ing)
	Quality Control Evaluation
QCI	Quality Counsellor to Industry
QCM	Quartz Crystal Microbalance
QCO	Quebec Construction Office [Canada]
QCONF	Quantity Conversion Factor
QCR	Quality Control Reliability
	Quick Change Response
QCSEE	Quiet Clean STOL (= Short Takeoff and Landing) Experimental Engine
QC&T	Quality Control and Test
QD	Quick Disconnect
QDC	Quick Dependable Communications
	Quick Disconnect Coupling
QDRI	Qualitative Development Requirements Information [US Army]
QE	Quiesce (Communication)
QEA	Queue Element Area
QEC	Quick Engine Change
QED	Quantum Electrodynamics
	Quick Text Editor
qed	(quod erat demonstrandum) - which was to be demonstrated
QEDC	Quebec Engineering Design Competition [Canada]
qef	(quod erat faciendum) - which was to be done
QEL	Quality Element
QEP	Quality Evaluation Package
QETE	Quality Engineering Test Establishment
QF	Qabel Foundation
	Quality Factor

	Quick Firing
QFIRC	Quick Fix Interference Reduction Capability
QFITC	Quinolizino-Substituted Fluorescein Isothiocyanate
QFL	Quebec Federation of Labour [Canada]
1/4 H	Quarter-Hard
QI	Quinoline-Insoluble
QIA	Quantitative Image Analysis
QIC	Quarter Inch Compatibility
QIRC	Quebec Industrial Research Center [Canada]
QIRI	Quebec Industrial Relations Institute [Canada]
QISAM	Queued Indexed Sequential Access Method
QIT	Quality Information and Test System
QIUF	Quebec Industrial Union Federation [Canada]
QL	Query Language
QLAP	Quick Look Analysis Program
QLDS	Quick Look Data Station
QLIT	Quick Look Intermediate Tape
QLS	Quebec Land Surveyor [Canada]
QLTY	Quality
QLY	Quality
QM	Quadrature Modulation
	Qualification Model
	Quality Management
	Quartermaster
	Quick-Make
QMDO	Qualitative Material Development Objective
QMF	Query Management Facility
QMI	Qualification Maintainability Inspection
	Quality Management Institute [US]
QMMA	Quebec Metals Mining Association [Canada]
QMQB	Quick-Make, Quick-Break
QMR	Qualitative Material Requirement
QMRP	Quality Management Registration Program
QMS	Quadrupole Mass Spectrometry
QN	Question
	Quotation
QNTY	Quantity
QO	Quinoline Oxide
QOD	Quick-Opening Device
Q1D	Quasi-One-Dimensional
QP	Quasi Peak
QPDM	Quad Pixel Dataflow Manager
QPL	Qualified Products List
QPP	Quantized Pulse Position
QPPM	Quantized Pulse Position Modulation
QPQ	Quench-Polish-Quench (Technique)
QPS	Quantitative Physical Science
QPSK	Quadra-Phase Shift Keyed
QQ	Quill and Quire
qqr	(quae vide) - which see
QR	Quality and Reliability
	Quantity Received
	Quarter
	Quasi-Random
	Quick Reaction
QRA	Quality Reliability Assurance
	Quick Reaction Alert
QRBHCA	Quebec Road Builders and Heavy Construction Association [Canada]

QRBM	Quasi-Random Band Model
QRC	Quick Reaction Capability
QRCC	Query Response Communications Console
1/4 RD	Quarter-Round
QRI	Qualitative Requirements Information
QRPG	Quebec Rubber and Plastics Group [Canada]
QRSS	Quasi-Random Signal Source
QRTLY	Quarterly
QRY	Quarry
QS	Quadrupole Splitting
QSAM	Queued Sequential Access Method
QSATS	Quiet Shorthaul Air Transportation System
QSE	Qualified Scientists and Engineers
QSEE	Quiet STOL (= Short Takeoff and Landing) Experimental Engine
QSG	Quasi-Stellar Galaxy
QSL	Queneau-Schuhmann-Lurgi (Process)
QSO	Quasi-Stellar Object
QSOP	Quadripartite Standing Operating Procedures
QSRA	Quiet STOL (= Short Takeoff and Landing) Research Aircraft
QSRS	Quasi-Stellar Radio Sources
QSS	Quasi-Stellar Sources
QSTOL	Quiet Short Takeoff and Landing
QT	Quadruple Thermoplastic (Wire)
	Quantity
	Quering Theory
	Quotient
qt	quart
QTA	Quebec Trucking Association [Canada]
QTAM	Queued Telecommunications Access Method
QTB	Quarry Tile Base
QTF	Quarry Tile Floor
QTM	Quantitative Television Microscope
QTO	Quarto
QTR	Quarry Tile Roof
	Quarter
QTRCD	Quarter Code
QTS	Quarter Turn Screw
QTY	Quantity
QTYOH	Quantity on Hand
QTYOO	Quantity on Order
QTYOR	Quantity Ordered
QTYSH	Quantity Shipped
QTZ	Quartz
QU	Quarterly
	Quasi
	Queen's University [Canada]
	Query
	Question
qu	quart
QUAD	Quadrant
	Quadruple
	Quadrangle
QUAL	Quality
QUAM	Quadrature Amplitude Modulation
QUANTRAS	Question Analysis Transformation and Search
QUAR	Quarry
	Quarter

QUASAR	Quasi-Stellar [also: quasar]
Que	Quebec [Canada]
QUEENSL	Queensland [Australia]
QUEN/TEMP	Quenched and Tempered
QUESAR	Quasi-Elliptical Self-Adaptive Rotation
QUEST	Quality Electrical Systems Test
QUIC	Quality Unit Inventory Control
QUICK	Queen's University Interpretative Coder, Kingston [Canada]
QUICO	Quality Improvement through Cost Optimization
QUIP	Query Interactive Processor
	Questionnaire Interpreter Program
QUOT	Quotation
qv	(quod vide) - which see
QVT	Quality Verification Testing
QW	Quantum Well
QWB	Quantum Well Box
QWL	Quality of Work Life
QWW	Quantum Well Wire
QY	Query
QZP	Quadruple Zooming Power

R

R	Arginine
	(Outside) Radius
	Radical
	Railway
	Rankine
	Rare Earth
	Ratio
	Read(er)
	Receiver
	Recommendation
	Record
	Register
	Report
	Relation
	Reliability
	Request
	Research
	Resistance
	Resistor
	Reaumur [unit]
	Ring
	Rockwell
	Roentgen
	Rubber
r	(inside) radius
	rare
	reset
	right
RA	Radar Altimeter
	Radioactive
	Random Access
	Rate of Application
	Read Amplifier

	Rear Access
	Record Address (File)
	Recrystallization Annealing
	Reducing Adapter
	Reduction in Area
	Replacement Algorithm
	Research Association
	Reserve Alkalinity
	Return Address
	Return Air
	Right Ascension
	Root Addressable (Record)
	Rosin-Activated
Ra	Radium
	Rayon
RAA	Random Access Array
	Remote Axis Admittance
RAAP	Residue Arithmetic Associative Processor
	Resource Allocation and Planning
RABPCVM	Research Association of British Paint, Color and Varnish Manufacturers
RAC	Railway Association of Canada
	Reliability Analysis Center [US]
	Remote Access and Control
	Remote Automatic Calibration
	Royal Automobile Club
	Rubber Association of Canada
RACC	Radiation and Contamination Control
RACE	Random Access Computer Equipment
	Random Access Control Equipment
	Rapid Automatic Checkout Equipment
RACEP	Random Access and Correlation for Extended Performance
RACES	Radio Amateur Civil Emergency Service
RACF	Resource Access Control Facility
RACIC	Remote Area Conflict Information Center [BMI]
RACON	Radar Beacon
RACS	Remote Access Computing System
	Remote Access Calibration System
	Remote Automatic Calibration System
RACT	Remote Access Computer Technique
RAD	Radial
	Radial Detector
	Radiation Absorbed Dose
	Radiator
	Radical
	Radio
	Radioactivity Detection
	Radius
	Random Access Data
	Random Access Disk
	Rapid Access Device
	Rapid Access Disk
	Ratio Analysis Diagram
	Relative Air Density
rad	radian [unit]
RADA	Random Access Discrete Address
RADAC	Radar Analog/Digital Data and Control
	Rapid Digital Automatic Computing
RADAN	Radar Doppler Automatic Navigator

RADAR	Radio Detecting and Ranging [also: Radar]
RADAS	Random Access Discrete Address System
RADATA	Radar Automatic Data Transmission and Assembly
RADC	Rome Air Development Center [US Air Force]
RADCM	Radar Countermeasures
RADCON	Radar Data Converter
RADEM	Random Access Delta Modulation
RADFAC	Radiation Facility
RADHAZ	Radiation Hazard
RADIAC	Radioactivity Detection, Identification and Computation
	Radioactivity Detection, Indication and Computation
RADIAN	Radial Detector for Ion Beam Analysis
RADIQUAD	Radio Quadrangle
RADIR	Random Access Document Indexing and Retrieval
RADM	Random Access Data Modulation
RADNOTE	Radio Note
RADOME	Radar Dome
RADOP	Radar Operator
RADOPWEAP	Radar Optical Weapons
RADOSE	Radiation Dosimeter Satellite [NASA]
RADPLANBD	Radio Planning Board
RADPROPCAST	Radio Propagation Forecast
RADSO	Radiological Survey Officer
RADTT	Radio Teletype
RADU	Radar Analysis and Detection Unit
RADVS	Radar Altimeter and Doppler Velocity Sensor
RAE	Radio Astronomy Explorer (Satellite)
	Range Azimuth Elevation
	Royal Aircraft Establishment [UK]
RAEN	Radio Amateur Emergency Network
RAER	Range, Azimuth, Elevation and Reproduction
RAES	Remote Access Editing System
	Royal Aeronautical Society [UK]
RAeS	Royal Aeronautical Society [UK]
RAF	Royal Air Force
RAFAX	Radar Facsimile Transmission
RAFT	Radially Adjustable Facility Tube
RAGU	Radio Receiving and Generally Useful
RAH	Radiation Anneal Hardening
RAI	Radioactive Interference
	Random Access and Inquiry
	Royal Architectural Institute [UK]
RAIC	Redstone Arsenal Information Center [US]
	Royal Architectural Institute of Canada
RAID	Remote Access Interactive Debugger
RAIDS	Rapid Availability of Information and Data for Safety
RAIL	Runway Alignment Indicator Lights
RAILS	Remote Area Instrument Landing Sensor
RAIR	Random Access Information Retrieval
RAJ	Reverse Air Jet
RAL	Radio Annoyance Level
	Resorcyclic Acid Lactone
	Riverbend Acoustical Laboratory [US]

RALI	Resource and Land Investigation
RALU	Register Arithmetic and Logic Unit
RALW	Radioactive Liquid Waste
RAM	Radar Absorbing Material
	Radioactive Material
	Radio Attenuation Measurement
	Random Access Memory
	Rapid Access Manual
	Real-Time Aerosol Monitor
	Right Ascension of the Meridan
RAMA	Rome Air Materiel Area [US]
RAMAC	Random Access Method of Accounting and Control [also: Ramac]
RAMAR	Random Access Memory Address Register
RAMARK	Radar Marker
RAMIS	Rapid Access Management Information System
	Rapid Automatic Malfunction Isolation System
RAMONT	Radiological Monitoring
RAMP	Radar Modernization Project
	Radiation Airborne Measurement Program
	Rapid Access of Manufactured Parts
	Raytheon Airborne Microwave Platform
	Rural Abandoned Mine Program
RAMPART	Radar Advanced Measurement Program for Analysis of Reentry Techniques
RAMPS	Resource Allocation and Multi-Project Scheduling
RAN	Reconnaissance/Attack Navigator
RANDAM	Random-Access Nondestructive Advanced Memory
RANDID	Rapid Alphanumerical Digital Indicating Device
RANN	Research Applied to National Needs [US]
RAO	Radio Astronomical Observatory
RAOB	Radio Observation
RAP	Reduced Air Pressure
	Redundancy Adjustment of Probability
	Regional Acceleratory Phenomenon
	Reliable Acoustic Path
	Rocket-Assisted Projectile
RAPCOE	Random Access Programming and Checkout Equipment
RAPCON	Radar Approach Control
RAPEC	Rocket-Assisted Personnel Ejection Catapult
RAPID	Reactor and Plant Integrated Dynamics
	Relative Address Programming Implementation Device
	Research in Automatic Photocomposition and Information Dissemination
RAPO	Resident Apollo Project Office [NASA]
RAPPI	Random Access Plan Position Indicator
RAPRA	Rubber and Plastic Research Association
RAPS	Retrieval Analysis and Presentation System
	Risk Appraisal of Programs System
RAPT	Reusable Aerospace Passenger Transport
RAPTUS	Rapid Thorium-Uranium-Sodium (Reactor)
RAR	Rapid Access Recording
RARC	Regional Administrative Radio Conference
RARDE	Royal Armament Research and Development Establishment [UK]

RARE	Ram Air Rocket Engine	**RBA**	Radar Beacon Antenna
RAREP	Radar Report		Recovery Beacon Antenna
RAS	Radar Advisory Service		Relative Byte Address
	Receptor Analysis System		Road Bitumen Association
	Reliability, Availability, Serviceability	**RBB**	Remazol Brilliant Blue
	Remote Analysis System	**RBC**	Red Blood Cell
	Row-Address Strobe	**RBDE**	Radar Bright Display Equipment
	Royal Aeronautical Society [UK]	**RBE**	Radiation Biological Effectiveness
	Royal Astronomical Society [UK]		Relative Biological Effectiveness
	Rutgers Annihilation Spectrometer	**RBEC**	Roller Bearing Engineers Committee
RASC	Rome Air Service Command [US]	**RBI**	Radar Bearing Indicator
RASPO	Resident Apollo Spacecraft Program Office [NASA]		Ripple-Blanking Input
			Roof-Bolt Inserter
RASS	Rapid Area Supply Support	**RBL**	Radiation Biology Laboratory
	Rotating Acoustic Stereo-Scanner	**RBM**	Real Batch Monitor
RASSR	Reliable Advanced Solid-State Radar		Rock Burst Monitor
RAST	Recover Assist, Secure and Traverse System	**RBMS**	Remote Bridge Management Software
RASTA	Radiation Special Test Apparatus	**RBO**	Rapid Burn-Off
RASTAC	Random Access Storage and Control		Ripple-Blanking Output
RASTAD	Random Access Storage and Display	**RBOT**	Rotating Bomb Oxidation Test
RASTI	Rapid Speech Transmission Index	**RBP**	Registered Business Programmer
RAT	Reactive Acidified Tailings	**RBPSSD**	Ruud-Barrett Position Sensitive Scintillation Detector
	Reliability Assurance Test		
RATA	Rankine Cycle Air Turboaccelerator	**RBR**	Radar Boresight Range
RATAC	Radar Analog Target Acquisition Computer		Roll-Bend-Roll
	Radar Target Acquisition		Rubber
RATAN	Radar and Television Aid to Navigation	**RBS**	Radar Bomb Scoring
RATC	Rate-Aided Tracking Computer		Random Barrage System
RATCC	Radar Air-Traffic Control Center		Rutherford (Ion) Backscattering Spectrometry
RATE	Remote Automatic Telemetry Equipment		
RATEL	Radio-Telephone	**RBSN**	Reaction-Bonded Silicon Nitride
RATER	Response Analysis Tester	**RBSS**	Recoverable Booster Support System
RATG	Radio-Telegraph	**RBT**	Resistance Bulb Thermometer
RATIO	Radio Telescope in Orbit	**RBV**	Relative Bioavailability
RATO	Rocket-Assisted Takeoff		Return Beam Vidicon
RATOG	Rocket-Assisted Takeoff Gear	**RC**	Radio Compass
RATS	Rate and Track Subsystem		Radix Complement
	Reactive Acidified Tailings Study		Range Control
RATSC	Rome Air Technical Service Command [US]		Range Correction
RATSCAT	Radar Target Scatter		Rapid Curing
RATT	Radio Teletype		Read and Compute
RAV	Rising Arc Voltage		Reinforced Concrete
RAVE	Radar Acquisition Visual-Tracking Equipment		Remote Control
			Research Center
RAVEN	Ranging and Velocity Navigation		Resistance - Capacitance
RAVIR	Radar Video Recording		Resolver Control
RAWARC	Radar and Warning Coordination		Revenue Canada
RAX	Remote Access Computing System		Reverse Current
RAYDAC	Raytheon Digital Automatic Computer	**R/C**	Radio Command
RAZEL	Range, Azimuth and Elevation		Radio Control
RAZON	Range and Azimuth Only		Range Clearance
RB	Radar Beacon		Rate of Climb
	Radio Beacon		Resistance-Capacitance
	Read Buffer	**RCA**	Reinforced Concrete Association
	Reducing Bushing		Research Council of Alberta [Canada]
	Resistance Brazing		Royal Canadian Academy
	Rest Button		Royal Canadian Artillery
	Return to Bias	**RCAC**	Royal Canadian Armored Corps
	Roller Bearing	**RCACA**	Royal Canadian Armored Corps Association
R&B	Ring and Ball	**RCAF**	Royal Canadian Air Force
Rb	Rubidium	**RCAG**	Remote Control Air/Ground

RCAT	Radio Code Aptitude Test	**RCS**	Radar Cross-Section
RCBC	Recycling Council of British Columbia [Canada]		Radio Command System
			Radio-Controlled Solar
RCC	Radio Common Channels		Reaction Control Subsystem
	Real-Time Computer Complex		Reaction Control System
	Recovery Control Center		Rearward Communications System
	Remote Center Compliance		Reentry Control System [NASA]
	Remote Communications Center		Registration Control System
	Remote Communications Console		Reloadable Control Storage
	Resistance-Capacitance Coupling		Remote Contact Sensor
	Routing Control Center		Remote Control Station
RCCE	Revenue Canada, Customs and Excise		Remote Control System
RCCM	Remote Carrier-Controlled Modem		Replacement Collision Sequences
RCCPLD	Resistance-Capacitance Coupled		Resin-Coated Sand
RCD	Record		Reverse Circulation System
	Received		Rigid Container Sheet
RCDC	Radar Course-Directing Center		Round-Cornered Square
RCDCD	Record Code	**RCSC**	Research Council on Structural Connections
RCDP	Remote Community Demonstration Program	**RCSS**	Resin-Coated Stainless Steel
RCE	Royal Canadian Engineers	**RCT**	Resolver Control Transformer
RCEI	Range Communications Electronics Instruction		Revenue Canada, Taxation
		RCTL	Rectangular Coaxial Transmission Line
RCEME	Royal Canadian Electrical and Mechanical Engineers		Resistor-Capacitor Transistor Logic
		RCTSR	Radio Code Test Speed on Response
RCF	Radial Centrifugal Force	**RCU**	Road Construction Unit
	Recall Finder	**RCVR**	Receiver
	Refrigeration Centrifuge	**RCWP**	Rural Clean Water Program
	Relative Centrifugal Force	**RCWV**	Rated Continuous Working Voltage
	Remote Call Forwarding	**RD**	Radar Data
	Reverse Column Flotation		Radar Display
	Rolling Contact Fatigue		Radiation Detection
RCG	Relative Cooling Gain		Random Drift
RCGS	Royal Canadian Geological Society		Read
R CHG	Reduced Charge		Read Direct
RCI	Radar Coverage Indicator		Readiness Date
	Rating Core Index		Red
RCIA	Royal Canadian Institute of Architects		Register Drive
RCJ	Reaction Control Jet		Restricted Data
RCL	Radiation Counting Laboratory		Road
	Robot Command Language		Rod
RCLS	Runway Centerline Light System		Rolling Direction
RCM	Radar Countermeasures		Roof Drain
	Radio Countermeasures		Root Diameter
	Royal Canadian Mint		Round
RCMP	Royal Canadian Mounted Police		Rural District
RCN	Royal Canadian Navy	**R&D**	Research and Development
RCO	Reactor Core	**RDA**	Reliability Design Analysis
	Remote Control Oscillator	**R-DAT**	Rotary Head - Digital Audio Tape
	Restoration Control Office	**RDB**	Radar Decoy Balloon
RCOC	Royal Canadian Ordnance Corps		Relational Database
RCOM	Regional Chapter Officers Meeting [ASM, US]		Research and Development Board
		RDC	Rail Diesel Car
RCP	Rapid Cooling Process		Recording Doppler Comparator
	Recognition and Control Processor		Regional Dissemination Center [NASA]
RCPB	Reactor Coolant Pressure Boundary		Reliability Data Center
RCPT	Receptacle		Remote Data Collection
RCR	Reader Control Relay		Rural District Council
	Recrystallization-Controlled Rolling	**RDE**	Radial Defect Examination
RCR	Required Carrier Return		Reactive Deposition Epitaxy
RCRA	Resource Conservation and Recovery Act		Receptor Destroying Enzyme
RCRC	Reinforced Concrete Research Council		Research and Development Effectiveness

	Rotating Disk Electrode
RDF	Radial Distribution Function
	Radio Direction Finder
	Radio Direction Finding
	Record Definition Field
	Refuse-Derived Fuel
	Repeater Distribution Frame
RDG	Resolver Differential Generator
RDH	Recirculating Document Handler
	Roundhead
RDI	Radio Doppler Inertial
RDIA	Regional Development Incentives Act
RDIP	Research Development Incentives Program
RDL	Radiological Defense Laboratory [US Navy]
	Rocket Development Laboratory [US Air Force]
RDM	Recording Demand Meter
	Random
RDMU	Range-Drift Measuring Unit
rDNA	Recombinant Deoxyribonucleic Acid
RDOS	Real-Time Disk Operating System
RDOUT	Readout
RDP	Radar Data Processing
	Research Data Publication
	Research, Development and Production
RD&P	Research, Development and Production
RDPS	Radar Data Processing System
RDRINT	Radar Intermittent
RDR	Radar
	Reader
RDR XMTR	Radar Transmitter
RDS	Relational Data System
	Remote Degassing Station
	Rendezvous and Docking Simulator
	Research Defense Society
	Robust Detection Scheme
RDT	Reactor Development and Technology
	Remote Data Transmitter
	Resource Definition Table
RDT&E	Research, Development, Testing and Evaluation [also: RDTE]
RDTL	Resistor Diode Transistor Logic
RDTR	Radiographic Dielectric Track Registration
	Research Division Technical Report
RE	Radiated Emission
	Ram Effect
	Raw End
	Rare Earth
	Rate Effect
	Reentry
	Reference Electrode
	Relative Efficiency
	Royal Engineer
Re	Real Part
	Reaumur (Temperature)
	Reynolds Number
	Rhenium
R&E	Research and Engineering
REA	Radar Echoing Area
	Rural Electrification Administration
REAC	Reactor

	Reaction
	Reactive
	Reeves Electronic Analog Computer
REACT	Radio Emergency Associated Citizens Teams [US]
	Real-Time Expert Analysis and Control System
	Register-Enforced Automated Control Technique
READ	Radar Echo Augmentation Device
	Real-Time Electronic Access and Display
	Remote Electronic Alphanumeric Display
READAC	Remote Environmental Automatic Data Acquisition Concept
READI	Rocket Engine Analysis and Decision Instrumentation
REALCOM	Real-Time Communication
REAP	Remote Entry Acquisition Package
	Roof Evaluation - Accident Prevention
REAR	Reliability Engineering Analysis Report
REASM	Reassemble
REBA	Relativistic Electron Beam Accelerator
REBAM	Rolling Element Bearing Activity Monitor
REBE	Recovery Beacon Evaluation
REC	Receiver
	Receipt
	Record
	Recorder
	Request for Engineering Change
RECEN	Receivable Entry (Code)
RECD	Received
REC'D	Received
RECFM	Record Format
RECIP	Reciprocate
RECIPE	Recomp Computer Interpretive Program Expediter
RECIRC	Recirculation
RECL	Reclosing
RECM	Recommend
RECMA	Radio and Electronic Component Manufacturers Association
RECOG	Recognition
RECOV	Recovery
RECP	Receptacle
RECPT	Receipt
RECSTA	Receiving Station
RECT	Receipt
	Rectangle
	Rectifier
RED	Radial Electron Distribution
	Reduce(r)
	Reduction
	Reflection Electron Diffraction
REDAP	Reentry Data Processing
REDSOD	Repetitive Explosive Device for Soil Displacement
REDTN	Reduction
REDUPL	Reduplicate(d)
REE	Rare Earth Element
	Reentrant
REECDP	Renewable Energy and Energy Conservation Demonstration Program

REED	Radio and Electrical Engineering Division [of NRCC]
REEP	Range Estimating and Evaluation Procedure
	Regression Estimation of Event Probabilities
REF	Refinery
	Reference
	Refund
REFL	Reflector
REF L	Reference Line
REFNY	Refinery
REFR	Refractory
REFRGN	Refrigeration
REFS	References
REG	Radioencephalogram
	Range Extender with Gain Register
	Region
	Register(ed)
	Registry
	Regular
	Regulator
	Rheoencephalography
REGAL	Range and Evaluation Guidance for Approach and Landing
REG'D	Registered
REGEN	Regeneration
REGIS	Register
REG TM	Registered Trademark
REHVA	Representatives of European Heating and Ventilating Association
REI	Relative Erodability
	Research-Engineering Interaction
REIC	Radiation Effects Information Center
REIL	Runway End Identification Lights
REINF	Reinforce(ment)
REINS	Requirements Electronic Input System
REL	Rapidly Extensible Language
	Rate of Energy Loss
	Relation
	Relay
	Release
	Relief
	Relocation
REM	Rapid Eye Movement
	Rare Earth Metals
	Reflection Electron Microscopy
	Reliability Engineering Model
	Remainder
	Remark
	Reminder
	Remote-Manual Bolter
	Removal
	Replacement Micrographs
	Replica Electron Microscopy
	Roentgen Equivalent Man
REMC	Resin-Encapsulated Mica Capacitor
REME	Royal Electrical and Mechanical Engineers
REMOS	Real-Time Event Monitoring System
REM-RED	Reflection Electron Microscopy - Reflection Electron Diffraction
REMSA	Railway Engineering Maintenance Suppliers Association [US]
REMSCAN	Remote Sensor Communication and Navigation
REN	Remote Enable
	Renewal
RENE	Rocket Engine Nozzle Ejector
RENM	Request for Next Message
REO	Rare Earth Oxide
	Regenerated Electrical Output
	Rocket Engine Operations
REON	Rocket Engine Operations - Nuclear
REP	Radar Evaluation Pad
	Range Error Probable
	Reentry Physics Program
	Rendezvous Evaluation Pad [NASA]
	Repair
	Report
	Representation
	Representative
	Republic
	Repulsion
	Request for Proposals
	Roentgen Equivalent Physical
	Rotating Electrode Process
REPL	Replace(ment)
REPLAB	Responsive Environment Programmed Laboratory
REPO	Remote Emergency Power Off
REPPAC	Repetitively Pulsed Plasma Accelerator
REPR	Representation
	Reprint(ed)
REPRO	Reproduction
REPROM	Reprogrammable Read-Only Memory
REPT	Report
REQ	Request
	Requirement
	Requisition
REQD	Required
REQT	Requirement
REQWQ	Requisition Word Queue
RERC	Rare Earth Research Conference
RERL	Residual Equivalent Return Loss
RES	Remote Entry Service
	Remote Entry Subsystem
	Reservation
	Reset
	Residence
	Residue
	Resistance
	Resistance Electroslag (Surfacing)
	Resistivity
	Resistor
	Restore
RESA	(Scientific) Research Society of America [US]
RESCAN	Reflecting Satellite Communication Antenna
RESCU	Radio Emergency Search Communications Unit
RESCUE	Remote Emergency Salvage and Clean-Up Equipment
RESD	Research and Engineering Support Division [of IDA, US]

RESER	Reentry Systems Evaluation Radar		Reducing Flame
RESG	Research Engineering Standing Group [USDOD]		Release Fraction
			Rice Flour
RESID	Residual		Roof
RESORS	Remote Sensing On-Line Retrieval System		Roof Fan
RESP	Regulated Electrical Supply Package		Rough Finish
	Respondent	**Rf**	Rutherfordium
RESRT	Restart	**RFA**	Redundant Force Analysis
RESS	Radar Echo Simulation System		Renewable Fuels Association
	Radar Echo Simulation Study		Retarding Field Analyzer
	Rapid Expansion of Supercritical Solutions		Radiographic Fluorescence Spectral Analysis
REST	Radar Electronic Scan Technique	**RFC**	Radio Frequency Chart
	Radar Electronic Scan Test		Radio Frequency Choke
	Reentry Environment and Systems Technology		Rosette Forming Cells
	Reentry System Test	**RFCP**	Radio Frequency Compatibility Program
RESTA	Reconnaissance, Security and Target Acquisition [US Army]	**RFD**	Ready for Data
			Reentry Flight Demonstration
RET	Reliable Earth Terminal	**RFDU**	Reconfiguration and Fault Detection Unit
	Retainer	**RFEI**	Request for Engineering Information
	Retardation	**RFF**	Remote-Fiber Fluorometry
	Return	**RFG**	Radar Field Gradient
	Retire(ment)		Rapid-Fire Gun
RETA	Refrigerating Engineers and Technicians Association [US]		Receive Format Generator
			Relative Growth Rate
			Roofing
RETAIN	Remote Technical Assistance and Information Network	**RFI**	Radio Frequency Interference
			Request for Information
RETC	Rapid Excavation and Tunneling Conference	**RFIT**	Radio Frequency Interference Test
	Regional Emergency Transportation Center [US]	**RFL**	Radio-Frequency Laboratories
			Requested Flight Level
RETEC	Regional Technical Conference		Resorcinol-Formaldehyde Latex
RETM	Rare Earth Transition Metal	**RFLP**	Restriction Fragment Length Polymorphism
RETMA	Radio Electronics Television Manufacturers Association [now: EIA]	**RFM**	Reactive Factor Meter
		RFMO	Radio Frequency Management Office
	Rare Earth Transition Metal Alloy	**RFNA**	Red Fuming Nitric Acid
RETR	Retraction	**RFNM**	Request for Next Message
RETSPL	Reference Equivalent Threshold Sound Pressure Level	**RFO**	Radio Frequency Oscillator
		RFP	Request for Proposal
REV	Reentry Vehicle	**RFPG**	Radio-Frequency Protection Guide
	Revenue	**RFQ**	Request for Quotation
	Review	**RFR**	Reduced Frequency Responses
	Revision		Reject Failure Rate
	Revolution	**RFS**	Radio Frequency Shift
rev/min	revolutions per minute		Regardless of Feature Size
REVOCON	Remote Volume Control		Remote File Sharing
REVS	Rotor Entry Vehicle System		Renormalized Forward Scattering
rev/s	revolutions per second	**RFSH**	Refresh
rev/sec	revolutions per second	**RF SQ**	Roof Squares
REV VER	Revised Version	**RFSTF**	Radio Frequency Systems Test Facility
REW	Rewind	**RFT**	Repeat Formation Tester
REWSONIP	Reconnaissance Electronic Warfare Special Operation and Naval Intelligence Processing		Revisable Form Text
		RFU	Reference Frequency Unit
		RG	Rate Gyroscope
REX	Real-Time Executive Routine		Red-Green
	Reduced Exoatmospheric Cross-Section		Reduction Gear
	Route Extension		Reset Gate
RF	Radio Frequency		Residual Gas
	Raised Face		Reticulated Grating
	Range Finder		Reverse Gate
	Rating Factor		Rolled Gold
	Reactive Factor	**RGA**	Rate Gyro Assembly

	Residual Gas Analyzer
RGB	Red/Green/Blue (Monitor)
RGCC	Roll Gas Combined Cooling System
RGH	Rough
RGH OPNG	Rough Opening
RGI	Residual Gas Ions
RGL	Report Generator Language
RGLET	Rise-Time Gated Leading Edge Trigger
RGLTR	Regulator
RGM	Recorder Group Monitor
RGN	Region
RGNCD	Region Code
RGO	Reference Gear Oil
RGP	Rate Gyro Package
	Rolled Gold Plate
RGS	Radio Guidance System
	Rate Gyro System
	Rocket Guidance System
	Royal Geographical Society
RGT	Resonant Gate Transistor
RH	Radiological Health
	Refrigeration Hardened
	Reheater
	Relative Humidity
	Request Header
	Request/Response Header
	Response Header
	Right-Hand
	Rockwell Hardness
	Roughness Height
	Ruhrstahl-Henrichshuette/Heraeus (Process)
Rh	Rhesus Factor
	Rhodium
R/h	Roentgen per Hour
RHA	Road Haulage Association
RHAW	Radar Homing and Warning
RHBDR	Rhombohedral
RHC	Reheat Coil
	Right-Hand Circular (Polarization)
	Rotation Hand Controller
RHCP	Right-Hand Circular Polarization
RHD	Right-Hand Drive
RHE	Radiation Hazard Effects
	Reversible Hydrogen Electrode
RHEED	Reflection High Energy Electron Diffraction
RHEL	Rutherford High Energy Laboratory
RHOMB	Rhombic
RHEO	Rheostat
RHI	Range-Height Indicator
RHIP	Radiation Health Information Project
RHN	Rockwell Hardness Number
RHOGI	Radar Homing Guidance
RHP	Reduced Hard Pressure
RHR	Reheater
	Rejectable Hazard Rate
	Roughness Height Rating
RHS	Rectangular Hollow Section
	Right-Hand Side
RI	Radar Input
	Radio Inertial

	Radio Influence
	Radio Interference
	Range Instrumentation
	Read-In
	Reflective Insulation
	Reliability Index
	Report of Investigation
	Reproducibility Index
	Resistance Inductance
	Rhode Island [US]
	Robotics International [of SME]
RIA	Radioimmunoassay
	Research Institute of America
	Robotics Industries Association
	Robot Institute of America
RIAA	Recording Industry Association of America
RIAS	Research Institute of Advanced Studies [US]
RIB	Request Indicator Byte
RIBA	Royal Institute of British Architects
RIBE	Reactive Ion Beam Etching
RIBS	Rutherford Ion Backscattering
RIC	Radar Indicating Console
	Radar Input Control
	Range Instrumentation Coordination
	Rare Earth Information Center [US]
	Reconstructed Ion Chromatography
	Relocation Instruction Counter
	Royal Institute of Chemistry [UK]
RICA	Research Institute for Consumer Affairs
RICASIP	Research Information Center and Advisory Service on Information Processing [US]
RICB	Reactive Ionized Cluster Beam
RICMT	Radar Inputs Countermeasures Technique
RICS	Range Instrumentation Control System
	Royal Institute of Chartered Surveyors [UK]
RID	Radar Input Drum
	Reset Inhibit Drum
RIDD	Range Instrumentation Development Division
RIDL	Ridge Instrument Development Laboratory [US Navy]
RIE	Reactive Ion Etching
	Royal Institute of Engineers [UK]
RIEF	Recirculating Isoelectric Focusing
RIETCOM	Regional Interagency Emergency Transportation Committee [US]
RIF	Radio Interference Field
	Reliability Improvement Factor
RIFI	Radio Interference Field Intensity
	Radio-Interference-Free Instrument
RIFS	Radioisotope Field Support
RI&FS	Remedial Investigation and Feasibility Study [EPA, US]
RIFT	Reactor-in-Flight Test
RIGS	Radio Inertial Guidance System
	Runway Identifiers with Glide Slope
RIIC	Rural Industries Innovation Center
RIISOM	Research Institute for Iron, Steel and other Metals
RIL	Radio Interference Level
	Red Indicating Lamp

RIM	Radar Input Mapper
	Radio Imaging Method
	Reaction Injection Molding
RIMS	Radiation Intensity Measuring System
	Resonance Ionization Mass Spectrometry
RIMtech	Research Institute for the Management of Technology
RIN	Regular Inertial Navigator
	Royal Institute of Navigation [UK]
RINA	Royal Institution of Naval Architects [UK]
RINAL	Radar Inertial Altimeter
RIND	Research Institute of National Defense [US]
RING	Ringing
RINS	Research Institute for the Natural Sciences
RINT	Radar Intermittent
RIO	Real-Time Input/Output
RIOMETER	Relative Ionospheric Opacity Meter
RIOT	Real-Time Input/Output Transducer
	Real-Time Input/Output Translator
RIP	Ring Index Pointer
	Rural Industrialization Program [US]
RIPCAM	Real-Time Interactive Process Control and Management
RIPPLE	Radioactive Isotope Powered Pulse Light Equipment
	Radioisotope Powered Prolonged Life Equipment
RIPS	Radio Isotope Power System
	Range Instrumentation Planning Study
RIR	Reliability Investigation Request
RIRS	Reflectance Infrared Spectroscopy
RIRTI	Recording Infrared Tracking Instrument
RIS	Radiology Information System
	Range Instrumentation Ship
	Receipt Inspection Segment
	Research Information Service
	Resonance Ionization Spectroscopy
	Retail Information System
	Revolution Indicating System
	Rotatable Initial Susceptibility
RISC	Radiology Information Systems Consortium
	Reduced Instruction Set Computer
RISE	Research in Supersonic Environment
RI/SME	Robotics International of the Society of Manufacturing Engineers [US]
RIT	Radar Input Test
	Radio Information Test
	Receiver Incremental Tuning
	Rochester Institute of Technology [US]
	Rocket Interferometer Tracking
RITA	Radio-Frequency Ion Thruster Assembly
RITC	Rhodamine Isothiocyanate
RITU	Research Institute of Temple University [US]
RIV	Radio-Influence Voltage
	River
	Rivet
RJE	Remote Job Entry
RJP	Remote Job Processing
RKHS	Reducing Kernel Hilbert Space
RKO	Range Keeper Operator

rkva	reactive kilovolt-ampere
RL	Radiation Laboratory
	Radio Link
	Radiolocation
	Record Length
	Reduced Level
	Relay Level
	Remote Loopback
	Research Laboratory
	Resistance-Inductance
	Resistor Logic
	Return Loss
	Rocket Launcher
RLA	Remote Loop Adapter
RLAP	Rural Land Analysis Program [Canada]
RLBM	Rearward Launched Ballistic Missile
RLC	Radio Launch Control System
	Resistance-Inductance-Capacitance
RLCS	Radio Launch Control System
RLD	Relocation Dictionary
RLE	Research Laboratory for Electronics [US]
	Runlength Encoder
RLG	Railing
RLHTE	Research Laboratory of Heat Transfer in Electronics
RLIN	Research Libraries Information Network
RLL	Recorded Lithology Logging
RLO	Restoration Liaison Officer
RLS	Radar Line-of-Sight
RLSE	Release
RLSD	Received Line Signal Detect
RLTS	Radio-Linked Telemetry System
RLY	Railway
	Relay
RM	Radar Mapper
	Radiation Measurement
	Radio Monitor(ing)
	Range Marks
	Ream
	Record Mark
	Research Memorandum
	Resistance Melting
	Resistance Monitor
	Resource Manager
	Room
	Rotating Machinery
R&M	Reports and Memoranda
RMA	Radio Manufacturers Association
	Reactive Modulation-Type Amplifier
	Rosin Mildly Activated
	Rubber Manufacturer's Association
	Royal Military Academy [UK]
R-MAD	Reactor Maintenance, Assembly, and Disassembly [USAEC]
RMARL	Rocky Mountain Analytical Research Laboratory [US]
RMAX	Range Maximum
RMC	Rod Memory Computer
	Royal Military College
RMCAO	Ready-Mixed Concrete Association of Ontario [Canada]

RMCS	Royal Military College of Science [UK]
RMEA	Rubber Manufacturing Employers Association
RMI	Radio Magnetic Indicator
	Reliability Maturity Index
RMIC	Research Materials Information Center [of ORNL]
RMICBM	Road Mobile Intercontinental Ballistic Missile
RMIN	Range Minimum
RML	Radar Mapper, Long-Range
	Radar Microwave Link
	Rocky Mountain Laboratory [NIH]
RMM	Radar Map Matching
	Read Mostly Memory
RMOS	Refractory Metal-Oxide Semiconductor
RMP	Reentry Measurement Program
	Royalty Management Program
RMS	Radar Mapper, Short-Range
	Record Management Service
	Recovery Management Support
	Remote Manipulator System
	Resource Management System
	Root Mean Square
	Royal Meteorological Society
	Royal Microscopical Society [UK]
RMSCC	Rock Mechanics and Strata Control Committee
RMSE	Root Mean Square Error
RMSS	Root Mean Square Strain
RMT	Remote Terminal
RMTE	Remote
RMU	Remote Maneuvering Unit
RMV	Reentry Measurement Vehicle
RN	Radar Navigation
	Random Number
	Reference Noise
	Removable Needle
	Research Note
	Reynolds Number
Rn	Radon
RNA	Ribonucleic Acid
RND	Round
RNase	Ribonuclease
RNFP	Radar Not Functioning Properly
RNIT	Radio Noise Interference Test
RNP	Ribonucleoprotein
RNS	Residue Number System
RNSS	Royal Naval Scientific Service [UK]
RNV	Radio Noise Voltage
RO	Radar Operator
	Range Operations
	Readout
	Receive Only
	Record 0 (= Zero)
	Reddish Orange
	Reference Oscillator
	Reverse Osmosis
	Royal Observatory [UK]
R/O	Read Only
	Receive Only

ROA	Return on Assets
ROAMA	Rome Air Materiel Area [US]
ROAT	Radio Operator's Aptitude Test
ROB	Radar Order of Battle
	Robotic Operating Buddy
ROBIC	Robotic Integrated Cell
ROBIN	Remote On-Line Business Information Network
ROBO	Rocket Bomber
ROBOMB	Robot Bomb
ROC	Range Operations Conference
	Rapid Omnidirectional Compaction
	Receiver Operating Characteristic
	Required Operational Capacity
	Remote Operator Console
	Return on Capital
	Reusable Orbital Carrier
ROCAPPI	Research on Computer Applications for the Printing and Publishing Industries
ROCE	Return on Capital Employed
ROCKET	Rand's Omnibus Calculator of the Kinetics of Earth Trajectories
ROCP	Radar Out of Commissions for Parts
ROCR	Remote Optical Character Recognition
ROD	Rate of Descent
	Recorder on Demand
	Required Operational Data
RODATA	Registered Organizations Database
ROE	Return on Equity
ROF	Royal Ordnance Factory [UK]
R of A	Reduction of Area
ROFT	Radar Off Target
ROI	Range Operations Instruction
	Region of Interest
	Return on Investment
ROIC	Return of Invested Capital
ROIP	Residual Oil in Place
ROIS	Radio Operational Intercom System
ROK	Republic of Korea
ROLF	Remotely Operated Longwall Face
ROLS	Recoverable Orbital Launch System
	Request On-Line Status
ROM	Read-Only Memory
	Readout Memory
	Return on Market Value
	Romania
	Romanian
	Rough Order of Magnitude
	Run-of-Mine (Ores)
ROMAC	Robotic Muscle Actuator
ROMBUS	Reusable Orbital Module Booster and Utility Shuttle
ROMON	Receiving-Only Monitor
ROMOTAR	Range-Only Measurement of Trajectory and Recording
ROM/RPROM	Read-Only Memory/Reprogrammable Read-Only Memory
ROMV	Return on Market Value
RON	Research Octane Number
RONT	Radar On Target
ROOFG	Roofing

ROOI	Return on Original Investment	**RPD**	Radar Planning Device
ROOST	Reusable One-Stage Orbital Space Truck		Retarding Potential Difference
ROOT	Relaxation Oscillator Optically Tuned	**RPE**	Radial Probable Error
ROP	Rate of Penetration		Registered Professional Engineer
	Record of Performance		Required Page-End (Character)
	Record of Production		Rocket Propulsion Establishment [UK]
ROPE	Ring of Prefetch Elements		Rotating Platinum Electrode
ROPP	Receive-Only Page Printer	**RPF**	Radiometer Performance Factor
ROPS	Range Operation Performance Summary	**RPG**	Random Pulse Generator
	Roll-Over-Protective Structure		Report Program Generator
ROPS-FOPS	Roll-Over-Protective and Falling-Object-Protective Structure		Rocket-Propelled Grenade
		RPH	Recommended Power Handling
RORC	Rolla Research Institute [US]		Relative Pulse Height
ROS	Read-Only Storage	**RPI**	Radar Precipitation Integrator
ROSE	Remotely Operated Special Equipment		Rensselaer Polytechnic Institute
	Retrieval by On-Line Search		Resource Policy Institute
RoSPA	Royal Society for the Prevention of Accidents [UK]		Rubber and Plastics Industry
		RPIE	Real Property Installed Equipment
ROT	Radar On Target	**RPL**	Radar Processing Language
	Reusable Orbital Transporter		Radiation Physics Laboratory [US]
	Rotary		Radiophysics Laboratory [of CSIRO]
	Rotate		Remote Program Loader
	Rotation		Request Parameter List
ROTCC	Receiver-Off-Hook Tone Connecting Circuit		Research Programming Language
ROTI	Recording Optical Tracking Instrument		Robot Programming Language
ROTR	Read-Only Typing Reperforator		Rocket Propulsion Laboratory [US Air Force]
	Receive-Only Tape Reperforator		Running Program Language
ROV	Remotely Operated Vehicle	**RPM**	Random Phase Model
ROVD	Remotely Operated Volume Damper		Rate per Minute
ROW	Right-of-Way		Reinforced Plastic Mortar
	Roll Welding		Reliability Performance Measure
ROWAP	Rotating Water Atomization Process		Resupply Provisions Module
ROWS	Robotic Winding System		Revolutions per Minute [also: rpm]
RP	Reader-Printer		Rollcast Planetary Mill
	Real Part	**RPMC**	Remote Performance Monitoring and Control
	Recommended Practice		
	Record Processor	**RPMI**	Revolutions-per-Minute Indicator
	Recovery Phase	**RPN**	Reverse Polish Notation
	Reddish Purple	**RPNA**	Reverse Pacific/North American
	Reference Pulse	**RPO**	Revolution per Orbit
	Refilling Point	**RPP**	Radar Power Programmer
	Reinforced Plastics		Radar Processing Plant
	Relative Pressure		Reductive Pentosephosphate
	Repair		Reinforced Pyrolyzed Plastic
	Research Paper	**RPQ**	Request for Price Quotation
	Rocket Projectile	**RPR**	Read Printer
	Rocket Propellant	**RPROM**	Reprogrammable Read-Only Memory
	Rocket Propulsion	**RPS**	Random Program Selection
	Round Pin		Radar Plotting Sheet
	Rust Preventive		Reactive Polystyrene
RP1D	Repeated Unidirectional		Real-Time Photogrammetry System
RPA	Radar Performance Analyzer		Refrigerator-Cooled Pump System
	Random-Phase Approximation		Reduced Pressure Spraying
	Rapid Pressure Application		Remote Processing Service
RPAO	Radium Plaque Adaptometer Operator		Revolutions per Second [also: rps]
RPC	Radar Planning Chart		Rotational Position Sensing
	Remote Position Control		Royal Photographic Society [UK]
	Research and Productivity Council	**RPSM**	Resources Planning and Scheduling Method
	Reversed Phase Chromatography	**RPT**	Repeat (Character)
	Row Parity Check		Report
RP/CI	Reinforced Plastics/Composites Institute [of SPI]		

RPU	Radio Phone Unit		**RRPTN**	Receiving Report Number
	Radio Propagation Unit		**RRRV**	Rate of Rise of Restriking Voltage
	Real-Time Photogrammetry Unit		**RRS**	Radiation Research Society [US]
RPV	Reactor Pressure Vessel			Radio Ranging System
	Remotely Piloted Vehicle			Radio Relay Station
	Remote Pilotless Vehicle			Radio Research Station [UK]
RPW	(Resistance) Projection Welding			Reaction Research Society
R/Q	Resolver/Quantizer			Required Response Spectrum
RQA	Recursive Queue Analyzer			Restraint Release System
R&QA	Reliability and Quality Assurance			Retrograde Rocket System
RQD	Rock Quality Designation		**RRT**	Relative Retention Time
RQE	Relative Quantum Efficiency			Rendezvous Radar Transponder
RQI	Request for Initialization		**RRTC**	Retractable Replaceable Thermocouple
RQL	Reference Quality Level		**RRU**	Radiobiological Research Unit
RQS	Rate Quoting System		**RS**	Radiated Susceptibility
RQMT	Requirement			Radio Simulator
RR	Radio Regulation			Range Selector
	Railroad			Range Safety
	Rapid Rectilinear (Lens)			Range Surveillance
	Readout and Relay			Rapidly Solidified
	Receive Ready			Rapid Solidification
	Recurrence Rate			Rapid Setting
	Register to Register (Instruction)			Reader Stop
	Rendezvous Radar			Real Storage
	Repetition Rate			Record Separator (Character)
	Research Reactor			Remote Station
	Research Report			Residual Stress
	Resonance Raman (Spectroscopy)			Resistance Soldering
	Return Rate			Rolled Shapes
	Retro Rocket		**R&S**	Research and Statistics
	Round Robin		**R-S**	Reset-Set
	Running Reverse		**RSA**	Radar Signature Analysis
	Rural Road			Related Scientific Activities
	Ruthenium Red			Remote Station Alarm
R/R	Readout and Relay			Rubber Shippers Association [US]
	Record/Retransmit		**RSAC**	Radiological Safety Analysis Computer
r/R	Radius Ratio			Reactor Safety Advisory Committee [Canada]
RRA	Restrictive Requirement Quality A		**RSB**	Reticulocyte Standard Buffer
	Retrogression and Reaging Treatment		**RSC**	Radar Set Control
RRB	Railroad Retirement Board [US]			Rapid Spinning Cup
	Restrictive Requirement Quality B			Remote Services Center
R&RC	Reactors and Reactor Controls [IEEE Committee]			Reversed Sigmoidal Curve
RRDS	Relative-Record Data Set			Royal Society of Canada
RRE	Radar Research Establishment [UK]		**RSCESFS**	Rapid Scanning Constant Energy Synchronous Fluorescence Spectrometry
	Receive Reference Equivalent		**RSCIE**	Remote Station Communication Interface Equipment
	Royal Radar Establishment [UK]		**RSCS**	Rate Stabilization and Control System
RREAC	Royal Radar Establishment Automatic Computer			Remote Spooling Communications Subsystem
RREG	Regional Rheoencephalography		**RSD**	Radar Signal Discrimination
RRI	Range-Rate Indicator			Relative Standard Deviation
	Rocket Research Institute [US]		**RSDP**	Remote Site Data Processor
RRIM	Reinforced Reaction Injection Molding		**RSDS**	Range Safety Destruct System
RRIS	Record Room Interrogation System		**RSE**	Reducing Street Elbow
	Remote Radar Integration Station			Register Signaling Equipment
RRL	Radio Relay Link		**RSEW**	Resistance-Seam Welding
	Radio Research Laboratories [Japan]		**RSF**	Relative Sensitivity Factor
	Road Research Laboratory [UK]			Remote Support Facility
RRN	Relative-Record Number		**RSFSR**	Russian Soviet Federated Socialist Republic
RRNS	Redundant Residue Number System			
RRPT	Receiving Report			

RSG	Resistance Strain Gauge		Receiver Transmitter
RSGB	Radio Society of Great Britain		Receptor Technology
RSI	Research Studies Institute		Recrystallization Time
	Reusable Surface Insulation		Reduction Table
RSIC	Radiation Shielding Information Center [of ORNL]		Reference Temperature
			Registered Technician
	Redstone Scientific Information Center [US Army]		Register Ton
			Register Transfer
RSID	Resource Identification (Table)		Released Time
RSIO	Reinforced Steel Institute of Ontario [Canada]		Remote Terminal
			Reperforator/Terminal
RSL	Radio Standards Laboratory [of NSB]		Reperforator/Transmitter
	Requirements Specifications Language		Research and Technology
RSLVR	Resolver		Resolution Tested
RSM	Rapidly Solidified Material		Return
	Real Storage Management		Road Tar
	Royal School of Mines		Room Temperature
RSME	Reversible Shape Memory Effect	**R/T**	Real Time
RSMPS	Romanian Society for Mathematics and Physical Sciences	**RTA**	Rapid Thermal Annealing
			Reliability Test Assembly
RSN	Radiation Surveillance Network		Remote Trunk Arrangement
	Record Sequence Number		Roads and Transportation Association
RSNA	Royal School of Naval Architects	**RTAC**	Roads and Transportation Association of Canada
RSORS	Remote Sensing On-Line Retrieval System		
RSP	Radio Switch Panel	**RTAM**	Remote Terminal Access Method
	Rapid-Solidified Product	**RTB**	Read Tape Binary
	Record Select Program		Response/Throughout Bias
	Required Space (Character)	**RTC**	Radar Tracking Center
	Respond		Radar Tracking Control
RSPA	Research and Special Programs Association [US]		Railway Transport Committee
			Range Telemetry Center
RS/PM	Rapid Solidification/Powder Metallurgy		Reader/Tape Contact
RSPT	Real Storage Page Table		Real-Time Command
RSR	Random Signal Reject		Real-Time Computer
	Rapid Solidification Rate		Reference Transfer Calibrator
RSRA	Rotor Systems Research Aircraft	**RTCA**	Radio-Technical Commission for Aeronautics [US]
RSRI	Regional Science Research Institute [US]		
RSRS	Radio and Space Research Station	**RTCC**	Real-Time Computer Complex
RSS	Random Selection for Service	**RTCF**	Real-Time Computer Facility
	Range Safety System	**RTCM**	Radio Technical Commission for Marines [US]
	Reactant Service System		
	Relaxed Static Stability	**RTCS**	Real-Time Computer System
	Residual Sum of Squares	**RTCU**	Real-Time Control Unit
	Ribbed Smoked Sheets	**RTD**	Range Time Decoder
	Root Sum Square		Read Tape Decimal
	Royal Statistical Society		Real-Time Display
RST	Rapid Solidification Technology		Research and Technology Division [US Air Force]
	Remote Station		
	Research Study Team		Residence Time Distribution
	Reset-Set Trigger		Resistance Temperature Detector
RSTN	Regional Seismic Test Network		Resistance Thermometer Detector
RSV	Revised Standard Version		Resistive Temperature Device
RSW	Resistance-Spot Welding	**RTDC**	Real-Time Data Channel
RT	Radiographic Testing	**RTDD**	Real-Time Data Distribution
	Radio Telegraphy	**RTDHS**	Real-Time Data Handling System
	Radio Telephony	**RTDS**	Real-Time Data System
	Raise Top	**RTE**	Real-Time Executive
	Range Tracking		Regenerative Turboprop Engine
	Rated Time		Reversible Temper Embrittlement
	Ratio Transformer	**RTECS**	Registry of Toxic Effects of Chemical Substances
	Real Time		

RTEM	Radar Tracking Error Measurement
RTF	Rich Text Format
RTG	Radioisotope Thermoelectric Generator
RTI	Real-Time Interface
	Referred to Input
	Research Triangle Institute [US]
RTIF	Rapid Test and Integration Facility [US Navy]
RTIFRAMP	Rapid Test and Integration Facility, Rapid Access to Manufactured Parts [US Navy]
RTI/OC	Real-Time Input/Output Controller
RTIRS	Real-Time Information Retrieval System
RTITB	Road Transport Industry Training Board
RTK	Range Tracker
RTL	Real-Time Language
	Resistor-Transistor Logic
RTM	Real-Time Monitor
	Recording Tachometer
	Register Transfer Module
	Research Technical Memoranda
	Resin-Transfer Molding
	Response Time Module
RTMA	Radio and Television Manufacturers Association [US]
RTMOS	Real-Time Multiprogramming Operating System
RTN	Return
	Registered Tradename
	Routine
RTO	Referred to Output
RTOL	Restricted Takeoff and Landing
RTOS	Real-Time Operating System
RTP	Real-Time Peripheral
	Reinforced Thermoplastic
	Remote Terminal Processor
	Remote Transfer Point
	Requirement and Test Procedures
	Research Triangle Park [US]
RTPC	Restrictive Trade Practices Commission
RTPH	Round Trips per Hour
RTR	Reinforced Thermosetting Resin
RTS	Radar Target Simulator
	Radar Tracking Station
	Range Time Signal
	Reactive Terminal Service
	Real-Time Subroutine
	Real-Time System
	Relaxation Time Spectrum
	Remote Targeting System
	Remote Testing System
	Request-to-Send
	Royal Television Society [UK]
	Rural Telephone System
RTSD	Resources and Technical Services Division [ALA, US]
RTSRS	Real-Time Simulation Research System
RTSS	Real-Time Scientific System
RTST	Radio Technician Selection Test
RTT	Radiation Tracking Transducer
	Radioteletypewriter
	Real-Time Tracer

	Recrystallization Time - Temperature (Curve)
	Reservoir and Tube Tunnel
RTTDS	Real-Time Telemetry Data System
RTTV	Real-Time Television
RTTY	Radio Teletypewriter
RTU	Remote Terminal Unit
RTV	Remote Television
	Room Temperature Vulcanizing
	Rotating Traveling Vehicle
RTWS	Raw Tape Write Submodule
RTX	Real-Time Executive
RTZ	Return-to-Zero
RU	Reproducing Unit
	Request Unit
	Request/Response Unit
	Response Unit
Ru	Ruthenium
RUB	Rubber
RuBP	Ribulosebiphosphate
RUD	Rudder
RUDI	Restricted Use Digital Instrument
RuDP	Ribulose Diphosphate
RUM	Remote Underwater Manipulator
	Rumania
	Rumanian
RuP	Ribulosephosphate
RUR	Rosum's Universal Robot
RUS	Russia(n)
RUSH	Remote Use of Shared Hardware
RUSS	Robotic Ultrasonic Scanning System
	Russia(n)
RV	Rated Voltage
	Rear View
	Recreational Vehicle
	Reentry Vehicle
	Relative Volume
	Relief Valve
	Rendezvous
	Restriking Voltage
	Revised Version
	Rough Vacuum
RVA	Reactive Volt-Ampere (Meter)
	Recorded Voice Announcement
	Reliability Variation Analysis
rva	reactive volt-ampere
RVC	Relative Velocity Computer
	Ribonucleoside Vanadyl Complex
RVCM	Residual Vinyl Chloride Monomer
RVDT	Rotary Variable Differential Transformer
RVI	Reverse Interrupt
RVM	Reactive Voltmeter
RVP	Reid Vapor Pressure
RVR	Runway Visual Range
RVS	Real-Time Visual Simulation
	Reverse
RVT	Resource Vector Table
RVV	Runway Visibility Value
RW	Raw Water
	Readily Weldable
	Read/Write

	Resistance Welding
R/W	Read/Write
	Right of Way
RWAA	Resistance Welding Alloy Association
RWC	Read, Write and Compute
RWCS	Report Writer Control System
RWG	Roebling Wire Gauge
RWI	Radio Wire Integration
	Read, Write and Initialize
RWM	Read-Write Memory
	Rectangular Wave Modulation
RWMA	Resistance Welder Manufacturers Association [US]
RWND	Rewind
RWP	Reaction Wave Polymerization
RWR	Relative Wear Resistance
RWS	Receiver Waveform Simulation
RWY	Railway
RX	Receive
	Resolver-Transmitter
RY	Railway
RZ	Reinheitszahl [biochemistry]
	Reset-to-Zero
	Return-to-Zero (Recording)
RZL	Return-to-Zero Level
RZM	Return-to-Zero Mark
	Root Zone Method
RZ(NP)	Nonpolarized Return-to-Zero (Recording)
RZ(P)	Polarized Return-to-Zero (Recording)

S

S	Saturday
	Scalar
	Scuttle
	Sea
	Section
	Segment
	September
	Serine
	Side
	Siemens
	Sign
	Silicate
	Silk
	Slate
	Society
	Software
	Soldering
	Solubility
	Source
	South
	Stack
	State
	Storage
	Stress
	Strain
	Submarine

	Sulphur [or: Sulfur]
	Sunday
	Supervisory
	Switch
	System
s	scruple [unit]
	second [unit]
	see
	set
	single
	soft
	solid
	soluble
	stere [unit]
	synchronous
SA	Sail Area
	Salt Added
	Scientific Authority
	Selected Area (Diffraction)
	Sense Amplifier
	Sequential Access
	Shaft Alley
	Shear Area
	Small Arms
	Snubber Adapter
	South Africa
	South America
	South Australia
	Special Application
	Spectrum Analyzer
	Spherical Aberration
	Spherical Agglomeration
	Stability Augmentation
	Stress Annealing
	Submerged Arc
	Successive Approximation
	Symbolic Assembler
	System Administrator
	System Analysis
	Systems Analyst
SAA	Solution Annealed and Aged
	Standards Association of Australia
	Storage Accounting Area
	Surface Active Agent
SAAC	Space Applications Advisory Committee
SAAEB	South African Atomic Energy Board
SAAL	Single-Axis Acoustic Levitator
SAAP	Saturn/Apollo Application Program [of NASA]
SAAS	Science Achievement Awards for Students
SAB	Scientific Advisory Board
	Secondary Application Block
	Silicon-Aluminum-Bronze
	Solar Alignment Bay
	Space Applications Board
	System Advisory Board
SAb	Serum Antibody
SABE	Society for Automation in Business Education [US]
SABIR	Semi-Automatic Bibliographic Information Retrieval

SABMIS	Seaborne Antiballistic Missile Intercept System [US Navy]
SABO	Sense Amplifier Blocking Oscillator
SABRE	Sales and Business Reservations done Electronically
	Secure Airborne Radar Equipment
	Self-Aligning Boost and Reentry System
SABS	South African Bureau of Standards
SAC	Semi-Automatic Coding
	Scientific Advisory Committee
	Shipbuilding Advisory Council [UK]
	Society for Analytical Chemistry
	Sound Absorption Coefficient
	Store and Clear Accumulator
	Strategic Advisory Committee [US]
	Strategic Air Command [NATO]
	Sulfacetamide
	Surface-Area-Center
SACCEI	Strategic Air Command Communications-Electronics Instruction
SACCOMNET	Strategic Air Command Communications Network
SACDE	System Analysis Control and Design Activity
SACE	Saskatchewan Association for Computers in Education [Canada]
SACEUR	Supreme Allied Commander, Europe [of NATO]
SACL	Space and Component Log
SACLANT	Supreme Allied Commander, Atlantic [of NATO]
SACMA	Suppliers of Advanced Composite Materials Association
SACMAPS	Selective Automatic Computational Matching and Positioning System
SACOM	Ships Advanced Communications Operational Model
SACP	Selected Area Channeling Pattern
SACSIR	South African Council for Scientific and Industrial Research
SACTTYNET	SAC (= Strategic Air Command) Teletype Network
SACVT	Society of Air Cushion Vehicle Technicians
SAD	Selected Area Diffraction
	Sentence Appraiser and Diagrammer
	Solar Array Drive
SADA	Seismic Array Data Analyzer
	Solar Array Drive Assembly
	Stand-Alone Data Acquisition
SADAP	Simplified Automatic Data Plotter
SADAPTA	Solar Array Drive and Power Transfer Assembly
SADAS	Sperry Airborne Data Acquisition System
SADC	Sequential Analog-Digital Computer
SADE	Solar Array Drive Electronics
SADIC	Solid-State Analog-to-Digital Computer
SADIE	Scanning Analog-to-Digital Input Equipment
	Semi-Automatic Decentralized Intercept Environment
	Sterling and Decimal Invoicing Electronically
SADL	Sterilization Assembly Development Laboratory [of NASA]
SADP	Selected Area Diffraction Pattern
SADR	Six Hundred Megacycles Air Defense Radar
SADSAC	Sampled Data Simulator and Computer
	Seiler ALGOL Digitally Simulated Analog Computer
SADSACT	Self-Assigned Descriptors from Self and Cited Titles
SADT	Self-Accelerating Decomposition Temperature
	Structured Analysis and Design Technique
SAE	Shaft-Angle Encoder
	Society of Automotive Engineers [US]
	Spiral Aftereffect
SAEH	Society for Automation in English and the Humanities
SAES	Scanning Auger Electron Spectroscopy
SAF	Safety
	Spacecraft Assembly Facility [of NASA]
SAFE	Safe Assessment and Facilities Establishment [of JAIF]
	Society for the Advancement of Fission Energy
	Society for the Application of Free Energy
	Space and Flight Equipment
SAFEA	Space and Flight Equipment Association [US]
SAFER	Stress and Fracture Evaluation of Rotors
SAFI	Semi-Automatic Flight Inspection
SAFOC	Semi-Automatic Flight Operations Center
S AFR	South Africa(n)
SAFSR	Society for the Advancement of Food Service Research
SAFT	Shortest Access First Time
	Synthetic Aperture Focussing Technique
SAG	Self-Aligned Gate
	Semi-Autogenous Grinding
	Standard Address Generator
SAGA	Short-Arc Geodetic Adjustment
SAGE	Semi-Automatic Ground Environment
	Stratospheric Aerosol and Gas Experiments (Satellite) [US]
SAHF	Semi-Automatic Height Finder
SAI	Science Applications International [US]
SAIB	Sucrose Acetate Isobutyrate
SAID	Speech Auto-Instructional Device
SAIL	Sea-Air Interaction Laboratory [US]
	Shuttle Avionics Integration Laboratory [of NASA]
	Stanford Artificial Intelligence Laboratory [US]
SAILS	Simplified Aircraft Instrument Landing System
SAIMS	Selected Acquisitions Information and Management System
SAINT	Satellite Interceptor
SAIS	South African Interplanetary Society
SAISAC	Ship's Aircraft Inertial System Alignment Console
SAKI	Solatron Automatic Keyboard Instructor

SAL	Salinometer
	Semantic Abstraction Language
	Short Approach Light
	Space Astronomy Laboratory [US]
	Supersonic Aerophysics Laboratory [US]
	Surface Airmail Lifted
	Symbolic Assembly Language
	Systems Assembly Language
SALE	Simple Algebraic Language for Engineers
SALEN	Salicylidene Ehtylenediamine
SALM	Society of Airline Meteorologists [US]
SALOPHEN	Salicylidene Phenylenediamine
SALORS	Structural Analysis of Layered Orthotropic Ring-Stiffened Shells
SALPN	Salicylidene Propanediamine
SALPS	Salicylideneamino Phenyl Disulfide
SALS	Ship Aircraft Locating System
	Short Approach Light System
	Solid-State Acoustoelectric Light Scanner
SALT	Strategic Arms Limitation Talks
SALV	Salvage
SAM	S-Adenosyl Methionine
	Scanning Acoustic Microscopy
	Scanning Auger Microprobe
	Scanning Auger Microscopy
	School of Aerospace Medicine [US Air Force]
	Script-Applier Mechanism
	Selective Automonitoring
	Semantic Analyzing Machine
	Semi-Automated Mathematics
	Sequential Access Memory
	Sequential Access Method
	Simulation of Analog Methods
	Society for Advancement of Management [US]
	Sort and Merge
	Strain Absorbent Module
	Strong Absorption Model
	Supply Analysis Model
	Surface-to-Air Missile
	Suspension Automatic Monitor
	Symbolic and Algebraic Manipulation
	System Activity Monitor
	Systems Analysis Module
S AM	South American
SAMA	Scientific Apparatus Makers Association [US]
SAMD	Surface-to-Air Missile Development
SAME	Society of American Military Engineers [US]
S AMER	South America(n)
SAMI	Service and Maintenance Indicator
	Socially Acceptable Monitoring Instrument
SAMIS	Structural Analysis and Matrix Interpretative System
SAMMIE	Scheduling Analysis Model for Mission Integrated Experiments
SAMOS	Satellite and Missile Observation System
SAMP	Sampling
	Succino Adenosine Monophosphate
SAMPE	Society for the Advancement of Materials and Process Engineering [US]
	Society of Aerospace Material and Process Engineers [US]
SAMS	Satellite Automonitor System
SAMSO	Space and Missile Systems Organization [US Air Force]
SAMSOM	Support-Availability Multisystem Operations Model
SAMSON	Strategic Automatic Message Switching Operational Network
	System Analysis of Manned Space Operations
SAMT	Swiss Association for Materials Testing
SAMTEC	Space and Missile Test Center [US]
SAN	Sanitary
	Styrene-Acrylonitrile
SANCAR	South African National Council for Antarctic Research
SAND	Shelter Analysis for New Designs
SANE	Solar Alternatives to Nuclear Energy
SANOVA	Simultaneous Analysis of Variance
SANS	Small-Angle Neutron Scattering
SAO	Smithsonian Astrophysical Observatory [US]
SAP	Semi-Armor Piercing
	Share Assembly Program
	Sintered Aluminum Powder
	Start of Active Profile
	Symbolic Address Program
	Symbolic Assembly Program
	Systems Assurance Program
s ap	scruple, apothecaries [unit]
SAPE	Society for Professional Education
	Solenoid Array Pattern Evaluator
SAPIR	System of Automatic Processing and Indexing of Reports
SAR	Search and Rescue Radio
	Sodium Absorption Ratio
	Storage Address Register
	Successive-Approximation Register
	Synthetic Aperture Radar
SARA	Society of American Registered Architects
SARAH	Search, Rescue and Homing
SARARC	Stable Auroral Red Arc
SARBE	Search and Rescue Beacon Equipment
SARCOM	Search and Rescue Communicator
SARISA	Surface Analysis by Resonance Ionization of Sputtered Atoms
SARP	Schedule and Report Procedure
	Sumitomo Alkali Refining Process
SARPS	Standard and Recommended Practices
SARS	Single-Axis Reference System
SARSAT	Search and Rescue Satellite
SARUC	Southeastern Association of Regulatory Utility Commissioners [US]
SAS	Saturated Ammonium Sulfate
	Segment Arrival Storage
	Small Angle Scattering
	Small Astronomy Satellite
	Society for Applied Spectroscopy

	Sodium Alkane Sulfonate
	Sodium Aluminum Sulfate
	Statistical Analysis System
	Support Amplifier Station
	Surface Active Substances
	Switched Access System
SASE	Self-Addressed and Stamped Envelope
	Statistical Analysis of a Series of Events
SASI	South African Standards Institute
	System on Automotive Safety Information
SASIDS	Stochastic Adaptive Sequential Information Dissemination System
Sask	Saskatchewan [Canada]
SASPL	Saturated Ammonium Sulfate Precipitation Limit
SASSY	Supported Activity Supply System
SASTU	Signal Amplitude Sampler and Totalizing Unit
SAT	Saturation
	Saturday
	Scholastic Aptitude Test
	Sine Acido Thymonucleico
	Society of Acoustic Technology
	Stabilization Assurance Test
	Stepped Atomic Time
	Symmetric Antitrapping
SATAN	Satellite Automatic Tracking Antenna
	Sensor for Airborne Terrain Analysis
SATANAS	Semi-Automatic Analog Setting
SATAR	Satellite for Aerospace Research [NASA]
SATCO	Signal Automatic Air Traffic Control
SATCOL	Satellite Network of Columbia
SATCOM	Scientific and Technical Communication Committee [of NAS/NAE, US]
	Satellite Communications
	Satellite Communication Agency [USDOD]
SATD	Saturated
SATIF	Scientific and Technical Information Facility
SATIN	SAGE Air Traffic Integration
SATIRE	Semi-Automatic Information Retrieval
SATO	Self-Aligned Thick Oxide
	Synthetic Aircraft Turbine Oil
SATRAC	Satellite Automatic Terminal Rendezvous and Coupling
SATS	Short Airfield and Tactical Support
	Solar Alignment Test Site
SATT	Shear Area Transistion Temperature
SAUCERS	Space and Unexplained Celestial Events Research Society [US]
	Saucer and Unexplained Celestial Events Research Society [US]
SAVA	Society for Accelerator and Velocity Apparatus
SAVE	Society of American Value Engineers
	System for Automatic Value Exchange
SAVES	Sizing of Aerospace Vehicle Structures
SAVITAR	Sanders Associates Video Input/Output Terminal Access Resource
SAVOR	Single-Actuated Voice Recorder
SAVS	Status and Verification System
SAW	Submerged Arc Welding

	Surface Acoustic Wave
SAWC	Special Air Warfare Center [US Air Force]
SAWE	Society of Allied Weight Engineers [US]
SAWMARCS	Standard Aircraft Weapons Management and Release Control System
SAWS	Small Arms Weapon System
SAW-S	Series Submerged Arc Welding
SAX	Strong Anion Exchange
SAXS	Small Angle X-Ray Scattering
SAYTD	Sales Amount Year-to-Date
SB	Secondary Battery
	Serial Binary
	Sideband
	Simultaneous Broadcast
	Sleeve Bearing
	Society for Biomaterials
	Soot Blower
	Splash Block
	Standby
	Straight Binary
	Stuffing Box
	Styrene-Butadiene
	Surface Barrier
	Synchronization Bit
Sb	(Stibium) - Antimony
SBA	Secondary Butyl Alcohol
	Security by Analysis
	Shared Batch Area
	Small Business Administration
	Standard Beam Approach
SBAC	Society of British Aerospace Companies
SBASI	Single-Bridge Apollo Standard Initiator
SBBNF	Ship and Boat Builders National Federation [UK]
SBC	Silicon Blue Cell
	Single Board Computer
SBCA	Sensor-Based Control Adapter
SBCC	Southern Building Code Congress [US]
SBD	Schottky Barrier Diode
	Single Button Dial
SBE	Supertwisted Birefringence Effect
SBH	Schottky Barrier Height
SBK	Single-Beam Klystron
SBLG	Small Blast Load Generator
SBM	Separate Bombardment Mode
	System Balance Measure
SBMA	Steel Bar Mills Association
SBMPL	Simultaneous Binaural Midplane Localization
SBN	Standard Book Number
SBO	Soy Bean Oil
	System for Business Operations
SBP	Shore-Based Prototype
	Special Boiling Point
	Stainless Ball Plunger
SBPO	Spun-Bounded Polyolefin
SBPT	Small-Modular-Weight Basic Protein Toxin
SBR	Styrene-Butadiene Rubber
SBRC	Santa Barbara Research Center [US]
SBS	Satellite Business System
	Serially Balanced Sequence

	Silicon Bilateral Switch		Selectivity Clear Accumulator
	Subscript (Character)		Sequence Control Area
SBSE	Scintillator Backscattered Electron Detector		Single Channel Analyzer
SBT	Submarine Bathythermograph		Subsidiary Communications Authorization
	Surface Barrier Transistor	SCADA	Supervisory Control and Data Acquisition
SBU	Station Buffer Unit	SCADS	Scanning Celestial Attitude Determination System
SBX	S-Band Transponder		Simulation of Combined Analog Digital Systems
	Subsea Beacon/Transponder		
SC	Satellite Computer		
	Saturable Core	SCALE	Space Checkout and Launch Equipment
	Scale	SCAM	Spectrum Characteristics Analysis and Measurement
	Science		
	Scottish	SCAMA	Station Conferencing and Monitoring Arrangement
	Scotland		Switching, Conference, and Monitoring Arrangement
	Screwed and Coupled		
	Search Control	SCAMP	Sectionalized Carrier and Multipurpose Vehicle
	Security Council [of UN]		
	Selector Channel	SCAMPS	Small Computer Analytical and Mathematical Programming System
	Self-Closing		
	Self-Contained	SCAN	Select Current Aerospace Notices
	Semiconductor		Self-Correcting Automatic Navigation
	Send Common		Semiconductor Component Analysis Network
	Set Clock		
	Simulation Council		Stock Control and Analysis
	Shaped Charge		Stock Market Computer Answering Network
	Shaping Circuit		Student Career Automated Network
	Short Circuit		Switched Circuit Automatic Network
	Shutdown Controller	SCAND	Scandinavia(n)
	Silver Copper (Wire)	SCANDOC	Scandinavian Documentation Center
	Simple Cubic	SCANIIR	Surface Compositional Analysis by Neutral and Ion Impact Radiation
	Single Column		
	Single Contact	SCANS	Scheduling and Control by Automated Network System
	Single Crystal		
	Slow Curing		System Checkout Automatic Network Simulator
	Society for Cryobiology		
	Solar Cell	SCAP	Silent Compact Auxiliary Power
	South Carolina [US]	SCAPE	Self-Contained Atmospheric Personnel Ensemble Suit
	Spacecraft		
	Special Committee		Self-Contained Atmospheric Protective Ensemble Suit
	Specific Conductivity		
	Splat-Cooled	SCAR	Satellite Capture and Retrieval
	Spreading Coefficient		Scandinavian Council for Applied Research
	Squeeze Cast		Scientific Committee on Antarctic Research
	Standing Committee		Submarine Celestial Altitude Recorder
	Statistics Canada	SCARA	Selective Compliance Assembly Robot Arm
	Steel Casting	SCARAB	Submersible Craft for Assisting Repair and Burial
	Steel Cover		
	Subcommittee	SCARF	Santa Cruz Acoustic Range Facility
	Super Composite		Side-Looking Coherent All-Range Focussed
	Superimposed Coding	SCARS	Software Configuration Accounting and Reporting System
	Superimposed Current		
	Suppressed Carrier	SCAS	Stability and Control Augmentation
	Swing Clamp	SCAT	School and College Ability Test
	Switch Cell		Sequentially-Controlled Automatic Transmitter
	Synchrocyclotron		
S/C	Short Circuit		Share Compiler-Assembler and Translator
	Spacecraft		Small Car Automatic Transit System
Sc	Scandium		Space Communication and Tracking
	Strato-Cumulus		Speed Command of Attitude and Thrust
sc	(scilicet) - namely		Supersonic Commercial Air Transport
SCA	Saskatchewan Construction Association [Canada]		

	Surface-Controlled Avalanche Transistor
SCATANA	Security Control of Air Traffic and Air Navigation Aids
SCATE	Stromberg-Carlson Automatic Test Equipment
SCATHA	Spacecraft Charging at High Altitudes
SCATS	Sequentially-Controlled Automatic Transmitter Start
	Simulation, Checkout and Training System
SCAV	Scavenge
SCAW	Shielded Carbon-Arc Welding
SCB	Segment Control Bit
	Selenite Cystine Broth
	Silicon Circuit Board
	Site Control Block
	Station Control Block
	String Control Byte
	System Control Block
SCC	Satellite Control Center
	Science Council of Canada
	Serial Communications Controller
	Short Circuit Current
	Simulation Control Center
	Single-Conductor Cable
	Single Cotton Covered (Wire)
	Society of Cosmetic Chemists
	Somatic Cell Concentration
	Specialized Common Carrier
	Standards Council of Canada
	Storage Connecting Circuit
	Stress-Corrosion Cracking
SCCIG	Stress Corrosion Crack Initiation and Growth
SCCM	Standard Cubic Centimeter per Minute
SCCS	Ship Command and Control System
	Source Code Control System
SCD	Satellite Control Department
	Screwed
	Source Control Drawing
	Space Control Document
	Specification Control Drawing
	Subcarrier Discriminator
ScD	Doctor of Science
SCDC	Steel Castings Development Center
SCDP	Society for Certified Data Processors
SCDSB	Suppressed-Carrier Double Sideband
SCE	Saturated Calomel Electrode
	Signal Conditioning Equipment
	Single Charge Exchange
	Single Cycle Execute
	Society of Christian Engineers
	Standard Calomel Electrode
SCEA	Signal Conditioning Electronic Assembly
SCEDR	Special Committee on Electronic Data Retrieval
SCEL	Signal Corps Engineering Laboratory [US]
SCEO	System Civil Engineering Office
SCEPC	Senior Civil Emergency Planning Committee [of NATO]
SCEPTRE	Systems for Circuit Evaluation and Prediction of Transient Radiation Effects
SCEPTRON	Spectral Comparative Pattern Recognizer
SCERT	Systems and Computers Evaluation and Review Technique
SCF	Satellite Control Facility
	Self-Consistent Field
	Sequential Compatibility Firing
	Short Chain Fat
	Single Crystal Ferrite
	Standard Cubic Feet
	Stress Concentration Factor
	System Control Facility
SCFEL	Standard COSMIC Facility Equipment List
SCFH	Standard Cubic Feet per Hour
SCFM	Standard Cubic Feet per Minute
	Subcritical Fracture Mechanics
SCFMO	Self-Consistent Field Molecular Orbital
SCG	Silicon Carbide Graphite
	Slow Crack Growth
	Solution Crystal Growth
SCH	Schedule(r)
	Socket Head
SCHEM	Schematic
SCHED	Schedule
SCHG	Supercharge(r)
S-CHG	Supercharge(r)
SCHM	Schematic
SCHMOO	Space Cargo Handler and Manipulator for Orbital Operations
SCI	Science
	Science Citation Index
	Scientific
	Serial Communications Interface
	Society of Chemical Industry
	Society of Computer Intelligence
	Switched Collector Impedance
SCIC	Semiconductor Integrated Circuit
	Single-Column Ion Chromatography
SCICON	Scientific Control
SCIAS	Society of Chemical Industry, American Section
SCIC	Saskatchewan Council for International Cooperation [Canada]
SCIF	Static Column Isoelectric Focusing
SCIGR	Standing Committee on International Geoscientific Relations
SCIM	Speech Communication Index Meter
SCIP	Scanning for Information Parameters
	Self-Contained Instrument Package
SCIR	Standing Committee for Installation Rebuilding [UK]
SCL	Space Charge Layer
SCLC	Space-Charge-Limited Current
SCM	Service Command Module
	Signal Conditioning Module
	Small Core Memory
	Software Configuration Management
	STARAN Control Module
SCMA	Southern Cypress Manufacurers Association [US]
	Systems Communications Management Association

SCMM	Semiconductor Memory Module	**SCR**	Scanning Control Register
SC-MOSFET	Surface-Channel Metal-Oxide		Screen
	Semiconductor Field-Effect Transistor		Screw
SCN	Satellite Control Network		Selective Catalytic Reduction
	Scanner		Selective Chopper Radiometer
	Sensitive Command Network		Semiconductor-Controlled Rectifier
	Shortest Connected Network		Silicon-Controlled Rectifier
	Succinonitrile		Short-Circuit Radio
SCNA	Sudden Cosmic-Noise Absorption	**SCRAM**	Selective Combat Range Artillery Missile
SCNR	Scientific Committee of National	**SCRAMJET**	Supersonic Combustion Ramjet
	Representatives [of SHAPE]	**SCRAP**	Super-Caliber Rocket-Assisted Projectile
SCO	Subcarrier Oscillator	**SCRATA**	Steel Casting Research and Trade
SCOF	Self-Contained Oil-Filled		Association [UK]
SCOMO	Satellite Collection of Meteorological	**SCRI**	Science Court and Research Institute
	Observations	**SCRIPT**	Scientific and Commercial Interpreter and
SCOOP	Scientific Computation of Optimal Programs		Program Translator
SCOOPS	Scheme Object-Oriented Programming	**SCRN**	Screen
	System	**SCRTD**	South California Rapid Transit District [US]
SCOPE	Schedule-Cost-Performance	**SCS**	Satellite Communication System
	Sequential Customer Order Processing		Scientific Computer System
	Electronically		SEM (= Scanning Electron Microscope)
	Special Committee on Paperless Entries		Cold Stage
	Specifiable Coordinating Positioning		Sensor Coordinate System
	Equipment		Silicon-Controlled Switch
	System for Coordination of Peripheral		Simulated Compton Scattering
	Equipment		Simulation Control Subsystem
SCOPEP	Steering Committee on the Performance of		SNA (= System Network Architecture)
	Electrical Products		Character String
SCOR	Scientific Computer On-Line Resource		Society for Computer Simulation
	Scientific Committee on Oceanographic		Southern Computer Service [US]
	Research		Space Cabin Simulator
	Self-Calibrating Omnirange		Speed Control Switch
SCORE	Satellite Computer Operated Readiness		Stabilization Cabin Simulator
	Equipment		Stabilization and Control System
	Selection Copy and Reporting		Standard Coordinate System
	Signal Communications by Orbiting Relay	**SC&S**	Strapped, Corded and Sealed
	Equipment	**SCSE**	South Carolina Society of Engineering [US]
SCORPI	Subcritical Carbon-Moderated Reactor	**SCSI**	Small Computer System Interface
	Assembly for Plutonium Investigations	**SCSP**	Supervisory Computer Software Package
SCOST	Special Committee on Space Technology	**SCSR**	Ship Construction Subsidy Regulations
SCOSTEP	Scientific Committee on Solar Terrestrial	**SCT**	Scanning Telescope
	Physics		Schottky Clamped Transistor
SCOT	Scotland		Single Cassette Tape Reader
	Scottish		Subroutine Call Table
SCOUT	Surface-Controlled Oxide Unipolar		Surface Charge Transistor
	Transistor		Surface-Crack Tension (Specimen)
SCOWR	Special Committee on Water Research [of	**SCTE**	Society of Cable Television Engineers
	ICSU]		Society of Carbide and Tool Engineers
SCP	Safety Control Program	**SCTL**	Short-Circuited Transmission Line
	Serial Character Printer	**SCTO**	Stalled Call Timed Out
	Single-Cell Protein	**SCTOC**	Satellite Communication Test Operations
	Spherical Candlepower		Center
	Surveillance Communication Processor	**SCTP**	Straight Channel Tape Print
	Symbolic Conversion Program	**SCTPP**	Straight Channel Tape Print Program
	System Control Program	**SCTTL**	Schottky Clamped Transistor-Transistor
SCPC	Single Channel per Carrier		Logic
SCPD	Scratch Pad	**SCU**	Scanner Control Unit
SCPI	Scientists' Committee for Public		Sensor Control Unit
	Information		Single Conditioning Unit
SCPRF	Structural Clay Products Research		Station Control Unit
	Foundation		Sulfur Coated Urea

SCUBA	Self-Contained Underwater Breathing Apparatus [also: Scuba]
SCUC	Satellite Communications Users Conference
SCUL	Simulation of the Columbia University Libraries [US]
SCUP	School Computer Use Plan
	Scupper
SCV	Subclutter Visibility
SCW	Supercritical Wing
SCWG	Satellite Communications Working Group [of NATO]
SCWIST	Society for Canadian Women in Science and Technology
SCWM	Special Commission on Weather Modification
SCX	Strong Cation Exchange
SD	Seasoned Dry
	Second Difference
	Selenium Diode
	Signal Distributor
	Soft-Drawn
	South Dakota [US]
	Speech Detector
	Square-Law Detector
	Standard Data
	Standard Deviation
	Standard Displacement
	Strength Differential
	Sweep Driver
	Synchronous Detector
SDA	Screen Design Aid
	Shaft Drive Axis
	Software Development Association
	Source Data Acquisition
	Source Data Automation
	Symbol Device Address
SDAD	Satellite Digital and Analog Display
SDADS	Satellite Digital and Analog Display System
S Dak	South Dakota [US]
SDAP	Systems Development Analysis Program
SDAS	Scientific Data Automation System
SDAT	Spacecraft Data Analysis Team [of NASA]
	Symbol Device Allocation Table
S-DAT	Stationary-Head Digital Audio Tape
SDB	Society for Developmental Biology
	Strength and Dynamics Branch [US Air Force]
SD BL	Sand Blast
SDBP	Small Database Package
SDC	Semiconductor Devices Council [of JEDEC]
	Signal Data Converter
	Society of Dyers and Colorists
	Stabilization Data Computer
	Submersible Decompression
	Submersible Diving Chamber
	Synchro-to-Digital Converter
SDCA	Society of Dyers and Colorists of Australia
SDCC	San Diego Computer Center [US]
SDCE	Society of Die Casting Engineers [US]
SDCR	Source Data Communication Retrieval
SDD	Selected Dissemination of Documents

	Synthetic Dynamic Display
SDDTTG	Stored Data Definition and Translation Task Group
SDE	Society of Data Educators
	Students for Data Education
	System Development Engine
SDF	Ship Design File
	Simplified Directional Facility
	Standard Data Format
	Supergroup Distribution Frame
SDFC	Space Disturbance Forecast Center [of ESSA]
SDG	Siding
	Simulated Data Generator
SDH	Self-Dumping Hopper
	Sorbitol Dehydrogenase
	Succinic Dehydrogenase
SDI	Selective Dissemination of Information
	Silt Density Index
	Source Data Information
	Standard Disk Interface
	Strategic Defense Initiative [US]
SDIO	Strategic Defense Initiative Organization [US]
SDL	Saddle
	Space Disturbances Laboratory [ESSA]
	System Descriptive Language
	System Directory List
SDLC	Synchronous Data Link Control
SDLC/BSC	Synchronous Data Link Control/Binary Synchronous Communication
SDM	Schwarz Differential Medium
	Standardization Design Memorandum
	STARAN Debug Module
	Statistical Delta Modulation
	Symmetric Dimer Model
SDMA	Space-Division Multiple Access
SDMS	Software Development and Maintenance System
SDN	Software Defined Network
SDP	Site Data Processor
	Signal Data Processor
	Slowing-Down Power
	Sputter Depth Profiling
	Stationary Diffraction Pattern
	Sulfonyldiphenol
SDPL	Servomechanisms and Data Processing Laboratory [US]
SDR	Search Decision Rule
	Sender
	Small Development Requirement
	Statistical Data Recorder
	System Design Review
SDRT	Slot Dipole Ranging Test
SDS	Safety Data Sheet
	Salzgitter-Dolomit-Schlacke (Process)
	Satellite Data System
	Scientific Data Systems
	Shared Data Set
	Simulation Data Subsystem
	Sodium Dodecylsulfate

	Space Documentation Service [of EURODOC]
	Supplier Delivery Schedule
	System Data Synthesizer
SDS-PAGE	Sodium Dodecylsulfate - Polyacrylamide Gel Electrophoresis
SDSW	Sense Device Status Word
SDT	Simulated Data Tape
	Single Delayed Trapping
	Start-Data-Traffic
	Step-Down Transformer
SDTA	Stress-Dependent Thermal Activation
SDW	Standing Detonation Wave
S+DX	Speech plus Duplex
SE	Secondary Electrode
	Secondary Electron
	Shielding Effect
	Slip End
	Small End
	Smoke Eliminator
	Society of Engineers
	Southeast
	Spectroscopic Ellipsometry
	Standard Error
	Starch Equivalent
	Steam Emulsion
	Street Elbow (Pipe Fitting)
	Sulfoethyl
	System Enhanced
	Systems Engineer
Se	Selenium
SEA	Software Engineering Architecture
	Southeast Asia
	Spherical Electrostatic Analyzer
	Statistical Energy Analysis
	Sudden Enhancement of Atmospherics
	Systems Effectiveness Analyzer
SEAC	Standard's Eastern Automatic Computer [of NBS]
SEACOM	Southeast Asia Commonwealth Cable
SEAISI	Southeast Asia Iron and Steel Institute
SEAL	Signal Evaluation Airborne Laboratory [of FAA]
	Standard Electronic Accounting Language
SEALS	Severe Environmental Air Launch Study
	Stored Energy Actuated Lift System
SEAM	Surface Environment and Mining Program
SEAN	Scientific Event Alert Network
SEAOC	Structural Engineers Association of California [US]
SEAPEX	Southeast Asia Petroleum Exploration Society
SEAS	(Committee for the) Scientific Exploration of the Atlantic Shelf
	School of Engineering and Applied Science
SEASCO	Southeast Asia Science Cooperation Office [India]
SEATO	Southeast Asia Treaty Organization
SEB	Source Evaluation Board
SEBS	Styrene-Ethylene-Butadiene-Styrene
SEC	Sanitary Engineering Center [US]

	Secondary Electron Conduction
	Secondary Emission and Conduction
	Secretary
	Section
	Simple Electronic Computer
sec	secant
	second
	secondary [chemicals]
	(secundum) - according to
SECA	Solar Energy Construction Association
SECAL	Selective Calling
SECAM	Sequential Color and Memory
SECAP	System Experience Correlation and Analysis Program
SECAR	Secondary Radar
	Structural Efficiency Cones with Arbitrary Rings
SECCAN	Science and Engineering Clubs of Canada
sec-ft	second-foot [unit]
sech	hyperbolic secant
SECNAV	Secretary of the Navy [US]
SECO	Self-Regulating Error-Correct Coder-Decoder
	Sequential Coding
	Sustainer-Engine Cutoff
SECON	Secondary Electron Conduction
SECOR	Sequential Collation of Range
SECPS	Secondary Propulsion System
SECS	Sequential Events Control System
	Solar Energy Conversion Systems
SECS	Sections
SECTAM	Southeastern Conference on Theoretical and Applied Mechanics [US]
SECURE	Systems Evaluation Code under Radiation Environment
SECT	Section
	Sectional
SECTL	Sectional
SECY	Secretary
SED	Sanitary Engineering Division
	Space Environment Division [of NASA]
	Spectral Energy Distribution
	Strong Exchange Degeneracy
SEDD	Systems Evaluation and Development Division [of NASA]
SEDIT	Sophisticated String Editor
SEDIX	Selected Dissemination of Indexes
SEDR	Systems Engineering Department Report
SEDS	Space Electronic Detection System
	Society for Educational Data Systems
SEE	Sabotage, Espionage and Embezzlement
	Secondary Electron Emission
	Secondary Emission Enhancement
	Society of Environmental Engineers [US]
	Society of Explosives Engineers
	Southeastern Electric Exchange
SEED	Self Electro-Optic Effect Device
SEEDS	Society, Environment and Energy Development Studies
SEEH	Super Energy-Efficient Home
SEEK	State of the Environment Education Kit [Canada]

	Systems Evaluation and Exchange of Knowledge
SEES	Secondary Electron Emission Sensing
SEF	Scanning Electron Fractograph
	Small End Forward
	Space Education and Foundation
SEFAR	Sonic End Fire for Azimuth and Range
SEFOR	Southwest Experimental Fast Oxide Reactor
SEG	Segment
	Society of Economic Geologists
	Society of Exploration Geophysicists [US]
	Standardization Evaluation Group
	Systems Engineering Group
SEGM	Segment
SEI	Secondary Electron Image
	Secondary Electron Imaging
	Society of Engineering Illustrators [US]
	Systems Engineering and Integration
SEIA	Solar Energy Industries Association
SEIAC	Science Education Information Analysis Center [of ERIC]
SEINAM	Solar Energy Institute of North America
SEIP	Smelter Environmental Improvement Project
	Systems Engineering Implementation Plan
SEIT	Satellite Educational and Informational Television
SEL	Selection
	Selector
	Space Environment Laboratory [US]
	Stanford Electronics Laboratories [US]
	Systems Engineering Laboratories [US]
SELDOM	Selective Dissemination of MARC
SELR	Saturn Engineering Liaison Request
SELS	Selsyn
SEM	Scanning Electron Micrograph
	Scanning Electron Microscopy
	Seminary
	Silylethoxymethyl
	Society for Experimental Mechanics
	Standard Error of Means
	Solar Equipment Manufacturers
SEMA	Specialty Equipment Manufacturers Association
	Spray Equipment Manufacturers Association
SEMCOR	Semantic Correlation
SEME	School of Electrical and Mechanical Engineering
SEMI	Semiconductor Equipment and Material Institute
SEMIRAD	Secondary-Electron Mixed Radiation Dosimeter
SEMLAM	Semiconductor Laser Amplifier
SEMLAT	Semiconductor Laser Array Technique
SEMPE	Socio-Economic Model of the Planet Earth
SEMS	Severe Environment Memory System
SEMY	Seminary
SEN	Senate
	Senator
	Senior
	Sense Command

	Single Edge Notch
	Steam Emulsion Number
SENB	Single End Notch Beam
SENL	Standard Equipment Nomenclature List
SENTOS	Sentinel Operating System
SEO	Satellite for Earth Observation [India]
SEOS	Synchronous Earth Observation Satellite
S/EOS	Standard Earth Observation Satellite
SEP	Search Effectiveness Probability
	Separation
	Separation Parameter
	September
	Solar Energy Program
	Space Electronic Package
	Standard Electronic Package
	Star Epitaxial Planar
	Strain Energized Powder
	Strain Energizing Process
SEPBOP	Southeastern Pacific Biological Oceanographic Program
SEPC	Space Exploration Program Council [of NASA]
SEPE	Single Escape Peak Efficiency
SEPM	Society of Economic Paleontologists and Mineralogists [US]
SEPOL	Settlement Problem-Oriented Language
	Soil-Engineering Problem-Oriented Language
SEPS	Service Module Electrical Power System
	Severe Environment Power System
	Solar Electric Propulsion System
SEPT	September
SEQ	Sequence
	Sequential
seq	(sequens) - the following
SER	Satellite Equipment Room
	Series
	Single Electron Response
	Single-Ended Radiant (Tube)
Ser	Serine
SERAPE	Simulator Equipment Requirements for Accelerating Procedural Evolution
SERB	Study of the Enhanced Radiation Belt [NASA]
SERC	Science and Engineering Research Council [UK]
SERDES	Serializer/Deserializer
SEREP	System Environment Recording and Editing Program
	System Error Recording Editing Program
SERI	Solar Energy Research Institute [US]
SERJ	Supercharged Ejector Ramjet
SERL	Services Electronics Research Laboratory
SERPS	Service Propulsion System
SERR	Serration
SERS	Surface-Enhanced Raman Scattering
	Surface-Enhanced Raman Spectroscopy
SERT	Society of Electronic and Radio Technicians [US]
	Space Electrical Rocket Test
SERV	Service

	Surface Effect Rescue Vehicle
SES	Society of Engineering Science [US]
	Socio-Economic Status
	Solar Energy Society [US]
	Standards Engineering Society [US]
	Strategic Engineering Survey
	Suffield Experimental Station [now: DRES]
	Superexcited Electronic States
	Surface Effect Ship
SE&S	Square Edge and Sound
SESA	Society for Experimental Stress Analysis [US]
SESC	Solar Energy Society of Canada
SESE	Secure Echo-Sounding Equipment
SESL	Space Environment Simulation Laboratory [of NASA]
SESO	Single End Shutoff
SESOC	Single End Shutoff Connector
	Surface Effect Ship for Ocean Commerce
SESOME	Service, Sort and Merge
SESP	Science and Engineering Student Program
SESS	Saskatoon Engineering Student Society [Canada]
	Session
SET	Self-Extending Translator
	Set Endpoint Titration
	Settling
	Single Exposure Technique
	Solar Energy Thermionics
	Systems Environment Team
SETA	Simplified Electronic Tracking Apparatus
SETAR	Serial Event Time and Recorder
SETBC	Society of Engineering Technologists of British Columbia [Canada]
SETE	Secretariat for Electronic Test Equipment [USDOD and NASA]
SETF	STARAN Evaluation and Training Facility
SETI	Search for Extraterrestrial Intelligence [NASA]
SETP	Society of Experimental Test Pilots [US]
SEU	Small End Up
	Source Entry Utility
SECURE	Systems Evaluation Code under Radiation Environment
SEV	Special Equipment Vehicle
SEVAS	Secure Voice Access System
SEW	Sewer
	Satellite Early Warning
	Sonar Early Warning
SEWS	Satellite Early Warning System
SEXAFS	Surface Extended X-Ray Absorption Fine Structure (Spectroscopy)
SF	Safety Factor
	Salt Film (Treatment)
	Sampled Filter
	Scale Factor
	Science Fiction
	Sea Flood
	Semifinished
	Separate Function
	Shift Forward

	Signal-Frequency
	Single Feeder
	Single Frequency
	Slow-Fast (Wave)
	Spontaneous Fission
	Spotface(d)
	Square Feet
	Stacking Fault
	Stopping Factor
	Subject Field
S/F	Store and Forward
SFA	Short Field Aircraft
	Scientific Film Association
	Sunfinder Assembly
SFACA	Solid Fuel Advisory Council of America
SFAPS	Spaceflight Acceleration Profile Simulator [of NASA]
SFAR	System Failure Analysis Report
SFB	Semiconductor Functional Block
SFC	Slow Furnace Cool
	Society of Flavour Chemists
	Solar Forecast Center [of US Air Force]
	Specific Fuel Consumption
	Supercritical Fluid Chromatography
SFCG	Spectrum Fatigue Crack Growth
SFD	Sudden Frequency Deviation
	System Function Description
SFE	Society of Fire Engineers
	Stacking Fault Energy
SFEA	Space and Flight Equipment Association [US]
SFF	Solar Forecast Facility [of US Air Force]
SFG	Signal Frequency Generator
SFI	Support for Innovation
SFIR	Specific Force Integrating Receiver
SFIT	Swiss Federal Institute of Technology
SFK	Step Fixture Key
SFL	Saskatchewan Federation of Labour [Canada]
SFM	Switching-Mode Frequency Multiplier
	Surface Feet per Minute
SFNA	Stabilized Fuming Nitric Acid
SFO	Service Fuel Oil
	Space Flight Operations
SFOC	Space Flight Operations Complex [of NASA]
SFOF	Space Flight Operations Facility [of NASA]
SFOM	Segregation Figures of Merit
SFP	Shop Floor Programming
	Slack Frame Program
SFPE	Society of Fire Protection Engineers [US]
SFPT	Society of Fire Protection Technicians
SFR	Signal Frequency Receiver
SFRC	Steel Fiber Reinforced Concrete
SFRTP	Short-Fiber-Reinforced Thermoplastics
SFS	Saybold Furol Second
	Sorption Filter System
S4S	Surfaced four Sides [lumber]
SFSA	Steel Founders' Society of America
SFSC	Shunt Feedback Schottky Clamped
SFSR	Soviet Federated Socialist Republic
SFSS	Satellite Field Service Station
SFT	Shaft

	Shift
	Simulated Flight Test
	Super-Fast Train
SFTE	Society of Flight Test Engineers
SFTP	Science for the People
SFTR	Shipfitter
SFTS	Standard Frequency and Time Signals
SFU	Simon Fraser University [Canada]
SFXD	Semi-Fixed
SG	Sawtooth Generator
	Scanning Gate
	Screen Grid
	Screw Gauge
	Secretary General
	Set Gate
	Silicon Dioxide Glazing
	Single Girder (Crane)
	Single Groove (Insulators)
	Solution Growth
	Solution of Glucose
	Specific Gravity
	Spheroidal Graphite
	Spin Glass
	Standing Group
	Steam Generator
	Strain Gauge
	Subsurface Geology
	Supergranule
	Supergroup
	Symbol Generator
S/G	Solid/Gas
SGBD	Secondary Grain Boundary Dislocation
SGC	Solicitor General of Canada
	Standard Gas Cycle
	Standard Geographical Classification
	Superior Geocentric Conjunction
SGDF	Supergroup Distribution Frame
SGE	Secondary Group Equipment
	Starch Gel Electrophoresis
SGF	Solution Growth Facility
SGHR	Steam Generating Heavy Water Moderated Reactor
SGHWR	Steam Generating Heavy Water Reactor
SGJP	Satellite Graphic Job Processor
SGL	Society of Gas Lighting
SGLS	Space-Ground Link System
SGN	Scan Gate Number
sgn	signum function
SGOT	Serum Glutamine Oxalacetic Transaminase
SGP	Strain Gauge Package
SGR	Short Growth Rate
	Sodium Graphite Reactor [USEAC]
SGRCA	Sodium Graphite Reactor Critical Assembly
SGS	Strain-Gauge System
	Symbol Generator and Storage
SGSI	Symposium on Gas-Surface Interactions
SGSP	Single Groove, Single Petticoat (Insulator)
SGSR	Society for General Systems Research
SGT	Segment Table
	Strain Gauge Technology
SGUR	Steering Group on Uranium Resources
SGW	Silanized Glass Wool
SGZ	Surface Ground Zero
SH	Scleroscope Hardness
	Scratch Hardness
	Sheet
	Shock
	Shower
	Shunt
	Start of Heading
	Steinhart and Hart (Thermistor)
	Superheater
S/H	Sample and Hold
SHA	Sidereal Hour Angle
SHAB	Soft and Hard Acids and Bases
SH ABS	Shock Absorber
SHADCOM	Shipping Advisory Committee
SHAP	Sintered Hydroxyapatite
SHAPE	Supreme Headquarters, Allied Powers Europe [of NATO]
SHARP	SHIPS Analysis and Retrieval Project
SH BL	Shot Blast
SHC	Superior Heliocentric Conjunction
SH CON	Shore Connection
SHD	Slant Hole Distance
SHE	Spacecraft Handling Equipment
	Spontaneous Hall-Effect
	Standard Hydrogen Electrode
SHEED	Scanning High-Energy Electron Diffraction
SHELV	Shelving
SHEP	Solar High-Energy Particles
SHF	Super-High Feed
	Super-High Frequency
	Super Hyperfine
SHG	Second Harmonic Generation
	Special High Grade
S-HI	System-Human Interaction
SHIEF	Shared Information Elicitation Facility
SHIELD	Sylvania High Intelligence Electronic Defense
SHINCOM	Ship's Integrated Communication System
SHINPADS	Shipboard Integrated Processing and Display System
SHINS	Shipping Instruction
SHIPG	Shipping
SHIPS	Bureau of Ships [US Navy]
SHIRAN	S-Band High-Accuracy Ranging and Navigation
SHIRTDIF	Storage, Handling and Retrieval of Technical Data in Image Formation
SHL	Shell
	Shellac
SHLD	Shield
	Shoulder
SHLMA	Southern Hardwood Lumber Manufacturers Association [US]
SHM	Simple Harmonic Motion
SHN	Scleroscope Hardness Number
SHODOP	Short Range Doppler
SHORAN	Short Range Navigation
SHOT	Society for the History of Technology [US]
SHOW	Scripps Institution - University of Hawaii - Oregon State University - University of Wisconsin [US]

SHP	Shaft Horsepower
SHPDT	Shipping Date
SHPE	Society of Hispanic Professional Engineers [US]
SHPG	Shipping
SHPT	Shipment
SHPNM	Ship-to Name
SHRAP	Shrapnel
SHR	Safety Hoist Ring
SHS	Self-Propagating High-Temperature Synthesis
	Small Hydro Society
	Swivel Head Screw
SHTB	Saskatchewan Highway Traffic Board [Canada]
SHTC	Short Time Constant
SHTG	Shortage
SHTHG	Sheathing
sh tn wt	short ton-weight [unit]
SHY	Syllable Hyphen (Character)
SI	Sample Interval
	Self-Interstitial
	Semi-Insulating (Sample)
	Serial In
	Shift-in (Character)
	Silicone
	Signal-to-Interference
	Signal-to-Intermodulation (Ratio)
	Single Instruction
	Smithsonian Institution [US]
	Space Industrialization
	Specific Inventory
	Station Identification
	Storage Immediate
	Surveyors Institute
	Swap-in
	System Integration
	International System of Units
Si	Silicon
SIA	Self-Interstitial Atom
	Semiconductor Industry Association [US]
	Standard Instrument Approach
	Stereo-Image Alternator
	Subminiature Integrated Antenna
	Systems Integration Area
SIAC	Specialized Information Analysis Center
SIAM	Shipborne Ice Alert and Monitoring
	Society for Industrial and Applied Mathematics [US]
SIAO	Smithsonian Institute Astrophysical Observatory [US]
SIALON	$Si_3Al_3O_3N_5$ [a ceramic]
SIAP	Shipbuilding Industry Assistance Program
SIAT	Single Integrated Attack Team
SIB	Satellite Ionospheric Beacon
	Screen Image Buffer
	Shipbuilding Industry Board [UK]
SIBL	Scanning Ion Beam Lithography
SIBM	Strain-Induced Boundary Migration
SIBS	Stellar-Inertial Bombing System
SIC	Science Information Council

	Semiconductor Integrated Circuit
	Standard Industrial Classification
	Sulphur Impregnated Concrete
SICC	Strain-Induced Corrosion Cracking
SICE	Society of Instrument and Control Engineers
SICEJ	Society of Instrument and Control Engineers of Japan
SiC-LAS	Silicon Carbide in Lithium Aluminosilicate
SICO	Switched in for Checkout
SICS	Semiconductor Integrated Circuits
SID	Seismic Intrusion Detector
	Silicon Imaging Device
	Society for Information Display [US]
	Solubilization by Incipient Development
	Standard Instrument Departure
	Sudden Ionospheric Disturbance
	Symbolic Instruction Debugger
	Synchronous Identification
	Syntax Improving Device
SIDA	Swedish International Development Agency
SIDASE	Significant Data Selection
SIDC	Strategy for International Development and Cooperation
SIDE	Suprathermal Ion Detector Experiment
SIDS	Speech Identification System
	Stellar Inertial Doppler System
SIE	Science Information Exchange
	Single Instruction Execute
SIF	Selective Identification Feature
	Sound Intermediate Frequency
	Stress Intensity Factor
SIFT	Share Internal FORTRAN Translator
SIG	Signal
	Special Interest Group
SIGARCH	Special Interest Group on Computer Architecture [of ACM]
SIGBDP	Special Interest Group on Business Data Processing [of ACM]
SIGBIO	Special Interest Group on Biomedical Computing [of ACM]
SIGCOSIM	Special Interest Group on Computer Systems Installation Management [of ACM]
SIGCSE	Special Interest Group on Computer Science Education [of ACM]
SIGGEN	Signal Generator
SIGI	System for Interactive Guidance and Information
SIGIR	Special Interest Group on Information Retrieval
SIGMA	Shielded Inert Gas Metal-Arc (Welding)
SIGMICRO	Special Interest Group on Microprogramming [of ACM]
SIGOP	Signal Operation
SIGOPS	Special Interest Group on Operating Systems [of ACM]
SIGPLAN	Special Interest Group on Programming Languages [of ACM]
SIGSAM	Special Interest Group for Symbolic and Algebraic Manipulation [of ACM]
SIGSCSA	Special Interest Group on Small Computing Systems and Applications [of ACM]

SIGSD	Special Interest Group for Systems Documentation [of ACM]
SIGUCC	Special Interest Group on University Computer Centers [of ACM]
SII	Standards Institute of Israel
SIL	Silence
	Silicone
	Speech Interference Level
SILS	Silver Solder
	Shipboard Impact Locator System
SIM	Scanning Ion Microscope
	Scientific Instrument Module
	Secondary Ion Microscope
	Selected Ion Monitoring
	Signal Isolator Module
	Simulator
	Society for Industrial Microbiology
	Stress-Induced Martensite
	Superconductor-Insulator-Metal
SIMA	Scientific Instrument Manufacturers Association
SIMAJ	Scientific Instrument Manufacturers Association of Japan
SIMCHE	Simulation and Checkout Equipment
SIMCOM	Simulator Compiler
	Simulator Computer
SIMCON	Scientific Inventory and Management Control
SIMD	Single-Instruction Multiple-Data
SIMDIS	Simulated Distillation Software
SIMICOR	Simultaneous Multiple Image Correlation
SIMILE	Simulator of Immediate Memory in Learning Experiments
SIMM	Single-in-Line Memory Module
	Symbolic Integrated Maintenance Manual
SIMNS	Simulated Navigation System
SIMP	Specific Impulse
SIMPAC	Simplified Programming for Acquisition and Control
	Simulation Package
SIMPP	Simple Image Processing Package
SIMS	Secondary Ion Mass Spectroscopy
	Single-Item, Multi-Source
	Symbolic Integrated Maintenance System
SIMULA	Simulation Language
SIMULCAST	Simultaneous Broadcast [also: simulcast]
SIN	Sensitive Information Network
	Simultaneous Interpenetrating Network
	Social Insurance Number
	Support Information Network
sin	sine
SINAD	Signal plus Noise and Distortion
SINADS	Shipboard Integrated Navigation and Display System
SINAP	Satellite Input to Numerical Analysis and Prediction
SINB	Southern Interstate Nuclear Board [US]
sinh	hyperbolic sine
SINS	Ship's Inertial Navigation System
SINSS	Shipboard Ice Navigation Support System
SIO	Scripps Institute of Oceanography [US]

	Serial Input/Output
	Staged in Orbit
SIOC	Serial Input/Output Channel
SIOP	Selector Input/Output Processor
	Single Integrated Operations Plan
SIOUX	Sequential Iterative Operation Unit X
SIP	Sheetmetal Insert Process
	Short Irregular Pulse
	Single-In-Line Package
	Sintering/Hot Isostatic Pressing
	Sputter Ion Plating
	Standard Information Package
	State Implementation Plan
	Stay-in-Place
	Submerged Injection Process
	Symbolic Input Program
SIPES	Society of Independent Professional Earth Scientists
SIPN	Semi-Interpenetrating Polymer Network [also: S-IPN]
	Simultaneous Interpenetrating Network
S-IPN	Semi-Interpenetrating Polymer Network [also: SIPN]
SIPO	Serial-in, Parallel-out
SIPOP	Satellite Information Processor Operational Program
SIPOS	Semi-Insulating Polycrystalline Silicon
SIPRE	Snow, Ice and Permafrost Research Establishment [US Army]
SIPROS	Simultaneous Processing Operating System
SIPS	Simulated Input Preparation System
	Sputter-Induced Photon Spectroscopy
	SAC (= Strategic Air Command) Intelligence Data Processing System
SIR	Selective Information Retrieval
	Semantic Information Retrieval
	Simultaneous Impact Rate
	Shuttle Imaging Radar
	Special Information Retrieval
	Statistical Information Retrieval
	Submarine Intermediate Reactor
	Supersonic Infantry Rocket
	Symbolic Input Routine
SIRA	Scientific Instrument Research Association [UK]
SIRC	Special Investment Research Contract
SIRS	Salary Information Retrieval System
	Satellite Infrared Spectrometer
SIRSA	Special Industrial Radio Service Association
SIRTF	Space Infrared Telescope Facility [US]
SIRU	Strapdown Inertial Reference Unit
SIS	Satellite Interceptor System
	Science Information Service
	Scientific Instruction Set
	Semiconductor-Insulator-Semiconductor
	Short-Interval Scheduling
	Siltstone
	Simulation Interface Subsystem
	Standards Information System
	Strategic Information System
SISD	Single-Instruction, Single-Data

SISO	Serial-in, Serial-out
SISS	Single-Item, Single-Source
	Submarine Integrated Sonar System
SISTM	Simulation Incremental Stochastic Transistion Matrices
SIT	Separation-Initiated Timer
	Silicon-Intensified Target
	Society of Instrument Technology
	Software Integration Test
	Spontaneous Ignition Temperature
	Stevens Institute of Technology [US]
	Structural Integrity Test
	Subarea Index Table
	System Initialization Table
SITC	Standard International Trade Classification [of UN]
	Standard International Trade Commodity
SITE	Satellite Instructional Television Experiment
	Search Information Tape Equipment
	Spacecraft Instrumentation Test Equipment
	Superfund Innovative Technology Evaluation
SITS	Acetamidoisothiocyanatostilbenedisulfonic Acid
	SAGE Intercept Target Simulator
SITVC	Secondary Injection Thrust Vector Control
SIU	Serial Interface Unit
	Southern Illinois University [US]
SIUE	Southern Illinois University, Edwardsville [US]
SIXPAC	System for Inertial Experiment Priority and Attitude Control
SIZ	Security Identification Zone
SJAC	Society of Japanese Aerospace Companies
SJCC	Spring Joint Computer Conference [US]
SJCM	Standing Joint Committee on Metrication
SJF	Shortest Job First
SJP	Safe Job Procedures
SK	Saskatchewan [Canada]
	Sketch
	Sink
	Skip
SKL	Skip Lister
SKOR	Sperry-Kalman Optimum Reset
SKP	Snoek-Koester Peak
SKS	Scanning Kinetic Spectroscopy
SKT	Socket
SKU	Stockkeeping Unit
SL	Subscriber's Lines
	Sea Level
	Searchlight
	Sensation Level
	Single-Loop
	Slate
	Slide
	Source Level
	Spacelab
	Space Lattice
	Standard Label
	Straight Line
	Streamline
	Stretcher-Levelled

S/L	Solid/Liquid
	Spacelab
sl	(sine loco) - without place
SLA	Spacecraft Lunar Module Adapter [NASA]
	Special Libraries Association [US]
SLAB	Small Amount of Bits [also: slab]
SLAC	Stanford Linear Accelerator Center [US]
SLAET	Society of Licenced Aircraft Engineers and Technicians
SLAG	Safe Launch Angle Gate
SLAM	Scanning Laser Acoustic Microscopy
	Simulation Language for Alternative Modeling
	Space-Launched Air Missile
	Strategic Low-Altitude Missile
	Supersonic Low-Altitude Missile
SLAMS	Simplified Language for Abstract Mathematical Structures
SLANG	Systems Language
SLANT	Simulator Landing Attachment for Night Landing Training
SLAP	Symbolic Language Assembler Program
S/LAP	Shiplap
SLAR	Side-Looking Airborne Radar
SLASH	Seiler Laboratory ALGOL Simulated Hybrid
SLAST	Submarine-Launched Anti-Surface Ship Torpedo
S LAT	South Latitude
SLATE	Small Lightweight Altitude Transmission Equipment
	Stimulated Learning by Automated Typewriter Environment
SLB	Scanning Laser Beam
	Side-Lobe Blanking
SLBM	Ship-Launched Ballistic Missile
	Submarine-Launched Ballistic Missile
SLBMDWS	Submarine-Launched Ballistic Missile Detection and Warning System
SLC	Selector Channel
	Side Lobe Cancellation
	Side Lobe Clutter
	Simulated Linguistic Computer
	Single Layer Ceramic
	Space Launch Complex
	Straight-Line Capacitance
	Sustained-Load Cracking
SLCA	Single Line Communications Adapter
SLCSAT	Submarine Laser Communications Satellite
SLCB	Single-Line Color Bar
SLCC	Saturn Launch Computer Complex
SLCD	Spring-Loaded Camming Device
SLCM	Sea-Launched Cruise Missile
SLCP	Saturn Launch Computer Program
SLD	Simulated Launch Demonstration
	Slim Line Diffuser
	Solder
SLDC	Single-Loop Digital Controller
SLDR	Solder
	System Loader
SLE	Society of Logistics Engineers [US]
	Superheat Limit Explosion

SLEAT	Society of Laundry Engineers and Allied Trades
SLEDGE	Simulating Large Explosive Detonable Gas Experiment
SLEP	Second Large ESRO Project
SLEW	Static Load Error Washout System
SLF	Stud Leveling Foot
	System Library File
SLFK	Sure Lock Fixture Key
SLI	Sea Level Indicator
	Suppress Length Indicator
SLIB	Source Library
	Subsystem Library
SLIC	Selective Listing in Combination
	Shear Longitudinal Inspection Characterization
	Simulation Linear Integrated Circuit
SLIP	Symmetric List Processor
SLIS	Shared Laboratory Information System
SLL	Satellite Line Link
SLM	Statistical Learning Model
SLO	Swept Local Oscillator
SLOCOP	Specific Linear Optimal Control Program
SLP	Segmented Level Programming
	Skip-Lot Plan
	Sound Level Pressure
	Source Language Processor
SLPH	Solid Photography
SLPP	Serum Lipophosphoprotein
SLQTM	Sales Quantity This Month
SLR	Side-Looking Radar
	Simple Left-to-Right
	Single Lens Reflex (Camera)
	Storage Limits Register
SLRN	Select Read Numerically
SLRV	Surveyor Lunar Roving Vehicle
SLRC	Salt Lake Research Center [US]
SL/RN	Stelco-Lurgi/Republic Steel-National Lead (Process)
SLS	Saskatchewan Land Surveyors [Canada]
	Segment Long Spacing
	Side-Lobe Suppression
	Side-Looking Sonar
	Strained-Layer Superlattice
	Superlarge Scale
	Superlattice Structure
SLSA	Saskatchewan Land Surveyors Association [Canada]
	St. Lawrence Seaway Authority
SLSI	Superlarge Scale Integration
SLT	Simulated Launch Test
	Solid-Logic Technique
	Solid-Logic Technology
SLTC	Society of Leather Trades Chemists
SLU	Secondary Logic Unit
SLV	Satellite Launch Vehicle [of NASA]
	Sleeve
SLW	Specific Leaf Weight
SLWL	Straight-Line Wavelength
SM	Scientific Memorandum
	Security Mechanism

	Service Module
	Sequence and Monitor
	Set Mode
	Shared Memory
	Shell Model
	Siemens-Martin (Steelmaking)
	Solid Modelling
	Special Memorandum
	Storage Mark
	Strategic Missile
	Student Manual
	Surface Measure
	Surface Mount
	System Mechanics
Sm	Samarium
SMA	Science Masters Association
	Semimajor Access
	Sequential Multiple Analyzer
	Shielded Metal-Arc
	Standard Methods Agar
	Styrene Maleic Anhydride
	Surface Modeling and Analysis
SMAB	Solid Motor Assembly Building
SMAC	Shielded Metal-Arc Cutting
	Special Mission Attack Computer
SMACNA	Sheetmetal and Air Conditioning Contractors National Association
SMACS	Simulated Message Analysis and Conversion Subsystem
SMAF	Specific Microphage Arming Factor
SMALGOL	Small Computer ALGOL
SMAP	Small Manufacturers Assistance Program
SMART	Sequential Mechanism for Automatic Recording and Testing
	Systems Management Analysis, Research and Test
	System Monitor Analysis and Response Technique
SMARTIE	Simple Minded Artificial Intelligence
SMATS	Source Module Alignment Test Site
SMAW	Shielded Metal-Arc Welding
	Short-Range Man Portable Antitank Weapon
SMAWT	Short-Range Man Portable Antitank Weapon Technology
SMC	Scientific Manpower Commission
	Sheet Molding Compound
	Small Magellanic Cloud
	Smooth Muscle Cell
	Supply and Maintenance Command [US Army]
	Surface Mounted Component
	Systems, Man and Cybernetics [IEEE Society]
SMCC	Simulation Monitor and Control Console
SMCRA	Surface Mining Control and Reclamation Act
SMD	Singular Multinomial Distribution
	Surface-Mounted Device
	Systems Manufacturing Division
	Systems Measuring Device

SMDC	Sodium Methyl Dithiocarbamate
	Superconductive Materials Data Center
SMDF	SAGE Main Distributing Frame
SME	School of Military Engineering
	Shape Memory Effect
	Small and Medium-Sized Enterprises
	Society of Manufacturing Engineers [US]
	Society of Mechanical Engineers
	Society of Military Engineers
	Society of Mining Engineers [of AIME]
SMEAT	Skylab Medical Experiments Altitude Test
SMED	Single-Minute Exchange of Die
SMEDP	Standard Methods for the Examination of Dairy Products
SMEK	Summary Message Enable Keyboard
S-MEM	Minimum Essential Medium for Suspension Cultures
SMF	Solar Magnetic Field
	System Management Facility
	System Measurement Facility
SMG	Spacecraft Meteorology Group
SMGP	Strategic Missile Group
SMI	Start Manual Input
SMIC	Study of Man's Impact on the Climate
SMIE	Solid-Metal-Induced Embrittlement
SMIEEE	Senior Member of the Institute of Electrical and Electronic Engineers
SMIL	Sawmill
SMILE	Significant Milestone Integration Lateral Evaluation
SMIP	Small Manufacturers Incentive Program
SMIS	Society for Management Information Systems
SMIT	Simulated Mid-Course Instruction Test [NASA]
SMITE	Simulation Model of Interceptor Terminal Effectiveness
Smith Inst	Smithsonian Institution
SMK	Smoke
SMKLS	Smokeless
SML	Symbolic Machine Language
SMLM	Simple-Minded Learning Machine
SMLS	Seamless
SMM	Standard Method of Measurement
	Start of Manual Message
	System Management Monitor
SMMC	System Maintenance Monitor Console
SMMIB	Surface Modification of Metals by Ion Beams
SMMP	Standard Methods of Measuring Performance
SMMT	Society of Motor Manufacturers and Trades
SMO	Small Magnetospheric Observatory
	Stabilized Master Oscillator
SMOBC	Solder Mask over Bare Copper
SMODOS	Self-Modulating Derivative Optical Spectrometer
SMOG	Special Monitor Output Generator
SMOI	Second Moment of Inertia
SMOW	Standard Mean Ocean Water
SMP	Scanning Microscope Photometer

	Spatial Mobilized Planes (Theory)
	Sucrose Monopalmitate
SMPL	Sample
SMPS	Simplified Message Processing Simulation
SMPTE	Society of Motion Picture and Television Engineers [US]
SMR	Solid Moderator Reactor
	Standard Malaysian Rubber
	Standard Mortality Ratio
	Static Maintenance Reactor
	Super Metal Rich
SMRD	Spin Motor Rate Detector
SMRE	Safety in Mines Research Establishment [UK]
SMRVS	Small Modular Recovery Vehicle System [USAEC]
SMS	Shuttle Mission Simulator
	Site-Mixed Slurry
	Small Magnetospheric Satellite [of NASA]
	Standard Modular System
	Strategic Missile Squadron
	Styrene Methylstyrene
	Surface Missile System
	Synchronous Meteorological Satellite [of NASA]
	Systems Maintenance Service
SMSA	Silica and Molding Sands Association
	Standard Metropolitan Statistical Area
SMSAE	Surface Missile System Availability Evaluation
SMSI	Strong Metal-Support Interaction
SMT	Service Module Technician
	Square Mesh Tracking
	Sulfamethazine
	Surface-Mount (Device) Technology
SMTA	Surface-Mount Technology Association [US]
SMTI	Selective Moving Target Indicator
	Southeastern Massachusetts Technological Institute [US]
SMU	Saint Mary University [US]
	Secondary Multiplexing Unit
	Self-Maneuvering Unit
	Southern Methodist University [US]
SMUF	Simulated Milk Ultrafiltrate
SMWG	Strategic Missile Wing
SMWIA	Sheet Metal Workers International Association
SMX	Submultiplexer Unit
SMYS	Specified Minimum Yield Stress
SN	Scientific Note
	Semiconductor Network
	Shipping Note
	Stress-Number (Curve)
	Supernova
	Swivel Nut
S/N	Signal-to-Noise (Ratio)
Sn	(Stannum) - Tin
sn	sine of the amplitude
SNA	Systems Network Architecture
SNACS	Share News on Automatic Coding Systems
	Single Nuclear Attack Case Study [USDOD]

SNAFU	Situation Normal, All Fouled Up
SNAME	Society of Naval Architects and Marine Engineers [US]
SNAP	Simplified Numerical Automatic Programmer
	Space Nuclear Auxiliary Power
	Structural Network Analysis Program
	Succinimidylnitroazidophenylaminoethyl-dithiopropionate
	Synchronous Nuclear Array Processor
	Systems for Nuclear Auxiliary Power
SNBU	Switched Network Backup
SND	Scientific Numeric Database
	Sound
SNDT	Society for Nondestructive Testing
SNEMSA	Southern New England Marine Sciences Association [US]
SNF	System Noise Figure
SNG	Substitute Natural Gas
	Synthetic Natural Gas
SNI	Sequence Number Indicator
SNJ	Switching Network Junction
SNL	Standard Nomenclature List
SNLA	Sandia National Laboratories, Albuquerque [US]
SNLC	Senior NATO Logistician Conference
SNM	Special Nuclear Material
SNME	Society of Naval Architects and Marine Engineers
SNMMS	Standard Navy Maintenance and Material Management System [US Navy]
SNMS	Secondary Neutral Mass Spectrometry
	Sputtered Neutral Mass Spectroscopy
SNORT	Supersonic Naval Ordnance Research Track [US Navy]
SNP	Sodium Nitroprusside
	Soluble Nucleoprotein
SNPM	Standard and Nuclear Propulsion Module
SNPO	Space Nuclear Propulsion Office [of NASA]
SNPS	Satellite Nuclear Power Station
SNR	Signal-to-Noise Ratio
	Supernova Remanent
SNS	Space Navigation System
	Spallation Neutron Source
SNT	Sign-on Table
SNX	Succinonitrile
SO	Serial Out
	Shift-out (Character)
	Shop Order
	Shutoff
	Signal Oscillator
	Slow Operating
	South
	Southern Oscillation
	Standoff
	Suboffice
	Switchover
SOA	State-of-the-Art
SOAC	System on a Chip
SOAP	Self-Optimizing Automatic Pilot
	Society of Office Automation Professionals
	Spectrometric Oil Analysis Program
	Symbolic Optimizer and Assembly Program
SOB	Shipped on Board
SOC	Satellite Operations Center [US]
	Separated Orbit Cyclotron
	Set Overrides Clear
	Simulation Operations Center
	Single Orbit Computation
	Society
	Socket
	Space Operations Center
	Specific Optimal Controller
	Superposition of Configuration
SOCC	Salvage Operational Control Center
SOCKO	Systems Operational Checkout
SOCM	Standoff Cluster Munitions
SOCO	Switched-out for Checkout
SOCOM	Solar Optical Communications System
SOCR	Sustained Operations Control Room
SOCRATES	Study of Complementary Research and Teaching in Engineering Science
	System for Organizing Content to Review and Teach Educational Subjects
SOCS	Spacecraft Orientation Control System
SOD	Small Object Detector
	Superoxide Dismutase
SODA	Source-Oriented Data Acquisition
SODAC	Society of Dyers and Colorists
SODAR	Sound Detecting and Ranging [also: Sodar]
SODAS	Structure-Oriented Description and Simulation
SODB	Science Organization Development Board [NAS, US]
S1E	Surfaced one Edge [lumber]
SOEP	Solar-Oriented Experimental Package [of NASA]
SOERO	Small Orbiting Earth Resources Observatory
SOF	Start of Format
	Storage Oscilloscope Fragment
SOFAR	Sound Fixing and Ranging
	Sound Fusing and Ranging
SOFC	Solid-Oxide Fuel Cell
SOFNET	Solar Observing and Forecasting Network [US Air Force]
SOFT	Simple Output Format Translator
	Software
SO GR	Soft Grind
SOH	Start of Heading
SOHC	Single Overhead Camshaft
SOHR	Solar Hydrogen Rocket Engine
SOI	Silicon-on-Insulator
	Southern Oscillation Index
	Specific Operating Instruction
	Standard Operating Instruction
SOIC	Small-Outline Integrated Circuit
SOIS	Shipping Operations Information System
SOJ	Small-Outlet J-Leaded Package
SOL	Simulation-Oriented Language
	Solenoid
	Solution
	System-Oriented Language

SOLAR	Serialized On-Line Automatic Recording
SOLARIS	Submerged Object Locating and Retrieving Identification System
SOLAS	Safety of Life at Sea
SOLE	Society of Logistics Engineers [US]
SOLID	Self-Organizing Large Information Dissemination System
SOLION	Solution Ion
SOLN	Solution
SOLO	Selective Optical Lock-On
SOLOGS	Standardization of Operations and Logistics
SOLRAD	Solar Radiation
SOL TRT	Solution Treatment
SOLV	Solenoid Valve
SOLZ	Second Order Laue Zone
SOM	Scanning Optical Microscopy
	Start of Message
	Space Oblique Mercator
SOMADA	Self-Organizing Multiple-Access Discrete Address
SOMED	School of Mines and Energy Development
SOMP	Start-of-Message Priority
SON	Supra-Optic Nucleus
SONAC	Sonar Nacelle
SONAR	Sound Navigation and Ranging [also: Sonar]
SONCM	Sonar Countermeasures
SONCR	Sonar Control Room
SONIC	System-Wide On-Line Network for Informational Control
SOP	Simulation Operations Plan
	Standard Operating Procedure
	Strategic Orbit Point
	Sum-of-Products
SOPHYA	Supervisor of Physics Analysis
SOPI	Serial-out, Parallel-in
SOPM	Standard Orbital Parameter Message
SOR	Society on Rheology
	Specific Operating Requirement
	Start-of-Record
	Steam/Oil Ratio
	Successive Overrelaxation
	Synchronous Orbital Resonance
SORD	Separation of Radar Data
SORTE	Summary of Radiation Tolerant Electronics
SORTI	Satellite Orbital Track and Intercept
SORTIE	Supercircular Orbital Reentry Test Integrated Environment
SOS	Share Operating System
	Silicon-on-Sapphire
	Scheduled Oil Sampling
	Start-of-Significance
	Symbolic Operating System
S1S	Surfaced one Side [lumber]
SOSI	Serial-out, Serial-in
	Shift-in, Shift-out
S1S1E	Surfaced one Side and one Edge [lumber]
SOSS	Shipboard Oceanographic Survey System
S1S2E	Surfaced one Side and two Edges [lumber]
SOSTEL	Solid-State Electronic Logic
SOT	Syntax-Oriented Translator
SOTA	State-of-the-Art

SOTIM	Sonic Observation of the Trajectory and Impact of Missiles
SOTUS	Sequentially-Operated Teletypewriter Universal Selector
SOUT	Swap-out
SOUTHEASTCON	Southeastern Convention [of IEEE]
SOV	Shutoff Valve
SOW	Standoff Weapon
	Start-of-Word
SOx	Solid Oxygen
SP	Scratch Pad
	Screw Pump
	Self-Propelled
	Sequential Phase
	Service Package
	Shear Plate
	Shift Pulse
	Shipping Port
	Shoulder Pin
	Signal Processor
	Single-Pole
	Single Programmer
	Small Pica
	Softening Point
	Soil Pipe
	Space (Character)
	Spain
	Spanish
	Spare
	Spare Part
	Special Paper
	Special Publication
	Speed
	Splash-Proof
	Sputter
	Stack Pointer
	Standpipe
	Static Pressure
	Structured Programming
	Sulphopropyl
	Summary Punch
	Supervisory Package
	Supervisory Printer
	Symbol Programmer
	System Processor
S&P	Systems and Programming
sp	(sine prole) - without issue
SPA	S-Band Power Amplifier
	Servo Power Assembly
	Servo Preamplifier
	Small Part Analysis
	Software Publishers Association
	Spectrum Analyzer
	Stereo Power Amplifier
	Sudden Phase Anomaly
	Systems and Procedures Association [US]
SP-A	Slightly Polishing-Abrasive
SPAC	Spacecraft Performance Analysis and Command [of NASA]
	Spatial Computer
SPACE	Self-Programming Automatic Circuit Evaluator

	Sequential Position and Covariance Estimation
	Sidereal Polar Axis Celestial Equipment
	Spacecraft Prelaunch Automatic Checkout Equipment
SPACECOM	Space Communications
SPACON	Space Control
SPAD	Satellite Position Predictor and Display
	Satellite Protection for Area Defense
	Simplified Procedures for Analysis of Data
SPADATS	Space Detection and Tracking System
SPADE	SCPC/PCM Multiple Access Demand Assigned Equipment
	Spare Parts Analysis, Documentation and Evaluation
	SPARTA Acquisition Digital Equipment
	Sperry Air Data Equipment
SPADES	Solar Perturbation and Atmospheric Density Measurement Satellite
SPADETS	Space Detection System
SPADNS	Sulfophenylazodihydroxynaphthalenedisulfonic Acid
SPAM	Satellite Processor Access Method
	Scanning Photoacoustic Microscopy
	Ship Position and Attitude Measurement
SPAN	Solar Particle Alert Network [of NASA]
	Statistical Processing and Analysis
	Stored Program Alphanumerics
SPANRAD	Superposed Panoramic Radar Display
SPAQUA	Sealed Package Quality Assurance
SPAR	Seagoing Platform for Acoustics Research
	Space Processing Applications Rocket
	Special Products and Applied Research
	Symbolic Program Assembly Routine
	Synchronous Position Altitude Recorder
SPARC	Space Air Relay Communications
	Space Program Analysis and Review Council [US Air Force]
	Space Research Conic [NASA]
SPARCS	Solar Pointing Aerobee Rocket Control System
SPARS	Space Prediction Attitude Reference System
	Society for Professional Audio Recording Studios
SPARSA	Sferics Pulse, Azimuth, Rate and Spectrum Analyzer
	Sferics, Position, Azimuth, Range Spectrum Analyzer
SPARTA	Spatial Antimissile Research Tests in Australia
	Special Antimissile Research Tests in Australia
SPARTECA	South Pacific Regional Trade and Economic Cooperation Agreement
SPAS	Shuttle Pallet Satellite [FRG]
SPASM	System Performance and Activity Software Monitor
SPASUR	Space Surveillance
SPAT	Silicon Precision Alloy Transistor
SPAYZ	Spatial Property Analyzer
SPBC	Snap Pack Battery Cartridge

SPBN	Sulfophenylbutylnitrone
SPC	Solid Proton Conductor
	Soy Protein Council
	Static Power Converter
	Statistical Process Control
	Stored Program Command
	Stored Program Control
SPD	Software Product Description
	Surge Protective Device
	Synchronous Phased Detector
SPDC	Specialized Products Distribution Center
SPDL	Spindle
SPDP	Society of Professional Data Processors
	Succinimidyl Pyridyldithiopropionate
SPDS	Safe-Practice Data Sheet
SPDT	Single-Pole, Double-Throw
SPDT DB	Single-Pole, Double-Throw, Double Break
SPDT SW	Single-Pole, Double-Throw Switch
SPE	Society of Petroleum Engineers [of AIME]
	Society of Plastics Engineers [US]
	Solid Phase Epitaxy
	Solid Phase Extraction
	Spherical Probable Error
	Stored Program Element
	Systems Performance Effectiveness
SPEARS	Satellite Photoelectric Analog Rectification System
SPEC	Society of Pollution and Environmental Control
	Specialty
	Specification
	Speech Predictive Encoding Communication
	Stock Precision Engineered Components
	Stored Program Educational Computer
SPECON	System Performance Effectiveness Conference
SPECS	Specifications
SPECT	Single-Photon Emission Computed Tomography
SPED	Supersonic Planetary Entry Decelerator
SPEDAC	Solid-State, Parallel, Expandable, Differential Analysis Computer
SPEDE	State System for Processing Educational Data Electronically
SPEDTAC	Stored Program Educational Transistorized Automatic Computer
SPEED	Selective Potentiostatic Etching by Electrolytic Dissolution
	Self-Programmed Electronic Equation Delinerator
	Subsistence Preparation by Electronic Energy Diffusion
SPEPD	Space Power and Electric Propulsion Division [of NASA]
SPERT	Schedule Performance Evaluation and Review Technique
	Special Power Excursion Reactor Test
SPES	Stored Program Element System
SPET	Solid Propellant Electric Thruster
SPF	Spectrofluorometer
	Superplastic Formability

	Superplastic Forming
SPFA	Steel Plate Fabricators Association [US]
SPF/DB	Superplastic Forming/Diffusion Bonding
SPFP	Single Point Failure Potential
SPFS	Small Payload Flight System
SPG	Scan Pattern Generator
	Single-Point Ground
	Sort Program Generator
	Spring
SP GR	Specific Gravity
SPGS	Secondary Power Generating Subsystem
SPGT	Serum Pyruvic Glutamic Transaminase
SPHE	Society of Packaging and Handling Engineers
SPHER	Spherical
SP HT	Specific Heat
SPI	Single Particle Impact
	Single Program Indicator
	Society of the Plastics Industry [US]
	Sonic Pulse-Echo Instrument
	Specific Productivity Index
	Symbolic Pictorial Indicator
SPIA	Solid Propellant Information Agency
SPIA-LPIA	Solid and Liquid Propellant Information Agency [of JHU]
SPIC	Society of Plastics Industry of Canada
	Spacelab Payload Integration and Coordination
SPICE	Spacelab Payload Integration and Coordination in Europe
SPIDAC	Specimen Input to Digital Automatic Computer
SPIDER	Sonic Pulse-Echo Instrument Designed for Extreme Resolution
SPIE	Scavenging-Precipitation-Ion Exchange
	Self-Programmed Individualized Education
	Simulated Problem, Input Evaluation
	Society of Photo-Optical Instrumentation Engineers [US]
SPIG	Symposium on the Physics of Ionized Gases
SPIN	Science Procurement Information Network
SPIR	Search Program for Infrared Spectra
SPIT	Selective Printing of Items from Tape
SPIW	Special Purpose Individual Weapon
SPKR	Speaker
SPL	Simple Programming Language
	Software Programming Language
	Sound Pressure Level
	Sound Pulse Level
	Spaceborne Programming Language
	Space Physics Laboratory [US]
	Space Programming Language
	Specialty
SPLC	Standard Point Location Code
SPLY	Supply
SPM	Self-Propelled Mount
	Sequential Processing Machine
	Shock Pulse Method
	Signal Processing Modem
	Small Perturbation Method
	Solar Proton Monitor

	Source Program Maintenance
	Strokes per Minute
	Sun Probe-Mars [NASA]
	System Performance Monitor
	Symbol Processing Machine
SPMC	Solid Polyester Molding Compound
SPMS	Solar Particle Monitoring System
	Special Purpose Manipulator System
	System Program Management Survey
SPO	System Program Office
SPOC	Simulated Processing of Ore and Coal
SPOCK	Simulated Procedure for Obtaining Common Knowledge
SPOOL	Simultaneous Peripheral Operations On-Line
SPOT	Satellite Positioning and Tracking
	Smithsonian Precision Optical Tracking
SPPF	Solid Phase Pressure Forming
SP PH	Split Phase
SPPO	Spacelab Payload Project Office [of NASA]
SPQR	Small Profits and Quick Returns
SP4T	Single-Pole, Quadruple-Throw
SP4T SW	Single-Pole, Quadruple-Throw Switch
SPR	Silicon Power Rectifier
	Simplified Practice Recommendation
	Spring
	Sprinkler
	Supervisory Printer Read
SPRA	Space Probe Radar Altimeter
SPRC	Self-Propelled Robot Craft
SPRDA	Solid Pipeline Research and Development Association
SPREADWARE	Spreadsheet Software [also: Spreadware]
SPRI	Scott Polar Research Institute [UK]
SPRITE	Solid Propellant Rocket Ignition Test and Evaluation
SPRS	Signal Processing Router/Scheduler
SPRT	Sequential Probability Ratio Test
	Standard Platinum Resistance Thermometer
SPRU	Science Policy Research Unit
SPS	Samples per Second
	School of Practical Science
	Secondary Power System
	Secondary Propulsion System
	Shrouded Plasma Spray
	Society of Physics Students
	Spark Plug Socket
	Standard Pipe Size
	Statistical Problem Solving
	String Process System
	Subsea Production System
	Super Proton Synchrotron
	Symbolic Program(ming) System
SPSA	Single Phase Statistical Analyzer
SP-SA	Slightly Polishing-Slightly Abrasive
SPSE	Society of Photographic Scientists and Engineers [US]
SPST	Single Pole, Single-Throw
SPSTNODM	Single-Pole, Single-Throw, Normally Open, Double-Make
SPST SW	Single-Pole Single-Throw Switch

SP SW	Single-Pole Switch		Semiregular
SPT	Shared Page Table		Shift Register
	Strength-Probability-Time (Diagram)		Short Range
	Star Point Transfer		Slip Range
	Symbolic Program Tape		Slow Release
	Symbolic Program Translator		Solid Rocket
SPTF	Sodium Pump Test Facility		Sorter Reader
SPTT	Single-Pole, Triple-Throw		Sound Ranging
SP3T	Single-Pole, Triple-Throw		Special Register
SPTT SW	Single-Pole, Triple-Throw Switch		Special Regulation
SP3T SW	Single-Pole, Triple-Throw Switch		Special Report
SPUR	Space Power Unit Reactor		Speed Recorder
SPURM	Special Purpose Unilateral Repetitive		Speed Regulator
	Modulation		Split Ring
SPURT	Spinning Unguided Rocket Trajectory		Spring-Return (Valve)
SPV	Surface Photo Voltage		Sputter Rate
SPW	Spring Plunger Wrench		Stateroom
SPWLA	Society of Professional Well Log Analysts		Status Register
SPWM	Single-Sided Pulsewidth Modulation		Storage Register
SPX	Simplex		Storage and Retrieval
SQ	Sequence		Styrene Rubber
	Square		Subroutine
	Structural Quality		Summary Report
	Superquick		Sweep Rate
sq	(sequens) - the following		Switch Register
SQA	Software Quality Assurance	**S/R**	Send and Receive
	Supplier Quality Assurance		Subroutine
	System Queue Area	**S-R**	Set-Reset
SQUAD	Squadron	**Sr**	Strontium
SQAPS	Supplier Quality Assurance Provisions	**sr**	steradian [unit]
SQC	Statistical Quality Control	**SRA**	Shop Replaceable Assembly
SQ CG	Squirrel Cage		Snubber Reducing Adapter
sq ch	square chain [unit]		Society of Research Administrators
sq cm	square centimeter		Spring Research Association [UK]
SQF	Subjective Quality Factor		Stress-Relief Annealing
sq ft	square foot		Sulfo-Ricinoleic Acid
sq in	square inch		Surveillance Radar Approach
sq km	square kilometer	**SRAM**	Short-Range Attack Missile
SQL	Structured Query Language		Static Random-Access Memory
sq m	square meter	**sRAM**	Static Random-Access Memory
sq mi	square mile	**SRARQ**	Selective Repeat-Automatic Repeat Request
sq mm	square millimeter	**SRATS**	Solar Radiation and Thermospheric Satellite
sq mu	square micron		[Japan]
sqq	(sequentia) - the following ones	**SRB**	Screw Rest Button
SQR	Sequence Relay		Sorter Reader Buffered
sq rd	square rod [unit]		Sulfate-Reducing Bacteria
SQUID	Sperry Quick Updating of Internal	**SRBM**	Short-Range Ballistic Missile
	Documentation	**SRBPP**	Saskatchewan River Basin Planning
	Superconducting Quantum Interference		Program [Canada]
	Device	**SRBSN**	Sintered Reaction-Bonded Si_3N_4
SQUIRE	Submarine Quickened Response	**SRC**	Sanitary Refuse Collectors
SQW	Single Quantum Well		Saskatchewan Research Council [Canada]
	Square Wave		Semiconductor Research Cooperative
sq yd	square yard		Science Research Council [UK]
SR	Saturable Reactor		Sound Ranging Control
	Scanning Radiometer		Source
	Scientific Report		Spokane Research Center [US]
	Selective Ringing		Standard Requirements Code
	Selenium Rectifier		Steel Research Center
	Self-Rectifying		Stimulus/Response Compare
	Semantic Reaction		Stored Response Chain

	Systems Research Center [US]
SRCL	Spectrally-Resolved Cathodoluminescence
SRCRA	Ship Owners Refrigeration Cargo Research Association
SRD	Secret Restricted Data
	Standard Reference Data
SRDAS	Service Recording and Data Analysis System
SRDE	Signals Research and Development Establishment [UK]
SRDL	Signals Research and Development Laboratory [UK]
SRDS	Standard Reference Data System
	Systems Research and Development Service [of FAA]
SRE	Sodium Reactor Experiment
	Speech Recognition Equipment
	Surveillance Radar Element
SRF	Self-Resonant Frequency
	Service Request Flag
	Software Recovery Facility
	Sorter Reader Flow
	Spacecraft Research Foundation
	Surface Roughness Factor
SRFB	Space Research Facilities Branch [NRC, Canada]
SRG	Shift-Register Generator
	Statistical Research Group
SRI	Southwest Research Institute [US]
	Space Research Institute
	Spalling Resistance Index
	Stanford Research Institute [US]
SRIF	Somatostatin Release Inhibiting Factor
SRIM	Structural Reaction Injection Molding
SRL	Savannah River Laboratory [of USAEC]
	Shift Register Latch
	System Reference Library
SRLY	Series Relay
SRM	Scrim-Reinforced Material
	Shock Remanent Magnetization
	Simple Reservoir Model
	Solid Rocket Motor
	Standard Reference Material
	Strategic Reconnaissance Missile
	Surface Reflection Microscope
	System Resources Manager
SRMA	Split-Channel Reservation Multiple-Access
SRME	Submerged Repeater Monitoring Equipment
SRMU	Space Research Management Unit [SRC, US]
SRO	Savannah River Operation [of USAEC]
	Short Range Order
	Standing Room Only
SRP	Signal Reference Point
	Solute Retarding Parameter
SRPES	Synchrotron Radiation Photoemission Spectroscopy
SRQ	Service Request
SRR	Search and Rescue Region
	Serially Reusable Resource
	Shift Register Recognizer

	Sound Recorder Reproducer
SRRB	Search and Rescue Radio Beacon
SRS	Selenium Rectifier System
	Simulated Remote Sites
	Simulated Remote Station
	Sodium Removal Station
	Specialized Rework System
	Statistical Reporting Service [of USDA]
	Strain-Rate Sensitivity
	Student Records System
	Subscriber Response System
	Supplemental Restraint System
	Synchrotron Radiation Source
SRSA	Scientific Research Society of America
SRSCC	Simulated Remote Station Control Center
SRSK	Short-Range Station Keeping
SRSS	Simulated Remote Sites Subsystem
SRT	Search Radar Terminal
	Secondary Ranging Test
	Self-Reinforcing Thermoplastic
	Slow Rise Time
	Society of Radiologic Technologists [US]
	Solids Retention Time
	Speech Reception Threshold
	Standard Radio Telegraph
	Supply Response Time
	Supporting Research and Technology
	Systems Readiness Test
SRTC	Scientific Research Tax Credit
SRTF	Shortest Remaining Time First
SRTS	Self-Regulating/Temperature Source [also: SR/TS]
SRU	Shop Replaceable Unit
SRV	Space Rescue Vehicle
SS	Sandstone
	Satellite Switching
	Secretary of State [US]
	Select Standby
	Semi-Steel
	Set Screw
	Ship Service
	Shoulder Screw
	Signal Strength
	Single Scan
	Single Shot
	Single Space
	Single Strength
	Slow Setting
	Slow-Slow (Wave)
	Sodium Salicylate
	Solid/Solid
	Solid Solution
	Solid State
	Space Simulator
	Space Station
	Spherical Symmetry
	Spin-Stabilized
	Spread Spectrum
	Stainless Steel
	Start-Stop (Character)
	Statistical Standards

	Steamship	SSCNS	Ships Self-Contained Navigation System
	Summing Selector	SSCPA	Single Site Coherent Potential
	Supersonics		Approximation
	Suspended Solids	SSCR	Set Screw
S/S	Steamship	SSCS	Safety and Security Communications
	Source/Sink		System
	Start/Stop (Character)	S&SCS	Scintillation and Semiconductor Counter
ss	(scilicet) - namely		Symposium
SSA	Seismological Society of America	SSCU	Special Signal Conditioning Unit
	Sequential Spectrometer Accessory	SSD	Single Station Doppler
	Serial Systems Analyzer		Solid-State Detector
	Smoke Suppressant Additive		Solid-State Disk
	Solid State Abstracts		Space Systems Division [of US Air Force]
	Spherical Sector Analyzer	SS&D	Synchronization Separator and Digitizer
	Sulfosalicylic Acid	SSDC	Semi-Submersible Drilling Caisson
SSAC	Society for the Study of Architecture in		Single Steel Drilling Caisson
	Canada	SSDD	Single-Sided Double-Density
	Space Science Analysis and Command	SSE	Safety Shutdown Earthquake
	[NASA]		Solid-State Electronics
SSAL	Simplified Short Approach Light		South-Southeast
SSALS	Simplified Short Approach Light System		Special Support Equipment
SSAR	Spin-Stabilized Aircraft Rocket		Sum of Squares Error
SSB	Serial Systems Bus	SSEAT	Surveyor Scientific Evaluation Advisory
	Single Sideband		Team [of NASA]
	Space Science Board [NRC, US]	SSEC	Selective Sequence Electronic Calculator
	Spring Stop Button	SSEP	System Safety Engineering Plan
SSBAM	Single Sideband Amplitude Modulation [also:	SSESM	Spent Stage Experimental Support Module
	SSB-AM]	SSF	Saybolt Seconds Furol
SSBFM	Single Sideband Frequency Modulation [also:		System Support Facility
	SSB-FM]	SSFF	Solid Smokeless Fuels Federation
SSBO	Single Swing Blocking Oscillator	SSFL	Steady-State Fermi Level
SSBSC	Single Sideband Suppressed Carrier [also:	SSFM	Single Sideband Frequency Modulation
	SSB-SC]	SSG	Search Signal Generator
SSBSCAM	Single Sideband-Suppressed Carrier		Second Stage Graphitization
	Amplitude Modulation [also: SSB-		Small Signal Gain
	SC-AM]	SSGS	Standard Space Guidance System
SSBSCASK	Single Sideband-Suppressed Carrier-	SSH	Solid Solution Hardening
	Amplitude Shift Keyed [also: SSB-	SSI	Sector Scan Indicator
	SC-ASK]		Small-Scale Integration
SSBSCOM	Single Sideband Suppressed-Carrier Optical		Storage-to-Storage Instruction
	Modulation [also: SSB-SC-OM]	SSIDA	Steel Sheet Information and Development
SSBSCPAM	Single Sideband-Suppressed Carrier Pulse-		Association
	Amplitude Modulation [also: SSB-	SSIE	Smithsonian Science Information Exchange
	SC-PAM]		[US]
SSC	Sensor Signal Conditioner	SSIG	Single Signal
	Settled Sludge Concentration	SSIP	Shuttle Student Involvement Project [US]
	Ship Structure Committee [US]	SSK	Soil Stack
	Single Specimen Compliance	SSL	Sodium Stearoyl Laclylate
	Sintered Silicon Carbide		Solid-State Lamp
	Solid-State Circuit		Source Statement Library
	Spectroscopy Society of Canada		Space Sciences Laboratory
	Station Selection Code	SSLO	Solid-State Local Oscillator
	Statistical Society of Canada	SSLV	Standard Space Launch Vehicle
	Stellar Simulation Complex	SSM	Satellite System Monitor
	Sulfide Stress Cracking		Single-Sideband Signal Multiplier
	Superconducting Super Collider		Special Sensor Microwave
	Supply and Services Canada		Surface-to-Surface Missile
	Swivel Screw Clamp	SSMA	Spread-Spectrum Multiple Access
SSCA	Strobed Single Channel Analyzer	SSMD	Silicon Stud-Mounted Diode
SSCC	Spin Scan Cloud Camera	SSME	Space Shuttle's Main Engine
SS/CF	Signal Strength/Center Frequency	SSM/I	Special Sensor Microwave/Imager

SSMS	Spark Source Mass Spectroscopy	
SSMT	Stress Survival Matrix Test	
SSMTG	Solid-State and Molecular Theory Group	
SSN	Sintered Silicon Nitride	
	Switched Services Network	
SSO	Steady-State Oscillation	
SSOG	Satellite System Operations Guide	
SSOP	Satellite System Operations Plan	
SSP	Scientific Subroutine Package	
	Signaling and Switching Processor	
	Space Station Program	
	Space Summary Program	
	Steam Service Pressure	
	Sum of Squares and Products	
SSPA	Solid-State Power Amplifier	
SSPC	Steel Structures Painting Council	
SSPF	Signal Structure Parametric Filter	
SSR	Secondary Surveillance Radar	
	Slow Strain Rate	
	Solid State Refining	
	Solid State Relay	
	Stretched Surface Recording	
	Sum of the Squared Residuals	
	Synchronous Stable Relaying	
SSRA	Spread-Spectrum Random Access	
SSRC	Structural Stability Research Council [US]	
	Swedish Space Research Committee	
SSRI	Swedish Silicate Research Institute	
SSRL	Stanford Synchrotron Radiation Laboratory [US]	
SSRS	Start-Stop-Restart System	
SSRT	Slow Strain Rate Technique	
	Slow Strain Rate Testing	
	Subsystem Readiness Test	
SSS	Scientific Subroutine System	
	Shipboard Survey Subsystem	
	Simulation Study Series	
	Small Scientific Satellite	
	Small Solar Satellite	
	Solid-State Switching	
	Steering and Suspension System	
SSSA	Soil Science Society of America	
SSSC	Surface Subsurface Surveillance Center	
SSSD	Single-Sided Single-Density	
SSSF	Self-Scanner Stop Failure	
SSSP	System Startup Service Package	
4S	Society for Social Studies of Science	
SSSV	Subsurface Safety Valve	
SST	Sea Surface Temperature	
	Simulated Structural Test	
	Simultaneous Self-Test	
	Stainless Steel	
	Subsystems Test	
	Supersonic Transport	
	Symmetric Single Trapping	
SSTC	Single Sideband Transmitted Carrier	
SS/TDMA	Satellite-Switched Time-Division Multiple Access	
SSTP	Subsystems Test Procedure	
SSTR	Solid-State Track Detector	
SSTV	Sea Skimming Test Vehicle [US Navy]	

	Slow-Scan Television
SSU	Saybolt Seconds Universal
	Secondary Sampling Unit
	Sequential Shunt Unit
SSUS	Spinning Solid Upper Stage
SSV	Settled Sludge Volume
	Supersonic Vehicle
SSW	Solid-State Welding
	South-Southwest
	Synchro Switch
SSWWS	Seismic Sea-Wave Warning System
SSX	Small System Executive
ST	Sawtooth
	Schmitt Trigger
	Scientific and Technical
	Segment Table
	Select Time
	Short Transverse
	Single-Throw
	Skin Temperature
	Sounding Tube
	Sound Trap
	Space Telescope
	Standard Time
	Start Signal
	Static Thrust
	Station
	Steam
	Steel
	Stone
	Strait
	Street
	Street Tee (Pipe)
	Sucrose Tallowate
S/T	Search/Track
	Specific Heat/Temperature (Curve)
S&T	Science and Technology
	Scientific and Technical
St	Stoke [unit]
STA	Saskatchewan Trucking Association [Canada]
	Science and Technology Agency [Japan]
	Shuttle Training Aircraft
	Slurry Transportation Association
	Solution Treated and Aged
	Station
	Stationary
	Supersonic Tunnel Association
STAAS	Surveillance and Target Acquisition Aircraft System
STAB	Stabilization
STAC	Science and Technology Advisory Committee [MSF]
STACO	Standing Committee
STADAC	Station Data Acquisition and Control
STADAN	Satellite Tracking and Data Acquisition Network [of NASA]
STAE	Specify Task Asynchronous Exit
STAF	Scientific and Technological Applications Forecast
STAFF	Stellar Acquisition Flight Feasibility

STAGG	Small Turbine Advanced Gas Generator
STAGS	Structural Analysis of General Shells
STAI	Subtask ABEND Intercept
STAIR	Structural Analysis Interpretive Routine
STALO	Stabilized Local Oscillator
STAMO	Stabilized Master Oscillator
STAMOS	SORTIE Turnaround Maintenance Operations Simulation
STAMP	Systems Tape Addition and Maintenance Program
STAMPS	Scientific and Technological Aspects of Materials Processing in Space
STAN	Stanchion
	Standard(ization)
STAR	Satellite Telecommunications with Automatic Routing
	Scientific and Technical Aerospace Report
	Self-Testing and Repair
	Space Thermionic Auxiliary Reactor
	Speed through Air Resupply
	Standard Terminal Approach Route
	Star and Stellar Systems Advisory Committee [of ESRO]
	Steerable Array Radar
	Stellar Attitude Reference
	Submarine Test and Research
STARAD	Starfish Radiation (Satellite) [of NASA]
STARAN	Stellar Attitude Reference and Navigation
STARE	Steerable Telemetry Antenna Receiving Equipment
STARFIRE	System to Accumulate and Retrieve Financial Information with Random Extraction
STARS	Satellite Telemetry Automatic Reduction System [of NASA]
	Satellite Transmission and Reception Specialist
START	Selections to Active Random Testing
	Short Term Aid for Research and Technology
	Spacecraft Technology and Advanced Reentry Test
	Strategic Arms Reduction Talks
	Summit Technology and Research Transfer Center
	Systematic Tabular Analysis of Requirements Technique
STAT	Shipping Transit Analysis Tabulation
	Statics
	Stationary
	Statistics
STATE	Simplified Tactical Approach and Terminal Equipment
STATLAB	Statistics Laboratory
STATPAC	Statistics Package
StatsCan	Statistics Canada
STB	Segment Tag Bit
	Subsystems Test Bed
STBD	Starboard
STBS	Stirred Tank Biological Reactor
STBY	Standby

STC	Satellite Test Center [US Air Force]
	Scientific and Technical Committee [of ESRO]
	Sensitive Time Control
	SHAPE Technical Center [NATO]
	Short Time Constant
	Sliding Twin-Crossbar
	Society for Technical Communication
	Sound Transmission Class
	Staff Training Center
	Standard Telephone Cable
	Standard Transmission Code
	System Test Complex
STCC	Spacecraft Technical Control Center
	Standards Council of Canada
STCW	System Time Code Word
STD	Salinity Temperature Depth
	Separations Technology Division [of Alcoa]
	Standard
	Subscriber Trunk Dialing
STDA	Solution Treated and Double Aged
S-TDA	Selenium-Tellurium Development Association
std atm	standard atmosphere [unit]
STD DEV	Standard Deviation
STDM	Synchronous Time-Division Multiplexing
STDN	Space Tracking and Data Network
STE	Society of Tractor Engineers
	Spacecraft Test Equipment
	Sterolester
	Supergroup Translating Equipment
STED	Scanning Transmission Electron Diffraction
STEER	Steering
STEL	Short Term Exposure Limit
STEM	Scanning Transmission Electron Microscopy
	Stay Time Extension Module [NASA]
	Stored Tubular Extendable Member
STEP	Scientific and Technical Exploitation Program
	Simple Transition Electronic Processing
	Space Technology Experiments Platform
	Standard for the Exchange of Product Model Data
	Standard Terminal Program
	Supervisory Tape Executive Program
	Support for Technology-Enhanced Productivity
STEPS	Solar Thermionic Electric Power System
STER	Sterilizer
STEREO	Stereotype
STET	Specialized Technique for Efficient Typesetting
STF	Satellite Tracking Facility [US Air Force]
STG	Space Task Group [of NASA]
	Storage
	Starting
STH	Somatotropin
STI	Scientific and Technical Information
	Speech Transmission Index
STIC	Scientific and Technical Intelligence Center [USDOD]

STID	Scientific and Technical Information Division [of NASA]
STIFF	Stiffener
STIN	Science and Technology Information Network
STINFO	Scientific and Technical Information
STINGS	Stellar Inertial Guidance System
STIP	Stipendiary
STIR	Stirring
STK	Stack
	Stock
	Strake
STL	Schottky Transistor Logic
	Sound Transmission Loss
	Space Technology Laboratory [US]
	Steel
	Studio Transmitter Link
STLO	Scientific and Technical Liaison Office
STL WG	Steel Wire Gauge
STM	Scanning Tunneling Microscopy
	Short-Term Memory
	Standard Thermal Model
	Structural Test Model
	Structural Thermal Model
	System Master Tape
STMIS	System Test Manufacturing Information System
STMT	Statement
STMU	Special Test and Maintenance Unit
STN	Satellite Tracking Network [of NASA]
	Stainless
	Station
	Stone
	Switched Telecommunication Network
STO	Stock-Tank Oil
	System Test Objectives
STOA	Solution Treated and Overaged
STOL	Short Takeoff and Landing
STOM	SPADE Terminal Operator's Manual
STOP	Stable Ocean Platform
	Statistical Operations Processor
	Supersonic Transport Optimization Program [of NASA]
STOPS	Shipboard Toxicological Operational Protective System
STOR	Storage
STORC	Self-Ferrying Trans-Ocean Rotary-Wing Crane
STORET	Storage and Retrieval
STORLAB	Space Technology Operations and Research Laboratory [US]
STORM	Statistically-Oriented Matrix Program
STOVL	Short Takeoff and Vertical Landing
STOW	System Takeoff Weight
	Stowage
STOX	Speech and Telegraphy in Voice Channel
STP	Scientifically Treated Petroleum
	Selective Tape Print
	Sewage Treatment Plant
	Signal Transfer Point
	Silver-Bearing Tough Pitch (Copper)

	Simultaneous Test Procedure
	Simultaneous Track Processor
	Space Test Program
	Special Technical Publication
	Stamp
	Standard Temperature and Pressure
	Stop (Character)
	System Test Plan
STPF	Stabilized Temperature Platform Furnace
STPO	Science and Technology Policy Office
STPP	Sodium Tripolyphosphate
STPSS	Science and Technology Program Support Section [Canada]
STPTC	Standardization of Tar Products Test Committee
STQ	Solution Treated and Quenched
STR	Short-Term Revitalization
	Status Register
	Steamer
	Strainer
	Strait
	Strip
	Stroke
	Structure
	Symbol Timing Recovery
	Synchronous Transmitter Receiver
STRAD	Signal Transmission, Reception and Distribution
	Switching, Transmitting, Receiving and Distributing
STRAP	Star Tracking Rocket Attitude Positioning
	Stellar Tracking Rocket Attitude Positioning
STRATCOM	Strategic Communications Command [US Army]
STRATWARM	Stratospheric Warming
STRAW	Simultaneous Tape Read and Write
STRD	Strand
STRESS	Structural Engineering Systems Solver
STRI	Sequential Tracking, Registration and Information
STRIPS	Standard Research Institute Problem-Solver
STRIVE	Standard Techniques for Reporting Information on Value Engineering
STRL	Structural
STROBES	Shared Time Repair of Big Electronic Systems
STRR	Statistical Treatment of Radar Returns
STRUC	Structure
STRUDL	Structural Design Language
STS	Satellite Tracking Station
	Scanning Tunnel Spectroscopy
	Science and Technology Society [US]
	Space Transportation System
	Special Treatment Steel
	Standard Test Signal
	Static Test Stand
	Structural Transition Section
S2S	Surfaced two Sides [lumber]
S2S1E	Surfaced two Sides and one Edge [lumber]
STT	Standard Tube Test
STU	Secure Telephone Unit

	Story Understander
	Submersible Test Unit
	Systems Test Unit
	System Timing Unit
STUD	Safety Training Update
STUFF	Systems to Uncover Facts Fast
STUMP	Story Understanding and Memory Program
STV	Separation Test Vehicle
	Suction Throttling Valve
	Surveillance Television
STVW	Symmetrical Triangle Voltage Waveform
ST W	Storm Water
STWP	Society of Technical Writers and Publishers [US]
ST WP	Steam Working Pressure
STWY	Stairway
STX	Saxitoxin
	Start-of-Text (Character)
SU	Selectable Unit
	Service Unit
	Set-up
	Sonics and Ultrasonics [IEEE Group]
	Stanford University [US]
	Station Unit
	Storage Unit
	Syracuse University [US]
SUA	State Universities Association [US]
SUAS	System for Upper Atmospheric Sounding
SUB	Submarine
	Submerge
	Subroutine
	Substitute (Character)
SUBCAL	Subcaliber
SUBCON	Subcontracting Industries Exhibition
SUBDIZ	Submarine Defense Identification Zone
SUBIC	Submarine Integrated Control
SUBL	Sublimes
SUBORD	Subordinate
SUBPROGRAM	Subordinate Program [also: Subprogram]
SUBROUTINE	Subordinate Routine [also: Subroutine]
SUBSID	Subsidiary
SUBSTA	Substation
SUBSTR	Substructure
SUC	Saskatchewan Universities Commission [Canada]
SUCT	Suction
SUDAA	Stanford University, Division of Aeronautics and Astronautics [US]
SUDIC	Sulfur Development Institute of Canada
SUDOSAT	Sudan Domestic Satellite
SUDT	Silicon Unilateral Diffused Transistor
SUEDE	Surface Evaluation and Definition
SUGI	SAS Users Group International
SUGG	Suggestion
SUHL	Sylvania Ultrahigh Level Logic
SUI	State University of Iowa [US]
SUIPR	Stanford University, Institute for Plasma Research [US]
SUM	Summary
	Surface-to-Underwater Missile

SUMMIT	Supervisor of Multiprogramming, Multiprocessing, Interactive Time-Sharing
SUMS	Sperry UNIVAC Material System
SUMT	Sequential Unconstrained Minimization Technique
SUN	Sunday
SUNI	Southern Universities Nuclear Institute
SUNY	State University of New York [US]
SUP	Superior
	Support
	Supplement
	Supply
SUPARCO	Space and Upper Atmosphere Research Committee [Pakistan]
SUPERSTR	Superstructure
SUPHTR	Superheater
SUPP	Supplement
SUPPL	Supplement
SUPR	Supervisor
SUPROX	Successive Approximation
SUPT	Superintendent
	Support
SUPV	Supervisor
SUPVR	Supervisor
SUR	Surface
	Surcharge
	Surplus
SURCAL	Surveillance Calibration
SURDD	Southern Utilization Research and Development Division [of USDA]
SURE	Symbolic Utilities Revenue Environment
SURF	Support of User Records and Files
	Surface
SURGE	Sorting, Updating, Report Generating
SURI	Syracuse University Research Institute [US]
SURIC	Surface Ship Integrated Control
SURSAT	Surveillance Satellite
SURSULF	Surface Hardening Sulfur Catalyst
SURV	Survey
SUS	Saybolt Universal Second
	Silicon Unilateral Switch
	Single Underwater Sound
SUSIE	Sequential Unmanned Scanning and Indicating Equipment
	Stock Updating Sales Invoicing Electronically
SUSOPS	Sustained Operations
SUSP	Suspension
SV	Safety Valve
	Sailing Vessel
	Self-Verification
	Simulated Video
	(Single) Silk Varnish
	Solenoid Valve
	Space Vehicle
	Synchronous Voltage
sv	(sub verbo) - under the following word or heading
SVA	Shared Vitual Area
SVC	Secure Voice Communication
	Society of Vacuum Coaters

	Supervisor Call (Instruction)
	Supervisory Cell
SVCS	Star Vector Calibration Sensor
SVER	Spatial Visual Evoked Response
SVH	Solar Vacuum Head
SVIC	Shock and Vibration Information Center
SVM	Silicon Video Memory
SVR	Slant Visual Range
	Supply-Voltage Rejection
SVS	Space Vision System
SVTL	Services Valve Test Laboratory
SVTP	Sound Velocity, Temperature and Pressure
SW	Salt Water
	Sandwich-Wound
	Shortwave
	Single Weight
	Software
	Southwest
	Specific Weight
	Spherical Washer
	Spot Weld(ing)
	Stud (Arc) Welding
	Surface Wave
	Sweden
	Swedish
	Switch
	Switchband-Wound
S/W	Software
SWA	Scheduler Work Area
	Single Wire Armored
SWAC	Standards Western Automatic Computer [of NBS]
SWAD	Special Warfare Aviation Detachment [US Army]
SWADS	Scheduler Work Area Data Set
SWAM	Standing Wave Area Monitor
SWAP	Stress Wave Analyzing Program
SWAT	Sidewinder IC Acquisition Track
	Special Warfare Armored Transporter
	Special Weapons and Tactics
	Stress Wave Analysis Technique
SWB	Single Weight Baryta
SWBD	Switchboard
SW BHD	Swash Bulkhead
SWBS	Ship Work Breakdown Structure
SWC	Stepwise Cracking
SWCH	Switch
SWCL	Sea Water Conversion Laboratory
SWD	Sliding Watertight Door
	Surface Wave Device
SWE	Society of Women Engineers [US]
SWEAT	Student Work Experience and Training
SWED	Sweden
	Swedish
SWEEP	Soil and Water Environmental Enhancement Program [Canada]
SWF	Stress Wave Factor
SWF/ISS	Stress Wave Factor/Interlaminar Shear Strength
SWFR	Slow Write, Fast Read
SWG	Society of Women Geographers

	Standard Wire Gauge
	Steel Wire Gauge
	Stubs (Steel) Wire Gauge
SWG BKT	Swinging Bracket
SWGR	Switchgear
SWIEEECO	Southwestern IEEE Conference and Exhibition [US]
SWIFT	Selected Words in Full Title
	Society for Worldwide Interbank Financial Telecommunications
	Software Implemented Friden Translator
	Strength of Wings including Flutter
SWIP	Standing Wave Impedance Probe
SWIR	Short Wave Infrared
SWIRA	Swedish Industrial Robot Association
SWITT	Surface Wave Independent Tap Transducer
SWL	Safe Working Load
	Shortwave Listener
	Specific Work Load
	Sulfide Waste Liquor
	Surface Wave Lines
SWOF	Switchover Operation Failure
SWOP	Structural Weight Optimization Program
SWOPS	Single Well Offshore Production System
SWP	Safe Working Pressure
SWPA	Steel Works Plant Association
SWR	Sine Wave Response
	Spin Wave Resonance
	Standing-Wave Ratio
	Steel Wire Rope
SWRI	Southwestern Research Institute [US]
SWS	Safe Working Stress
	Shift Word, Substituting
SWSA	Southern Wood Seasoning Association [US]
SWSI	Single Width, Single Inlet
SWST	Society of Wood Science and Technology
SWT	Supersonic Wind Tunnel
SWTL	Surface-Wave Transmission Line
SWW	Severe Weather Warning
SX	Simplex
	Solvent Extraction
SXAPS	Soft X-Ray Appearance Potential Spectroscopy
SXB	Spring Extension Bar
SXPS	Soft X-Ray Photoelectron Spectroscopy
SXS	Step by Step
SXT	Spacecraft Sextant
SY	Stripping Yield
	Symbol
	System
SYCOM	Synchronous Communications
SYD	Sum of the Year's Digits
SYDAS	System Data Acquisition System
SYL	Syllabus
SYM	Symbol
	Symmetry
SYMAN	Symbol Manipulation
SYMAP	Synagraphic Mapping System
SYMPAC	Symbolic Program for Automatic Control
SYN	Synchronous (Idle Character)
	Synthesizer

	Synthetic
	Synonym
SYNC	Synchronization
	Synchronizer
SYNCOM	Synchronous-Orbiting Communications Satellite
SYNCHROMESH	Synchronous Mesh [also: synchromesh]
SYNSCP	Synchroscope
SYNSEM	Syntax and Semantics
SYNT	Synthetic
SYS	System
SYSADMIN	System Administration
SYSCAP	System of Circuit Analysis Programs
SYSCTLG	System Catalogue
SYSGEN	System Generation
SYSIN	System Input
SYSLOG	System Log
SYSPOP	System Programmed Operator
SYSOUT	System Output
SYSRC	System Reference Count
SYST	System
SYSTRAN	Systems Analysis Translator
SZVR	Silicon Zener Voltage Regulator

T

T	Tee
	Temperature
	Terminal
	Tesla
	Test
	Threonine
	Thymine
	Thymidine
	Thermoplastic
	Timer
	Toll
	Tooth
	Torque
	Track
	Transaction
	Translation
	Triode
	Tritium
	Truss
	Tuesday
t	metric ton
	thickness
	time
	transmit
	troy weight [unit]
TA	Tape Address
	Telegraphic Address
	Thermal Analysis
	Time Analyzer
	Transverse Acoustical
	Triacetic

	Turbocharged and Aftercooled
	Turbulence Amplifier
Ta	Tantalum
TAA	Transportation Association of America
	Turbine-Alternator Assembly
TAALS	Tactical Army Aircraft Landing System
TAAR	Target Area Analysis Radar
TAAS	Three Axis Attitude Sensor
TAB	Tablet
	Tabular Language
	Tabulate
	Tape Automated Bonding
	Technical Activities Board [of IEEE]
	Technical Advisory Bureau
	Technical Analysis Branch
	Technical Assistance Board [UN]
TABL	Tablet
TABS	Terminal Access to Batch Service
	Total Automated Broker System
	Total Aviation Briefing Service
TABSAC	Targets and Backgrounds Signature Analysis Center [US]
TABSIM	Tabulating Equipment Simulator
TABSOL	Tabular Systems-Oriented Language
TABSTONE	Target and Background Signal-to-Noise Evaluation
TAC	Technical Advisory Committee
	Telemetry and Command
	Thermostatically-Controlled Air Cleaner
	Time Amplitude Converter
	Total Available Carbohydrates
	TRANSAC Assembler Compiler
	Transformer Analog Computer
	Transistorized Automatic Control
	Translator - Assembler - Compiler
	Transport Advisory Council
	Trapped Air Cushion
	Triallylcyanurate
	Tunneling Association of Canada
TACAMO	Take Charge and Move out
TACAN	Tactical Air Navigation
TACAN-DME	TACAN Distance Measuring Equipment
TACCAR	Time-Averaged Clutter-Coherent Airborne Radar
TACCO	Tactical Coordinator
TACDACS	Target Acquisition and Data Collection System
TACE	Turbine Automatic Control Equipment
TACH	Tachometer
TACL	Time and Cycle Log
TACMAR	Tactical Multifunction Array Radar
TACNAV	Tactical Navigation
TACODA	Target Coordinate Data
TACOL	Thinned Aperture Computed Lens
TACOM	Tank-Automotive Command [US Army]
TACPOL	Tactical Procedure Oriented Language
TACR	Time and Cycle Record
TACRV	Tracked Air Cushion Research Vehicle
TACS	Tactical Air Control System
TACSAT	Tactical Communications Satellite [US]
TACSATCOM	Tactical Satellite Communications

TACT	Technological Aids to Creative Thought
	Transistor and Component Tester
	Transsonic Aircraft Technology
TACV	Tracked Air Cushion Vehicle
TAD	Target Acquisition Data
	Technical Approach Document
	Telephone Answering Device
	Tensile Axis Direction
	Thrust-Augmented Delta [NASA]
	Top Assembly Drawing
tAD	Metric Tons, Air Dried
TADOG	Towed Array Deep Operating Gear
TADS	Teletypewriter Automatic Dispatch System
TADSS	Tactical Automatic Digital Switching System
TAEC	Thiolated Aminoethylcellulose
TAERS	Transportation Army Equipment Record System
TAF	Torque Amplification Factor
TAG	Technical Advisory Group
	Technical Assistance Group
	Transient Analysis Generator
TAI	Time to Auto-Ignition
TAIC	Tokyo Atomic Industrial Consortium [Japan]
TAID	Thrust Augmented Improved Delta [NASA]
TAL	Trans-Alpine [Europe]
TALA	Tactical Landing Approach
TALAR	Tactical Landing Approach Radar
TALS	Transfer Airlock Section
TAM	Telephone Answering Machine
	Terminal Access Method
	Total Available Market
TAMA	Methylanilinium Trifluoroacetate
TAMCO	Training Aid for MOBIDIC Console Operations
TAME	Tosylarginine Methylester
TAMIS	Telemetric Automated Microbial Identification System
TAMS	Tunnel Air Monitoring System
TAMU	Texas A & M University [US]
tan	tangent
tanh	hyperbolic tangent
TANS	Tactical Air Navigation System
TAO	Technical Assistance Order
TAP	Tactical Area Positioning
	T-Angle Plate
	Tape Automatic Positioning
	Tape Automatic Preparation
	Tapping
	Technological Assessment and Planning [Canada]
	Technical Advisory Panel
	Technology Assessment Program [Canada]
	Terminal Applications Package
	Time-Sharing Assembly Program
	Test Article Protector
	Triisoamyl Phosphate
TAPAC	Tape Automatic Positioning Control
TAPE	Tape Automatic Preparation Equipment
	Technical Advisory Panel for Electronics [US Air Force]

TAPP	Tarapur Atomic Power Project [India]
	Two-Axis Pneumatic Pickup
TAPPI	Technical Association of the Pulp and Paper Industry [US]
TAPRE	Tracking in an Active and Passive Radar Environment
TAPS	Tactical Area Positioning System
	Terminal Area Positive Separation [FAA]
	Trimethylaminopropanesulfonic Acid
	Turboalternator Power System
TAPSO	Trimethylaminohydroxypropanesulfonic Acid
TAR	Tactical Air Reconnaissance
	Tactical Attack Radar
	Technical Assistance Request
	Terminal Area Radar
	Terrain-Avoidance Radar
	Trajectory Analysis Room
TARABS	Tactical Air Reconnaissance and Aerial Battlefield Surveillance System
TARAN	Tactical Attack Radar and Navigation
TARC	The Army Research Council [US]
	Trace Analysis Research Center
TARE	Telegraph Automatic Routing Equipment
	Telemetry Automatic Reduction Equipment
	Transistor Analysis Recording Equipment
TARGA	TrueVision Advanced Raster Graphics Adapter
TARGET	Thermal Advanced Reactor Gas-Cooled Exploiting Thorium
TARIF	Technical Apparatus for Rectification of Indifferent Films
TARMAC	Terminal Area Radar Moving Aircraft
TARP	Tarpaulin
TARS	Terrain Analog Radar Simulator
	Three-Axis Reference System
TART	Twin Accelerator Ring Transfer
TAS	Tactical Analysis System
	Target Acquisition System
	Telephone Answering System
	Terminal Address Selector
	True Air Speed
TASC	Tactical Articulated Swimmable Carrier
	Terminal Area Sequencing and Control
TASCC	Tandem Accelerator Superconducting Cyclotron
TASCON	Television Automatic Sequence Control
TASES	Tactical Airborne Signal Exploration System
TAS-F	Trisdimethylaminosulfur Difluoride
TASI	Time-Assignment Speech Interpolation
TAS-PAC	Tactical Analysis System for Production, Accounting and Control
TASR	Terminal Area Surveillance Radar
	Thermal Activation Strain Rate
TASRA	Thermal Activation Strain Rate Analysis
TASS	Tactical Avionics System Simulator
TAT	Target Aircraft Transmitter
	Thrust-Augmented Thor
	Transatlantic Telephone
TATC	Terminal Air Traffic Control
	Transatlantic Telephone Cable

TATCS	Terminal Air Traffic Control System	**TC**	Tab Card
TAV	Transatmosphere Vehicle		Tactical Computer
	Transatmospheric Vehicle		Tantalum Capacitor
TAVE	Thor-Agena Vibration Experiment		Tape Command
TAWCS	Tactical Air Weapons Control System		Tariff Commission
TAXIR	Taxonomic Information Retrieval		Task Control
TB	T-Bolt		Technical College
	Technical Bulletin		Technical Committee
	Terminal Board		Technical Control
	Thermal Barrier		Technical Cooperation
	Tight-Binding		Telecommunications
	Time Base		Telemetry and Command
	Title Block		Telephone Channel
	Torch Brazing		Telescoping Collar
	Torpedo Boat		Television Control
	Transborder		Temperature Coefficient
	Transmitter Blocker		Temperature Compensation
T/B	Title Block		Temperature Control
Tb	Terbium		Temporary Council
TBA	Testbed Aircraft		Terminal Computer
	Tetrabutylammonium		Terminal Control
	Torsional Braid Analysis		Terra Cotta
TBAH	Tetrabutylammonium Hydroxide		Test Conductor
TBA-OH	Tetrabutylammonium Hydroxide		Test Console
TBAX	Tube Axial		Test Control(ler)
TBBF	Top Baseband Frequency		Tetracycline
TBC	Tensile Bolting Cloth		Tetrahedral Cubic
	Thermal Barrier Coating		Thermal Conditioning
	Toss Bomb Computer		Thermal Conductivity
TBD	Target-Bearing Designator		Thermal Control
TBDF	Transborder Data Flow		Thermal Cutting
TBDP	Transborder Data Processing		Thermocompression
TBE	Tetrabromoethane		Thermocouple
TBF	Butylformamidine		Thermocurrent
	Tail Bomb Fuse		Thermoplastic Composite
	Transmitted Bright-Field		Thiocarbamyl
TBHP	Tertbutylhydroperoxide		Threshold Circuit
TBHQ	Tertiary Butylhydroquinone		Thrust Chamber
TBI	Tape and Buffer Index		Thrust Control
	Target-Bearing Indicator		Time Code
TBL	Terminal Ballistics Laboratory [US Army]		Time Constant
TBM	Temporary Bench Mark		Timed Closing
	Terabit Memory		Time to Computation
	Thyssen Basic-Oxygen Metallurgy		Timing Channel
	Tunnel Boring Machine		Tissue Culture
TBME	Tactical Ballistic Missile Experiment		Toggle Clamp
TBMX	Tactical Ballistic Missile Experiment		Total Carbon
TBN	Total Base Number		Total Cholesterol
TBO	Time between Overhauls		Tracking Camera
TBP	Tributylphosphate		Traffic Control
	Tributylphosphoric Acid		Transfer Channel
	True Boiling Point		Transistorized Carrier
TBPO	Tertiary Butylperoctoate		Transmission Control (Character)
TBR	Timber		Transmission Controller
TBRC	Top-Blown Rotary Converter		Transmitting Circuit
TBRI	Technical Book Review Index		Transportation Commodity
TBS	Tight Building Syndrome		Transport Canada
TBT	Tight-Binding Theory		Trip Coil
TBTUP	Tight-Binding Theory with Universal Parameters		True Complement
		Tc	Technetium
TBU	Transmit Baseband Unit	**TCA**	Teach Cable Assembly

	Terminal Control Area		Trillion Cubic Feet
	Thrust Chamber Assembly	TCFTD	Triple Constant Fraction Time
	Tissue Culture Association		Discriminator
	Transmission Control Area	TCG	Transponder Control Group
	Tricarboxylic Acid		Tune-Controlled Gain
	Trichloroacetic Acid	TCH	Thiocarbohydrazide
	Turbulent Contact Absorber	TCI	Telemetry Components Information
TCAB	Tetrachloroazobenzene		Terrain Clearance Indicator
	Twin-Carbon Arc Brazing		Theoretical Chemistry Institute [US]
TCAE	Thermal Coefficient of Area Expansion	TCID	Tissue Culture Infectious Dose
TCAI	Tutorial Computer-Assisted Instruction	TCL	Time and Cycle Log
TCAM	Telecommunication Access Method		Transistor-Coupled Logic
TCAW	Twin-Carbon Arc Welding	TCLE	Thermal Coefficient of Linear Expansion
TCB	Task Control Block	TCM	Telemetry Code Modulation
	Technical Coordinator Bulletin		Temperature Control Model
	Terminal Control Block		Terminal Capacity Matrix
	Trusted Computing Base		Terminal-to-Computer Multiplexer
TCBM	Transcontinental Ballistic Missile		Thermal Conduction Module
TCBOC	Trichlorobutoxycarbonyl		Thermochemical Machining
TCBS	Tesla Coil Builders Society		Thermoplastic Cellular Molding
TCBV	Temperature Coefficient of Breakdown	TCMF	Touch-Calling Multifrequency
	Voltage	TCMP	Toxic Chemicals Management Program
TCC	Technical Coordination Committee	TC-NBT	Thiocarbamyl Nitro Blue Tetrazolium
	Technical Computing Center	TCNE	Tetracyanoethylene
	Telecommunications Coordinating	TCNQ	Tetracyanoquinodimethane
	Committee	TCO	Thermal Cutoff
	Television Control Center		Trunk Cutoff
	Temperature Coefficient of Capacitance	TCP	Technical Cooperation Program [Canada,
	Temporary Council Committee		US, UK and Australia]
	Test Control Center		Test Checkout Procedure
	Test Controller Console		Thrust Chamber Pressure
	Thermal Controlled Coating		Tool Center Point
	Thermofor Catalytic Cracking		Topologically Close-Packed
	Traffic Control Center		Traffic Control Post
	Transfer Channel Control		Transmission Control Protocol
	Transportation Commodity Classification		Transport Control Protocol
	Trichlorocarbanilide		Trichlorophenoxyacetic Acid
TCCSR	Telephone Channel Combination and		Tricresylphosphate
	Separation Racks	TCPC	Tab Card Punch Control
TCD	Telemetry and Command Data	TCP/IP	Transport Control Protocol/Internet
	Temperature Coefficient of Decay		Protocol
	Thyratron Core Driver	TCPL	Trans-Canada Pipeline
	Time Code Division	TCPO	Trichlorophenyloxalate
	Thermal Conductivity Detector	TCR	Telemetry Compression Routine
	Transistor-Controlled Delay		Temperature Coefficient of Resistance
TCDD	Tetrachlorinated Dibenzo-P-Dioxin		Temperature Coefficient of Resistivity
	Tetrachloro-Dibenzo-P-Dioxin		Tracer
TCE	Telemetry and Command Equipment		Transfer Control Register
	Telemetry Checkout Equipment	TC&R	Telemetry Command and Ranging
	Terrace	TCRA	Telegraphy Channel Reliability Analyzer
	Tetrachloroethane	TCRC	Twin Cities Research Center
	Thermal Coefficient of Expansion	TCS	Terminal Communications System
	Total Composite Error		Terminal Countdown Sequencer
	Transmission Control Element		Terminal Count Sequence
TCEA	Training Center for Experimental		The Classification Society
	Aerodynamics [NATO]		The Constant Society
TCED	Thrust Control Exploratory Development		Thermal Conditioning Service
TCENM	Triscyanoethylnitromethane		Thermal Control Subsystem
TCEP	Triscyanoethoxypropane		Tool Coordinate System
TCF	Technical Control Facility		Total Control System
	Terminal Configuration Facility		Traffic Control Station

	Transmission-Controlled Spark	**TDAL**	Tetradecenal
	Transportable Communication System	**TDB**	Traffic Database
	Transportation Communication Service	**TDBI**	Test During Burn-In
	True Chemical Shift	**TDC**	Target Data Collection
TCSC	Trainer Control and Simulation Computer		Telegraphy Data Channel
TCST	Trichlorosilanated Tallow		Time Delay Closing
TCT	Terminal Control Table		Time-to-Digital Converter
	Thrombin Clotting Time		Top Dead Center
	Translator and Code Treatment (Frame)		Track Detection Circuit
TCTS	Trans-Canada Telephone System		Transportation Development Center [Canada]
TCTTE	Terminal Control Table Terminal Entry	**TDCC**	Transportation Data Coordinating Committee
TCU	Tape Control Unit		
	Teletypewriter Control Unit	**TDCM**	Technology Development for the Communications Market
	Terminal Control Unit		
	Transmission Control Unit		Transistor Driver Core Memory
TCVC	Tape Control via Console	**TDCO**	Torpedo Data Computer Operator
TCVE	Thermal Coefficient of Volume Expansion	**TDCTL**	Tunnel-Diode Charge-Transformer Logic
TCW	Time Code Word	**TDD**	Target Detection Device
TCWG	Telecommunications Working Group		Technical Data Digest
TCXO	Temperature-Compensated Crystal Oscillator		Telecommunication Device for the Deaf
	Temperature-Controlled Crystal Oscillator		Telemetry Data Digitalizer
TCZD	Temperature-Compensated Zener Diode	**TDDA**	Tetradecadienyl Acetate
TD	Tabular Data	**TDDL**	Time-Division Data Link
	Tank Destroyer	**TDDR**	Technical Data Department Report
	Tape Drive	**TDE**	Tetrachlorodiphenylethane
	Technical Data	**TDEC**	Technical Division and Engineering Center
	Technical Division		Telephone Line Digital Error Checking
	Technical Document		
	Technology Development	**2DEG**	Two Dimensional Electron Gas
	Telemetry Data	**TDEP**	Tracking Data Editing Program [NASA]
	Temperature Differential	**TDF**	Total Dietary Fiber
	Terminal Distributor		Transmitted Dark-Field
	Testing Device		Two Degrees of Freedom
	Thor-Delta (Satellite)	**TDFA**	Total Dietary Fiber Assay
	Thoria-Dispersed	**TDFAB**	Total Dietary Fiber Assay Bulletin
	Thorium Dioxide	**TDG**	Test Data Generator
	Time Delay		Transmit Data Gate
	Time Difference	**TDI**	Textile Dye Institute [US]
	Time Domain		Toluene Diisocyanate
	Timing Device		Tool and Die Institute
	Top Down	**TDIA**	Transient Data Input Area
	Toxic Dose	**TDIO**	Tuning Data Input-Output
	Toyota Diffusion (Process)	**TDL**	Tunnel-Diode Logic
	Track Data	**TDLAS**	Tunable Diode Laser Absorption Spectrometry
	Track Display		
	Transmitter-Distributor	**TDM**	Tandem
	Transverse Direction		Tertiary Dodecyl Mercaplan
	Tunnel Diode		Time Division Multiplex(er)
	Turbine Drive		Time Division Multiplexing
T&D	Transmission and Distribution		Torpedo Detection Modification
3D	Three Dimensional [also: 3-D]	**TDMA**	Time Division Multiple Access
2D	Two Dimensional [also: 2-D]	**TDMS**	Telegraph Distortion Measurement Set
TDA	Target Docking Adapter		Terminal Data Management System
	Terminal Diode Amplifier		Thermal Desorption Mass Spectroscopy
	Tetracarboxylic Dianhydride		Time-Shared Data Management System
	Tetradecenylacetate	**TDN**	Target Doppler Nullifier
	Textile Distributors Association [US]		Total Digestible Nutrient
	Titanium Development Association	**T-DNA**	Tumor-Desoxyribonucleic Acid
	Tracking and Data Acquisition	**TDO**	Tallow Diaminopropanedioleate
	Transport Distribution Analysis		Time Delay Opening
	Tunnel Diode Amplifier	**TDOL**	Tetradecenol

TDOS	Tape/Disk Operating System		Trailing Edge
TDP	Thiodiphenol		Transversal-Electric
	Technical Data Package		Transverse-Electric
	Technical Development Plan		Trunk Equalizer
	Teledata Processing		Turbo-Electric
	Tracking Data Processor	**T&E**	Test and Evaluation
TDPAC	Time Differential Perturbed Angular Correlation	**Te**	Tellurium
		TEA	Technical Engineers Association
TDR	Target Discrimination Radar		Technical Exchange Agreement
	Technical Data Relay		Tetraetylammonium
	Technical Data Report		Transferred Electron Amplifier
	Technical Documentary Report		Transversely Excited Atmosphere
	Temperature-Dependent Resistor		Triethanolamine
	Time-Delay Relay		Triethyl Aluminum
	Time Domain Reflectometry		Triethylamine
	Torque Differential Receiver		Triethylaminoethyl
TDRP	Treatment Development Research Project		Triethylammonium
TDRSS	Tracking and Data Relay Satellite System		Tunnel-Emission Amplifier
TDS	Tape Data Selector		Tyrethylaluminum
	Target Designation System	**TEAE**	Triethylaminoethyl
	Technical Data System	**TEAM**	Technique for Evaluation and Analysis of Maintainability
	Tertiary Data Set		
	Test Data Sheet		Test and Evaluation of Air Mobility
	Test Development Series	**TEAMS**	Test, Evaluation and Monitoring System
	Test Development Station	**TEAOH**	Tetraethylammonium Hydroxide
	Testing Data System	**TEC**	Tactical Electromagnetic Coordinator
	Thermal Desorption Spectroscopy		Total Electron Content
	Time Distance Speed		Transearth Coast
	Time-Division Switching		Transient Electron Current
	Titanium Descaling Salt	**TECH**	Technical
	Thermal Diffuse Scattering		Technician
	Total Dissolved Solids		Technological
	Totally Dissolved Solids		Technologist
	Track Data Simulator		Technology
	Track Data Storage	**TECHNOL**	Technology
	Tracking and Data System	**TECOM**	Test and Evaluation Command [US Army]
	Translation and Docking Simulator	**TED**	Technology Evaluation and Development
TDSF	Triangle-Dimer Stacking Fault		Test Engineering Division [US Navy]
TDSS	Time Dividing Spectrum Stabilization		Threshold Extension Demodulator
TDT	Target Designation Transmitter		Translation Error Detector
	Task Dispatch Table		Transmission Electron Diffraction
TDTA	Toluoyl-D-Tartaric Acid		Triethylenediamine
TDTL	Tunnel-Diode Transistor Logic		Turbine Engine Division [US Air Force]
TDU	Target Detection Unit	**TE/DC**	Traffic Enforcement/Driver Control
TDX	Thermal Demand Transmitter	**TEE**	Telecommunications Engineering Establishment
	Time Division Exchange		
	Torque Differential Transmitter		Torpedo Experimental Establishment [UK]
TDW	Tons Deadweight		Trans Europe Express
TE	Technical Exchange	**TEEOC**	Turret Electronics and Electrooptical Console
	Television Electronics		
	Terminal Emulator	**TEFA**	Total Esterified Fatty Acid
	Terminal Equipment	**TEFC**	Totally Enclosed Fan-Cooled (Motor)
	Terminal Exchange	**TEG**	Thermo-Electric Generator
	Test Equipment		Triethylene Glycol
	Thermal Element		Triethyl Gallium
	Thermoelastic(ity)	**TEGa**	Triethyl Gallium
	Thermoelectric(ity)	**TEIn**	Triethyl Indium
	Threshold Extension	**TEIC**	Tissue Equivalent Ionization Chamber
	Tokamak Experiment	**TEJ**	Transverse Expansion Joint
	Totally Enclosed	**TEL**	Telegram
	Traffic Enforcement		Telegraph

	Telephone
	Tetraethyl Lead
TELEC	Telecommunication
TELECOM	Telecommunication
TELEDAC	Telemetric Data Converter
TELENET	Telecommunication Network
TELEPAC	Telemetering Package
TELESAT	Telecommunications Satellite
TELEX	Teletype Exchange
	Teleprinter Exchange
TELOPS	Telemetry On-Line Processing System
TELSCOM	Telemetry-Surveillance-Communications
TELSIM	Teletypewriter Simulator
TELTIPS	Technical Effort Locator and Technical Interest Profile System [US Army]
TELUS	Telemetric Universal Sensor
TEM	Thermal Expansion Molding
	Transmission Electron Microscopy
	Transverse Electromagnetic Mode
	Transversal Electromagnetic
	Triethylenemelamine
TEMA	Telecommunication Engineering and Manufacturing Association
	Tubular Exchanger Manufacturers' Association
TEM-BF	Transmission Electron Microscopy - Bright Field
TEM-DF	Transmission Electron Microscopy - Dark Field
TEMED	Tetramethylethylenediamine
TEMO	Topological Effect on Molecular Orbitals
TEMP	Temperature
	Template
	Temporary
	Total Energy Management Professionals
	Transportation Energy Management Program
TEMPO	Tetramethylpiperidinyloxy
	Total Evaluation of Management and Production Output
TEMS	Test Equipment Maintenance Set
	Toyota Electronically Modulated Suspension
TEM-TED	Transmission Electron Microscopy - Transmission Electron Diffraction
TENE	Total Estimated Net Energy
Tenn	Tennessee [US]
TENS	Tension
	Transcutaneous Electronic Nerve Stimulator
TEO	Transferred Electron Oscillator
TEOM	Transformer Environment Overcurrent Monitor
TEOS	Tetraethoxysilane
	Tetraethylorthosilicate
TEP	Transmitter Experimental Package
	Transportation Energy Panel
TEPA	Tetraethylenepentamine
TEPG	Thermionic Electrical Power Generator
TEPIAC	Thermophysical and Electronic Properties Information Analysis Center [US]
TEPRSSC	Technical Electronic Product Radiation Safety Standards Committee
TER	Terazzo
	Tergitol
	Terrace
	Territory
	Tertiary
	Thermal Expansion Rubber
	Thyssen-Extrem-Rechtkant (Process)
	Transmission Equivalent Resistance
	Triple Ejector Rack
TERAC	Tactical Electromagnetic Readiness Advisory Council
TEREC	Tactical Electromagnetic Reconnaissance
TERLS	Thumba Equatorial Rocket Launching System [India]
TERM	Terminal
	Termination
	Terminology
TERP	Terrain Elevation Retrieval Program
TERPS	Terminal Planning System
TERR	Terrace
	Territory
TERS	Tactical Electronic Reconnaissance System
tert	tertiary [chemicals]
TES	Thermal Evaluation Spectroscopy
	Transportable Earth Station
	Trisethanesulfonic Acid
TESA	Television Electronics Service Association
TESb	Triethyl Antimon
TESS	Tactical Electromagnetic Systems Study
	Thermocouple Emergency Shipment Service
TESTG	Testing
TET	Tetrachloride
	Total Elapsed Time
	Transportation Engineering Technology
TETA	Triethylenetetramine
TETR	Test and Training Satellite [US]
	Tetragonal
TETRA	Terminal Tracking (Telescope)
TETRAG	Tetragonal
TETRAHED	Tetrahedral
TETROON	Tethered Meteorological Balloon
TEU	Twenty-Foot Equivalent Unit
TEV	Thermoelectric Voltage
TEVROC	Tailored Exhaust Velocity Rocket
TEW	Tactical Electronic Warfare
TEWS	Tactical Electronic Warfare System
TEX	Target Excitation
	Teletype Exchange
	Telex
Tex	Texas [US]
TEXTIR	Text Indexing and Retrieval
TEXTOR	Tokamak Experiment for Technology-Oriented Research
TEZG	Tribological Experiments in Zero Gravity
TF	Task Force
	Technological Forecasting
	Test Fixture
	Thin Film
	Threshold Factor
TFA	Thin Film Analysis
	Timing Filter Amplifier

	Trifluoroacetic Acid
TFAA	Trifluoroacetic Anhydride
TFB	Towed Flexible Barge
TFC	Telefilm Canada
	Traffic
TFCG	Thin Film Crystal Growth
TFCX	Tokamak Fusion Core Experiment
TFD	Television Feasibility Demonstration
	Thin Film Deposition
	Thin Film Detector
	Transflective Device
TFDU	Thin Film Deposition Unit
TFE	Tetrafluoroethylene
	Turbofan Engine
TFEL	Thin Film Electroluminescence
TFFET	Thin Film Field-Effect Transistor
TFG	Transmit Format Generator
TFI	Thick Film Ignition
TFL	Threaded Full Length
TFM	Trifluoromethane
TFMC	Trifluoromethylhydroxymethylenecamphorato
TFR	Terrain Following Radar
TFRS	Tungsten Fiber-Reinforced Superalloy
TFS	Thin Film Solid
	Tin-Free Steel
	Transfer Standard
TFSA	Thin Film Spreading Agent
TFSUS	Task Force on the Scientific Uses of Space Stations
TFT	Thin-Film Technology
	Thin-Film Transistor
	Threshold Failure Temperature
	Time-to-Frequency Transformation
TFTR	Tokamak Fusion Test Reactor
TFU	Timing and Frequency Unit
TFX	Tactical Fighter Experiment
	Transverse Flux
TFZ	Transfer Zone
TG	Telegraph
	Terminal Guidance
	Thermogravimetry
	Top Grille
	Torpedo Group
	Total Graph
	Transgranular
	Triglyceride
	Tuned Grid
T&G	Tongue and Groove
TGA	Thermogravimetric Analysis
	Thermogravimetric Analyzer
	Thioglycolic Acid
TGB	Tongued, Grooved and Beaded
TGC	Transmit Gain Control
TGCA	Transportable Ground Control Approach
TGF	Through Group Filter
TGFA	Triglyceride Fatty Acid
TGG	Third Generation Gyro
TGI	Target Intensifier
TGID	Transmission Group Identifier
TGL	Toggle
TGM	Toroidal Gate Monochromator

TGMDA	Tetraglycidylated Methylene Dianiline
TGDMA/DDS	Tetraglycidylated Methylene Dianiline modified with Diaminodiphenylsulfone
TGID	Transmission Group Identifier
TGS	Tactical Ground Support
	Telemetry Ground Station
	Translator Generator System
	Triglycine Sulfate
TGSE	Tactical Ground Support Equipment
	Telemetry Ground Support Equipment
TGSO	Tertiary Groups Shunt Operation
TGSS	Triglycine Sulfate Selenate
TGT	Target
TGWU	Transport and General Workers Union
TH	Thursday
	Thyristor
	Transformation Hardened
	Transmission Header
Th	Thorium
THAM	Trishydroxymethylaminomethane
THB	Temperature-Humidity Bias
THC	Tetrahydracannabinol
	Thermal Converter
	Thrust Hand Controller
THD	Thread
	Total Harmonic Distortion
THE	Tetrahydrocortisone
	Thunderstorm Event
THEED	Transmission High Energy Electron Diffraction
THEO	Theoretical
THEOR	Theoretical
	Theory
	Theorem
THERM	Thermometer
THERMO	Thermostat
THERP	Technique for Human Error Rate Prediction
THF	Tetrahydrocortisol
	Tetrahydrofolic Acid
	Tetrahydrofuran
THFC	Trisheptafluoropropylhydroxymethylene-camphorato
THFTDA	Tetrahydrofuran Tetracarboxylic Dianhydride
THHP	Tetrahydrohomopteroic Acid
THI	Temperature Humidity Index
	Total Height Index
THIR	Temperature Humidity Infrared Radiometer
THK	Thick
THL	True Heavy Liquid
THM	Tonnes Hot Metal
THN	Tetrahydronaphthalene
THOMIS	Total Hospital Operating and Medical Information System
THOPS	Tape Handling Operational System
THOR	Tape-Handling Option Routine
	Transistorized High-Speed Operations Recorder
THPFB	Treated Hard Pressed Fiberboard
THQ	Thermionic Quadrupole
Thr	Threonine

THRM	Thermal
THROT	Throttle
THRU	Through
THS	Tetrabutylammonium Hydrogen Sulfate
THSP	Thermal Spraying
THT	Tetrahydrothiophene
THTR	Thorium High Temperature Reactor
THTRA	Thorium High Temperature Reactor Association
THUR	Thursday
THURS	Thursday
THWM	Trinity High-Water Mark
THYMOTRO	Thyratron Motor Control
THz	Terahertz
TI	Tape Inverter
	Target Identification
	Technical Information
	Temperature Index
	Thermal Insulation
	Threaded Insert
	Time Index
	Time Interval
	Track Identifier
	Track Identity
	Track Initiator
	Transfer Impedance
Ti	Titanium
	Tumor-Induced
TIA	Time Interval Analyzer
TIAA	Timber Importers Association of America
TIAC	Thermal Insulation Association of Canada
TIARA	Target Illumination and Recovery Aid
TIAS	Target Identification and Acquisition System
TIB	Technical Information Bureau [UK]
TIBA	Triiodobenzoic Acid
	Triisobutyl Aluminum
TIBBS	Trace Integrated Bare Board System
TIBC	Total Iron-Binding Capacity
TIBER	Tokamak Ignition/Burn Experimental Research
TIBI	Training in Business and Industry
TIBOE	Transmitting Information by Optical Electronics
TIC	Tape Intersystem Connection
	Target Intercept Computer
	Task Interrupt Control
	Technical Information Center
	Telemetry Instruction Conference
	Temperature Indicating Controller
	Total Inorganic Carbon
	Transfer in Channel
	Transducer Information Center [of BMI]
TICA	Technical Information Center Administration
TICAS	Taxonometric Intra-Cellular Analytic System
TICCIT	Time-Shared Interactive Computer-Controlled Informational Television
TICE	Time Integral Cost Effectiveness
TICS	Teacher Interactive Computer System
	Tidal Current System
TID	Technical Information Division

	Test Instrument Division
	Total Ion Detector
TIDAR	Time Delay Array Radar
TIDB	Tester-Independent Database
TIDDAC	Time in Deadband Digital Attitude Control
TIDE	Transponder Interrogation and Decoding Equipment
TIDES	Time Division Electronic Switching System
TIDY	Track Identity
TIE	Technical Integration and Evaluation
TIES	Total Integrated Engineering System
	Transmission and Information Exchange System
TIF	Telephone Influence Factor
	Telephone Interference Factor
	True Involute Form [gears]
TIFF	Tag Image File Format
TIFR	Tata Institute of Fundamental Research [India]
TIFS	Total In-Flight Simulator
TIG	Tungsten Inert-Gas (Welding)
TIIF	Tactical Image Interpretation Facility
TIIPS	Technically Improved Interference Prediction System
TILS	Technical Information and Library Services
	Tumor-Infiltrating Lymphocytes
TIM	Teeth in Mesh [gears]
	Test Instrumented Missile
	Time Meter
	Total Ion Monitoring
	Transistor Information Microfile
TIME	Test, Inspection, Measurement and Evaluation
TIMM	Thermionic Integrated Micromodules
TIMS	The Institute of Management Services
	Thermal Ionization Mass Spectroscopy
	Transmission Impairment Measuring Set
TIN	Temperature Independent
TINCT	Tincture
TINS	Thermal Imaging Navigation Set
TINT	Track in Track
TIO	Time Interval Optimization
TIOA	Terminal Input/Output Area
TIP	Tape Input
	Technical Information Processing
	Technical Information Program
	Technical Information Project
	Teletype Input Processing
TIPA	Tank and Industrial Plant Association
	Tetraisopropylpyrophosphoramide
TIPACS	Texas Instruments Planning and Control System
TIPI	Tactical Information Processing and Interpretation
TIPL	Teach Information Processing Language
TIPP	Time-Phasing Program
TIPS	Technical Information Processing System
	Triisoprophylsilyl
TIPS	Technical Information Programming System
	TrueVision Image Processing Software
TIPTOP	Tape Input/Tape Output

TIPTOM	Towards Improved Performance of Tool Materials
TIR	Technical Information Release
	Technical Information Report
	Technical Intelligence Report
	Total Indicator Reading
	Total Internal Reflection
	Transient Impulse Resonance
	Transient Impulse Response
TIRAM	Taper Insulated Random-Access Memory
TIRF	Traffic Injury Research Foundation
TIRH	Theoretical Indoor Relative Humidity
TIROS	Television and Infrared Observation Satellite
TIRP	Total Internal Reflection Prism
TIRS	Thermal Infrared Scanner
TIS	Target Information Sheet
	Target Information System
	Technical Information Service
	Total Information System
	Total Ionic Strength
	Transponder Interrogation Sonar
TISAB	Total Ionic Strength Adjustment Buffer
TISSS	Tester-Independent Support Software System
TIT	Title
	Turbine Inlet Temperature
TIU	Tape Identification Unit
	Terrestrial Interface Unit
	Typical Information Use
TIUPIL	Typical Information Use per Individual
TIV	Total Indicator Variation
TJC	Trajectory Chart
TJD	Trajectory Diagram
TJID	Terminal Job Identification
TK	Thymide Kinase
	Track
	Transmission Kossel (Technique)
TKN	Total Kjehldal Nitrogen
TKS	Tessman-Kahn-Shockley
TKT	Ticket
TL	Tape Library
	Target Language
	Target Loss
	Test Link
	Thermoluminescence
	Thin Layer
	Tie Line
	Time Limit
	Total Lipid
	Transmission Level
	Transmission Line
	Transmission Loss
	Truck Load
Tl	Thallium
TLAS	Tactical, Logical and Air Stimulation
TLB	Translation Lookaside Buffer
TLC	Tank Landing Craft
	Thin Layer Chromatography
TLCK	Tosyl Lysine Chloromethyl Ketone
TLD	Thermoluminescent Dosimeter
	Trapped Lattice Dislocation
TLE	Temperature-Limited Emission
	Theoretical Line of Escape
	Tracking Light Electronics
TLG	Telegraph
TLI	Telephone Line Interface
TLK	Test Link
TLL	Teflon Luer Lock
	Thin-Layer Leaching
TLM	Tape-Laying Machine
	Telemeter
	Thin Lipid Membrane
	Transition Line Model
TLO	Tracking Local Oscillator
TLP	Tension Leg Platform
	Threshold Learning Process
	Total Language Processor
	Transient Liquid Phase
	Transient Lunar Phenomena
	Transmission Level Point
TLRV	Tracked Levitated Research Vehicle
TLS	Telecommunication Liaison Staff
	Telemetry Listing Schedule
	Terminal Landing System
	Three-Stage Least Squares
TL/SX/EW	Thin-Layer Leaching followed by Solvent Extraction and Electrowinning
TLT	Terminal List Table
TLTA	Toluoyl-L-Tartaric Acid
TLTR	Translator
TLU	Table Lookup
TLV	Threshold Limit Value
TLV-C	Threshold Limit Value - Ceiling
TLV-TWA	Threshold Limit Value - Time-Weighted Average
TLZ	Titanium-Lead-Zinc
	Transfer on Less than Zero
TM	Tactical Missile
	Tapemark
	Technical Manual
	Technical Memo(randum)
	Technical Monograph
	Telemetry
	Temperature Meter
	Test Mode
	Thematic Mapper
	Thermomagnetic
	Thermomechanical
	Thermoplastic Molding
	Thickness Measurement
	Time Modulation
	Tone Modulation
	Trademark
	Transition Metal
	Transmission Matrix
	Transversal-Magnetic
	Transverse-Magnetic
	Trench Mortar
	Turing Machine
	Twist Multiplier
Tm	Thulium
TMA	Tape Motion Analyzer

	Testability Measure Analyzer
	Test Module Adapter
	Tetrahydroaminacrine
	Tetramethylammonium
	Thermomechanical Analysis
	Thermomechanical Analyzer
	Trimellitic Anhydride
	Trimethyl Aluminum
	Trimethylamine
	Trimethylanilinium
TMAC	Tetramethylammonium Carbonate
TMAH	Trimethylanilinium Hydroxide
TMAl	Trimethyl Aluminum
TMAMA	Textile Machinery and Accessory Manufacturers Association
TMAO	Trimethylamine Oxide
TMB	Tetramethylbenzidine
	Trimethoxybenzoic Acid
TMBA	Thickness Measurement by Beam Alignment
TMBR	Timber
TMC	The Maintenance Council
	Thick Molding Compound
	Total Molding Concept
TMCB	Tetramethoxycarbonyl Benzophenone
TMCC	Time-Multiplexed Communication Channel
TMCOMP	Telemetry Computation
TMCP	Thermomechanical Control Process
TMCS	Trimethylchlorosilane
TMD	Tensor Meson Dominance
	Theoretical Maximum Density
TME	Telemetric Equipment
	Tempered Martensite Embrittlement
TMEDA	Tetramethylethylenediamine
TMF	Thermomechanical Fatigue
TMG	Methylthiogalactoside
	Thermal Meteoroid Garment
	Trimethyl Gallium
TMGa	Trimethyl Gallium
TMGE	Thermo-Magnetic-Galvanic Effect
TMH	Tons per Man-Hour
TMI	Test, Measurement and Inspection (Conference and Exhibition)
	Timing Measurement Instrument
	Trimethyl Indium
TMIn	Trimethyl Indium
TMIS	Technical and Management Information System [NASA]
	Television Measurement Information System
	Television Metering Information System
TML	Terminal
	Tetramethyl Lead
	Total Mass Loss
TMM	Test Message Monitor
TMN	Technical and Management Note
	Trimethylamine Nitrogen
TMO	Telegraph Money Order
	Time out
TMOS	T-Type Metal Oxide Semiconductor
TMP	Terminal Monitor Program
	Thermomechanical Process(ing)

	Thymidine Monophosphate
	Transparent Multiprocessing
	Transmembrane Potential
	Turbomolecular Pump
TMPAH	Trimethylphenylammonium Hydroxide
TMPTA	Trimethylolpropane Triacrylate
TMPTMA	Trimethylolpropane Trimethacrylate
TMR	Triple Modular Redundancy
TMRBM	Transportable Medium Range Ballistic Missile
TMS	Tactical Missile Squadron
	Telephone Management System
	Temperature Measurements Society [US]
	Test Management System
	Tetramethylsilane
	The Masonry Society
	The Metallurgical Society [of AIME]
	Time-Shared Monitor System
	Transmission Measuring Set
	Trimethylsilyl
TMSb	Trimethyl Antimon
TMSDEA	Trimethylsilyldiethylamine
TMSi	Tetramethylsilane
TMT	Turbine Motor Train
TMTC	Through-Mode Tape Converter
TMTD	Tetramethylthiuram Disulfide
TMTSF	Tetramethyltetraselenafulvalene
TMTU	Tetramethythiourea
TMU	Time Measurement Unit
TMV	Tobacco Mosaic Virus
TMX	Telemeter Transmitter
	Terminal Multiplexer
TMXO	Tactical Miniature Crystal Oscillator
TN	Technical Note
	Tennessee [US]
	Terminal Node
	Thermonuclear
	Thyssen-Niederrhein (Steelmaking Process)
	Total Nitrogen
	Track Number
tn	ton [unit]
TNA	Transient Network Analyzer
TNAA	Thermal Neutron Activation Analysis
TNBS	Trinitrobenzenesulfonic Acid
TNBT	Tetrabutyltitanate
	Tetranitro Blue Tetrazolium
TNC	Total Nonstructural Carbohydrates
	Trinitrocellulose
TNDC	Thai National Documentation Center [Thailand]
TNG	Training
TNIP	Terrestrial Network Interface Processor
TNL	Tunnel
TNLDIO	Tunnel Diode
TNM	Tetranitromethane
TNP	Trinitrophenol
TNPAL	Trinitrophenylaminolauryl
TNR	Thermonuclear Reaction
TNS	Toluidinylnaphthylenesulfonate
	Toluidinonaphthalenesulfonic Acid
TNT	Terminal Name Table

	Thallium Nitrate Trihydrate
	Total Network Test (System)
	Transient Nuclear Test
	Trinitrotoluene
TNXCD	Transaction Type Code
TNZ	Transfer to Non-Zero
TO	Takeoff
	Telegraph Office
	Transistor Outline
	Transverse Optical
	Time out
	Turnover
T&O	Test and Operation
TOA	Takeoff Angle
	Time of Arrival
TOB BRZ	Tobin Bronze
TOC	Table of Contents
	Television Operating Center
	Time of Contact
	Total Optical Color
	Total Organic Carbon
TOCC	Technical and Operational Control Center
TOD	Technical Objective Directive
	Technical Objective Documents
	Time-of-Day
	Top of Duct
	Total Oxygen Demand
TODS	Test-Oriented Disk System
	Transaction on Database System
TOE	Total Operating Expenses
TOEFL	Test of English as a Foreign Language
TOF	Time-of-Flight
	Top of File
	Top of Form
TOFF	Thin Overlay for Friction
TOFISS	Time-of-Flight Ion Scattering Spectroscopy
TOGA	Tropical Oceans and Global Atmosphere
TOH	Tyrosine Hydroxylase
TOHM	Terohmmeter
TOHP	Takeoff Horsepower
TOI	Technical Operation Instruction
TOIRS	Transfer Orbit Infrared Earth Sensor
TOJ	Track on Jamming
TOL	Test-Oriented Language
	Tolerance
TOLIP	Trajectory Optimization and Linearized Pitch
TOM	Tool Material
	Translator Octal Mnemonic
	Typical Ocean Model
TOMCAT	Telemetry On-Line Monitoring Compression and Transmission
tonf	ton-force [unit]
TONLAR	Tone-Operated Net Loss Adjuster
TONN	Tonnage
TOOL	Test Oriented Operation Language
TOOLS	Total Operating On-Line System
TOP	Tape Output
	Technical (and) Office Protocol
	Total Obscuring Power
	Transaction-Oriented Programming

TOPIC	Time Ordered Programmer Integrated Circuit
TOPICS	Transport Operations Program for Increasing Capacity and Safety
TOPO	Trioctylphosphine Oxide
TOPOG	Topography
TOPP	Terminal Operated Production Program
TOPS	Telemetry Operations
	Telephone Order Processing System
	Teletype Optical Projection System
	Total Operations Processing System
	Transducer Operated Pressure System
TOPSI	Topside Sounder, Ionosphere
TOPSY	Test Operations and Planning System
	Thermally Operated Plasma System
TOPTS	Test-Oriented Paper-Tape System
TOR	Technical Operations Research
	Technology-Oriented Research
	Torque
TORC	Traffic Overload Reroute Control
TORP	Torpedo
TOS	Tactical Operations System
	Tape Operating System
	Terminal-Oriented Software
	Terminal-Oriented System
	Test Operating System
	TIROS Operational Satellite [NASA]
	Top of Steel
	Transfer Orbit Stage
TOSBAC	Toshiba Scientific and Business Automatic Computer
TosMIC	Tosylmethylisocyanide
TOSS	Test Operation Support Segment
	TIROS Operational Satellite System
TOSSA	Transfer Orbit Sun Sensor Assembly
TOST	Turbine Oxidation Stability Test
TOT	Time-of-Tape
	Time of Transmission
	Total
TOTE	Terminal On-Line Test Executive
TOW	Tube-Launched, Optically-Tracked, Wire-Guided (Missile)
TOWA	Terrain and Obstacle Warning and Avoidance
TP	Tandem Propeller
	Tank Pressure
	Target Practice
	Technical Pamphlet
	Technical Paper
	Technical Program
	Technical Publication
	Technological Properties
	Telemetry Processing
	Teleprinter
	Teleprocessing
	Teletype Printer
	Terphenyl
	Testosterone Propionate
	Test Point
	Test Procedure
	Thermal Printer

	Thermal Properties
	Thermophysics
	Thermoplastics
	Thiophosphamide
	Throttle Positioner
	Time Pulse
	Timing Pulse
	Title Page
	Torpedo Part of Beam
	Torque Pressure
	Total Pressure
	Total Protein
	Township
	Transactions Paper
	Triple Pole
	Tryptophan Pyrrolase
	Tuned Plate
3P	Triple-Pole
TPA	Tape Pulse Amplifier
	Technical Publications Association
	Terephthalic Acid
	Tetradecanoylphorbol
	Tissue Plasminogen Activator
	Two-Point Approximation
TPB	Tetraphenylbutadiene
TPC	Telecommunications Planning Committee
	Thymolphthalein Complexone
	Time Pickoff Control
	Tire Performance Criteria
	Totally Pyrolytic Cuvette
	Trans-Pacific Cable
	Trifluoroacetylprolylchloride
TPCK	Tosylaminophenylethylchloromethylketone
TPCU	Thermal Precondition Unit
TPCV	Turbine Power Control Valve
TPD	Temperature-Programmed Desorption
	Time Pulse Distributor
	Tons per Day
TPDT	Triple-Pole, Double-Throw
3PDT	Triple-Pole, Double-Throw
3PDT SW	Triple-Pole, Double-Throw Switch
TPE	Teleprocessing Executive
	Tetraphenylethylene
	Thermoplastic Elastomer
	Two Pion Exchange
TPF	Taper per Foot
	Terminal Phase Finalization
	Track Pick Fragments
	Two Photon Fluorescence
TPG	Time Pulse Generator
	Tin Plate Gauge
	Transmission Project Group [of CEGB]
	Trypticase Peptone Glucose
TPH	Tons per Hour
2PH	Two-Phase
TPHC	Time to Pulse Height Converter
TPI	Tape Phase Inverter
	Taper per Inch
	Target Position Indicator
	Teeth per Inch
	Terminal Phase Initiate

	Thermoplastic Polyimide
	Threads per Inch
	Tracks per Inch
	Triosephosphate Isomerase
	Turns per Inch
TPL	Test Parts List
	Tritium Process Laboratory
	Total Phospholipid
TPLAB	Tape Label
TPM	Tape Preventive Maintenance
	Telemetry Processor Module
	Total Productive Maintenance
	Transmission and Processing Model
	Transport Planning Mobilization
TPMA	Thermodynamic Properties of Metals and Alloys
	Timber Products Manufacturers Association [US]
TPmm	Taper per Millimeter
TPN	Triphosphopyridine Nucleotide
TPNH	Triphosphopyridine Nucleotide, Reduced Form
TPO	Thermoplastic Olefin
	Tryptophan Pyrrolase
TPP	Test Point Pace
	Tetraphenylporphine
	Thiamine Pyrophosphatase
TPPC	Total Package Procurement Concept
TP-PCB	Teleprocessing Program Communication Block
TPPD	Technical Program Planning Division
TPQ	Two Piston Quenching
TPR	Tape Programmed Row
	Taper
	Telescopic Photographic Recorder
	Thermoplastic Recording
	Total Peripheral Resistance
TPRC	Thermophysical Properties Research Center [US]
TPS	Task Parameter Synthesizer
	Technical Problem Summary
	Telecommunications Programming System
	Telemetry Processing Station
	Terminals per Station
	Test Program Set
	Thermal Protection System
	Triisoprophylbenzenesulfonyl
TPSI	Torque Pressure in Pounds per Square Inch
TPST	Triple-Pole, Single-Throw
3PST	Triple-Pole, Single-Throw
3PST SW	Triple-Pole Single-Throw Switch
3P SW	Triple-Pole Switch
TPT	Tetraisopropyl Titanate
	Transmission Path Translator
TP-T	Target Practice with Tracer
TpT	Thymidylyl Thymidine
TPTG	Tuned Plate, Tuned Grid
TpTpT	Thymidylyl Thymidylyl Thymidine
TPTZ	Tripyridyl-S-Triazine
TPU	Tape Preparation Unit
	Thermoplastic Polyurethane

	Time Pickoff Unit
TPWB	Three Program Wire Broadcasting
TPV	Thermophotovoltaic
	Thermoplastic Vulcanizate
TPX	Polymethylpenetene [also: PMP]
TQA	Total Quality Assurance
TQC	Total Quality Control
TQCA	Textile Quality Control Association [US]
TR	Tape Recorder
	Technical Report
	Temperature Rate
	Temper Rolled
	Test Request
	Top Register
	Top Running (Crane)
	Torque Receiver
	Trace(r)
	Track
	Train
	Transfer
	Transformer-Rectifier
	Transient Response
	Transistor
	Transistor Radio
	Transition
	Translation
	Translator
	Transmit-Receive
	Transmitter-Receiver
	Transport
	Transposition
TRA	Technical Requirement Analysis
	Transfer
TRAACS	Transit Research and Attitude Control Satellite
TRAC	Text Reckoning and Compiling
	Thermally Regenerative Alloy Cell
	Transient Radiation Analysis by Computer
TRACALS	Traffic Control, Approach and Landing System
TRACE	Tactical Readiness and Checkout Equipment
	Tape-Controlled Reckoning and Checkout Equipment
	Teleprocessing Recording for Analysis by Customer Engineers
	Time-Shared Routines for Analysis, Classification and Evaluation
	Tolls Recording and Computing Equipment
	Tracking and Communications, Extraterrestrial
	Transistor Radio Automatic Circuit Evaluator
	Transportable Automated Control Environment
TRACOM	Tracking Comparison
TRACON	Terminal Radar Approach Control
TRACS	Test and Repair Analysis/Control System
TRADA	Timber Research and Development Association
TRADEX	Target Resolution and Discrimination Experiment

TRADIC	Transistor Digital Computer
TRAIN	Telerail Automated Information Network [AAR]
TRAM	Target Recognition Attack Multisensor
TRAMP	Time-Shared Relational Associative Memory Program
TRAN	Transaction
	Transit
	Transmit
TRANDIR	Translation Director
TRANET	Tracking Network
TRANPRO	Transaction Processor
TRANS	Transaction
	Transfer
	Transformer
	Translation
	Translator
	Transportation
	Transposition
TRANSAC	Transistorized Automatic Computer
TRANSF	Transfer(red)
TRANSIF	Transient State Isoelectric Focusing
TRANSIM	Transportation Simulator
TRANSL	Translate
	Translation
	Translator
TRANSM	Transmission
TRANSP	Transportation
TRANSPT	Transport
TRANSV	Transverse
TRAP	Terminal Radiation Airborne Program
	Tracker Analysis Program
TRAPATT	Trapped Avalanche Triggered Transit
TRAWL	Tape Read and Write Library
TRAX	Total Reflection-Angle X-Ray Spectrometer
TRB	Transportation Research Board [US]
TRC	Tape Record Coordinator
	Technical Resources Center [US]
	Technology Reports Center
	Tracking, Radar-Input and Correlation
TR&D	Transport Research and Development
TRDTO	Tracking Radar Takeoff
TRE	Telecommunications Research Establishment [UK]
	Transmit Reference Equivalent
TREAT	Transient Radiation Effects Automated Tabulation
	Transient Reactor Test
TREC	Tethered Remote Camera
TREE	Transient Radiation Effects on Electronics
TREES	Turbine Rotors Examination and Evaluation System
TREND	Tropical Environment Data
TRF	Transportation Research Foundation [US]
	Tuned Radio Frequency
TRFC	Traffic
TRFCS	Temperature Rate Flight Control System
TRG	Technical Research Group
	Training
TRH	Thyrotropin Releasing Hormone
TRI	Technical Research Institute

	Tin Research Institute
	Triode
TRIAC	Triode AC Semiconductor [also: Triac]
TRIAL	Technique for Retrieving Information from Abstracts of Literature
TRICE	Transistorized Real-Time Incremental Computer Equipment
TRICINE	Trismethylglycine
TRICL	Triclinic
TRID	Track Identity
TRIDOP	Tri-Doppler
TRIG	Trigonal
	Trigonometric
	Trigonometry
TRIGON	Trigonometric
	Trigonometry
TRIM	Test, Rework and Inspection Management
	Trimetric
	Trimmer
TRIP	Transformation-Induced Plasticity
	Transformation Induced by Plastic Deformation
TRIS	Trishydroxymethylaminomethane
TRITC	Tetramethylrhodamine Isothiocyanate
TRIUMF	Tri-University Meson Facility [Canada]
TRK	Track
	Trunk
TRL	Thermodynamics Research Laboratory
	Transistor-Resistor Logic
	Transuranium Research Laboratory [USAEC]
TRM	Test Request Message
	Thermal Remanent Magnetization
	Thermoremanent Magnetization
	Transmit-Receive Module
TRMA	Time Random Multiple Access
TRMDT	Transmission Date
TRMS	TDMA Reference and Monitor Station
	True Root Mean Square
TRN	Technical Research Note
	Television Relay Network
TRNA	Transfer Ribonucleic Acid
t-RNA	Transfer Ribonucleic Acid
TRNBKL	Turnbuckle
TROCA	Tangible Reinforcement Operant Conditioning Audiometry
TROLL	Technion Robotics Laboratory Language
TROMEX	Tropical Meteorological Experiment
TROO	Transponder On-Off
TROP	Tropics
	Tropic(al)
TROS	Tape Resident Operating System
tr oz	troy ounces [unit]
TRP	Television Remote Pickup
	Thermal Ribbon Printer
	Trap
	Tuition Reimbursement Plan
Trp	Tryptophan
TRPGDA	Tripropylene Glycol Diacrylate
TRR	Target Ranging Radar
	Teaching and Research Reactor

TRRS	Total Reflection Raman Scattering
TRS	Test Response Spectrum
	Time Reference System
	Total Reducing Sugar
	Tough Rubber Sheathed
	Transmission Regulated Spark
	Transposition
	Transverse Rupture Strength
	Tree-Ring Society
TRSB	Time Reference Scanning Beam
TRSSM	Tactical Range Surface-to-Surface Missile
TRSSSV	Tubing-Retrievable Subsurface Safety Valve
TRTL	Transistor-Resistor-Transistor Logic
TRU	Total Recycle Unit
	Transmit-Receive Unit
	Transportable Radio Unit
TRUMP	Teller Register Unit Monitoring Program
	Total Revision and Upgrading of Maintenance Procedures
TRV	Transient Recovery Voltage
TRX	Transaction
TS	Tape Status
	Taper Shank
	Tape System
	Target Strength
	Target System
	Technical Specification
	Technical Service
	Telegraph Service
	Temporary Storage
	Tensile Strength
	Test Set
	Thermal Shock
	Time Sharing
	Time Switch
	Tool Steel
	Torch Soldering
	Total Solids
	Transformer Station
	Transition Set
	Transmission Service
	Transmitting Station
	Transverse Section
	Triple Space
TSA	Time Series Analysis
	Total Scan Area
	Transportation Standardization Agency [USDOD]
TSAZ	Target Seeker-Azimuth
TSB	Technical Support Building
	Twin Sideband
TSC	Tape System Calibrator
	Technical Service Contractor
	Technical Subcommittee
	Test Set Connection
	Thermally Stimulated Current
	Thiosemicarbazide
	Time Sharing Control
	Transmitter Start Code
	Transportation Safety Committee [of SAE]
	Transportation Systems Center

TSCA	Timing Single Channel Analyzer
	Toxic Substances Control Act [US]
TSCC	Telemetry Standards Coordination Committee
TSCF	Textronix Standard Codes and Formats
TSCLT	Transportable Satellite Communications Link Terminal
TSD	Thermally-Stimulated Depolarization
	Touch Sensitive Digitizer
	Traffic Situation Display
TSDD	Temperature-Salinity-Density-Depth (Relationship)
TSDF	Target System Data File
TSDM	Time-Shared Data Management (System)
TSDOS	Time-Shared Disk Operating System
TSDU	Target System Data Update
TSE	Tactical Support Equipment
	Test of Spoken English
	Twist Setting Efficiency
TSEE	Thermally-Stimulated Emission of Exoelectrons
TSEQ	Time Sequence
TSF	Ten Statement FORTRAN
	Tetraselenafulvalene
	Thin Solid Film
	Through Supergroup Filter
TSFO	Transportation Support Field Office [US]
TSG	Time Signal Generator
TSH	Thyroid Stimulating Hormone
TSI	Task Status Index
	Terminal Specific Interface
	Threshold Signal-to-Interference (Ratio)
	Tons per Square Inch
	Transmitting Station Identification
TSIM	Trimethylsilylimidazole
TSIMS	Telemetry Simulation Submodule
TSIOA	Temporary Storage Input/Output Area
TSK	Task
TSLS	Two-Stage Least Squares
TSM	Telephony Signaling Module
	Time-Shared Monitor System
TSN	Tryptone Sulfite Neomycine
TSO	Time-Sharing Option
TSOS	Time-Sharing Operating System
TSO/VTAM	Time-Sharing Option/Virtual Telecommunications Access Method
TSP	Titanium Sublimation Pump
	Triple Superphosphate
	Trisodium Phosphate
TSPC	Thermally-Stimulated Polarization Current
TSPC/DC	Thermally-Stimulated Polarization/Depolarization Current
TSPS	Time Sharing Programming System
	Traffic Service Position System
TSQ	Time and Superquick
TSR	Test Schedule Request
	Technical Summary Report
	Terminate-and-Stay Resident
	Thyroid Secretion Rate
	Tunnel Stress Relaxation
TSRO	Topological Short-Range Order

TSS	Tactical Strike System
	Telecommunication Switching System
	Teletype Switching Subsystem
	Time-Sharing System
	Total Suspended Solids
TSSA	Test Scorer and Statistical Analyzer
TST	Test
TSTA	Transmission, Signaling and Test Access
	Tritium Systems Test Assembly
TSU	Technical Service Unit
	Time Standard Unit
	Texas Southern University [US]
TSVD	Thermally-Stimulated Voltage Decay
TSW	Test Switch
	Tube Socket Weld
TSX	Time Sharing Executive
TT	Technical Translation
	Telegraphic Transfer
	Teletype(writer)
	Terminal Timing
	Timing and Telemetry
	Total Time
	Tracking and Telemetry
	Tracking Telescope
	Transformation-Toughened
	Transformation Twin
	Triple Thermoplastic (Wire)
T/T	Telegraphic Transfer
TTA	Time-Temperature Austenitization
	Transformation-Toughened Alumina
TTAC	Telemetry Tracking and Command
TTB	Troop Transport Boat
	Trunk Test Buffer
TTBWR	Twisted Tape Boiling Water Reactor
TTC	Technical Transfer Council [Australia]
	Tracking, Telemetry and Command
	Tape-to-Card
	Thiatricarboxyanine
	Triphenyltetrazolium Chloride
TT&C	Tracking, Telemetry and Command
TTCM	Tracking, Telemetry, Command and Monitoring Station
TTCP	Tripartite Technical Cooperation Program
TTD	Temporary Text Delay
TTE	Telephone Terminal Equipment
	Time to Event
TTeF	Tetratellurafulvalene
TTEGDA	Tetraethylene Glycol Diacrylate
TTF	Tetrathiofulvalene
	Timber Trade Federation
TTFC	Textile Technical Federation of Canada
TTG	Technical Translation Group
	Time to Go
TTHA	Triethylenetetraminehexaacetic Acid
TTHDCM	Thin and Thick Film High-Density Ceramic Module
TTI	Teletype Test Instruction
	Time-Temperature Indicator
TTK	Tie Trunk
TTL	Through-the-Lens
	Transistor-to-Transistor Link

	Transistor-Transistor Logic
TTL/LS	Transistor-Transistor Logic/Large Scale
TTMA	Truck Trailer Manufacturers Association [US]
TTMS	Telephoto Transmission Measuring Set
TTO	Transmitter Turn-Off
TTP	Tape-to-Print
	Thermal Transfer Printer
	Trunk Test Panel
	Thymidine Triphosphate
TTPB	Trunk Test Panel Buffer
TTR	Target Tracking Radar
	Time-Transfer Receiver
TTRC	Thrust Travel Reduction Curve
TTRM	Transition Thermoremanent Magnetization
TTS	TDMA Terminal Simulation
	Tearing Topography Surface
	Telecommunications Terminal System
	Teletypesetter
	Teletypesetting
	Temporary Threshold Shift
	Test and Training Satellite
	Transmission Test Set
	Transportable Transformer Substation
T²S	Technology Transfer Society
TTSP	Time-Temperature Superposition Principle
TTT	Time-Temperature-Transformation (Diagram)
TTU	Terminal Time Unit
	Through-Transmission Ultrasonics
TTW	Teletypewriter
TTX	Tetrodotoxin
TTY	Teletype(writer)
TTYBS	Teletype(writer), Backspace
TTZ	Transformation-Toughened Zirconia
TU	Tape Unit
	Temple University [US]
	Thermal Unit
	Timing Unit
	Top Up
	Trade Union
	Transfer Unit
	Tuesday
	Tulsa University [US]
	Turbopump Unit
TUB	Tubing
TUC	Trade Union Congress
TUCC	Triangle Universities Computing Center [US]
TUD	Technology Utilization Division [of NASA]
TUES	Tuesday
TUG	Towed Universal Glider
	TRANSAC Users Group
TUM	Tuning Unit Member
TUN	Tuning
TUNS	Technical University of Nova Scotia [Canada]
TUP	Technology Utilization Program
TUR	Test Uncertainty Ratio
	Traffic Usage Recorder
	Turbine

	Turret
TURB	Turbine
TURBO GEN	Turbo-Generator
TURK	Turkey
	Turkish
TURP	Turpentine
TURPS	Terrestrial Unattended Reactor Power System
TUT	Transistor Under Test
TV	Television
	Terminal Velocity
	Test Vector
	Test Vehicle
	Thermal Vacuum
	Throttle Valve
	Thrust Vector
TVA	Tennessee Valley Authority [US]
	Thrust Vector Alignment
TVC	Tag Vector Control
	Thrust Vector Control
TVCS	Thrust Vector Control System
TVDC	Test Volts, Direct Current
TVDP	Terminal Vector Display Unit
TVDR	Tag Vector Display Register
TVEL	Track Velocity
TVF	Tape Velocity Fluctuation
TVG	Time-Varied Gain
TVI	Television Interference
	Tomasetti Volatile Indicator
TVIST	Television Information Storage Tube
TVL	Tenth-Value Layer
TVM	Tachometer Voltmeter
	Track via Missile
	Transistor Voltmeter
TVN	Total Volatile Nitrogen
TVOC	Television Operations Center
TVOR	Terminal VHF (= Very High Frequency) Omnirange
TVP	True Vapor Pressure
TVPPA	Tennessee Valley Public Power Association [US]
TVR	Tag Vector Response
TVRO	Television Receive-Only
TVS	Thermostatic Vacuum Switch
TV-SAT	Television Satellite [FRG]
TVSS	Transient Voltage Surge Suppressor
TVT	Television Typewriter
TVW	Tag Vector Word
TW	Tape Word
	Thermal Wire
	Thermit Welding
	Travelling Wave
	Truncated Wave
	Twist
	Typewriter
TWA	Task Work Area
	Time-Weighted Average
	Travelling Wave Accelerator
	Travelling Wave Amplifier
TWB	Typewriter Buffer
TWC	(American) Truncated Whitworth Coarse (Thread)

TW-C	Truncated Wave with Compressive Dwell
TWCRT	Travelling-Wave Cathode Ray Tube
TWERLE	Tropical Wind Energy Conservation and Reference Level Experiment
TWF	(American) Truncated Whitworth Fine (Thread)
TWG	Telemetry Working Group
TWI	Thermal Wave Imaging
TWINAX	Twinaxial Cable [also: twinax]
TWK	Typewriter Keyboard
TWL	Total Weight Loss
TWM	Travelling-Wave Maser
TWMBK	Travelling-Wave Multiple-Beam Klystron
TWMR	Tungsten Water-Moderated Reactor
TWP	Township
TWR	Trans-World Radio
TWS	Track while Scan
	(American) Truncated Whitworth Special (Thread)
TWSB	Twin Sideband
TWT	Travelling-Wave Tube
TW-T	Truncated Wave with Tensile Dwell
TWTA	Travelling-Wave Tube Amplifier
TW-TC	Truncated Wave with Tensile and Compressive Dwell
TWX	Teletypewriter Exchange (Service)
	Time Wire Transmission
TX	Task Extension
	Temperature - Concentration (Diagram)
	Texas [US]
	Torque Transmitter
	Transmit(ter)
TXA	Task Extension Area
TXE	Telephone Exchange Electronics
TXRX	Transmitter-Receiver
TXT	Text
TY	Territory
3Y	Three Times Yield Strength
TYDAC	Typical Digital Automatic Computer
TYP	Typographical
	Typography
TYPSG	Typesetting
Tyr	Tyrosine
TZD	True Zenith Distance
TZM	Titanium-Zirconium-Molybdenum
TZP	Tetragonal Zirconia Polycrystalline

U

U	Unit
	University
	Uranium
	Uracile
	User
u	unidirectional
	update
	upper
UA	Ultra-Audible
	University of Alabama [US]
	University of Alaska [US]
	University of Arizona [US]
	User Area
UAA	University Aviation Association [US]
	Utility Arborists Association
UAB	University of Alabama, Birmingham [US]
UACN	University of Alaska Computer Network [US]
UADPS	Uniform Automatic Data Processing System
UADS	User Attribute Data Set
UAE	United Arab Emirates
UAGI	University of Alaska Geophysical Institute [US]
UAH	University of Alabama, Huntsville [US]
UAIDE	Users of Automatic Information Display Equipment
UAIS	Universal Aircraft Information System
UAM	Underwater-to-Air Missile
UAN	Unified Automatic Network
UAO	Unexplained Aerial Object
UAP	Unidentified Atmospheric Phenomena
	Upper Air Project
UAR	United Arab Republic
	Upper Atmosphere Research
UARAEE	United Arab Republic Atomic Energy Establishment
UARI	University of Alabama Research Institute [US]
UARS	Upper Atmosphere Research Satellite
UART	Universal Asynchronous Receiver-Transmitter
UARTO	United Arab Republic Telecommunication Organization
UAS	Unit Approval System [FAA]
UAUM	Underwater-to-Air-to-Underwater Missile
UAW	United Automobile Workers [US]
UB	Unbalance
	Upper Bound
U/B	Unbalance
UBC	Universal Buffer Controller
	University of British Columbia [Canada]
UBCW	United Brick and Clay Workers of America
UBD	Utility Binary Dump
UBFF	Urey-Bradley Force Field
UBHR	User Block Handling Routine
UBJ	Union Ball Joint
UBK	Unblock
UBM	Unit Bill of Materials
UBS	Unit Backspace (Character)
UBT	Universal Book Tester
UC	Unit Call
	Unit Cooler
	Universal Crown (Rolling Mill)
	University of California [US]
	University of Chicago [US]
	Upper Case
U/C	Up-Converter
UCA	Uncommitted Component Array
UCAR	University Corporation for Atmospheric Research

UCB	Unit Control Block
	University of California, Berkeley [US]
UCC	Uniform Commercial Code
UCCA	Universities Central Council on Admissions
UCCRS	Underwater Coded Command Release System
UCCB	University College of Cape Breton [Canada]
UCCS	Universal Camera Control System
UCD	Universal Control Drive
UCDP	Uncorrected Data Processor
UCEA	University Council for Educational Administration
UCF	Utility Control Facility
UCI	Ultrasonic Contact Impedance
	Utility Card Input
UCIS	Uprange Computer Input System
UCL	Upper Confidence Limit
	Upper Control Limit
UCLA	University of California, Los Angeles [US]
UCNI	Unified Communications Navigation and Identification
UCO	Utility Compiler
UCON	Utility Control
UCORC	University of California/Operations Research Center [US]
UCOS	Uprange Computer Output System
UCOWR	Universities Council on Water Research
UCP	Utility Control Program
UCPTE	Union for the Coordination of the Production and Transfer of Electric Energy
UCRI	Union Carbide Research Institute [US]
UCRL	University of California Research Laboratory [US]
	University of California Radiation Laboratory [US]
UCS	Uniform Chromaticity Scale
	Universal Camera Site
	Universal Character Set
	Universal Classification System
UCSB	Universal Character Set Buffer
UCSD	Universal Communications Switching Device
	University of California, San Diego [US]
UCSEL	University of California Structural Engineering Laboratory [US]
UC/SSL	University of California/Space Sciences Laboratory [US]
UCSTR	Universal Code Synchronous Transmitter-Receiver
UCT	Universal Continuity Tester
	Universal Coordinated Time
UCTE	Union of Canadian Transport Employees
UCW	Unit Control Word
UCWE	Underwater Countermeasures and Weapons Establishment
UD	University of Denver [US]
UDAR	Universal Digital Adaptive Recognizer
UDAS	Unified Direct Access System
UDB	Up Data Buffer
UDC	Ultrasonic Doppler Cardioscope

	Unidirectional Composite
	Universal Decimal Classification
	Universal Digital Controller
	Urban District Council
	User Defined Code
UDEC	Unitized Digital Electronic Calculator
UDET	Unsymmetrical Diethylenetriamine
UDF	Unducted Fan
	User Defined Function
UDI	Urban Development Institute
UDK	User Define Key
UDL	Uniform Data Link
	Up Data Link
UDM	Unified Defect Model
UDMH	Unsymmetrical Dimethylhydrazine
UDOFET	Universal Digital Operational Flight Trainer
UDOP	UHF (= Ultrahigh Frequency) Doppler
UDP	Uniform Distribution Pattern
	United Data Processing
	Uridine Diphosphate
UDPAG	Uridine Diphosphoacetylglucosamine
UDPG	Uridine Diphosphoglucose
UDPGA	Uridine Diphosphoglucuronic Acid
UDP-GAL	Uridine Diphosphogalactose
UDPGD	Uridine Diphosphoglucose Dehydrogenase
UDPGDH	Uridine Diphosphoglucose Dehydrogenase
UDR	Universal Document Reader
UDT	Universal Data Transcriber
	Universal Document Transport
UDTI	Universal Digital Transducer Indicator
UEP	Underwater Electric Potential
UERD	Underwater Explosives Research Division [US Navy]
UESA	Ukrainian Engineers' Society of America
UET	United Engineering Trustees
	Universal Engineered Tractor
UF	Ultrafiltration
	Ultrasonic Frequency
	Urea-Formaldehyde
	Urethane Foam
UFA	Unesterified Fatty Acids
UFB	Unfit for Broadcast
UFC	Uniform Freight Classification
UFCS	Underwater Fire Control System
UFFI	Urea-Formaldehyde Foam Insulation
UFO	Unidentified Flying Object
UFR	Underfrequency Relay
UGA	Uncommitted Gate Array
UGC	University Grants Committee
UGCW	United Glass and Ceramic Workers (of America)
UGLIAC	United Gas Laboratory Internally Programmed Automatic Computer
UH	Unit Heater
UHB	Ultrahigh Bypass (Engine)
UHC	Ultrahigh Carbon
UHF	Ultrahigh Frequency
UHFDF	Ultrahigh Frequency Direction Finder
UHM	Ultrahigh-Modulus
UHML	Upper-Half Mean-Length
UHMW	Ultrahigh Molecular Weight

UHMWPE	Ultrahigh Molecular Weight Polyethylene	**UMF**	Urea Melamine Formaldehyde
UHP	Ultrahigh Power	**UMI**	University Microfilms International
UHPFB	Untreated Hard Pressed Fiberboard	**UML**	Universal Machine Language
UHR	Ultrahigh Resistance	**UMLER**	Universal Machine Language Equipment
UHT	Ultrahigh Temperature		Register
UHTREX	Ultrahigh Temperature Reactor Experiment	**UMM**	Universal Measuring Microscope
UHV	Ultrahigh Vacuum	**UMP**	Upper Mantle Project [of ICSU]
UI	Unemployment Insurance		Uridine Monophosphate
	University of Illinois [US]	**UMR**	Unipolar Magnetic Region
UIA	Union of International Associations		University of Missouri-Rolla [US]
UIC	Unemployment Insurance Commission		Upper Maximum Range
UIEO	Union of International Engineering	**UMS**	Universal Maintenance Standards
	Organizations	**UMT**	Ultrasonic Machine Tool
UIF	Unrestricted Industrial Funds	**UMTA**	Urban Mass Transit Authority
UIL	UNIVAC Interactive Language		Urban Mass Transportation Administration
UIM	Ultra-Intelligent Machine		[US]
UIP	University Interactions Program	**UMTRA**	Uranium Mill Tailings Radiation Control
UIPT	International Union of Public Transport		Act [US]
UJM	Uncorrected Jet Model	**UMWA**	United Mineworkers of America
UJT	Unijunction Transistor	**UN**	Union
UK	United Kingdom		United Nations
UKAC	United Kingdom Automation Council		University of Nevada [US]
UKAEA	United Kingdom Atomic Energy Authority		Urban Network
UKPO	United Kingdom Post Office	**UNAC**	United Nations Association of Canada
UKR	Ukraine	**UNADS**	UNIVAC Automated Documentation System
	Ukrainian	**UNAMACE**	Universal Automatic Map Compilation
UKRAS	United Kingdom Railway Advisory Service		Equipment
UKSM	United Kingdom Scientific Mission	**UNB**	University of New Brunswick [Canada]
UL	Underwriters Laboratories	**UNC**	Unified National Coarse (Thread)
	Upper Level		University of North Carolina [US]
	Upper Limit	**UNCAST**	United Nations Conference on the
ULA	Uncommitted Logic Array		Applications of Science and Technology
ULB	Universal Logic Block	**UNCOL**	Universal Computer-Oriented Language
ULC	Underwriters Laboratories of Canada	**UNCSTD**	United Nations Conference on Science and
	Universal Logic Circuit		Technology for Development
ULCB	Ultralow Carbon Bainitic (Steel)	**UNCTAD**	United Nations Conference on Trade and
ULCC	Ultralarge Crude Carrier		Development
ULE	Ultralow Expansion	**UNDP**	United Nations Development Program
ULF	Ultralow Frequency	**UNDV**	Under Voltage
ULICP	Universal Log Interpretation Computer	**UNDW**	Underwater
	Program	**UNEF**	Unified Screw Thread Extra Fine
ULM	Ultrasonic Light Modulator		United Nations Emergency Force
ULMS	Underseas Long-Range Missile System	**UNEP**	United Nations Environmental Program
ULO	Unmanned Launch Operations	**UNESCO**	United Nations Educational, Scientific and
ULOW	Unmanned Launch Operations Western		Cultural Organization
	(Test Range)	**UNETAS**	United Nations Emergency Technical Aid
ULPR	Ultralow Pressure Rocket		Service
ULSI	Ultralarge-Scale Integration	**UNEXC**	Unexcavated
ULSV	Unmanned Launch Space Vehicle	**UNF**	Unified National Fine (Thread)
ULT	Ultimate(ly)	**UNICAT/TELECAT**	Union Catalogue/Telecommuni-
	Unique Last Term		cations Catalogue
ULTRA	Universal Language for Typographic	**UNICIS**	University of Calgary Information System
	Reproduction Applications		[Canada]
ULV	Ultra Low Volume	**UNICOM**	Universal Integrated Communications
UM	Universal Mill (Plate)	**UNICOMP**	Universal Compiler
	University of Michigan [US]	**UNIDO**	United Nations Industrial Development
UMA	Universal Measuring Amplifier		Organization
UMASS	Unlimited Machine Access from Scattered	**UNIFET**	Unipolar Field-Effect Transistor
	Sites	**UNIFREDI**	Universal Flight Range and Endurance Data
UMC	Unidirectional Molding Compound		Indicator
UMES	University of Michigan Executive System		

UNIPOL	Universal Procedure-Oriented Language
UNISAP	UNIVAC Share Assembly Program
UNISOR	University Isotope Separator at Oak Ridge [US]
UNITAR	United Nations Institute for Training and Research
UNITEL	Universal Teleservice
UNITRAC	Universal Trajector Compiler
UNIV	University
	Universal
UNIVAC	Universal Automatic Computer
UNIX	Universal Executive
UNM	University of New Mexico [US]
UNNSAD	Unit Neutral Normalized Spectral Analytical Density
UNO	United Nations Organization [now: UN]
UNOQ	Unit of Quantity
UNCPICPUNE	United Nations Conference for the Promotion of International Cooperation in the Peaceful Uses of Nuclear Energy
UNPS	Universal Power Supply
UNR	University of Nevada, Reno [US]
UNRR	Unidirectional Non-Reversing Relay
UNS	Unified National Special (Thread)
	Unified Numbering System
	Unsymmetrical
UNSC	United Nations Security Council
UNSCEAR	United Nations Scientific Committee on the Effects of Atomic Radiation
UNU	United Nations University [Japan]
UNWMG	Utility Nuclear Waste Management Group
U/O	Used on
UOC	Ultimate Operating Capability
UOD	Units of Optical Density
U of C	University of Calgary [Canada]
U of I	University of Idaho [US]
U of M	University of Michigan [US]
U of Nfld	University of Newfoundland [Canada]
U of O	University of Ottawa [Canada]
U of Penn	University of Pennsylvania [US]
U of T	University of Toronto [Canada]
U of V	University of Virginia [US]
UON	Unless Otherwise Noted
UOS	Underwater Ordnance Station
UOV	Unit of Variance
UOW	University of Waterloo [Canada]
UP	Uniprocessing
	United Press
	University of Pittsburgh [US]
	Unsolicited Proposal
	Urea Phosphate
	Utility Path
UPA	University Photographers Association (of America)
UpA	Uridylyladenosine
UpApA	Uridylyladenylyladenosine
UPB	Upper Bound
UPC	Universal Product Code
UPCS	Universal Process Control Software
UPDATE	Unlimited Potential Data through Automation Technology in Education
UPE	Unnatural Parity Exchange
UPEI	University of Prince Edward Island [Canada]
UpG	Uridylylguanosine
UPHS	Underground Pumped Hydro Storage
UPL	Universal Programming Language
UPLIFTS	University of Pittsburgh Linear File Tandem System
UPOS	Utility Program Operating System
UPP	United Papermakers and Paperworkers
UPR	Ultrasonic Paramagnetic Resonance
UPS	Ultraviolet Photoelectron Spectroscopy
	Ultraviolet Photoemission Spectroscopy
	Uninterrupted Power Supply
	Uninterruptible Power Supply
	Uninterruptible Power System
UPSE	Universal Power Service Equipment
UPU	Universal Postal Union [of UN]
UpU	Uridylyluridine
UpUpG	Uridylyluridylylguanosine
UPVC	Unplasticized Polyvinyl Chloride
UQ	University of Quebec [Canada]
UQAC	University of Quebec at Chicoutami [Canada]
UQAM	University of Quebec at Montreal [Canada]
UQAR	University of Quebec at Rimouski [Canada]
UQTR	University of Quebec at Trois-Riviére [Canada]
UR	Unattended Repeater
	Under Running (Crane)
	Unit Record
	University of Rochester [US]
U/R	Up Range
URA	Ultrared Absorption
	Universities Research Association [US]
URAG	Uranium Resources Appraisal Group [Canada]
URBM	Ultimate Range Ballistic Missile
URC	Utilities Research Commission
URD	Underground Residential Distribution (Cable)
URG	Universal Radio Group
URI	University of Rhode Island [US]
URIF	University Research Incentive Fund [Canada]
URIPS	Undersea Radioisotope Power Supply
URIR	Unified Radioactive Isodromic Regulator
URISA	Urban and Regional Information Systems Association
URL	Underground Research Laboratory
URLL	University of Rochester Laser Laboratory [US]
UROP	Undergraduate Research Opportunities
URPA	University of Rochester, Department of Physics and Astronomy [US]
URR	Ultrared Reflection
	Unidirectional Reversing Relay
URRI	Urban Regional Research Institute [US]
URS	Uniform Reporting System
	United Research Service
	Universal Reference System

	Universal Regulating System
	Unmanned Repeater Station
URT	Uranium Research Technology
URU	Uruguay
URV	Undersea Rescue Vehicle
	Undersea Research Vehicle
US	Ultrasonic(s)
	Underside
	United Services
	United States (of America)
	Unit Separator (Character)
	Universal System
USA	United States of America
USAAML	United States Army Aviation Materiel Laboratories
USAASO	United States Army Aeronautical Services Office
USAAVLABS	United States Army Aviation Materiel Laboratories
USAAVNS	United States Army Aviation School
USAAVNTA	United States Army Aviation Test Activity
USABAAR	United States Army Board for Aviation Accident Research
USABRL	United States Army Ballistics Research Laboratories
USACA	United States Advanced Ceramics Association
USACDA	United States Arms Control and Disarmament Agency
USACDC	United States Army Combat Developments Command
USACDCAVNA	United States Army Combat Developments Command Aviation Agency
USACDCCBRA	United States Army Combat Developments Command Chemical-Biological-Radiological Agency
USACSC	United States Army Computer Systems Command
USACSSC	United States Army Computer Systems Support (and Evaluation) Command
USADATCOM	United States Army Data Support Command
USADC	United States Army Data Support Command
USADSC	United States Army Data Services Command
USAEC	United States Atomic Energy Commission
USAECOM	United States Army Electronics Command
USAEPG	United States Army Electronic Proving Grounds
USAERDA	United States Army Electronic Research and Development Agency
USAERDL	United States Army Engineering Research and Development Laboratories
USAF	United States Air Force
USAFB	United States Air Force Base
USAFETAC	United States Air Force Environmental Technical Applications Center
USAFI	United States Armed Forces Institute [USDOD]
USAFSC	United States Air Force Systems Command

USAID	United States Agency for International Development
USAIDSC	United States Army Information and Data Systems Command
USALMC	United States Army Logistics Management Center
USAM	Unified Space Applications Mission [NASA]
	User Spool Access Method
USAMC	United States Army Materiel Command
USANDL	United States Army Nuclear Defense Laboratory
USANWSG	United States Army Nuclear Weapon Surety Group
USAOMC	United States Army Ordnance Missile Command
USAPC	United States Army Petroleum Center
USAPHS	United States Army Primary Helicopter School
USAPO	United States Antarctic Projects Office
USAREPG	United States Army Electronic Proving Grounds
USARIEM	United States Army Research Institute of Environmental Medicine
USARP	United States Antarctic Research Program [NSF, US]
	United States Atlantic Research Program
USARPA	United States Army Radio Propagation Agency
USAS	United States of America Standards
USASCAF	United States Army Service Center for Armed Forces
USASCC	United States Army Strategic Communications Command
USASCII	United States of America Standard Code for Information Interchange
USASI	United States of America Standards Institute [now: ANSI]
USASMSA	United States Army Signal Missile Support Agency
USATACOM	United States Army Tank-Automotive Command
USATEA	United States Army Transportation Engineering Agency
USATECOM	United States Army Test and Evaluation Command
USATIA	United States Army Transportation Intelligence Agency
USB	Unified S-Band
	Upper Sideband
	Upper Surface Blowing
USBE	Unified S-Band Equipment
USBM	United States Bureau of Mines
USBR	United States Bureau of Reclamation
USBS	Unified S-Band System
	United States Bureau of Standards
USC	Ultrasupercritical
	United States Code
	United States Congress
	University of Southern California [US]
USCAL	University of Southern California Aeronautical Laboratory [US]

USCEA	United States Committee for Energy Awareness
USCEC	University of Southern California Engineering Center [US]
USCEE	University of Southern California, Department of Electrical Engineering [US]
USCG	United States Coast Guard
USC&GS	United States Coast and Geodetic Survey [USDC]
USCMI	United States Commission on Mathematical Instruction
USD	Ultimate Strength Design
USDA	United States Department of Agriculture
USDC	United States Department of Commerce
USDI	United States Department of the Interior
USDL	United States Department of Labour
USDOC	United States Department of Commerce
USDOD	United States Department of Defense
USDR&E	Undersecretary of Defense for Research and Engineering [US]
USE	Unified S-Band Equipment
	Unit Support Equipment
	UNIVAC Scientific Exchange
USEC	United System of Electronic Computers
USERC	United States Environment and Resources Council
USERID	User Identification
USES	United States Employment Service
USF	United States Form Thread
USFCC	United States Federation for Culture Collections
USFED	United States Federal Specifications Board [also: USFed]
USFS	United States Frequency Standard
USFSS	United States Fleet Sonar School
USG	United States (Standard) Gauge
	United States Gallon
USGA	Ultrasonic Gas Atomization
USGPM	United States Gallons per Minute
USGPO	United States Government Printing Office
USGR	United States Government Report
USGRDR	United States Government Research and Development Report [NBS]
USGS	United States Geological Society
	United States Geological Survey
USGW	Undersea Guided Weapon
USI	User System Interface
USIA	United States Information Agency
USIB	Unsaturated Iron-Binding Capacity
USIC	United States Information Center
USIG	Ultrasonic Impact Grinding
USIS	United States Information Service
USISC	United States International Service Carriers
USITA	United States Independent Telephone Association
USITC	United States International Trade Commission
USIU	United States International University
USJPRS	United States Joint Publication Research Service

USL	Underwater Sound Laboratory [US Navy]
USM	Ultrasonic Machining
	Underwater-to-Surface Missile
USMA	United States Metric Association
	United States Military Academy
USMAC	United States Management Advisory Committee
USMT	United States Military Transport
USN	United States Navy
USNBS	United States National Bureau of Standards
USNC	United States National Committee
USNCSCOR	United States National Committee for the Scientific Committee on Oceanic Research
USNC-TAM	United States National Committee on Theoretical and Applied Mathematics
USNC-URSI	United States National Committee for the International Union of Radio Science
USNCWEC	United States National Committee of the World Energy Conference
USNEL	United States Navy Electronics Laboratory [now divided into NUWC and NCCCC]
USNO	United States Naval Observatory
USNR	United States Naval Research
USNRDL	United States Naval Radiological Defense Laboratory
USNRRC	United States Nuclear Reactor Regulatory Committee
USNUSL	United States Navy Undersea Laboratory
USNUWL	United States Navy Underwater Laboratory
USO	Unmanned Seismic Observatory [USDOD]
USOE	United States Office of Education
USP	United States Pharmacopoeia
	Usage Sensitive Pricing
USPO	United States Patent Office
	United States Post Office
USRA	Undergraduate Students Research Awards
	United States Railway Association
	Universities Space Research Association [US]
USRL	Underwater Sound Reference Laboratory [US Navy]
USRS	United States Rocket Society
USS	Unformatted System Service
	Unique Support Structure
	United States Ship
	United States Standard
	United States Steel
USSA	Underground Security Storage Association
USSG	United States Standard Gauge
USSR	Union of Soviet Socialist Republics
USSTL WG	United States Steel Wire Gauge [also: USStl WG]
USSWG	United States Steel Wire Gauge
UST	Ultrasonic Transducer
	Unblock SPADE Terminal (Command)
	Underground Storage Tank
USTAG	United States Technical Advisory Group
USU	Utah State University [US]
USW	Ultrasonic Welding
	Undersea Warfare
	United Steelworkers (of America)

USWB	United States Weather Bureau
UT	Ultrasonic Testing
	Umbilical Tower
	Universal Time (Scale)
	University of Texas [US]
	Urban Transport(ation)
	User Terminal
	Utah [US]
Ut	Utah [US]
UTA	Urban Transportation Administration
UTANG	University of Toronto Antinuclear Group
UTAP	Unified Transportation Assistance Program [US]
	Urban Transport Assistance Program
UTC	United Technology Center
	Universal-Test Communicator
UTCS	Urban Traffic Control System
UTD	United
	Universal Transfer Device
	University of Texas, Dallas [US]
UTEC	Utah University College of Engineering [US]
UTEP	University of Texas, El Paso [US]
UTF	Underground Test Facility
UTH	Universal Test Head
UTIA	University of Toronto Institute of Aerophysics [Canada]
UTIAS	University of Toronto Institute of Aerospace Studies [Canada]
UTK	Ultrasonic Trim Knife
UTLAS	University of Toronto Library Automation System
UTM	Universal Testing Machine
	Universal Test Message
	Universal Transverse Mercator
UTP	Universal Tape Processor
	Upper Turning Point
	Uridine Triphosphate
UTQGS	Uniform Tire Quality Grading System
UTS	Ultimate Tensile Strength
	Underwater Technology School
	Underwater Telephone System
	Unified Transfer System
	Unit Test Station
	Universal Test Station
	Universal Time-Sharing (System)
UTTAS	Utility Tactical Transport Aircraft System
UTTC	Universal Tape-to-Tape Converter
UTU	United Transportation Union
UTW	Ultrathin Window
	United Telegraph Workers
UTWA	United Textile Workers of America
UUA	UNIVAC Users Association [US]
UUM	Underwater-to-Underwater Missile
UUT	Unit Under Test
UV	Ultraviolet
	Ultravisible
	Undervoltage
UVA	Ultraviolet A [320-400 nm]
UVASER	Ultraviolet Amplification by Stimulated Emission of Radiation
UVB	Ultraviolet B [290-320 nm]
UVC	Ultraviolet C [below 290 nm]
	Unidirectional Voltage Converter
UVD	Undervoltage Device
UVFL	Ultraviolet Fluorescence
UVic	University of Victoria [Canada]
UVPROM	Ultraviolet Programmable Read-Only Memory
UVR	Undervoltage Relay
UV/VIS	Ultraviolet/Visible
UW	Ultrasonic Wave
	Unique Word
	Upset Welding
	University of Washington [US]
	University of Wisconsin [US]
UWAL	University of Washington Aeronautical Laboratory [US]
UWEX	University of Wisconsin-Extension [US]
UWI	University of the West Indies [Jamaica]
UWM	University of Wisconsin, Milwaukee [US]
UWO	University of Western Ontario [Canada]
UXB	Unexploded Bomb
UXD	Ultimate X-Ray Detector

V

V	Valine
	Value
	Valve
	Vanadium
	Vapor
	Variable
	Vector
	Verification
	Vitrified
	Volt(age)
	Voltmeter
	Volume
v	velocity
	verify
	virtual
VA	Value Analysis
	Vickers Armstrong Gun
	Video Amplifier
	Virginia [US]
	Virtual Address
	Visual Aid
	Volt-Ammeter
	Voltampere
V/A	Volume/Area
Va	Virginia [US]
VAA	Viscum Album Agglutinin
VAAC	Vanadyl Acetylacetonate
VAB	Vertical Assembly Building
	Voice Answerback
VAC	Vacuum
	Variable Amplitude Correction
	Vector Analog Computer
	Vertical Assembly Component

	Video Amplifier Chain
	Volts, Alternating Current
VACR	Variable Amplitude Correction Rack
VACTL	Vertical Assembly Component Test Laboratory
VAD	Vacuum Arc Decarburization
	Vacuum Arc Degassing
	Vapor Axial Deposition
	Voltmeter Analog-to-Digital (Converter)
VADC	Voltmeter Analog-to-Digital Converter
VADE	Versatile Auto Data Exchange
VADER	Vacuum Arc Double Electrode Remelting
VAE	Vinyl Acetate-Ethylene
VAEP	Variable Attributes Error Propagation
VAL	Value
	Vehicle Authorization List
Val	Valine
VALOR	Variable Locale and Resolution
VALSAS	Variable Length Word Symbolic Assembly System
VAM	Vasicular-Arbuscular Mycorrhizae
	Vector Airborne Magnetometer
	Virtual Access Method
	Vogel's Approximation Method
	Voltammeter
VAMAS	Versailles Project on Advanced Materials and Standards
VAMP	Vector Arithmetic Multiprocessor
	Vincristine Amethopterin
	Visual-Acoustic-Magnetic Pressure
VAN	Value-Added Network
VAOR	VHF (= Very High Frequency) Aural Omnirange
VAPI	Visual Approach Path Indicator
VAPP	Vector and Parallel Processor
VAP PRF	Vapor-Proof
VAPS	VSTOL Approach System
VAR	Vacuum Arc Remelting
	Value-Added Reseller
	Variable
	Variance
	Variant
	Variation
	Variety
	Variometer
	Video-Audio Range
	Visual and Aural Range
	Volt-Ampere Reactive
VARAD	Varying Radiation
VARHM	Var-Hour Meter
VARISTOR	Variable Transistor [also: Varistor]
VARITRAN	Variable-Voltage Transformer
VARN	Varnish
VARR	Variable Range Reflector
	Visual Aural Radio Range
VARS	Vertical Azimuth Reference System
VAS	Value-Added Service
	Vibration Absorption System
VASCA	(Electronic) Valve and Semiconductor Manufacturers Association
VASCAR	Visual Average Speed Computer and Recorder

VASI	Visual Approach Slope Indicator
VASIS	Visual Approach Slope Indicator System
VAST	Versatile Automatic Specification Tester
	Versatile Avionics Ship Test
	Vibration and Strength Analysis
VAT	Value-Added Tax
	Virtual Address Translator
VATE	Versatile Automatic Test Equipment
VATLS	Visual Airborne Target Location System
VAV	Variable Air Volume
	Ventricular Assist Valve
VA/VE	Value Analysis/Value Engineering
VAWT	Vertical Axis Wind Turbine
VAX	Virtual Address Extension
VB	Valence Band
	Valve Box
	Variable Block
	Voice Band
VBAS	Von Braun Astronomical Society [US]
VBD	Voice Band Data
VBDOS	Valance Band Density of States
VBE	Valinolbutylether
	Volumetric Balance Equation
VBI	Vertical Blanking Interval
VBL	Voyager Biological Laboratory [NASA]
VBM	Valence Band Maximum
VBO	Valence Band Offset
VBS	Valence Band Spectrum
VC	Varnished Cambric (Wire)
	Vector Control
	Versatility Code
	Vertical Circle
	Video Correlator
	Virtual Circuit
	Visual Capacity
	Voice Coil
	Voltage Comparator
	Volt-Coulomb
VCA	Voltage-Controlled Amplifier
VCBA	Variable Control Block Area
VCC	Visual Communications Congress
	Voice Control Center
	Voltage-Controlled Capacitor
VCCS	Voltage-Controlled Current Source
VCD	Vacuum Carbon Deoxidation
	Valve Coil Driver
	Variable-Capacitance Diode
	Variable Center Distance
	Vibrational Circular Dichroism
	Voltage Charge Device
VCE	Collective Emitter Voltage
VCF	Voltage-Controlled Frequency
VCG	Vapor Crystal Growth
	Vectorcardiogram
	Vertical Location of the Center of Gravity
VCGS	Vapor Crystal Growth System
VCI	Vehicle Cone Index
	Volatile Corrosion Inhibitor
VCL	Vertical Center Line
VCM	Vacuum Condensible Material
	Vehicle Condition Monitor

	Vibrating Coil Magnetometer
	Vinyl Chloride Monomer
	Voltage Controlled Multivibrator
VCNR	Voltage-Controlled Negative Resistance
VCO	Voltage-Controlled Oscillator
VCR	Video Cartridge Recorder
	Video Cassette Recorder
VCS	Video Cassette System
	Visually Coupled System
	Voltage Calibration Set
VCSR	Voltage-Controlled Shift Register
VCT	Voltage Control Transfer
VCVS	Voltage-Controlled Voltage Source
VCXO	Voltage-Controlled Crystal Oscillator
VD	Vandyke
	Void
	Voltage Detector
	Voltage Drop
	Volume Damper
V/D	Voice Data
VDA	Vertical Danger Angle
VDAS	Vibration Data Acquisition System
VDC	Versatile Digital Controller
	Vinylidene Chloride
	Volts, Direct Current
VDCT	Direct Current Test Volts
VDCU	Videograph Display Control Unit
VDCW	Direct Current Working Volts
VDD	Visual Display Data
	Voice Digital Display
VDDI	Voyager Data Detailed Index
VDDL	Voyager Data Distribution List
VDDS	Voyager Data Description Standards
VDE	Variable Display Equipment
VDET	Voltage Detector
VDF	VHF (= Very High Frequency) Direction Finding
	Video Frequency
VDFG	Variable Diode Function Generator
VDG	Variable Drive Group
VDI	Virtual Device Interface
VDL	Vienna Definition Language
VDM	Video Display Module
VDP	Van der Pauw (Structure)
	Vertical Data Processing
VDPI	Voyager Data Processing Instruction
VDR	Voice and Data Recording
	Voltage Dependent Resistor
VDRA	Voice and Data Recording Auxiliary
VDS	Variable Depth Sonar
	Visual Docking Simulator
VDT	Video Display Terminal
VDU	Video Display Unit
	Visual Display Unit
VE	Value Engineer(ing)
	Vernier Engine
	Vibration Eliminator
	Viscoelasticity
VEA	Value Engineering Association
VEB	Variable Elevation Beam
VECI	Vehicular Equipment Complement Index

VECO	Vernier Engine Cutoff
VECOS	Vehicle Checkout Set
VECP	Value Engineering Change Proposal
	Value Engineering Cost Proposal
VECU	Vacuum Pump Exhaust Cleanup (System)
VEDAR	Visible Energy Detection and Ranging
VEDS	Vehicle Emergency Detection System [NASA]
VEG	Vegetable
VEH	Vehicle
VEL	Velocity
VELCOR	Velocity Correction
VELF	Velocity Filter
VELOC	Velocity
VEM	Vasoexcitor Material
	Virtual Electrode Model
VENEZ	Venezuela
VENT	Ventilate
	Ventilation
	Ventilator
VEP	Visual Evoked Potential
VEPIS	Vocational Education Program Information System
VER	Verify
	Vernier
	Version
	Visual Evoked Response
	Voluntary Export Restraint
VERA	Versatile Experimental Reactor Assembly
VERDAN	Versatile Differential Analyzer
VERLORT	Very Long-Range Tracking
VERNITRAC	Vernier Tracking by Automatic Correlation
vers	versed sine
versin	versed sine
VERT	Vertical
VES	Variable Elasticity of Substitution
VESC	Vehicle Equipment Safety Commission
VESIAC	Vela Seismic Information Analysis Center [U of M]
VEST	Volunteer Engineers Scientists and Technicians
VEV	Voice Excited Vocoder
VEWS	Very Early Warning System
VF	Variable Frequency
	Vector Field
	Video Frequency
	Voice Frequency
V/F	Voltage-to-Frequency
VFA	Volatile Fatty Acid
VFB	Flatband Voltage
VFC	Vertical Forms Control
	Video Film Converter
	Video Frequency Carrier
	Voice Frequency Carrier
	Voltage-to-Frequency Converter
VFCT	Voice Frequency Carrier Teletype
VFFT	Voice Frequency Facility Terminal
VFL	Variable Field Length
VFO	Variable-Frequency Oscillator
VFR	Visual Flight Rules
VFS	Vertical Full Scale

VFT	Voice Frequency Telegraph
VFU	Vertical Format Unit
VFVC	Vacuum Freezing Vapor Compression
VG	Vector Generator
	Vertical Grain
	Very Good
	Viscosity Grade
	Voice Grade
VGA	Vapor Generation Accessory
	Variable Gain Amplifier
	Video Graphics Array
VGC	Vacuum Gauge Control(ler)
VGPI	Visual Glide Path Indicator
VGS	Vision Guidance System
VGSI	Visual Glide Slope Indicator
VHAA	Very High Altitude Abort
VHD	Video High Density
VHDL	Very High Density Lipoprotein
VHF	Very High Frequency
VHM	Virtual Hardware Monitor
VHN	Vickers Hardness Number
VHO	Very High Output
VHP	Vacuum Hot Pressing
	Very High Performance
	Very High Pressure
VHPIC	Very-High-Performance Integrated Circuit
VHRR	Very High Resolution Radiometer
VHS	Video Home System
VHS-C	Video Home System - Compact
VHSIC	Very High Speed Integrated Circuit
VI	Vibration Institute
	Video Integrator
	Virgin Islands [US]
	Viscosity Index
	Volume Indicator
VIA	Visual Interactive Access
VIAS	Voice Interference Analysis Set
VIB	Vertical Integration Building
	Vibrate
	Vibration
	Vibrator
VIC	Variable Instruction Computer
VICARS	Visual Integrated Crime Analysis and Reporting System
VICI	Velocity Indicating Coherent Integrator
VICOM	Visual Communication
VID	Video
VIDAC	Visual Data Acquisition
	Visual Information Display and Control
VIDAMP	Video Amplifier
VIDF	Video Frequency
VIDIAC	Vidicon Input-to-Automatic Computer
VIEW	Visual Information Enhanced Workstation
VIFC	VTOL Integrated Flight Control
VIFCS	VTOL Integrated Flight Control System
VIFI	Voyager Information Flow Instruction
VIG	Video Integrating Group
	Visual Integrating Group
VIGS	Virtual-Induced Gap States
	Visual Glide Slope
VIH	Voltage-Input High

VIL	Village
	Voltage-Input Low
VILP	Vector Impedance Locus Plotter
VIM	Vacuum Induction Melting
	Vibrational Microlamination
VIMIS	Vertically-Integrated Metal-Insulator-Semiconductor
VIMP	Vancouver Island Mainland Pipeline [Canada]
VIMS	Virginia Institute of Marine Science [US]
VIN	Vehicle Identification Number
VIND	Vicarious Interpolations Not Desired
VINS	Velocity Inertia Navigation System
	Very Intense Neutron Source
VINT	Video Integration
VIO	Virtual Input/Output
VIOC	Variable Input-Output Code
VIOLET	Voice Input/Output Lexically Endowed Terminal
VIOS	Versatile Instrument Operating Software
VIP	Variable Information Processing
	Variable Input Phototypesetting
	Very Important Person
	Vision Integrated with Positioning
	Visual Image Processor
	Visual Information Projection
VIPER	Video Processing and Electronic Reduction
VIPP	Variable Information Processing Package
VIPS	Voice Interruption Priority System
VIR	Vertical Interval Retrace
VIRNS	Velocity Inertia Radar Navigation System
VIS	Videotex Information System
	Visible
VISC	Viscosity
	Viscous
VISSR	Visible and Infrared Spin Scan Radiometer
VISTA	Verbal Information Storage and Text Analysis
	Visual Interpretation System for Technical Applications
VIT	Vertical Interval Test
	Vitreous
VITAL	Variably Initialized Translator for Algorithmic Languages
	VAST Interface Test Application Language
VITEAC	Video Transmission Engineering Advisory Committee [US]
VITS	Vertical Interval Test Signal
viz	(videlicet) - namely
VKIFD	Von Karman Institute for Fluid Dynamics [US]
VL	Vertical Ladder
	Video Logic
	Visual Learning
V/L	Vapor-to-Liquid
VLA	Very Large Array
	Very Low Altitude
VLB	Very Long Baseline
VLBI	Very Long Baseline Interferometry
VLC	Video Level Controller
VLCC	Very Large Crude Carrier

VLCR	Variable Length Cavity Resonance
VLCS	Voltage-Logic Current-Switching
VLDL	Very Low Density Lipoprotein
VLED	Visible Light-Emitting Diode
VLF	Vertical Launch Facility
	Very Low Frequency
VLFS	Variable Low Frequency Standard
VLIW	Very Long Instruction Word
VLP	Video Long Play
VLPE	Very Long Period Experiment
VLR	Very Long Range
VLS	Vacuum Loading System
	Vapor-Liquid-Solid
	Very Large Scale
	Vapor Feed Gases, Liquid Catalyst, Solid Crystalline Whisker Growth
VLSI	Very Large-Scale Integrated (Circuit)
	Very Large-Scale Integration
VLSIC	Very Large-Scale Integrated Circuit
VLSIIC	VLSI Implementation Center [Canada]
VLVS	Voltage-Logic Voltage Switching
VM	Vacuum Melting
	Vector Message
	Velocity Modulation
	Virtual Machine
	Virtual Memory
	Volatile Matter
	Voltmeter
VMA	Valve Manufacturers Association
	Vanillomandelic Acid
	Vehicle Maintenance Area
	Virtual Memory Allocation
VMC	Vertical Motion Carriage
	Virtual Machine Control
	Visual Meteorological Conditions
VMCB	Virtual Machine Control Block
VMD	Vector Meson Dominance
	Vertical Magnetic Dipole
VMGSE	Vehicle Measuring Ground Support Equipment
VMID	Virtual Machine Identifier
VMJ	Vertical Multijunction
VMM	Variable Mission Manufacturing
	Virtual Machine Monitor
VMOS	Vertical Metal-Oxide Semiconductor
VM&P	Varnishmakers and Painters
VMS	Volcanogenic Massive Sulfide
VMT	Video Matrix Terminal
VMTAB	Virtual Machine Table
VMTSS	Virtual Machine Time-Sharing System
VN	Volatile Nitrogen
VNA	Very Narrow Aisle
VNL	Via Net Loss
VNLF	Via Net Loss Factor
VO	Voice-Over
VOC	Voice-Operated Coder
	Volatile Organic Chemical
	Volatile Organic Compound
VOCODER	Voice Coder
VOCOM	Voice Communications
VOD	Vacuum-Oxygen Decarburization

	Velocity of Detonation
	Voice-Operated Device
VODACOM	Voice Data Communication
VODAS	Voice-Operated Device, Anti-Sing
VODAT	Voice-Operated Device for Automatic Transmission
VODC	Vacuum-Oxygen Decarburization Converter
VODER	Voice Decoder
VOGAD	Voice-Operated Gain Adjusting Device
VOH	Voltage-Output High
VOIS	Visual Observation Instrumentation Subsystem
	Visual Observation Instrumentation System
VOL	Voltage-Output Low
	Volcano
	Volume
VOL%	Volume Percent
VOLCAS	Voice-Operated Loss Control and Suppressor
VOLS	Volumes
VOLSER	Volume/Serial Number
VOLT	Volatilize
VOLTAN	Voltage Amperage Normalizer
VOM	Volt-Ohm Meter
VOR	VHF (= Very High Frequency) Omnidirectional Range
	VHF (= Very High Frequency) Omnirange
	Voice-Operated Recording
VORDAC	VOR/DME (= VHF Omnirange/Distance Measuring Equipment) for Average Coverage
VOR/DMET	VHF Omnirange/Distance Measuring Equipment Compatible with TACAN
VORTAC	VHF (= Very High Frequency) Omnirange TACAN
VOS	Virtual Operating System
	Voice-Operated Switch
VOSC	VAST Operating System Code
VOT	Voice-Operated Transmit
VOX	Voice-Operated Transmission
VP	Velocity Pressure
	Vent Pipe
	Vertical Polarization
	Vice-President
	Virtual Processor
	Vulnerable Point
VPAM	Virtual Partitioned Access Method
VPC	Vapor Phase Chromatography
VPCA	Video Prelaunch Command Amplifier
VPD	Vacuum Products Division
	Vapor Phase Deacidification
VPE	Vapor Phase Epitaxy
VPF	Vertical Processing Facility
VPG	Vapor Growth
VPI	Vacuum Pressure Impregnation
	Virginia Polytechnic Institute [US]
VPI&SU	Virginia Polytechnic Institute and State University [US]
VPL	Vanishing Point Left
VPLCC	Vehicle Propellant Loading Control Center
VPM	Vehicles per Mile
	Vibrations per Minute

	Volts per mil
VPR	Vanishing Point Right
VPPA	Variable Polarity Plasma Arc
V PRES	Vice-President
VPRF	Variable Pulse Repetition Frequency
VPR	Vapor-Phase Reflow
VPS	Vacuum Plasma Spraying
	Vapor Phase Soldering
	Vibrations per Second
	Video Program System
VPSP	Vinylpolystyrylpyridine
VPSW	Virtual Program Status Word
VPT	Viscosity-Pressure-Temperature
VR	Vacuum Residual (Oil)
	Variety Reduction
	Velocity Ratio
	Vibrational Relaxation
	Virtual Route
	Voltage Regulator
	Voltage Relay
	Vulcanized Rubber
V-R	Voltage Regulation
VRB	VHF (= Very High Frequency) Recovery Beacon
VRBL	Variable
VRC	Vertical Reciprocating Conveyor
	Vertical Redundancy Check
	Visual Record Computer
VRCI	Variable Resistive Components Institute [US]
VRD	Vacuum Tube Relay Driver
VRE	Voice Recognition Equipment
VREW	Vegetative Rehabilitation and Equipment Workshop
VRFI	Voice Reporting Fault Indicator
VRFWS	Vehicle Rapid Fire Weapon System
VRID	Virtual Route Identifier
VRL	Vertical Reference Line
	Vibration Research Laboratory
VRM	Variable Range Marker
	Vertical Retreat Mining
	Viscous Remanent Magnetization
VRMS	Volts Root-Mean-Square
VRPS	Voltage-Regulated Power Supply
VRR	Visual Radio Range
VRS	Vertical Raster Scan(ning)
	Volatile Reducing Substances
VRSA	Voice Reporting Signal Assembly
VRSS	Voice Reporting Signal System
VRT	Variable Reactance Transformer
	Vessel Residence Time
	Volume-Rendering Technique
VRU	Voltage Readout Unit
VS	Variable Speed
	Variable Store
	Variable Sweep
	Vent Stack
	Vertical Sounding
	Virtual Storage
	Voltmeter Switch
	Volumetric Solution

vs	(versus) - against
VSA	Vacuum Swing Adsorption
	Vibrating String Accelerometer
VSAM	Virtual Sequential Access Method
	Virtual Storage Access Method
VSAT	Very Small-Aperture Terminal
VSB	Vestigial Sideband
	VME Subsystem Bus
VSC	Vacuum Slag Suction
	Variable Speed Control
	Variable Speech Control
	Vibration Safety Cutoff
	Video System Controller
	Voltage-Saturated Capacitor
VSCF	Variable-Speed Constant-Frequency
VSD	Variable Speed Drive
VSE	Virtual Storage Equipment
VSG	Vertical Sweep Generator
VSI	Vacuum-Super-Insulation
	Vertical Speed Indicator
	Video Sweep Integrator
VSM	Vestigial Sideband Modulation
	Vibrating Sample Magnetometer
	Vibrating Space Modulator
	Virtual Storage Management
VSMF	Virtual Search Microfilm File
VSMS	Video Switching Matrix System
VSO	Very Stable Oscillator
VSP	Vision Statistical Processor
VSPC	Virtual Storage Personal Computer
VSR	Validation Summary Report
	Vertical Storage and Retrieval
VSRBM	Very Short Range Ballistic Missile
VSS	Variable Stability System
	Voice Signaling System
	Volatile Suspended Solids
VST	Variable Speed (Friction) Tester
	Variable Speed Transmission
VSTOL	Vertical and/or Short Takeoff and Landing
VSW	Very Short Wave
	Voltage Standing Wave
VSWR	Voltage Standing-Wave Ratio
VT	Vacuum Tube
	Vacuum Technology
	Variable Time
	Vehicular Technology
	Vertical Tab(ulation Character)
	Vertical Tail
	Video Tape
	Video Terminal
	Visual Testing
	Voice Tube
	Vermont [US]
Vt	Vermont [US]
VTAB	Vertical Tab(ulation Character)
VTAM	Virtual Telecommunications Access Method
VTCS	Vehicular Traffic Control System
VTD	Vertical Tape Display
VTDC	Vacuum Tube Development Committee
VTF	Vertical Test Fixture
	Via Terrestrial Facilities

VTL	Variable Threshold Logic
VTM	Voltage-Tunable Magnetron
VTMS	Vessel Traffic Management System
VTO	Voltage-Tuned Oscillator
	Volumetric Top-Off
VTOC	Volume Table of Contents
VTOHL	Vertical Takeoff and Horizontal Landing
VTOL	Vertical Takeoff and Landing
VTOVL	Vertical Takeoff Vertical Landing
VTPR	Vertical Temperature Profile Radiometer
VTR	Video Tape Recorder
VTRS	Video Tape Response System
VTS	Vertical Test Stand
VTVM	Vacuum Tube Voltmeter
VU	Vehicle Unit
	Voice Unit
	Volume Unit
VULC	Vulcanizing
VUTS	Verification Unit Test Set
VUV	Vacuum Ultraviolet
vv	(vice versa) - conversely
V/V	Volume by Volume
VVA	Vicia Villosa Agglutinin
VVC	Voltage-Variable Capacitor
VVDS	Video Vertex Decision Storage
VVE	Vertical Vertex Error
VVR	Variable Voltage Rectifier
VWDI	Vinyl Window and Door Institute
VWL	Variable Word Length
VWP	Variable Width Pulse
VWPI	Vacuum Wood Preservers Institute [US]
VWSS	Vertical Wire Sky Screen
VX	Volume - Concentration (Diagram)
VXO	Variable(-Frequency) Crystal Oscillator

W

W	Tryptophan
	Watt
	Wattmeter
	Wednesday
	Weight
	West
	Width
	Wire
	Wood
	Word
	Work
	(Wolfram) - Tungsten
w	wide
WA	Washington (State) [US]
	Waveform Analyzer
	Western Australia
	Woolwich Armstrong (Gun)
	Wrong Answer
WAA	Water Authorities Association
WAAG	Waveform Acquisition and Arbitrary Generator

	Waveform Analysis and Arbitrary Generator
WAAS	World Academy of Arts and Sciences
WABE	Western Association of Broadcast Engineers
WAC	Weighted Average Cost
	World Aeronautical Chart
	Write Address Counter
WACHO	Western Association of Canadian Highway Officials
WACK	Wait before Transmit Positive Acknowledgement
WACM	Western Association of Circuit Manufacturers [US]
WACS	Workshop Attitude Control System
WADC	Wright Aeronautical Development Center [US Air Force]
WADD	Wright Air Development Division [US Air Force]
WADEX	Word and Author Index
WADS	Wide Area Data Service
WAF	Width across Flats
	Wiring around Frame
WAGR	Western Australian Government Railways
	Windscale Advanced Gas-Cooled Reactor
WAIT	What Alloy Is That (Test)
WAITRO	World Association of Industrial and Technological Research Organizations
WALDO	Wichita Automatic Linear Data Output
WALOPT	Weapons Allocation and Desired Ground Zero Optimizer [US Military]
WAM	Words a Minute
	Worth Analysis Model
WAMDII	Wide Angle Michelson Doppler Imaging Interferometer
WAML	Wright Aero-Medical Laboratory [US]
WAMOSCOPE	Wave Modulated Oscilloscope
WAN	Wide Area Network
WANEF	Westinghouse Astronuclear Experimental Facility [US]
WANL	Westinghouse Astronuclear Laboratory [US]
WAP	Work Assignment Procedure
WARC	World Administrative Radio Conference
WARC/MT	World Administrative Radio Conference on Maritime Telecommunications
WARC/ST	World Administrative Radio Conference on Space Telecommunications
WARHD	Warhead
WARLA	Wide Aperture Radio Location Array
WASH	Washer
Wash	Washington (State) [US]
WASHO	Western Association of State Highway Officials [US]
WASP	Weightless Analysis Sounding Probe [NASA]
	Workshop Analysis and Scheduling Program
WASSP	Wire Arc Seismic Section Profiler
WAT	Weight Altitude and Temperature
WATS	Wide Area Telecommunications Service
	Wide Area Telephone Service
	Wide Area Telephone System
WB	Water Board
	Weak Beam (Image)
	Weather Bureau

	Weldbrazing
	Wet Bulb
	Wheel Base
	Wideband
W/B	Waybill
WBAN	Weather Bureau Air Force and Navy [US]
WBCO	Waveguide below Cutoff
WBCT	Wideband Current Transformer
WBCV	Wideband Coherent Video
WBD	Wideband Data
	Wire Bound
WBDL	Wideband Data Link
WBDF	Weak Beam Dark Field
WBFM	Wideband Frequency Modulation
WBGT	Wet Bulb Globe Thermometer
WBIF	Wideband Intermediate Frequency
WBL	Wideband Limiting
WBLC	Waterborne Logistics Craft
WBNL	Wideband Noise Limiting
WB/NWRC	Weather Bureau/National Weather Records Center [US]
WBP	Weather- and Boilproof
WBPA	Wideband Power Amplifier
WBRS	Wideband Remote Switch
WBTS	Wideband Transmission System
WC	Wood Cover
	Work Cell
	Worsted Count
	Write and Compute
WCA	Workmen's Compensation Act
WCAP	Westinghouse Commercial Atomic Power
WCB	Workmen's Compensation Board
WCC	Work Cell Control
	Workmen's Compensation Commission
	Write Control Character
WCDB	Wing Control during Boost
WCEE	World Conference on Earthquake Engineering
WCF	White Cathode Follower
WCG	Weakly Compactly Generated
WCGM	Writable Character Generation Module
WCL	World Confederation of Labour
WCLIB	West Coast Lumber Inspection Bureau
WCOS	Work Cell Operator Station
WCM	Wired Core Matrix
	Word Combine and Multiplex
WCPSC	Western Conference of Public Services Commissioners
WCR	Water Cooler
	Word Control Register
WCRA	Weather Control Research Association
WCS	Whatman Compression Screw
	Work Center Supervisor
	World Coordinate System
	Writable Control Store
WCSI	World Center for Scientific Information
WD	War Department
	Watt Demand Meter
	Width
	Wind
	Wood

	Word
	Work Distance
	Works Department
	Write Direct
WDB	Wideband
WDC	World Data Center [NAS, US]
WDF	Waste Derived Fuel
	Woodruff
WDG	Winding
WDM	Wavelength Division Multiplexer
WDP	Women in Data Processing
WDS	Wavelength Dispersive Spectrometry
	Wavelength Dispersive X-Ray Spectroscopy
WDST	Western Daylight Saving Time
WDWKG	Woodworking
WDX	Wavelength Dispersive X-Ray Analysis
WE	Women in Energy
	Working Electrode
	Write Enable
WEA	Workers' Education Association
WEC	World Energy Conference
	World Environment Center
WECOM	Weapons Command [US Army]
WED	Weak Exchange Degeneracy
	Wednesday
WEDC	Western Engineering Design Competition [Canada]
WEFAX	Weather Facsimile [of ESSA]
WEMA	Western Electronic Manufacturers Association
	Winding Engine Manufacturers Association
WER	Worth Estimating Relationship
WES	Waterways Experiment Station [US Army]
	Women's Engineering Society
WESCON	Western Electronics Show and Convention [US]
WESRAC	Western Research Application Center [US]
W/E&SP	With Equipment and Spare Parts
WESTAC	Western Transportation Advisory Council
WESTAR	Waterways Experiment Station Terrain Analyzer Radar
WESTEC	Western Metal and Tool Exposition and Conference
WETAC	Westinghouse Electronic Tubeless Analog Computer
WEU	Western European Union
	Wide End Up (Mold)
WF	Weighting Function
	Widefield
WFA	Wisteria Floribunda Agglutinin
WFAIT	Western Foundation for Advanced Industrial Technology
WFCMV	Wheeled Fuel-Consuming Motor Vehicle
WFEO	World Federation of Engineering Organizations
WF&Eq	Wave Filters and Equalizers [IEEE Committee]
WFF	Well-Formed Formula
WFMI	World Federation for the Metallurgical Industry
WFNA	White Fuming Nitric Acid

WF/PC	Widefield/Planetary Camera
WFR	Wafer
W-F	Wiedemann-Franz (Ratio)
WFS	World Future Society
WFTU	World Federation of Trade Unions
WG	Waveguide
	Water Gauge
	Wire Gauge
	Working Group
W/G	Waveguide
WGA	Wheatgerm Agglutinin
WGBC	Waveguide Operating below Cutoff
W GER	West German
	West Germany
WGS	Waveguide Glide Slope
WGSC	Wide Gap Spark Chamber
WGT	Weight
WH	Water Heater
	White
Wh	Watthour [unit]
WHC	Watthour Meter with Contact Device
WHDM	Watthour Demand Meter
WHIMS	Wet High Intensity Magnetic Separation
WHIMS-CF	Wet High Intensity Magnetic Separation and Cationic Flotation
WHL	Watthour Meter with Loss Compensator
WHM	Watthour Meter
WHMIS	Workplace Hazardous Materials Information System
WHO	World Health Organization [of UN]
WHOI	Woods Hole Oceanographic Institute [US]
Whr	Watthour [unit]
WHRA	Welwyn Hall Research Association [UK]
WHSE	Warehouse
WHT	Watthour Meter Thermal Type
	White
WI	Welding Institute
	Wisconsin [US]
	Wrought Iron
W&I	Weighing and Inspection
WIC	Welding Institute of Canada
WICB	Women in Cell Biology
WIDE	Wiring Integration Design
WIDJET	Waterloo Interactive Direct Job Entry Terminal (System) [Canada]
WIF	Water Immersion Facility
WIL	White Indicating Lamp
WIMP	Wisconsin Interactive Molecule Processor
WINCON	Winter Conference on Aerospace and Electronic Systems [US]
	Winter Convention [IEEE]
WIND	Weather Information Network and Display
WINDII	Wind Imaging Interferometer
WINDS	Weather Information Network and Display System
WIP	Women in Information Processing
	Work-in-Process
WIPE	Waste Immobilization Process Experiment
WIPP	Waste Isolation Pilot Plant
WIPS	Weather Image Processing System
WIRA	Wool Industries Research Association [UK]

WIRDS	Weather Information Remoting and Display System
WIRTC	Western Industrial Research and Training Center
WIS	Weizmann Institute of Science [US]
Wis	Wisconsin [US]
Wisc	Wisconsin [US]
WISP	Waves in Space Plasma Program [NRC, Canada]
WISTIC	Wisconsin University Theoretical Chemistry Institute [US]
WITS	(University of) Waterloo Interactive Terminal System [Canada]
WJCC	Western Joint Computer Conference [US]
WK	Week
	Work
WKB	Wentzel-Kramers-Brillouin (Theory)
WKBJ	Wentzel-Kramers-Brillouin-Jordan (Theory)
WKG	Working
WKNL	Walter Kidde Nuclear Laboratories
WDQD	Word Queue Directory
WL	Water Line
	Wavelength
	Wheel
	Wind Load
	Wire List
	Working Level
WLDG	Welding
WLF	Williams, Landel and Ferry (Equation)
WLI	Wavelength Interval
WLM	Working Level Month
WLN	Wiswesser Line Notation
W LONG	West Longitude
WL-RDCB	Wedge-Loaded Rectangular Double Cantilever Beam
WLU	Wilfred Laurier University [Canada]
WM	Wattmeter
	Word Mark
W&M	Washburn and Moen (Gauge)
WMC	World Materials Congress
	World Mining Congress
WMEC	Western Military Electronics Center [US]
WMHS	Wall-Mounted Handling System
WMO	World Meteorological Organization [of UN]
WMS	Warehouse Management System
	Waste Management System
	Wire Mesh Screen
	World Magnetic Survey
WMSC	Weather Message Switching Center
WMSI	Weak Metal-Support Interaction
	Western Management Science Institute [US]
WMSO	Wichita Mountains Seismological Observatory [US]
WMU	Western Michigan University [US]
W&M WIRE G	Washburn & Moen Wire Gauge
WNRC	Washington National Records Center [US]
WNRE	Whiteshell Nuclear Research Establishment [Canada]
WNW	West-Northwest
WO	Work Order
W/O	Water/Oil (Emulsion)

	Without
	Write-Off
WOC	Water/Oil Contact
W/O E&SP	Without Equipment and Spare Parts
WOF	Work of Fracture
WOG	Water Oil Gas
WOL	Wedge-Opening Load(ing)
WOM	Write Optional Memory
WORB	Work Roll Bending
WORC	Washington Operations Research Council [US]
WORCRA	Worner/Conzinc Riotinto of Australia (Process)
WORM	Write Once, Read Many Times
WOSAC	Worldwide Synchronization of Atomic Clocks
WOT	Wide-Open Throttle
WP	Weatherproof
	White Phosphorus
	Word Processing
	Word Processor
	Working Point
	Working Pressure
	Write Protection
WPAFB	Wright Patterson Air Force Base [US]
WPB	Write Printer Binary
WPC	Water Pillow Cooling
	Watts per Candle
	Wheat Protein Concentrate
	Wood Plastic Combination
	Wood Plastic Composite
	World Petroleum Congress
	World Power Conference
WPCA	Water Pollution Control Administration [US]
WPCF	Water Pollution Control Federation [US]
WPD	Write Printer Decimal
WPESS	Width-Pulse Electronic Sector Scanning
WPF	World Productivity Forum
WPI	Wholesale Price Index
	Worcester Polytechnic Institute [US]
WPL	Waste Pickle Liquor
WPM	Words per Minute
WPP	Waterproof Paper Packing
WPRL	Water Pollution Research Laboratory
WPRT	Write Protect
WPS	Water Phase Salt
	Welding Procedure Specification
	Word Processing Specialist
	Words per Second
WQ	Water Quenching
WQM	Weld Quality Monitor
WQMP	Water Quality Management Project
WQRC	Water Quality Research Council
WR	Wilson Repeater
	Write
WRA	Water Research Association [UK]
	Weapon Replaceable Assembly
WRAC	Willow Run Aeronautical Center [US]
WRAIR	Walter Reed Army Institute of Research [US]
WRAP	Weighted Regression Analysis Program

WRBC	Weather Relay Broadcast Center
WRC	Water Resource Congress
	Weather Relay Center
	Welding Research Council
WRD	Water Resource Division [of USGS]
WRE	Weapons Research Establishment [Australia]
WRESAT	WRE Satellite
WRL	Willow Run Laboratory [US]
WRMS	Watts Root-Mean-Square
WRPC	Weather Records Processing Center
WRP/CI	Worldwide Reinforced Plastics/Composites Institute
WRQ	Westinghouse Resolver/Quantizer
WRRC	Willow Run Research Center [US]
WRRS	Wire Relay Radio System
WRSIC	Water Resources Scientific Information Center [US]
WRSSSV	Wireline-Retrievable Subsurface Safety Valve
WRT	Wrought
WRU	Western Reserve University [US]
WS	Water Solid
	Wave Soldering
	Weather Stripping
	Weapon System
	Wetted Surface
	Wire Shear
	Working Space
	Work Space
	Work Support
	Workstation
Ws	Wattsecond [unit]
WSAD	Weapon Systems Analysis Division [US Navy]
WSB	Wheat Soy Blend
WSCS	Wide Sense Cyclo-Stationary
WSD	Working Stress Design
WSE	Washington Society of Engineers [US]
	Water Saline Extract
	Weapon System Efficiency
	Western Society of Engineers [US]
WSED	Weapon Systems Evaluation Division [of IDA]
WSEIAC	Weapon Systems Effectiveness Industry Advisory Committee [USDOD]
WSEP	Waste Solidification Engineering Prototype Plant [USAEC]
WSI	Water Solidity Index
WSIA	Water Supply Improvement Association
WSIT	Washington State Institute of Technology [US]
WSL	Warren Spring Laboratory [UK]
WSMR	White Sands Missile Range [US]
WSP	Water Supply Point
WSR	Weather Search Radar
	Weather Surveillance Radar
WSTD CT	Worsted Count
WSTF	White Sands Test Facility [US]
WSU	Washington State University [US]
WSW	West-Southwest
WSW/RS	Wind Shear Warning/Recovery System

WSZ	Wrong Signature Zero
WT	Wait Time
	Watertight
	Weight
	Wide Track (Tractor)
	Wireless Telegraphy
W/T	Wireless Telegraphy
WT%	Weight Percent
WTM	Wind Tunnel Model
	Write Tape Mask
WTMS	World Trade in Minerals (Database) System
WTO	Write to Operator
WTOR	Write to Operator with Reply
WTP	Width Table Pointer
WTR	Western Test Range [NASA]
WTS	World Terminal Synchronous
WTT	Weapons Tactical Trainer
WU	Washington University [US]
	Western Union
WUE	Water Use Efficiency
WUI	Western Union International
WUIS	Work Unit Information System
WUS	Word Underscore (Character)
WUSC	World University Service of Canada
WV	West Virginia [US]
	Working Voltage
W/V	Weight by Volume
W Va	West Virginia [US]
WVE	Water Vapor Electrolysis
WVDC	Working Voltage Direct Current
WVT	Water Vapor Transmission
WVTR	Water Vapor Transmission Rate
WVU	West Virginia University [US]
WW	Wide Woods
	Wire Way
	Wire-Wound
	Wire Wrap (Board)
WWCC	World Wide Cost Comparison
WWF	Whole Wheat Flour
WWG	Warrington (Iron) Wire Gauge
	Worldwide Guide
WWMCCS	Worldwide Military Command Control System
WWMP	Western Wood Molding Producers
WWP	Working Water Pressure
WWPA	Western Word Processing Association
	Western Wood Products Association [US]
WWSN	Worldwide Seismology Network [NBS]
WWW	World Weather Watch [of WMO]
WXTRN	Weak External Reference
WY	Wyoming [US]
Wyo	Wyoming [US]
WYSIAWYG	What you see is almost what you get
WYSIWYG	What you see is what you get

X

X	Computer
	Experiment(al)
	Extra
	Index
	Transmission
	Transmitter
	X-Unit
XACT	X (= Computer) Automatic Code Translation
XAES	X-Ray Excited Auger Electron Spectroscopy
XAFS	X-Ray Absorption Fine Structure
XANES	X-Ray Absorption Near-Edge Spectroscopy
	X-Ray Absorption Near-Edge Structure
XARM	Cross Arm
XBAR	Crossbar
XBF	Experimental Boundary File
XBT	Expendable Bathythermograph
XCLB	X-Ray Compositional Line Broadening
XCON	Expert Configurer
XCONN	Cross Connection
XCVR	Transceiver
XDH	Xanthine Dehydrogenase
XDP	Xanthosine Diphosphate
XDCR	Transducer
XDP	X-Ray Density Probe
XDS	Xerox Data System
XE	Experimental Engine
Xe	Xenon
XEC	Execute
	X-Ray Elasticity Constants
XECF	Experimental Engine - Cold Flow
XEDS	X-Ray Energy Dispersive Spectroscopy
XEG	X-Ray Emission Gauge
XEQ	Execute
XES	X-Ray Energy Spectrometry
XESD	X-Ray Induced Electron-Stimulated Ion Desorption
XET	Transparent End-of-Transmission
XETB	Transparent End-of-Transmission Block
XETX	Transparent End-of-Text
XF	Extra Fine
XFER	Transfer
XFMR	Transformer
XFS	X-Ray Fluorescence
XGAM	Experimental Guided Aircraft Missile
XHAIR	Cross Hair
XHFR	Experimental High-Frequency Radar
XHMO	Extended Hueckel Molecular Orbital
XHV	Extreme High Vacuum
XIC	Transmission Interface Converter
XIM	X-Ray Inspection Module
XIO	Execute Input/Output
XITB	Transparent Intermediate Text Block
XL	Execution Language
	Extra Long
XLBIB	Extra Large Burn-In Bath
XLDP	Xylose-Lysine-Deoxycholatepeplon
XLPE	Cross-Linked Polyethylene
XLR	Experimental Liquid Rocket
XM	Experimental Missile
XMA	X-Ray Microanalysis
	X-Ray Microanalyzer

XMAS	Extended Mission Apollo Simulation [NASA]
XMC	Directionally-Reinforced Molding Compound
XMIT	Transmit
XMP	Xanthosine Monophosphate
XMR	X-Ray Micro-Radiography
XMSN	Transmission
XMT	Transmit
XMTD	Transmitted
XMTG	Transmitting
XMTL	Transmittal
XMTR	Transmitter
XN	Execution Node
XNBR	Carboxylated Nitrile-Butadiene Rubber
XO	Crystal Oscillator
	Xanthine Oxidase
XOFF	Transmitter Off
XON	Transmitter On
XOP	Extended Operation
XOR	Exclusive Or
XP	Explosionsproof
XPD	Cross-Polarization Discrimination
	X-Ray Photoelectron Diffraction
XPDA	X-Ray Powder Diffraction Analysis
XPDR	Transponder
XPI	Cross-Polar Interference
XPL	Explosive
XPN	External Priority Number
X-POL	Cross-Polarization
XPS	X-Ray Photoelectron Spectroscopy
XPT	Cross-Point
	External Page Table
XR	External Reset
	Index Register
XRC	X-Ray Rocking Curves
XRD	X-Ray Diffraction
XREF	Cross-Reference
XRF	X-Ray Fluorescence
XRITC	Quinolizino-Substituted Fluorescein Isothiocyanate
XRLM	Extended Range Lance Missile
XRM	X-Ray Microanalysis
	X-Ray Microanalyzer
XRPD	X-Ray Powder Diffraction
XRPM	X-Ray Projection Microscope
XRS	X-Ray Spectroscopy
XS	Extra Strong
XSA	X-Ray Stress Analysis
XSCE	X-Ray Spectroscopy from Channeled Electrons
XSE	X-Ray Stress Evaluation
XSECT	Cross-Section
XSM	Experimental Strategic Missile
XSONAD	Experimental Sonic Azimuth Detector
XSPT	External Shared Page Table
XSPV	Experimental Solid Propellant Vehicle
XSTR	Transistor
XSTX	Transparent Start-of-Text
XSWIS	X-Ray Standing Wave Interference Spectroscopy
XSYN	Transparent Synchronous

XT	Cross Talk
XTA	X-Ray Texture Analysis
XTAL	Crystal
XTASI	Exchange of Technical Apollo Simulation Information
XTEL	Cross Tell
XTEM	Cross-Sectional Transmission Electron Microscopy
XTL	Crystal
XTLO	Crystal Oscillator
XTP	Xanthosine Triphosphate
XTS	Cross-Tell Simulator
XTSI	Extended Task Status Index
XTTD	Transparent Temporary Text Delay
XU	X-Unit
XUV	Extreme Ultraviolet
XVR	Transceiver
XWAVE	Extraordinary Wave
XX	Double Extra
XXL	Double Extra Long
XXS	Double Extra Strong
XXX	Triple Extra
XXXL	Triple Extra Long
XXXS	Triple Extra Strong

Y

Y	Tyrosine
	Yttrium
YAG	Yttrium Aluminum Garnet
YAP	Yttrium Aluminum Perovskite Oxide
Yb	Ytterbium
YBT	Yoshida Buckling Test
yd	yard [unit]
YDT	Yukon Daylight Time [Canada]
YEA	Yale Engineering Association [US]
YECIP	Yukon Energy Conservation Incentive Program [Canada]
YEL	Yellow
YES	Yeast Extract Sucrose
YG	Yellowish Green
YIG	Yttrium Iron Garnet
YIL	Yellow Indicating Lamp
YM	Yellow Metal
YMBA	Yacht and Motor Boat Association
YMC	Yeast Mold Count
YP	Yellow Pine
	Yield Point
YR	Year
YRS	Years
YS	Yield Strength
YSF	Yield Safety Factor
YSLF	Yield Strength Load Factor
YST	Yukon Standard Time [Canada]
YSZ	Yttria-Stabilized Zirconia
YT	Yukon Territory [Canada]
YTD	Year to Date
YU	Yale University [US]

YUDC York University [Canada]
York University Development Center
[Canada]

Z

Z	Atomic Number
	Impedance [symbol]
	Zero
	Zone
ZA	Zero Adjusted
	Zero and Add
	Zinc-Aluminum (Alloy)
ZADCA	Zinc Alloy Die Casters Association
ZAF	Atomic Number, Absorption and Fluorescence Correction
ZAMS	Zero Age Main Sequence
ZAP	Zone Axis Pattern
ZAR	Zeus Acquisition Radar
ZB	Zinc-Blende
ZCS	Zone Communication Station
ZD	Zenith Distance
	Zero Defects
ZDA	Zinc Development Association
ZDC	Zeus Defense Center
ZDCTBS	Zeus Defense Center Tape and Buffer System
ZDP	Aminoimidazolecarboxamideribofuranosyldiphosphate
	Zero Delivery Pressure
	Zinc Dithiophosphate
ZDR	Zeus Discrimination Radar
ZDT	Zero-Ductility Transition
ZEA	Zero Energy Assembly
ZEBRA	Zero Energy Breeder Reactor Assembly [UK]
ZEEP	Zero Energy Experimental Pile
ZENITH	Zero Energy Nitrogen-Heated Thermal Reactor
ZER	Zero Energy Reactor
ZERLINA	Zero Energy Reactor for Lattice Investigations and Study of New Assemblies
ZERT	Zero Reaction Tool
ZES	Zero Energy System
ZETA	Zero Energy Thermonuclear Apparatus
ZETR	Zero Energy Thermal Reactor
ZEUS	Zero Energy Uranium System
ZF	Zero Frequency
	Zone Finder
ZFC	Zero Failure Criterion
ZFS	Zero Field Splitting
ZG	Zero Gravity
ZGE	Zero Gravity Effect
ZGS	Zero Gradient Synchrotron
ZHS	Zero Hoop Stress
ZI	Zinc Institute
ZIF	Zero Insertion Force
ZIP	Zero Inventory Purchasing

	Zig-Zag In-Line Package
	Zinc Impurity Photodetector
ZIPCD	Zip Code
ZK	Magnesium-Zinc-Zirconium (Alloy)
ZkW	Zero Kilowatt
ZM	Zone Melting
ZMAR	Zeus Multifunction Array Radar
ZMAR/MAR	Zeus Multifunction Array Radar/Multifunction Array Radar
ZMB	Zero Moisture Basis
ZMR	Zone Melting Recrystallization
ZN	Zone
Zn	Zinc
ZnDTP	Zinc Dialkyl Dithiophosphate
ZODIAC	Zone Defense Integrated Active Capability
ZOE	Zero Energy
ZOLZ	Zero Order Laue Zone
ZP	Zero Point
ZPA	Zeus Program Analysis
ZPCK	Carbobenzyloxyphenylalanyl Chloromethylketone
ZPEN	Zeus Project Engineer Network
ZPO	Zeus Project Office
ZPPR	Zero Power Plutonium Reactor [USAEC]
ZPR	Zero Power Reactor
ZPRF	Zero Power Reactor Facility
ZR	Zone Refined
Zr	Zirconium
ZRA	Zero Range Approximation
	Zero Resistance Ammeter
ZS	Zero and Subtract
	Zero Slope
ZTA	Zirconia-Toughened Alumina
ZTC	Zirconia-Toughened Ceramic
ZTP	Aminoimidazolecarboxamideribofuranosyltriphosphate
ZTS	Zoom Transfer Scope
ZURF	Zeus Up-Range Facility
ZVD	Zinc Vapor Deposition